Lecture Notes in Computer Science 14610

The series Lecture Notes in Computer Science (LNCS), including its subseries Lecture Notes in Artificial Intelligence (LNAI) and Lecture Notes in Bioinformatics (LNBI), has established itself as a medium for the publication of new developments in computer science and information technology research, teaching, and education.

LNCS enjoys close cooperation with the computer science R & D community, the series counts many renowned academics among its volume editors and paper authors, and collaborates with prestigious societies. Its mission is to serve this international community by providing an invaluable service, mainly focused on the publication of conference and workshop proceedings and postproceedings. LNCS commenced publication in 1973.

Nazli Goharian · Nicola Tonellotto · Yulan He ·
Aldo Lipani · Graham McDonald ·
Craig Macdonald · Iadh Ounis
Editors

Advances in Information Retrieval

46th European Conference on Information Retrieval, ECIR 2024
Glasgow, UK, March 24–28, 2024
Proceedings, Part III

Editors
Nazli Goharian
Georgetown University
Washington, WA, USA

Nicola Tonellotto
University of Pisa
Pisa, Italy

Yulan He
King's College London
London, UK

Aldo Lipani
University College London
London, UK

Graham McDonald
University of Glasgow
Glasgow, UK

Craig Macdonald
University of Glasgow
Glasgow, UK

Iadh Ounis
University of Glasgow
Glasgow, UK

ISSN 0302-9743 ISSN 1611-3349 (electronic)
Lecture Notes in Computer Science
ISBN 978-3-031-56062-0 ISBN 978-3-031-56063-7 (eBook)
https://doi.org/10.1007/978-3-031-56063-7

This Springer imprint is published by the registered company Springer Nature Switzerland AG
The registered company address is: Gewerbestrasse 11, 6330 Cham, Switzerland

Paper in this product is recyclable.

Preface

The 46th European Conference on Information Retrieval (ECIR 2024) was held in Glasgow, Scotland, UK, during March 24–28, 2024, and brought together hundreds of researchers from the UK, Europe and abroad. The conference was organised by the University of Glasgow, in cooperation with the British Computer Society's Information Retrieval Specialist Group (BCS IRSG) and with assistance from the Glasgow Convention Bureau.

These proceedings contain the papers related to the presentations, workshops, tutorials, doctoral consortium and other satellite tracks that took place during the conference. This year's ECIR program boasted a variety of novel work from contributors from all around the world. In addition, we introduced a number of novelties in this year's ECIR. First, ECIR 2024 included for the first time a new "Findings" track, which was offered to some full papers that were deemed to be solid, but which could not make the main conference track. Second, ECIR 2024 ran a new special IR4Good track that presented high-quality, high-impact, original IR-related research on societal issues (such as algorithmic bias and fairness, privacy, and transparency) at the interdisciplinary level (e.g., philosophy, law, sociology, civil society), which go beyond the purely technical perspective. Third, ECIR 2024 featured a new innovation called the "Collab-a-thon", intended to provide an opportunity for participants to foster new collaborations that could lead to exciting new research, and forge lasting relationships with like-minded researchers. Finally, ECIR 2024 introduced a new award to encourage and recognise researchers who have made significant contributions in using theory to develop the information retrieval field. The award was named after Professor Cornelis "Keith" van Rijsbergen (University of Glasgow), a pioneer in modern information retrieval, and a strong advocate of the development of models and theories in information retrieval.

The ECIR 2024 program featured a total of 578 papers from authors in 61 countries in its various tracks. The final program included 57 full papers (23% acceptance rate), an additional 18 finding papers, 36 short papers (24% acceptance rate), 26 IR4Good papers (41%), 18 demonstration papers (56% acceptance rate), 9 reproducibility papers (39% acceptance rate), 8 doctoral consortium papers (57% acceptance rate), and 15 invited CLEF papers. All submissions were peer-reviewed by at least three international Program Committee members to ensure that only submissions of the highest relevance and quality were included in the final ECIR 2024 program. The acceptance decisions were further informed by discussions among the reviewers for each submitted paper, led by a Senior Program Committee member. Each track had a final PC meeting where final recommendations were discussed and made, trying to reach a fair and equal outcome for all submissions.

The accepted papers cover the state-of-the-art in information retrieval and recommender systems: user aspects, system and foundational aspects, artificial intelligence & machine learning, applications, evaluation, new social and technical challenges, and other topics of direct or indirect relevance to search and recommendation. As in previous years, the ECIR 2024 program contained a high proportion of papers with students as first authors, as well as papers from a variety of universities, research institutes, and commercial organisations.

In addition to the papers, the program also included 4 keynotes, 7 tutorials, 10 workshops, a doctoral consortium, an IR4Good event, a Collab-a-thon and an industry day. Keynote talks were given by Charles L. A. Clarke (University of Waterloo), Josiane Mothe (Université de Toulouse), Carlos Castillo (Universitat Pompeu Fabra), and this year's Keith van Rijsbergen Award winner, Maarten de Rijke (University of Amsterdam). The tutorials covered a range of topics including explainable recommender systems, sequential recommendation, social good applications, quantum for IR, generative IR, query performance prediction and PhD advice. The workshops brought together participants to discuss narrative extraction (Text2Story), knowledge-enhanced retrieval (KEIR), online misinformation (ROMCIR), understudied users (IR4U2), graph-based IR (IRonGraphs), open web search (WOWS), technology-assisted review (ALTARS), geographic information extraction (GeoExT), bibliometrics (BIR) and search futures (SearchFutures).

The success of ECIR 2024 would not have been possible without all the help from the strong team of volunteers and reviewers. We wish to thank all the reviewers and meta-reviewers who helped to ensure the high quality of the program. We also wish to thank: the reproducibility track chairs Claudia Hauff and Hamed Zamani, the IR4Good track chairs Ludovico Boratto and Mirko Marras, the demo track chairs Giorgio Maria Di Nunzio and Chiara Renso, the industry day chairs Olivier Jeunen and Isabelle Moulinier, the doctoral consortium chairs Yashar Moshfeghi and Gabriella Pasi, the CLEF Labs chair Jake Lever, the workshop chairs Elisabeth Lex, Maria Maistro and Martin Potthast, the tutorial chairs Mohammad Aliannejadi and Johanne R. Trippas, the Collab-a-thon chair Sean MacAvaney, the best paper awards committee chair Raffaele Perego, the sponsorship chairs Dyaa Albakour and Eugene Kharitonov, the proceeding chairs Debasis Ganguly and Richard McCreadie, and the local organisation chairs Zaiqiao Meng and Hitarth Narvala. We would also like to thank all the student volunteers who worked hard to ensure an excellent and memorable experience for participants and attendees. ECIR 2024 was sponsored by a range of learned societies, research institutes and companies. We thank them all for their support. Finally, we wish to thank all of the authors and contributors to the conference.

March 2024

Nazli Goharian
Nicola Tonellotto
Yulan He
Aldo Lipani
Graham McDonald
Craig Macdonald
Iadh Ounis

Organization

General Chairs

Craig Macdonald	University of Glasgow, UK
Graham McDonald	University of Glasgow, UK
Iadh Ounis	University of Glasgow, UK

Program Chairs – Full Papers

Nazli Goharian	Georgetown University, USA
Nicola Tonellotto	University of Pisa, Italy

Program Chairs – Short Papers

Yulan He	King's College London, UK
Aldo Lipani	University College London, UK

Reproducibility Track Chairs

Claudia Hauff	Spotify & TU Delft, Netherlands
Hamed Zamani	University of Massachusetts Amherst, USA

IR4Good Chairs

Ludovico Boratto	University of Cagliari, Italy
Mirko Marras	University of Cagliari, Italy

Demo Chairs

Giorgio Maria Di Nunzio	Università degli Studi di Padova, Italy
Chiara Renso	ISTI - CNR, Italy

Industry Day Chairs

Olivier Jeunen	ShareChat, UK
Isabelle Moulinier	Thomson Reuters, USA

Doctoral Consortium Chairs

Yashar Moshfeghi	University of Strathclyde, UK
Gabriella Pasi	Università degli Studi di Milano Bicocca, Italy

CLEF Labs Chair

Jake Lever	University of Glasgow, UK

Workshop Chairs

Elisabeth Lex	Graz University of Technology, Austria
Maria Maistro	University of Copenhagen, Denmark
Martin Potthast	Leipzig University, Germany

Tutorial Chairs

Mohammad Aliannejadi	University of Amsterdam, Netherlands
Johanne R. Trippas	RMIT University, Australia

Collab-a-thon Chair

Sean MacAvaney	University of Glasgow, UK

Best Paper Awards Committee Chair

Raffaele Perego	ISTI-CNR, Italy

Sponsorship Chairs

Dyaa Albakour	Signal AI, UK
Eugene Kharitonov	Google, France

Proceeding Chairs

Debasis Ganguly	University of Glasgow, UK
Richard McCreadie	University of Glasgow, UK

Local Organisation Chairs

Zaiqiao Meng	University of Glasgow, UK
Hitarth Narvala	University of Glasgow, UK

Senior Program Committee

Mohammad Aliannejadi	University of Amsterdam, Netherlands
Omar Alonso	Amazon, USA
Giambattista Amati	Fondazione Ugo Bordoni, Italy
Ioannis Arapakis	Telefonica Research, Spain
Jaime Arguello	The University of North Carolina at Chapel Hill, USA
Javed Aslam	Northeastern University, USA
Krisztian Balog	University of Stavanger & Google Research, Norway
Patrice Bellot	Aix-Marseille Université CNRS (LSIS), France
Michael Bendersky	Google, USA
Mohand Boughanem	IRIT University Paul Sabatier Toulouse, France
Jamie Callan	Carnegie Mellon University, USA
Charles Clarke	University of Waterloo, Canada
Fabio Crestani	Università della Svizzera italiana (USI), Switzerland
Bruce Croft	University of Massachusetts Amherst, USA
Maarten de Rijke	University of Amsterdam, Netherlands
Arjen de Vries	Radboud University, Netherlands
Tommaso Di Noia	Politecnico di Bari, Italy
Carsten Eickhoff	University of Tübingen, Germany
Tamer Elsayed	Qatar University, Qatar

Liana Ermakova	HCTI/Université de Bretagne Occidentale, France
Hui Fang	University of Delaware, USA
Nicola Ferro	University of Padova, Italy
Norbert Fuhr	University of Duisburg-Essen, Germany
Debasis Ganguly	University of Glasgow, UK
Lorraine Goeuriot	Université Grenoble Alpes (CNRS), France
Marcos Goncalves	Federal University of Minas Gerais, Brazil
Julio Gonzalo	UNED, Spain
Jiafeng Guo	Institute of Computing Technology, China
Matthias Hagen	Friedrich-Schiller-Universität, Germany
Allan Hanbury	TU Wien, Austria
Donna Harman	NIST, USA
Claudia Hauff	Spotify, Netherlands
Jiyin He	Signal AI, UK
Ben He	University of Chinese Academy of Sciences, China
Dietmar Jannach	University of Klagenfurt, Germany
Adam Jatowt	University of Innsbruck, Austria
Gareth Jones	Dublin City University, Ireland
Joemon Jose	University of Glasgow, UK
Jaap Kamps	University of Amsterdam, Netherlands
Jussi Karlgren	SiloGen, Finland
Udo Kruschwitz	University of Regensburg, Germany
Jochen Leidner	Coburg University of Applied Sciences, Germany
Yiqun Liu	Tsinghua University, China
Sean MacAvaney	University of Glasgow, UK
Craig Macdonald	University of Glasgow, UK
Joao Magalhaes	Universidade NOVA de Lisboa, Portugal
Giorgio Maria Di Nunzio	University of Padua, Italy
Philipp Mayr	GESIS, Germany
Donald Metzler	Google, USA
Alistair Moffat	The University of Melbourne, Australia
Yashar Moshfeghi	University of Strathclyde, UK
Henning Müller	HES-SO, Switzerland
Julián Urbano	Delft University of Technology, Netherlands
Marc Najork	Google, USA
Jian-Yun Nie	Université de Montreal, Canada
Harrie Oosterhuis	Radboud University, Netherlands
Iadh Ounis	University of Glasgow, UK
Javier Parapar	University of A Coruña, Spain
Gabriella Pasi	University of Milano Bicocca, Italy
Raffaele Perego	ISTI-CNR, Italy

Benjamin Piwowarski	CNRS/ISIR/Sorbonne Université, France
Paolo Rosso	Universitat Politècnica de València, Spain
Mark Sanderson	RMIT University, Australia
Philipp Schaer	TH Köln (University of Applied Sciences), Germany
Ralf Schenkel	Trier University, Germany
Christin Seifert	University of Marburg, Germany
Gianmaria Silvello	University of Padua, Italy
Fabrizio Silvestri	University of Rome, Italy
Mark Smucker	University of Waterloo, Canada
Laure Soulier	Sorbonne Université-ISIR, France
Torsten Suel	New York University, USA
Hussein Suleman	University of Cape Town, South Africa
Paul Thomas	Microsoft, USA
Theodora Tsikrika	Information Technologies Institute/CERTH, Greece
Suzan Verberne	LIACS/Leiden University, Netherlands
Marcel Worring	University of Amsterdam, Netherlands
Andrew Yates	University of Amsterdam, Netherlands
Shuo Zhang	Bloomberg, UK
Min Zhang	Tsinghua University, China
Guido Zuccon	The University of Queensland, Australia

Program Committee

Amin Abolghasemi	Leiden University, Netherlands
Sharon Adar	Amazon, USA
Shilpi Agrawal	Linkedin, USA
Mohammad Aliannejadi	University of Amsterdam, Netherlands
Satya Almasian	Heidelberg University, Germany
Giuseppe Amato	ISTI-CNR, Italy
Linda Andersson	Artificial Researcher IT GmbH TU Wien, Austria
Negar Arabzadeh	University of Waterloo, Canada
Marcelo Armentano	ISISTAN (CONICET - UNCPBA), Argentina
Arian Askari	Leiden University, Netherlands
Maurizio Atzori	University of Cagliari, Italy
Sandeep Avula	Amazon, USA
Hosein Azarbonyad	Elsevier, Netherlands
Leif Azzopardi	University of Strathclyde, UK
Andrea Bacciu	Sapienza University of Rome, Italy
Mossaab Bagdouri	Walmart Global Tech, USA

Ebrahim Bagheri	Ryerson University, Canada
Seyedeh Baharan Khatami	University of California San Diego, USA
Giacomo Balloccu	Università degli Studi di Cagliari, Italy
Alberto Barrón-Cedeño	Università di Bologna, Italy
Alvaro Barreiro	University of A Coruña, Spain
Roberto Basili	University of Roma Tor Vergata, Italy
Elias Bassani	Independent Researcher, Italy
Christine Bauer	Paris Lodron University Salzburg, Austria
Alejandro Bellogin	Universidad Autonoma de Madrid, Spain
Alessandro Benedetti	Sease, UK
Klaus Berberich	Saarbruecken University of Applied Sciences, Germany
Arjun Bhalla	Bloomberg LP, USA
Sumit Bhatia	Adobe Inc., India
Veronika Bogina	Tel Aviv University, Israel
Alexander Bondarenko	Friedrich-Schiller-Universität Jena, Germany
Ludovico Boratto	University of Cagliari, Italy
Gloria Bordogna	National Research Council of Italy - CNR, France
Emanuela Boros	EPFL, Switzerland
Leonid Boytsov	Amazon, USA
Martin Braschler	Zurich University of Applied Sciences, Switzerland
Pavel Braslavski	Ural Federal University, Russia
Timo Breuer	TH Köln (University of Applied Sciences), Germany
Sebastian Bruch	Pinecone, USA
Arthur Câmara	Zeta Alpha Vector, Netherlands
Fidel Cacheda	Universidade da Coruña, Spain
Sylvie Calabretto	LIRIS, France
Cesare Campagnano	Sapienza University of Rome, Italy
Ricardo Campos	University of Beira Interior, Portugal
Iván Cantador	Universidad Autónoma de Madrid, Spain
Alberto Carlo Maria Mancino	Politecnico di Bari, Italy
Matteo Catena	Siren, Italy
Abhijnan Chakraborty	IIT Delhi, India
Khushhall Chandra Mahajan	Meta Inc., USA
Shubham Chatterjee	The University of Edinburgh, UK
Despoina Chatzakou	Information Technologies Institute, Greece
Catherine Chavula	University of Texas at Austin, USA
Catherine Chen	Brown University, USA
Jean-Pierre Chevallet	Grenoble Alpes University, France
Adrian-Gabriel Chifu	Aix Marseille University, France

Evgenia Christoforou	CYENS Centre of Excellence, Cyprus
Abu Nowshed Chy	University of Chittagong, Bangladesh
Charles Clarke	University of Waterloo, Canada
Stephane Clinchant	Naver Labs Europe, France
Fabio Crestani	Università della Svizzera Italiana (USI), Switzerland
Shane Culpepper	The University of Queensland, Australia
Hervé Déjean	Naver Labs Europe, France
Célia da Costa Pereira	Université Côte d'Azur, France
Maarten de Rijke	University of Amsterdam, Netherlands
Arjen De Vries	Radboud University, Netherlands
Amra Deli	University of Sarajevo, Bosnia and Herzegovina
Gianluca Demartini	The University of Queensland, Australia
Danilo Dess	Leibniz Institute for the Social Sciences, Germany
Emanuele Di Buccio	University of Padua, Italy
Gaël Dias	Normandie University, France
Vlastislav Dohnal	Masaryk University, Czechia
Gregor Donabauer	University of Regensburg, Germany
Zhicheng Dou	Renmin University of China, China
Carsten Eickhoff	University of Tübingen, Germany
Michael Ekstrand	Drexel University, USA
Dima El Zein	Université Côte d'Azur, France
David Elsweiler	University of Regensburg, Germany
Ralph Ewerth	Leibniz Universität Hannover, Germany
Michael Färber	Karlsruhe Institute of Technology, Germany
Guglielmo Faggioli	University of Padova, Italy
Fabrizio Falchi	ISTI-CNR, Italy
Zhen Fan	Carnegie Mellon University, USA
Anjie Fang	Amazon.com, USA
Hossein Fani	University of Windsor, UK
Henry Field	Endicott College, USA
Yue Feng	UCL, UK
Marcos Fernández Pichel	Universidade de Santiago de Compostela, Spain
Antonio Ferrara	Polytechnic University of Bari, Italy
Komal Florio	Università di Torino - Dipartimento di Informatica, Italy
Thibault Formal	Naver Labs Europe, France
Eduard Fosch Villaronga	Leiden University, Netherlands
Maik Fröbe	Friedrich-Schiller-Universität Jena, Germany
Giacomo Frisoni	University of Bologna, Italy
Xiao Fu	University College London, UK
Norbert Fuhr	University of Duisburg-Essen, Germany

Petra Galuščáková	University of Stavanger, Norway
Debasis Ganguly	University of Glasgow, UK
Eric Gaussier	LIG-UGA, France
Xuri Ge	University of Glasgow, UK
Thomas Gerald	Université Paris Saclay CNRS SATT LISN, France
Kripabandhu Ghosh	ISSER, India
Satanu Ghosh	University of New Hampshire, USA
Daniela Godoy	ISISTAN (CONICET - UNCPBA), Argentina
Carlos-Emiliano González-Gallardo	L3i, France
Michael Granitzer	University of Passau, Germany
Nina Grgic-Hlaca	Max Planck Institute for Software Systems, Germany
Adrien Guille	Université de Lyon, France
Chun Guo	Pandora Media LLC, USA
Shashank Gupta	University of Amsterdam, Netherlands
Matthias Hagen	Friedrich-Schiller-Universität Jena, Germany
Fatima Haouari	Qatar University, Qatar
Maram Hasanain	Qatar University, Qatar
Claudia Hauff	Spotify, Netherlands
Naieme Hazrati	Free University of Bozen-Bolzano, Italy
Daniel Hienert	Leibniz Institute for the Social Sciences, Germany
Frank Hopfgartner	Universität Koblenz, Germany
Gilles Hubert	IRIT, France
Oana Inel	University of Zurich, Switzerland
Bogdan Ionescu	Politehnica University of Bucharest, Romania
Thomas Jaenich	University of Glasgow, UK
Shoaib Jameel	University of Southampton, UK
Faizan Javed	Kaiser Permanente, USA
Olivier Jeunen	ShareChat, UK
Alipio Jorge	University of Porto, Portugal
Toshihiro Kamishima	AIST, Japan
Noriko Kando	National Institute of Informatics, Japan
Sarvnaz Karimi	CSIRO, Australia
Pranav Kasela	University of Milano-Bicocca, Italy
Sumanta Kashyapi	University of New Hampshire, USA
Christin Katharina Kreutz	Cologne University of Applied Sciences, Germany
Abhishek Kaushik	Dublin City University, Ireland
Mesut Kaya	Aalborg University Copenhagen, Denmark
Diane Kelly	University of Tennessee, USA

Jae Keol Choi	Seoul National University, South Korea
Roman Kern	Graz University of Technology, Austria
Pooya Khandel	University of Amsterdam, Netherlands
Johannes Kiesel	Bauhaus-Universität, Germany
Styliani Kleanthous	CYENS CoE & Open University of Cyprus, Cyprus
Anastasiia Klimashevskaia	University of Bergen, Italy
Ivica Kostric	University of Stavanger, Norway
Dominik Kowald	Know-Center & Graz University of Technology, Austria
Hermann Kroll	Technische Universität Braunschweig, Germany
Udo Kruschwitz	University of Regensburg, Germany
Hrishikesh Kulkarni	Georgetown University, USA
Wojciech Kusa	TU Wien, Austria
Mucahid Kutlu	TOBB University of Economics and Technology, Turkey
Saar Kuzi	Amazon, USA
Jochen L. Leidner	Coburg University of Applied Sciences, Germany
Kushal Lakhotia	Outreach, USA
Carlos Lassance	Naver Labs Europe, France
Aonghus Lawlor	University College Dublin, Ireland
Dawn Lawrie	Johns Hopkins University, USA
Chia-Jung Lee	Amazon, USA
Jurek Leonhardt	TU Delft, Germany
Monica Lestari Paramita	University of Sheffield, UK
Hang Li	The University of Queensland, Australia
Ming Li	University of Amsterdam, Netherlands
Qiuchi Li	University of Padua, Italy
Wei Li	University of Roehampton, UK
Minghan Li	University of Waterloo, Canada
Shangsong Liang	MBZUAI, UAE
Nut Limsopatham	Amazon, USA
Marina Litvak	Shamoon College of Engineering, Israel
Siwei Liu	MBZUAI, UAE
Haiming Liu	University of Southampton, UK
Yiqun Liu	Tsinghua University, China
Bulou Liu	Tsinghua University, China
Andreas Lommatzsch	TU Berlin, Germany
David Losada	University of Santiago de Compostela, Spain
Jesus Lovon-Melgarejo	Université Paul Sabatier IRIT, France
Alipio M. Jorge	University of Porto, Portugal
Weizhi Ma	Tsinghua University, China

Georgios Peikos	University of Milano-Bicocca, Italy
Gustavo Penha	Spotify Research, Netherlands
Marinella Petrocchi	IIT-CNR, Italy
Aleksandr Petrov	University of Glasgow, UK
Milo Phillips-Brown	University of Edinburgh, UK
Karen Pinel-Sauvagnat	IRIT, France
Florina Piroi	Vienna University of Technology, Austria
Alessandro Piscopo	BBC, UK
Marco Polignano	Università degli Studi di Bari Aldo Moro, Italy
Claudio Pomo	Polytechnic University of Bari, Italy
Lorenzo Porcaro	Joint Research Centre European Commission, Italy
Amey Porobo Dharwadker	Meta, USA
Martin Potthast	Leipzig University, Germany
Erasmo Purificato	Otto von Guericke University Magdeburg, Germany
Xin Qian	University of Maryland, USA
Yifan Qiao	University of California, USA
Georges Quénot	Laboratoire d'Informatique de Grenoble CNRS, Germany
Alessandro Raganato	University of Milano-Bicocca, Italy
Fiana Raiber	Yahoo Research, Israel
Amifa Raj	Boise State University, USA
Thilina Rajapakse	University of Amsterdam, Netherlands
Jerome Ramos	University College London, UK
David Rau	University of Amsterdam, Netherlands
Gábor Recski	TU Wien, Austria
Navid Rekabsaz	Johannes Kepler University Linz, Austria
Zhaochun Ren	Leiden University, Netherlands
Yongli Ren	RMIT University, Australia
Weilong Ren	Shenzhen Institute of Computing Sciences, China
Chiara Renso	ISTI-CNR, Italy
Kevin Roitero	University of Udine, Italy
Tanya Roosta	Amazon, USA
Cosimo Rulli	University of Pisa, Italy
Valeria Ruscio	Sapienza University of Rome, Italy
Yuta Saito	Cornell University, USA
Tetsuya Sakai	Waseda University, Japan
Shadi Saleh	Microsoft, USA
Eric Sanjuan	Avignon Université, France
Javier Sanz-Cruzado	University of Glasgow, UK
Fabio Saracco	Centro Ricerche Enrico Fermi, Italy

Harrisen Scells	Leipzig University, Germany
Philipp Schaer	TH Köln (University of Applied Sciences), Germany
Jörg Schlötterer	University of Marburg, Germany
Ferdinand Schlatt	Friedrich-Schiller-Universität Jena, Germany
Christin Seifert	University of Marburg, Germany
Giovanni Semeraro	University of Bari, Italy
Procheta Sen	University of Liverpool, UK
Ismail Sengor Altingovde	Bilkent University, Türkiye
Vinay Setty	University of Stavanger, Norway
Mahsa Shahshahani	Accenture, Netherlands
Zhengxiang Shi	University College London, UK
Federico Siciliano	Sapienza University of Rome, Italy
Gianmaria Silvello	University of Padua, Italy
Jaspreet Singh	Amazon, USA
Sneha Singhania	Max Planck Institute for Informatics, Germany
Manel Slokom	Delft University of Technology, Netherlands
Mark Smucker	University of Waterloo, Canada
Maria Sofia Bucarelli	Sapienza University of Rome, Italy
Maria Soledad Pera	TU Delft, Germany
Nasim Sonboli	Brown University, USA
Zhihui Song	University College London, UK
Arpit Sood	Meta Inc, USA
Sajad Sotudeh	Georgetown University, USA
Laure Soulier	Sorbonne Université-ISIR, France
Marc Spaniol	Université de Caen Normandie, France
Francesca Spezzano	Boise State University, USA
Damiano Spina	RMIT University, Australia
Benno Stein	Bauhaus-Universität, Germany
Nikolaos Stylianou	Information Technologies Institute, Greece
Aixin Sun	Nanyang Technological University, Singapore
Dhanasekar Sundararaman	Duke University, UK
Reem Suwaileh	Qatar University, Qatar
Lynda Tamine	IRIT, France
Nandan Thakur	University of Waterloo, Canada
Anna Tigunova	Max Planck Institute, Germany
Nava Tintarev	University of Maastricht, Germany
Marko Tkalcic	University of Primorska, Slovenia
Gabriele Tolomei	Sapienza University of Rome, Italy
Antonela Tommasel	Aarhus University, Denmark
Helma Torkamaan	Delft University of Technology, Netherlands
Salvatore Trani	ISTI-CNR, Italy

Giovanni Trappolini	Sapienza University, Italy
Jan Trienes	University of Duisburg-Essen, Germany
Andrew Trotman	University of Otago, New Zealand
Chun-Hua Tsai	University of Omaha, USA
Radu Tudor Ionescu	University of Bucharest, Romania
Yannis Tzitzikas	University of Crete and FORTH-ICS, Greece
Venktesh V	TU Delft, Germany
Alberto Veneri	Ca' Foscari University of Venice, Italy
Manisha Verma	Amazon, USA
Federica Vezzani	University of Padua, Italy
João Vinagre	Joint Research Centre - European Commission, Italy
Vishwa Vinay	Adobe Research, India
Marco Viviani	Università degli Studi di Milano-Bicocca, Italy
Sanne Vrijenhoek	Universiteit van Amsterdam, Netherlands
Vito Walter Anelli	Politecnico di Bari, Italy
Jiexin Wang	South China University of Technology, China
Zhihong Wang	Tsinghua University, China
Xi Wang	University College London, UK
Xiao Wang	University of Glasgow, UK
Yaxiong Wu	University of Glasgow, UK
Eugene Yang	Johns Hopkins University, USA
Hao-Ren Yao	National Institutes of Health, USA
Andrew Yates	University of Amsterdam, Netherlands
Fanghua Ye	University College London, UK
Zixuan Yi	University of Glasgow, UK
Elad Yom-Tov	Microsoft, USA
Eva Zangerle	University of Innsbruck, Austria
Markus Zanker	University of Klagenfurt, Germany
Fattane Zarrinkalam	University of Guelph, Canada
Rongting Zhang	Amazon, USA
Xinyu Zhang	University of Waterloo, USA
Yang Zhang	Kyoto University, Japan
Min Zhang	Tsinghua University, China
Tianyu Zhu	Beihang University, China
Jiongli Zhu	University of California San Diego, USA
Shengyao Zhuang	The University of Queensland, Australia
Md Zia Ullah	Edinburgh Napier University, UK
Steven Zimmerman	University of Essex, UK
Lixin Zou	Wuhan University, China
Guido Zuccon	The University of Queensland, Australia

Additional Reviewers

Pablo Castells
Ophir Frieder
Claudia Hauff
Yulan He
Craig Macdonald
Graham McDonald
Iadh Ounis
Maria Soledad Pera
Fabrizio Silvestri
Nicola Tonellotto
Min Zhang

Contents – Part III

Full Papers

Short Papers

Full Papers

Knowledge Graph Cross-View Contrastive Learning for Recommendation

Zeyuan Meng(✉), Iadh Ounis, Craig Macdonald, and Zixuan Yi

University of Glasgow, Glasgow, UK
{z.meng.2,z.yi.1}@research.gla.ac.uk,
{iadh.ounis,craig.macdonald}@glasgow.gla.ac.uk

Abstract. Knowledge Graphs (KGs) are useful side information that help recommendation systems improve recommendation quality by providing rich semantic information about entities and items. Recently, models based on graph neural networks (GNNs) have adopted knowledge graphs to capture further high-order structural information, such as shared preferences between users and similarities between items. However, existing GNN-based methods suffer from two challenges: (1) Sparse supervisory signal, where a large amount of information in the knowledge graph is non-relevant to recommendation, and the training labels are insufficient, thereby limiting the recommendation performance of the trained model; (2) Valuable information is discarded whereby the use by the existing models of edge or node dropout strategies to obtain augmented views during self-supervised learning could lead to valuable information being discarded in recommendation. These two challenges limit the effective representation of users and items by existing methods. Inspired by self-supervised learning to mine supervision signals from data, in this paper, we focus on exploring contrastive learning based on knowledge graph enhancement, and propose a new model named Knowledge Graph Cross-view Contrastive Learning for Recommendation (KGCCL) to address the two challenges. Specifically, to address supervision sparseness, we perform contrastive learning between graph views at different levels and mine graph feature information in a self-supervised learning manner. In addition, we use noise augmentation to enhance the representation of users and items, while retaining all triplet information in the knowledge graph to address the challenge of valuable information being discarded. Experimental results on three public datasets show that our proposed KGCCL model outperforms existing state-of-the-art methods. In particular, our model outperforms the best baseline performance by 10.65% on the MIND dataset.

1 Introduction

Recommendation systems have evolved into an essential tool for consumers to access the content they are interested in. Collaborative filtering (CF) [12,14] is a commonly used recommendation framework for extracting user preferences from

N. Goharian et al. (Eds.): ECIR 2024, LNCS 14610, pp. 3–18, 2024.
https://doi.org/10.1007/978-3-031-56063-7_1

past interaction data. However, the effectiveness of CF-based recommender systems still faces the challenge of data sparsity. Most CF-based models – such as Factorisation Machine (FM) [29] and xDeepFM [22] – consider user and item interactions as independent information and neglect any links between interactions. This can prevent the resulting model from extracting collaboration signals from similar user interactions. Recently, knowledge graphs have been widely adopted as useful external information (also known as side information) to cope with data sparsity [2,9,16,35]. Knowledge graphs provide recommendation systems with semantic information about objects in order to improve the representation of users and items, thus enabling knowledge graph-enhanced recommendation. Existing enhancement solutions based on knowledge graphs have mainly experienced three stages of development. Early recommendation systems [28,29,38] focused on using different embedding models to learn the representation of triplet information in knowledge graphs, which link user-item interaction and knowledge graph learning. Later, the path-based methods [6,8,23] enhanced the recommendation system's learning of high-order information in the knowledge graph by capturing the multi-hop paths from the user to the item. Recently, the graph neural network (GNN)-based methods [7,14,44] have captured high-order information in the knowledge graph (KG) by incorporating multi-hop neighbours into the representation of nodes. However, while overall effective, these GNN-based models are still challenged by sparse supervisory signals. Indeed, sparse supervision signal refers to the phenomenon where the GNN approaches amplify a large amount of information in the KG that is non-relevant to recommendation, thereby limiting the accurate expression of the users' preferences.

Recently, as a form of self-supervised learning, contrastive learning [10,24,36] has been proposed to solve the problem of sparse supervision signals. This method realises the mining of representation features from the data itself without labels to improve the model's discriminative learning of the data embeddings. Existing self-supervised recommender models focus on exploring graph node/edge discarding (dropout) strategies to generate augmented *views* for contrastive learning. Contrastive learning compares the similarity of samples between two views to discover the feature information contained in the graph. For example, SGL [36] uses random edge dropout and node dropout to obtain augmented views, while KGCL [39] adopts edge dropout on the user-item interaction graph guided by the consistency of items in the knowledge graph (KG). However, these dropout-based models focus on using a specific method to delete the edge or node information of the input graph to generate augmented subgraphs (or views), which causes some valuable recommendation information to be discarded (the so-called valuable information loss problem). To solve the problem of data sparsity and that of valuable information being possibly discarded, we propose a new model, called Knowledge Graph Cross-View Contrastive Learning Recommendation (KGCCL), to fully exploit the rich semantic and collaborative information in both the user-item interaction graph and the knowledge graph. Specifically, we use the knowledge graph to enrich information representation, treat the user-item interaction graph and the entity-item interaction graph (knowledge graph) as two local-level graphs, and build a user-item-entity graph as a global-level graph to retain more complete structural information in both local graphs. Building on self-supervised learning, we propose a cross-view contrastive learning mechanism that performs contrastive learning (CL) in the local and global views (corresponding to the local and global graphs), which preserves the complete path information and mines the similarity between items. Furthermore, in order to cope with the loss of valuable information due to the dropout strategy used by traditional contrastive learning models in constructing contrastive views, we use noise augmentation to generate views for contrastive learning, which retains all information of the input graph thereby improving the robustness and generalisation of the model.

Overall, the contributions of our work are summarised as follows: (1) We propose a generic KGCCL model that improves representation learning for users and items by applying contrastive learning on the local views and the constructed global view; (2) In terms of data augmentation, we replace the traditional dropout strategy with noise augmentation, which retains all valuable information for recommendation; (3) We conduct extensive experiments on three public datasets. Our results show that KGCCL consistently outperforms the existing state-of-the-art baselines, demonstrating the effectiveness of KGCCL's cross-view contrastive learning and noise augmentation in recommendation.

2 Related Work

In this section, we briefly introduce two families of related work, namely KG-enhanced recommendation and CL-based Recommendation, and position our work with respect to the literature.

Knowledge Graph-Enhanced Recommendation: Depending on the encoding method, the existing effective knowledge graph-enhanced recommenders mainly focus on the embedding-based and GNN-based methods. First, the embedding-based methods [28,29,38] preprocess the input knowledge graph, and then apply the learned entity embeddings and the relations between entities embeddings for recommendation. The CKE model [45] is a typical example of this method, which obtains knowledge embedded information from the knowledge graph (KG), as well as the available text and picture information to represent items. This enriches the semantic information embedded in the items, leading to an increase in prediction accuracy. Another embedding-based model is DKN [32] for news entity recommendation. DKN captures entity, content and word embeddings in news headlines and feeds them through three channels of a CNN [23].

On the other hand, GNN-based methods [15,20,26,31,41] can enrich the representation of nodes through information from their neighbourhood, which has been claimed to be more effective than embedding-based methods [6,9]. Such methods can capture node features, and discover the relations between long-connected nodes by integrating multiple neighbours of nodes in the graph into the node representation. The early LightGCN model [13] used graph convolutional networks (GCNs) to aggregate information from user-item interaction data. Later, the GCN-based methods were combined with knowledge graphs for recommendation. KGAT [33] used a GCN-based method as a knowledge graph encoder to obtain node representations. KGIN [34] explored the user's intention under the guidance of a KG by modeling the user's intention in the user-item interaction, and combined it with the KG to perform GNN on the interaction information. However, these GNN-based methods, while effective, suffer from sparse interaction information and missing data labels, which require obtaining additional supervision signals to enhance recommendation performance. Therefore, in this paper, we propose to adopt contrastive learning based on self-supervised learning to further enhance GNN-based recommendation systems through supervisory signals generated from unlabeled data. Differently from existing approaches, we integrate the user-item graph and the KG into a unified user-item-entity graph, and introduce noise in them in order to enhance the generalisation of embeddings, thus providing effective supervision signals to enhance recommendation performance.

Contrastive Learning-Based Recommendation: Recently, recommendation models based on contrastive learning [10,24,36,40,42] have received a lot of attention in the recommendation field. Contrastive learning (CL) [18] aims to learn useful feature representations by comparing the similarities and differences between samples. The main idea is to compare positive samples (similar samples) with negative samples (dissimilar samples) and to train the model by maximising the similarity between the positive samples and minimising the similarity between the positive and negative samples. The auxiliary supervision of positive samples promotes the consistency between different views of the same node to make predictions, while the supervision of negative samples strengthens the differences among different nodes. For example, SGL [36] obtains the augmented graph by randomly deleting edge/node information (dropout), and performs contrastive learning with the original graph. The recently proposed KACL model [31] performs contrastive learning between the user-item interaction graph and a knowledge graph. However, these models only consider the contrastive learning of one level of graph view (the user-item graph or the KG), which cannot fully use the semantic information and collaborative information for recommendation. Different from previous work, we perform cross-level contrastive learning between the local-level graph (the user-item interaction graph) and the global-level graph (composed of the interaction graph and the KG) to mine information in the graphs for recommendation. In addition, existing models focus on exploring different augmentation methods to improve the recommendation performance. The SGL model [36] uses random edge/node dropout to augment the contrasting view. KGCL [39] uses the KG-guided dropout strategy to augment the user-item interaction graph. However, applying different graph augmentation modalities has been considered unnecessary due to possible negative effects on recommendation [37]. Yu et al. [43] pointed out that adopting graph augmentation severely distorts the interaction graph of users and items. Hence, in SimGCL, they adopted a more effective noise-based augmentation for the user-item graph. Compared with the graph augmentation method rep-

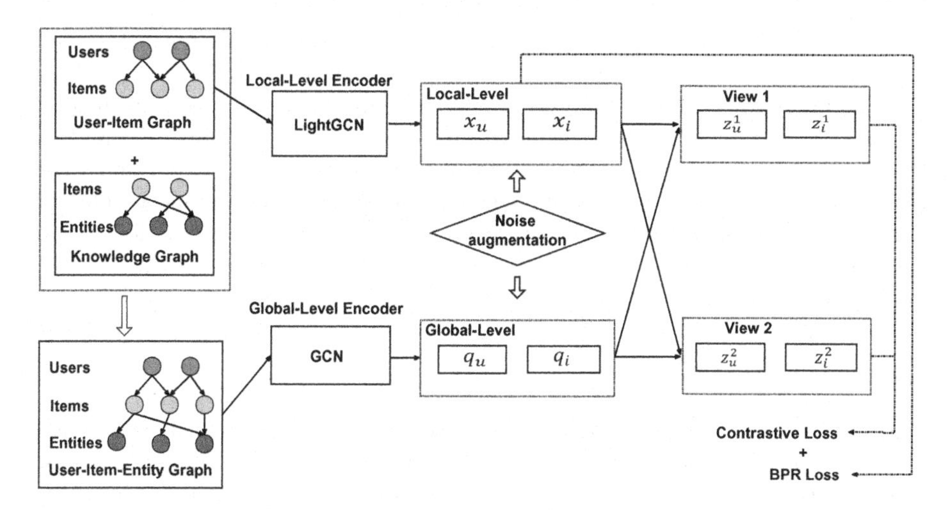

Fig. 1. The architecture of our proposed KGCCL model.

resented by the dropout strategy, our noise augmentation method is designed to retain all information in the KG and interaction graph, and only generates different views for contrastive learning at the representation level. Therefore, our proposed KGCCL model uses noise augmentation that can make full use of the knowledge and interaction graphs to enhance the user and item representations. Differently from prior work, we adopt noise augmentation on the obtained user-item-entity graph and perform cross-view contrastive learning. The latter addresses the problem of sparse supervision signals, while the former addresses the potential discarding of valuable information.

3 Model Architecture

In this section, we first introduce the recommendation task and the used notations (Sect. 3.1). We describe our proposed KGCCL model and its architecture (see Fig. 1). In Sects. 3.2 and 3.3, we introduce the local-level learning and the global-level learning of the model. Finally, Sect. 3.4 presents our cross-view contrastive learning, and describes how we jointly optimise our KGCCL model using the Bayesian personalised ranking loss and the contrastive learning loss.

3.1 Problem Formulation

In this paper, we focus on the ranking-based recommendation task. Two types of necessary structural data that are required by knowledge-enhanced recommender systems are the user-item interaction and the knowledge graphs. The user set and item set are expressed as $\mathcal{U} = \{u_1, u_2, \ldots, u_S\}$ and $\mathcal{I} = \{i_1, i_2, \ldots, i_T\}$, respectively. We define the user-item interaction matrix as $\mathbf{Y} \in \mathbb{R}^{S \times T}$, where S and T represent the number of users and items, respectively. In the interaction matrix, if $y_{u,i} = 1$, then the user has purchased or clicked on the item, otherwise $y_{u,i} = 0$. We first construct the user-item interaction graph $\mathcal{G}_u = \{\mathcal{V}, \mathcal{E}\}$, where $\mathcal{V} = \mathcal{U} \cup \mathcal{I}$ represents the node set, and $\mathcal{E}$ represents edges generated in $\mathcal{G}$ if $y_{u,i} = 1$. Then we build a knowledge graph $\mathcal{G}_k = \{(h, r, t)\}$, where h represents the head entity, r represents the relation between entities, and t represents the tail entity. Specifically, each entity-relation-entity triple represents a different semantic relationship between entities. Given the user-item interaction graph and a knowledge graph, our task is to learn the users's preferences through the model, and to recommend the top-K items for each user.

3.2 Local-Level Learning

Existing CL-based models focus on contrastive learning between local levels (e.g., SGL [36]) or between global levels (e.g., SimCLR [5]), which cannot fully use collaborative information (i.e., items preferred by the same users) and semantic information (i.e., entities related to the same items). Therefore, we propose to establish graph views between the local-level graph (the interaction graph) and the global-level graph (the constructed user-item-entity graph) for contrastive learning to avoid the waste of information caused by coarse-grained contrastive learning (i.e., using only global or local contrastive learning). As input graphs, we consider the user-item interaction graph as a local-level view graph, which focuses on the collaborative relations between users and items. Given a knowledge graph (KG) containing relations between items and entities, we also treat it as a local-level view graph, which aims to explore the semantic similarity between items. Unlike models that augment the input graph in the first step, we retain all interaction information in the user-item graph and define a local-level encoder to obtain node embeddings. As shown in Fig. 1, we use LightGCN to aggregate the node embeddings from the user-item graph:

$$\mathbf{x}_u^{(d+1)} = \sum_{i \in \mathcal{N}_u} \frac{\mathbf{x}_i^{(d)}}{\sqrt{|\mathcal{N}_u||\mathcal{N}i|}}, \qquad \mathbf{x}_i^{(d+1)} = \sum_{u \in \mathcal{N}_i} \frac{\mathbf{x}_u^{(d)}}{\sqrt{|\mathcal{N}_u||\mathcal{N}i|}}, \tag{1}$$

where d is the layer, $\mathbf{x}_u^{(d)}$ ($\mathbf{x}_i^{(d)}$) represents the embedding of the user (item) in layer d, and $\mathcal{N}_u$ ($\mathcal{N}_i$) represents the neighbours of the user (item).

3.3 Global-Level Learning

On the other hand, we regard the user-item-entity graph as a global-level view graph, which preserves the complete path information of long-distance connections from users to items to entities. We follow the method in [33] and represent the information in the interaction graph as $(u, relation, i)$, where $relation$ indicates whether the user interacted with the item. Then, according to the corresponding relationship between items and entities in the KG, the interaction and knowledge graphs are combined into a global-level graph $\mathcal{G}_g = \{(h, r, t)\}$, where h and t represent nodes in the interaction graph and KG, while r represents the relations in the two local graphs. Similarly, as shown in Fig. 1, we define a global-level encoder to obtain the node embeddings on the global graph. Inspired by [47], we adopt a GNN to aggregate the neighbour information to the nodes:

$$\mathbf{q}_u^{(k+1)} = \frac{\sum_{i \in \mathcal{N}_u} \mathbf{q}_i^{(k)}}{|\mathcal{N}_u|}, \tag{2}$$

$$\mathbf{q}_i^{(k+1)} = \frac{\sum_{(r,q) \in \mathcal{N}_i} \alpha(i,r,e) \mathbf{q}_r \odot \mathbf{q}_e^{(k)}}{|\mathcal{N}_i|}, \tag{3}$$

$$\alpha(i,r,e) = \frac{\exp\left((\mathbf{q}_i || \mathbf{q}_r)^T \cdot (\mathbf{q}_e || \mathbf{q}_r)\right)}{\sum_{(e',r) \in \hat{\mathcal{N}}(i)} \exp\left((\mathbf{q}_i || \mathbf{q}_r)^T \cdot (\mathbf{q}_{e'} || \mathbf{q}_r)\right)}, \tag{4}$$

where k denotes the layer, $\mathbf{q}_u$ and $\mathbf{q}_i$ represent the embedding of the user and item, respectively, (i, r, e) represents the triplet connecting the item and entity in the global graph, $\odot$ denotes the element-wise product, and $\hat{\mathcal{N}}(i)$ represents the neighbouring entities $\mathcal{N}_i$ and item i itself.

3.4 Cross-View Contrastive Learning

Augmentation techniques aim to increase the diversity of training data (graph information) by maximising the mutual information between views after augmentation, thus the model can learn the similarity between views through contrastive learning. As discussed in Sect. 2, existing models [31,36] obtain contrastive views by using edge/node dropout graph augmentation on the input graph, which can distort the user-item interaction graph information and could lead to a loss of valuable information used for recommendations. Therefore, we follow the method of SimGCL [43] and adopt noise augmentation at the representation level of the node embeddings in the local and global graphs to avoid the problem of valuable information being discarded. As we discussed in Sect. 3.2, we obtain the embeddings of the user and item in the interaction graph through the local encoder, represented by x_u and x_i, respectively. Similarly, the user and item embeddings obtained through the global-level encoder are represented by q_u and q_i, respectively. As shown in Fig. 1, we employ noise augmentation to represent node embeddings both on the local and global views. For example, given an item node i and an item representation q_i in the global-level graph in a p-dimensional embedding space, we can augment it using the following formulae:

$$\mathbf{q}_i' = \mathbf{q}_i + \Delta_i', \mathbf{q}_i'' = \mathbf{q}_i + \Delta_i'', \tag{5}$$

$$\text{with } \Delta_i = \Delta \odot \operatorname{sign}(\mathbf{q}_i) \odot \epsilon, \Delta \in \mathbb{R}^p \sim U(0,1) \tag{6}$$

where Δ_i' and Δ_i'' are noise vectors, $\mathbf{q}_i'$ and $\mathbf{q}_i''$ are augmented item embedding representations, ϵ controls the strength of additive noise in the range [0, 1], and sign is a function used to represents the positive and negative sign of the signal. Similarly, we can use this noise augmentation method to obtain the representations of the users and items on both local and global graphs. We employ noise augmentation at both local and global-level graphs to obtain augmented representations of nodes without deleting any information in the graphs, so as to avoid discarding valuable information. Then, we perform contrastive learning between the local-level graph and global-level graph. For each user or item node, we create positive pairs by considering the two view-specific embeddings of the user or item, taking advantage of the self-discrimination capabilities of the node. Conversely, negative pairs are formed by using the representations of distinct nodes present in both graph views. The contrastive loss of our KGCCL model is defined based on InfoNCE [3]:

$$\mathcal{L}_c = \sum_{n \in \mathcal{V}} -\log \frac{\exp(s(\mathbf{z}_n^1, \mathbf{z}_n^2)/\tau)}{\sum_{n' \in \mathcal{V}_{n' \neq n}} \exp(s(\mathbf{z}_n^1, \mathbf{z}_{n'}^2)/\tau)} \tag{7}$$

where $\mathbf{z}_n^1$ and $\mathbf{z}_n^2$ represent the embedding of user/item before and after augmentation, respectively, and τ represents the temperature parameter that controls the smoothness of the distribution in the loss function. The KGCCL model uses a cosine function $s(\cdot)$ to calculate the similarity between the positive and negative pairs. Through the minimisation of the contrastive objective loss $\mathcal{L}_c$, we aim to enhance the concordance among the positive pairs in contrast to the negative pairs. Furthermore, we combine the recommendation task with a self-supervised task through joint training to optimise the KGCCL model. The model uses a combination of a contrastive learning loss and the Bayesian Personalised Ranking (BPR) loss for joint training. The formula for joint learning is as follows:

$$\mathcal{L} = \sum_{u \in \mathcal{U}} \sum_{i \in \mathcal{N}_u} \sum_{i' \notin \mathcal{N}_u} -\log \sigma(\hat{y}_{u,i} - \hat{y}_{u,i'}) + \lambda \mathcal{L}_c + \lambda_1 \|\Theta\|_2^2, \tag{8}$$

where $\mathcal{N}_u$ represents items that have interacted with users, and we select items $i' \notin \mathcal{N}_u$ that are not connected to users as negative samples. Θ is the set of all model parameters. The balance between the BPR loss and $\mathcal{L}_c$ is achieved through the adjustment of the hyperparameter λ, and λ_1 is used to control the regularisation strength. In summary, our KGCCL model obtains the augmented view through noise augmentation for contrastive learning, which alleviates the valuable information discarding problem caused by the dropout augmentation strategy. In addition, we adopt cross-level contrastive learning to cope with the impact of the lack of supervision signals on the recommendation performance.

4 Experiments

To evaluate the performance of our proposed KGCCL model, we conduct experiments on three public datasets and compare with several classical and existing state-of-the-art models, addressing three research questions:

RQ1: How does KGCCL perform compared to existing recommendation models?

RQ2: How do different levels of contrastive learning and augmentation affect the recommendation performance?

RQ3: How do the settings of different hyperparameters (noise addition ratio Δ, contrastive loss balance parameter λ) affect KGCCL's performance?

4.1 Datasets and Evaluation Protocol

We conduct experiments to evaluate our KGCCL[1] model on three public datasets collected from different domains, namely Amazon-book [11], Yelp2018 [25] and MIND [1]. Among these three datasets, we follow similar settings to [33,34,39], where we only collect entities within two hops. In addition, we remove the user and item information with less than 10 interactions in the historical interaction information to improve the quality of the dataset. Amazon-book is used for book product recommendation, Yelp2018 is used for commercial site recommendation, and MIND is from the field of news recommendation. The salient statistics of the datasets are shown in Table 1. For the Amazon-book dataset, we map items to the Freebase entities via title matching [4,46]. For Yelp2018, we extract item information from the business information network as knowledge graph data [33]. The MIND dataset follows the data preprocessing of [39], and uses the spacy-entity-linker tool [17] and Wikidata [21,30] to build a knowledge graph. For the three used datasets, we randomly select 80% of the interaction information of each user as the training set, while the remaining 20% of the interaction information is the test set. In the training set, we randomly select 10% of all user-item interaction information as the validation set to adjust the hyperparameters. To evaluate effectiveness, we use two representative measures, Recall@20 and NDCG@20, to evaluate the performance of the KGCCL model in terms of top-K recommendation. We use Adam [19] to optimise our KGCCL model and baselines. We follow the same setting as [39], where the embedding size for all models is fixed to 64. Moreover, the training process involves using a learning rate of 0.001 and a batch size of 2048. To prevent overfitting, we apply early stopping while training, and stop training when the loss does not decrease for 20 epochs. We employ a grid search to determine the optimal hyperparameters for the model. For the noise addition ratio Δ, we search in the range $\{0.03, ..., 0.1, ..., 0.15\}$ with a step size of 0.01. For the contrastive loss balance parameter λ, we search in the range $\{0.02, ..., 0.1, ..., 0.2\}$ with a step size of 0.01.

4.2 Baselines

We compare the performance of KGCCL with different types of baselines. Among them, the baseline models that make use of a KG for recommendation include KGAT, KGIN, and KGCL, while the models BPR, LightGCN, and SGL are strong baselines that do not use any KG for recommendation:

[1] Source code for KGCCL is available at: https://github.com/terrierteam/KGCCL.

Table 1. Statistics of datasets.

	Yelp2018	Amazon-book	MIND
Users	45,919	70,679	300,000
Items	45,538	24,915	48,957
Interactions	1,183,610	846,434	2,545,327
Interaction density (%)	0.00057	0.00048	0.00017
Entities	47,472	29,714	106,500
Relations	42	39	90
Triples	869,603	686,516	746,270

(1) **BPR** [27]: This is a well-known recommendation method that ranks item candidates via a pairwise ranking loss function; (2) **LightGCN** [13]: LightGCN is an advanced GCN-based recommendation method that simplifies the design of feature propagation components by removing nonlinear activation and transformation matrices. This approach simplifies the convolution operation in the information transfer between users and items; (3) **CKE** [45]: CKE uses the widely adopted TransR method [22] to encode the semantic information of items. This method uses the knowledge base embedded component to obtain structural information from the knowledge graph. (4) **SGL** [36]: SGL is a state-of-the-art recommendation method based on self-supervised learning. It uses LightGCN as the encoder of users and items, and uses two different augmentation operators, edge dropping and node dropping, for the feature enhancement of users and items; (5) **KGAT** [33]: KGAT constructs a knowledge collaboration graph, and designs an effective information transfer scheme in the graph. It refines each node's embedding by recursively propagating the embeddings of its neighbours. It also distinguishes the importance of node neighbours through an attention mechanism; (6) **KGIN** [34]: KGIN is a knowledge graph enhancement model. It introduces a user intent mechanism to identify the users' latent intents and performs a perceptual aggregation of relational paths between users through intents; (7) **KGCL** [39]: KGCL is a model that applies different dropout strategies to generate views on the user-item interaction graphs and knowledge graphs, and then performs contrastive learning. We use two forms of KGCL for comparison, KGCL-Operator and KGCL-Random. KGCL-Operator is the original version of KGCL, which generates an augmented view on the interaction graph through a dropout operator, guided by the consistency of items in the KG. KGCL-Random denotes the use of only a random dropout to generate the augmented views.

4.3 Performance Comparison with Baselines (RQ1)

Table 2 compares the performance of our proposed KGCCL model with all used baselines. The best performance is indicated in bold, while the second best performance is underlined. From the table, we observe the following. First, our proposed KGCCL model outperforms all baselines in both measures on the three

Table 2. Performance comparison results between KGCCL and other baselines on three datasets. The best result is bolded and the second best result is underlined. The superscript * indicates that the results are significantly different from the KGCCL results using the Holm-Bonferroni corrected paired t-test with p-value < 0.05.

Dataset	Amazon-Book		Yelp2018		MIND	
Methods	Recall@20	NDCG@20	Recall@20	NDCG@20	Recall@20	NDCG@20
BPR	0.1151*	0.0629*	0.0583*	0.0381*	0.0867*	0.0379*
LightGCN	0.1395*	0.0737*	0.0680*	0.0439*	0.1025*	0.0519*
SGL	0.1448*	0.0768*	0.0727*	0.0477*	0.1021*	0.0530*
CKE	0.1375*	0.0685*	0.0597*	0.0395*	0.0871*	0.0402*
KGAT	0.1280*	0.0666*	0.0636*	0.0410*	0.0878*	0.0412*
KGIN	0.1436*	0.0748*	0.0712*	0.0462*	0.1044*	0.0527*
KGCL-Dropout	<u>0.1485*</u>	<u>0.0788*</u>	0.0744*	0.0486*	0.1066*	0.0547*
KGCL-Operator	0.1480*	0.0785*	<u>0.0756*</u>	<u>0.0493*</u>	<u>0.1070*</u>	<u>0.0549*</u>
KGCCL	**0.1537**	**0.0828**	**0.0794**	**0.0521**	**0.1184**	**0.0625**
%Improve.	3.50%	5.08%	5.03%	5.68%	10.65%	13.84%

datasets. The paired t-test with a Holm-Bonferroni correction shows that the performance improvements over all baselines are statistically significant, on all three datasets. Specifically, KGCCL has achieved significant improvements of 3.50%, 5.03%, and 10.65% over the strongest baselines in terms of Recall@20 on Amazon-book, Yelp2018, and MIND, respectively. This result demonstrates the effectiveness of adopting cross-level contrastive learning and noise augmentation to enrich the local and global graphs. Since the datasets we use vary due to factors such as sparsity and knowledge graph characteristics, the observed improvements by KGCCL demonstrate its flexibility and generalisation. Such improvements by KGCCL are mainly due to the applied cross-level contrastive learning and noise augmentation methods. Through cross-level contrastive learning, KGCCL effectively captures the user and item information to address the problem of sparse supervision signals. KGCCL also uses noise augmentation to effectively obtain the enhanced representations of the user and item, which can alleviate the phenomenon of discarding valuable information. Table 2 also shows that the recommendation models using a KG (KGAT, KGIN and KGCL) overall achieve a better performance than those models without a KG (BPR, LightGCN, SGL). In particular, KGCL, which augments the representation of users and items guided by the KG structure, performs the best among the baselines. These results overall indicate that the use of the knowledge graph can effectively improve the sparse interactive information problem.

4.4 Ablation Study (RQ2)

To investigate the contribution of the main components of KGCCL to its final performance, we consider two KGCCL variants. KGCCL$_{w/o\ CL}$ is a variant of

Table 3. Effects of the ablation study. The superscript * indicates that the results are significantly different from the KGCCL results using the Holm-Bonferroni corrected paired t-test with p-value < 0.05.

Dataset	Amazon-Book		Yelp2018		MIND	
Methods	Recall@20	NDCG@20	Recall@20	NDCG@20	Recall@20	NDCG@20
$\text{KGCCL}_{w/o\ CL}$	0.1523*	0.0810*	0.0784*	0.0514*	0.1152*	0.0603*
$\text{KGCCL}_{w/o\ NA}$	0.1516*	0.0819*	0.0785*	0.0516*	0.1127*	0.0610*
KGCCL	**0.1537**	**0.0828**	**0.0794**	**0.0521**	**0.1184**	**0.0625**

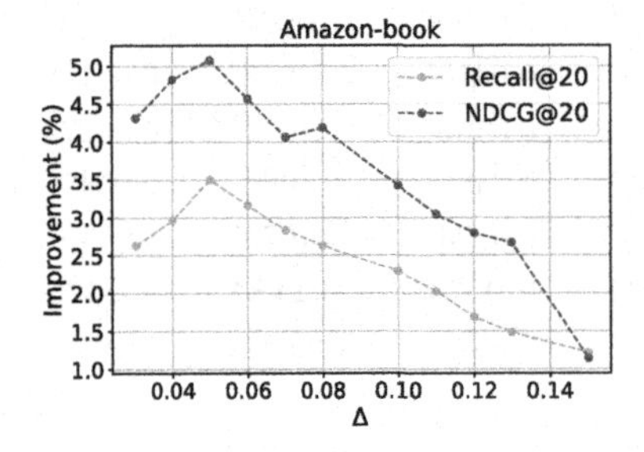

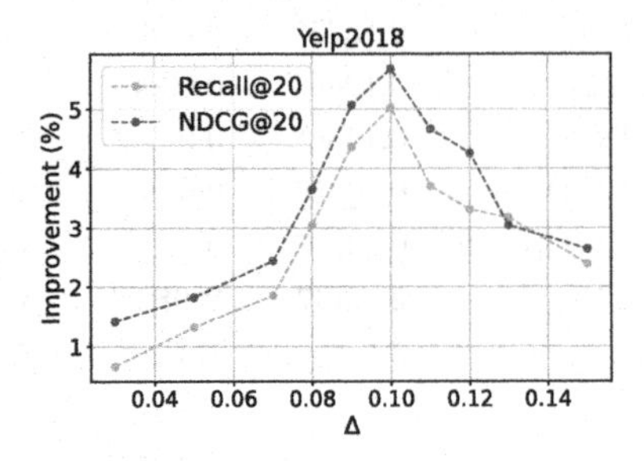

Fig. 2. Impact of noise addition ratio Δ.

KGCCL that does not apply a scheme for contrastive learning between the global-level graph and the local-level graph, while it uses contrastive learning between the two local-level graphs of the user-item interaction graph and the knowledge graph. On the other hand, $\text{KGCCL}_{w/o\ NA}$ is a variant of KGCCL where the noise augmentation component is removed, without adding noise to each embedding. Instead, this variant uses random edge dropout to obtain augmented views when obtaining the contrastive views. Table 3 presents the results. From the table, we make the following observations. First, our KGCCL model exhibits the best performance across all performance metrics on the the three datasets. The decrease in performance observed in the variant model that omits the global-level contrastive learning strategy highlights the effectiveness of this method in harnessing graph structural insights for KG-based recommendation. This shows that the application of contrastive learning between the global-level and local-level graphs can effectively alleviate the impact of sparse supervision signals on the model performance. Overall, these observations highlight the importance of exploring graph structure information for knowledge graph-aware recommendation. In addition, in all datasets, the variant without noise augmentation is overall the least competitive model, especially in the case of large datasets. This suggests that noise augmentation is more effective as a representation-level enhancement than graph structural information pruning. Indeed, noise augmentation helps to avoid losing the graph structure information that may be later useful for recommendation, and effectively improves the representation of users and items while retaining all graph information.

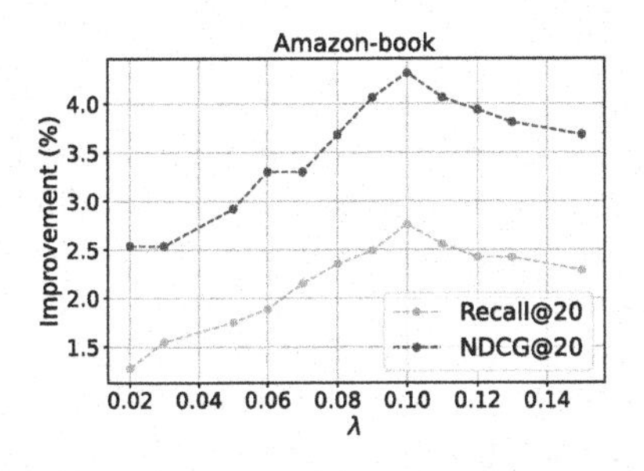

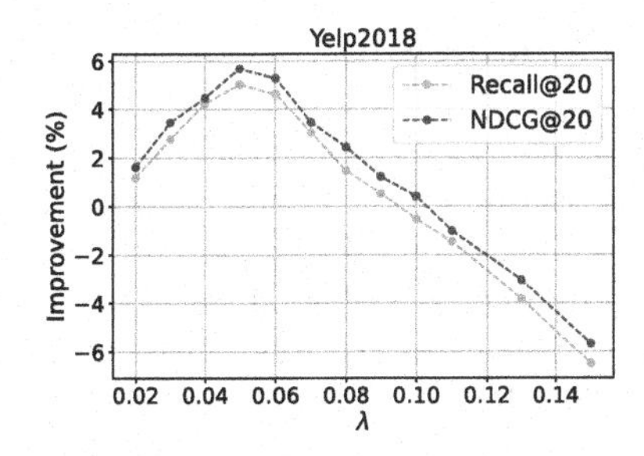

Fig. 3. Impact of the contrastive loss balance parameter λ.

4.5 Sensitivity Analysis (RQ3)

Impact of Noise Addition Ratio $\boldsymbol{\Delta}$. To investigate the effect of adding noise scales in noise augmentation, we vary Δ in the range $\{0.03, ..., 0.1, ..., 0.15\}$ with a step size of 0.01. Figure 2 shows the performance impact of adjusting the noise addition ratio Δ on the Amazon-book and Yelp-2018 datasets[2]. According to the results in Fig. 2, we observe that our KGCCL model achieves the best performance on Amazon-book and Yelp-2018, when $\Delta = 0.05$ and 0.10, respectively. The performance drops for markedly higher values of Δ. It is of note that the optimal noise addition ratio varies from a dataset to another. Overall, these results suggest that adding an appropriate proportion of noise can effectively enhance the user and item embedding representations, while excessively adding noise will have a negative impact on the model.

Impact of Contrastive Loss Balance Parameter $\boldsymbol{\lambda}$. The contrastive loss balance parameter λ indicates the importance of the contrastively learned loss in the joint training. In order to study the influence of the contrastive loss balance parameter on the KGCCL model, we vary λ in the range of $\{0.02, 0.03, ..., 0.1, ..., 0.2\}$ with a step size of 0.01. According to the results of Fig. 3, we can observe that when $\lambda = 0.10, 0.05$, KGCCL performs best on Amazon-book and Yelp-2018, respectively. The performance drops for markedly higher values of λ, especially on the Yelp2018 dataset. This result suggests that the setting of the λ parameter depends on the collection, and needs to be carefully chosen to match the loss of the recommendation task to improve the performance of the model.

Note that for the results in Table 2, KGCCL learned effective values for Δ and λ ($\Delta = 0.05$, $\lambda = 0.1$ on Amazon-book and $\Delta = 0.1$, $\lambda = 0.05$ on Yelp-2018), using our experimental setup in Sect. 4.1, demonstrating that adequate values (as per Fig. 2 and 3) for these parameters can be automatically learned through an adequate training of KGCCL.

[2] Results on the MIND dataset follow similar trends. We omit them because of space constraints.

5 Conclusions

In this work, we explored KG-based contrastive learning to improve the quality of the user and item representations in a self-supervised manner, so as to address the problems of data sparsity and the possible discarding of valuable information. We proposed the KGCCL model, which achieves a better user and item representation learning through cross-level contrastive learning and noise augmentation. Specifically, we performed contrastive learning on local-level and global-level graphs to mine more effective graph structural information and features in a self-supervised manner. We augmented the node embedding with noise augmentation to learn richer user and item representations. Our experimental results on three public datasets showed that our KGCCL model significantly outperforms various existing state-of-the-art baselines. Furthermore, we performed ablation experiments on the model to study the impact of the main components of our KGCCL model on its performance. Our study demonstrated the particularly added-value of the applied noise augmentation method on the recommendation performance. In addition, the conducted parameter sensitivity analysis showed that KGCCL can improve the recommendation performance by selecting appropriate value ranges for the noise addition ratio and the contrasting loss balance parameters.

References

1. Abdulhussein, N.A., Obaid, A.J.: User recommendation system based on mind dataset. arXiv preprint arXiv:2209.06131 (2022)
2. Ai, Q., Azizi, V., Chen, X., Zhang, Y.: Learning heterogeneous knowledge base embeddings for explainable recommendation. Algorithms (2018)
3. Aitchison, L.: InfoNCE is a variational autoencoder. arXiv preprint arXiv:2107.02495 (2021)
4. Bollacker, K., Evans, C., Paritosh, P., Sturge, T., Taylor, J.: Freebase: a collaboratively created graph database for structuring human knowledge. In: Proceedings of SIGMOD (2008)
5. Chen, T., Kornblith, S., Norouzi, M., Hinton, G.: A simple framework for contrastive learning of visual representations. In: Proceedings of ICML (2020)
6. Chen, X., Jia, S., Xiang, Y.: A review: knowledge reasoning over knowledge graph. Expert Syst. Appl. (2020)
7. Fan, W., et al.: Graph neural networks for social recommendation. In: Proceedings of WWW (2019)
8. Fensel, D., et al.: Introduction: what is a knowledge graph? Knowledge graphs (2020)
9. Guo, Q., et al.: A survey on knowledge graph-based recommender systems. Trans. Knowl. Data Eng. (2020)
10. Hayou, S., Doucet, A., Rousseau, J.: On the impact of the activation function on deep neural networks training. In: Proceedings of ICML (2019)
11. He, R., McAuley, J.: Ups and downs: modeling the visual evolution of fashion trends with one-class collaborative filtering. In: Proceedings of WWW (2016)
12. He, X., Chua, T.S.: Neural factorization machines for sparse predictive analytics. In: Proceedings of SIGIR (2017)

13. He, X., Deng, K., Wang, X., Li, Y., Zhang, Y., Wang, M.: LightGCN: simplifying and powering graph convolution network for recommendation. In: Proceedings of SIGIR (2020)
14. He, X., Liao, L., Zhang, H., Nie, L., Hu, X., Chua, T.S.: Neural collaborative filtering. In: Proceedings of WWW (2017)
15. Hu, B., Shi, C., Zhao, W.X., Yu, P.S.: Leveraging meta-path based context for top-N recommendation with a neural co-attention model. In: Proceedings of SIGKDD (2018)
16. Huang, C., et al.: Knowledge-aware coupled graph neural network for social recommendation. In: Proceedings of AAAI (2021)
17. Jo, Y., Yoo, H., Bak, J., Oh, A., Reed, C., Hovy, E.: Knowledge-enhanced evidence retrieval for counterargument generation. In: Proceedings of EMNLP Findings (2021)
18. Khosla, P., et al.: Supervised contrastive learning. In: Proceedings of NeurIPS (2020)
19. Kingma, D.P., Ba, J.: Adam: a method for stochastic optimization. arXiv preprint arXiv:1412.6980 (2014)
20. Kipf, T.N., Welling, M.: Semi-supervised classification with graph convolutional networks. In: Proceedings of ICML (2017)
21. Lai, T.M., Ji, H., Zhai, C.: Improving candidate retrieval with entity profile generation for Wikidata entity linking. In: Proceedings of ACL Findings (2022)
22. Lin, Y., Liu, Z., Sun, M., Liu, Y., Zhu, X.: Learning entity and relation embeddings for knowledge graph completion. In: Proceedings of AAAI (2015)
23. Liu, C., Li, L., Yao, X., Tang, L.: A survey of recommendation algorithms based on knowledge graph embedding. In: Proceedings of CSEI (2019)
24. Liu, S., Ounis, I., Macdonald, C.: An MLP-based algorithm for efficient contrastive graph recommendations. In: Proceedings of SIGIR (2022)
25. Ma, T., et al.: Social network and tag sources based augmenting collaborative recommender system. Trans. Inf. Syst. (2015)
26. Mancino, A.C.M., Ferrara, A., Bufi, S., Malitesta, D., Di Noia, T., Di Sciascio, E.: KGTORe: tailored recommendations through knowledge-aware GNN models. In: Proceedings of RecSys, pp. 576–587 (2023)
27. Rendle, S., Freudenthaler, C., Gantner, Z., Schmidt-Thieme, L.: BPR: Bayesian personalized ranking from implicit feedback. In: Proceedings of UCAI (2009)
28. Shi, C., Hu, B., Zhao, W.X., Philip, S.Y.: Heterogeneous information network embedding for recommendation. Trans. Knowl. Data Eng. (2018)
29. Sánchez-Moreno, D., Moreno-García, M.N., Sonboli, N., Mobasher, B., Burke, R.: Using social tag embedding in a collaborative filtering approach for recommender systems. In: Proceedings of WIC (2020)
30. Vrandečić, D.: Wikidata: a new platform for collaborative data collection. In: Proceedings of WWW (2012)
31. Wang, H., et al.: Knowledge-adaptive contrastive learning for recommendation. In: Proceedings of WSDM (2023)
32. Wang, H., Zhang, F., Xie, X., Guo, M.: DKN: deep knowledge-aware network for news recommendation. In: Proceedings of the WWW (2018)
33. Wang, X., He, X., Cao, Y., Liu, M., Chua, T.S.: KGAT: knowledge graph attention network for recommendation. In: Proceedings of SIGKDD (2019)
34. Wang, X., et al.: Learning intents behind interactions with knowledge graph for recommendation. In: Proceedings of WWW (2021)

35. Wang, Y., Liu, Z., Fan, Z., Sun, L., Yu, P.S.: DSKReG: differentiable sampling on knowledge graph for recommendation with relational GNN. In: Proceedings of CIKM (2021)
36. Wu, J., et al.: Self-supervised graph learning for recommendation. In: Proceedings of SIGIR (2021)
37. Xia, J., Wu, L., Chen, J., Hu, B., Li, S.Z.: SimGRACE: a simple framework for graph contrastive learning without data augmentation. In: Proceedings of WWW (2022)
38. Yang, L., Yin, X., Long, J., Chen, T., Zhao, J., Huang, W.: Spatio-temporal aware knowledge graph embedding for recommender systems. In: Proceedings of ISPA (2022)
39. Yang, Y., Huang, C., Xia, L., Li, C.: Knowledge graph contrastive learning for recommendation. In: Proceedings of SIGIR (2022)
40. Yi, Z., Ounis, I., Macdonald, C.: Contrastive graph prompt-tuning for cross-domain recommendation. Trans. Inf. Syst. **42** (2023)
41. Yi, Z., Ounis, I., Macdonald, C.: Graph contrastive learning with positional representation for recommendation. In: Proceedings of ECIR (2023)
42. Yi, Z., Wang, X., Ounis, I., Macdonald, C.: Multi-modal graph contrastive learning for micro-video recommendation. In: Proceedings of SIGIR (2022)
43. Yu, J., Yin, H., Xia, X., Chen, T., Cui, L., Nguyen, Q.V.H.: Are graph augmentations necessary? Simple graph contrastive learning for recommendation. In: Proceedings of SIGIR (2022)
44. Yu, J., Yin, H., Xia, X., Chen, T., Li, J., Huang, Z.: Self-supervised learning for recommender systems: a survey. Trans. Knowl. Data Eng. (2023)
45. Zhang, F., Yuan, N.J., Lian, D., Xie, X., Ma, W.Y.: Collaborative knowledge base embedding for recommender systems. In: Proceedings of SIGKDD (2016)
46. Zhao, W.X., et al.: KB4Rec: a data set for linking knowledge bases with recommender systems. Data Intell. (2019)
47. Zou, D., et al.: Multi-level cross-view contrastive learning for knowledge-aware recommender system. In: Proceedings of SIGIR (2022)

Mu2STS: A *Mu*ltitask *Mu*ltimodal *S*arcasm-Humor-Differential *T*eacher-*S*tudent Model for Sarcastic Meme Detection

Gitanjali Kumari(✉), Chandranath Adak, and Asif Ekbal

Department of Computer Science and Engineering, Indian Institute of Technology Patna, Bihta, India
{gitanjali_2021cs03,chandranath,asif}@iitp.ac.in

Abstract. Memes, a prevalent form of online communication, often express opinions, emotions, and creativity concisely and entertainingly. Amidst the diverse landscape of memes, the realm of sarcastic memes holds a unique position with its foundation in irony, mockery, satire, and messages that diverge from literal meanings. Detecting sarcasm in memes is challenging due to the intricate interplay between sarcasm and humor. While prior research has primarily concentrated on leveraging the relationship between sarcasm and humor for identifying sarcastic memes, our goal in this paper extends beyond establishing a fundamental connection between the two; instead, we aspire to unravel their distinct characteristics and nuances that differentiate sarcasm from humor. To accomplish this, we introduce a novel deep learning model, i.e., Mu2STS (***Mu****ltitask* ***Mu****ltimodal* ***S****arcasm-Humor-Differential* ***T****eacher*-***S****tudent*), for sarcasm detection in memes, with a special focus on humor. To bolster Mu2STS, we have developed the SHMH (WARNING: This paper contains meme samples that are offensive in nature.) (*Sarcasm-with-Humorous-Meme-in-Hindi*) dataset, designed for detecting sarcasm and humor in memes written in the Hindi language, which is the first of its kind to the best of our knowledge. Our empirical evaluation, which includes both qualitative and quantitative analyses conducted on the SHMH dataset and some benchmark meme datasets, clearly illustrates the effectiveness of Mu2STS, which outperformed major state-of-the-art models. (The dataset and codes are available at https://www.iitp.ac.in/~ai-nlp-ml/resources.html.)

Keywords: Sarcasm Detection · Humor Detection · Memes · Teacher-Student model · Knowledge Distillation · Perturbation

1 Introduction

Social media platforms like Facebook, Twitter, and Instagram play a significant role in shaping society, but they also contribute to the spread of hate speech, offensive content, and misinformation. Memes, a popular form of expression on social media, have gained attention due to their influence on public discourse [20,48,50,51,57]. Many memes, despite being humorous, use sarcasm and dark humor to promote societal harm [27,30,31]. Among the various types of memes, sarcastic memes hold a unique place.

N. Goharian et al. (Eds.): ECIR 2024, LNCS 14610, pp. 19–37, 2024.
https://doi.org/10.1007/978-3-031-56063-7_2

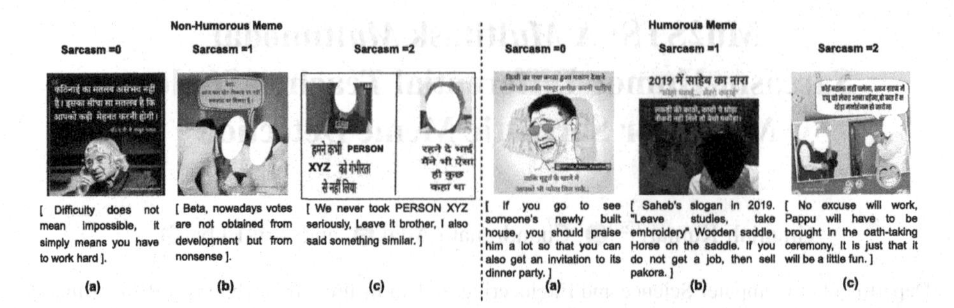

Fig. 1. Meme samples from the dataset to show the relationship between sarcasm and humor. Text enclosed in brackets [] is the English translation of the meme text in Hindi for better readability. To maintain the anonymity of the famous individual, we blurred out the face and replaced the name with "Person XYZ" throughout the paper. Also, zoom in on the figure for better readability.

Detecting sarcasm is challenging due to the intricate interconnection between sarcasm and humor [7,23,38]. Although Stieger et al. [55] categorize sarcasm as a variant of aggressive humor, the key differentiator that sets sarcasm apart from conventional humor lies in its inclination to mock or ridicule a specific target. While sarcastic memes often employ a blend of irony, satire, and humor to engage the audience, it is essential to note that sarcasm does not always intend to be humorous. Additionally, detecting sarcasm in memes is particularly challenging due to the relatively weak correlation between their textual and visual modalities, exacerbated by contextual complexities and subjectivity [7,50]. To illustrate more, in Fig. 1, we categorize sarcastic memes based on the presence of humor. It becomes apparent that humor often lessens the direct aggression aimed at a specific target when used in sarcastic memes. This delicate balance is achieved by incorporating elements such as wordplay, incongruity, and amusing expressions and emoticons within the meme (Fig. 1: Humorous memes (a)–(c)). However, it is worth noting that the inclusion of humor is not an obligatory component of sarcasm every time (Fig. 1: Non-humorous memes (a)–(c)).

Understanding and capturing this interplay between sarcasm and humor in memes requires robust automated models. Previous research in sarcasm detection has primarily focused on understanding the relationship between sarcasm and humor through multi-tasking learning [9,42]. Despite promising progress made by existing models, there is a significant gap in the literature, as there has been limited exploration into identifying the nuanced distinctions between humor and sarcasm. This paper addresses this gap by adopting a comprehensive approach that combines a teacher-student model with a multitask learning framework. To achieve this, we leverage the similarities between sarcasm and humor, with the teacher model serving as a multi-task learning model. However, our novel approach involves the student model not only mimicking the teacher model but also actively learning to discern the disparities between sarcasm and humor. This enhanced learning process incorporates the concept of perturbation, enabling a deeper understanding of the subtle nuances that differentiate these two complex phenomena. In addition to this approach, we have extended the existing sarcastic meme dataset [7] by including humor labels. To the best of our knowledge, previous literature has hardly delved into detecting sarcastic memes using humor in a low-resourced language like Hindi. The main **contribution** of this paper is as follows: ***(i) Dataset:*** We have created the first corpus of humorous memes in Hindi, building upon an existing sarcastic meme dataset [7]. This extended dataset is referred to as "SHMH" (*Sarcastic Humorous Meme in Hindi*). ***(ii) Model:*** We introduce a new model, say "Mu2STS" (*Multitask Multimodal Sarcasm-Humor-Differential Teacher-Student"*), designed for detecting sarcasm in memes. This model combines a multitask teacher-student approach with a perturbation-based model. Additionally, we devise a novel fusion mechanism, referred to as "M3F-CmU" (*Multimodal Multisegment Multihead Fusion with Cross-modal Understanding*), to integrate textual and visual features from memes effectively. ***(iii) Analysis:*** Through an extensive empirical study conducted on the SHMH dataset and various benchmark datasets, we illustrate the effectiveness of our Mu2STS model, surpassing major state-of-the-art models.

2 Related Works

Sarcasm Detection in Unimodal Data: In the realm of sarcasm detection, researchers have exhibited substantial interest, particularly in text-based analysis. Early studies primarily employed feature engineering approaches to discern congruity within the text [2,6,24,37,61]. Building upon these foundations, a series of works [41,62,64] delved into the realm of deep learning networks, leveraging architectures like CNN, LSTM, and self-attention for sarcasm detection. Recent advancements have introduced attention-based models, incorporating external resources that exploit inter- and intra-sentence relationships in texts to enhance sarcasm identification [2,4,35,36,54,59,63].

Sarcasm Detection in Multimodality: The advent of social media has driven researchers to incorporate images into their analyses to better understand sarcasm and its intentions. With the rapid growth of multi-modality posts on modern social media, detecting sarcasm for text and image modalities has increased research attention [23,49,60,65]. Schifanella et al. [49] first tackled this task as a multimodal classification problem by concatenating the visual and textual features and employing SVM and softmax layers to detect sarcasm. Similarly, Zhao et al. [65] presented a multi-modal sarcasm generation (MSG) task to understand sarcasm by generative models.

Other Major Studies on Memes: With the widespread proliferation of memes and their growing influence on online communication, the field of Natural Language Processing (NLP) has recently seen an upsurge in research related to meme analysis. However, much of this research has been centered around identifying memes that are hateful, offensive, harmful, or propaganda techniques [15,20,27,48,50,51,57]. There has been limited attention given to the identification of sarcastic or humorous memes. Furthermore, most of the existing research on memes is either conducted in English or in code-mixed settings. Identifying sarcastic memes using humor, particularly in low-resource languages like Hindi, remains largely unexplored due to a lack of necessary resources and tools.

Consequently, there is a significant research gap in identifying sarcastic memes. This paper aims to bridge this gap by focusing specifically on the task of identifying sarcastic memes and exploring their prevalence on social media platforms, primarily in the context of the Hindi language.

3 Meme Corpus Creation

In this research, we initially utilized a sarcastic meme dataset in Hindi [7], one of the most widely spoken languages globally. This dataset [7] comprises 7,416 meme samples, each meticulously labeled with sarcasm and emotion tags, which we chose due to its open access and diverse content encompassing various domains such as politics, religion, and social issues (e.g., terrorism, racism, and sexism). To further enrich the dataset's usefulness for humor analysis, we conducted manual annotations to introduce humor labels, indicating the presence of humorous elements within the memes. This augmentation allows a more comprehensive exploration of the humor context embedded within the memes (Fig. 1 for selected meme samples from the dataset). Detecting humor in memes can be a subjective task. However, prior research [22] has suggested that providing clear guidelines is crucial when addressing subjectivity during annotation.

To address this challenge, we developed comprehensive gold standard guidelines based on existing research [5,8,11,50]. These guidelines offer valuable insights for identifying humor in a given sample. It encompasses the presence of various humorous attributes, including wordplay, adaptations of poetry or song lyrics, satire, amusing emoticons, comical facial expressions, parody, absurdity, irony, hyperbole, and unexpected or incongruous elements either in the textual and visual components of the meme to be humorous (Table 1 for examples of such memes). In establishing our annotation guidelines, we adopted a similar strategy to [7], which commenced with an initial training phase for annotators, wherein annotators were provided with 200 pre-annotated samples during a dry run. Following this dry-run phase, the actual annotation process was initiated, where every meme was annotated by two annotators. Our annotators, comprising AI professionals and linguists, covered a wide age range (20 to 45 years) and had a balanced gender representation. They were compensated at local rates and were explicitly instructed to remain politically and religiously neutral to ensure objectivity and avoid biases. Their task involved categorizing each meme as either "*humorous* (1)" or "*non-humorous* (0)" following the provided guidelines, drawing upon their cultural knowledge and contextual understanding. Using Cohen's *kappa* coefficient [1], we found significant inter-rater agreement among annotators of 0.8623% for humorous/non-humorous labeling. In Table 1, we have mentioned the statistics of the *Sarcastic Humorous Meme in Hindi* dataset (henceforth referred to as SHMH) for sarcasm and humor labels.

Table 1. Data statistics of SHMH Dataset for sarcasm and humor labels

	Sarcasm			Humor	
classes	Non-Sarcastic (0)	Mildly Sarcastic (1)	Highly Sarcastic (2)	Non-humorous (0)	Humorous (1)
instance	1798	2770	2848	3,088	4,328
distribution %	24.25	37.35	38.4	41.64	58.36

4 Methodology

In this section, we illustrate our proposed *Mu2STS* model to solve two tasks in the end-to-end multi-task learning framework: (i) sarcastic meme detection and (ii) humorous meme detection. Our proposed system employs perturbation-based teacher-student learning to differentiate sarcasm from humor to identify sarcastic memes efficiently.

4.1 Problem Formulation

We are given a set of meme samples $S \in \{T, I\}$, where each sample S_i includes text T_i and RGB image $I_i \in \mathbb{R}^{224\times224\times3}$. Our goal is to predict the correct label of each task, i.e., $\hat{y}_{t1} \subseteq$ {Highly Sarcastic, Sightly Sarcastic, Non Sarcastic} and $\hat{y}_{t2} \subseteq$ {Humorous, Non-Humorous} for each S_i. The respective optimizing goal is then to learn the model weights θ and get the optimum loss $\mathcal{L}((\hat{y}_{t1}, \hat{y}_{t2}) \mid S_i, \theta)$. The overall workflow of our proposed *Mu2STS* model is shown in Fig. 2, and its components are discussed below.

4.2 Encoding of Meme

A meme sample S_i comprises of meme text $T_i = (t_{i_1}, t_{i_2}, \ldots, t_{i_k})$, which is tokenized into sub-word units and projected into high-dimensional feature vectors, where k is the number of tokens in the meme text, and image I_i with regions $r_i = \{r_{i_1}, r_{i_2}, ..., r_{i_N}\}$; for $r_{i_j} \in R^N$, where N is the number of regions. These are then fed into a multimodal CLIP-based [45] pre-trained model designed to extract features by understanding text and images at a semantic level.

$$ft_i, fv_i = CLIP(t_i, r_i)\,; \tag{1}$$

ft_i and fv_i are k and m dimensional textual and visual feature vectors, respectively.

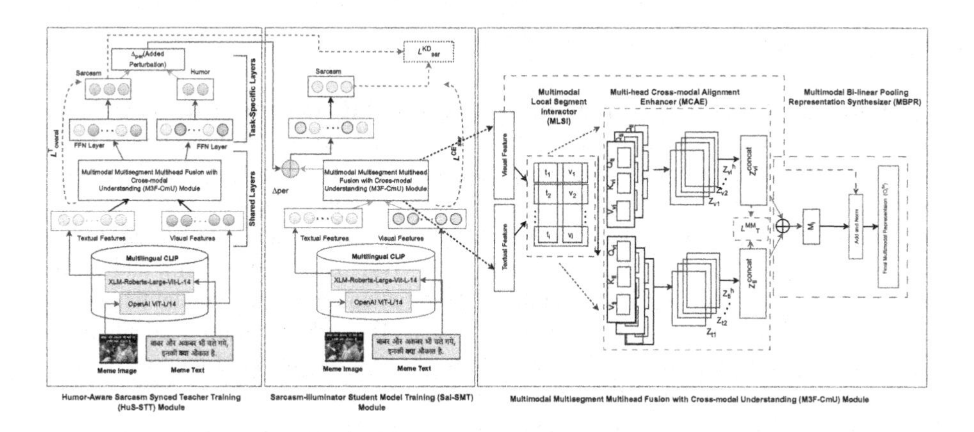

Fig. 2. Our proposed model *Mu2STS* architecture

4.3 Multimodal Multisegment Multihead Fusion with Cross-Modal Understanding (M3F-CmU) Module

To encapsulate the idea of harmonizing both textual and visual features of each meme S_i for achieving a unified multimodal representation, we devise a hierarchical fusion approach that leverages segmented multi-head co-attention mechanism by facilitating local and global interaction. Within our M3F-CmU module, we incorporate three essential sub-modules, i.e., (i) Multimodal Local Segment Interactor (MLSI), (ii) Multi-head Cross-modal Alignment Enhancer (MCAE) and (iii) Multimodal Bi-linear Pooling Representation Synthesizer (MBPR), as discussed below in detail.

(i) Multimodal Local Segment Interactor (MLSI) Sub-module: We employ a sliding window that simultaneously traverses the feature vectors ft_i and fv_i obtained as described in Eqn. 1. At each step of this operation, we perform a process of local fusion for the segments of feature vectors residing within the window. This approach facilitates a comprehensive interaction between features from both modalities within the same window. Consequently, it allows for localized interactions that are finely tuned and specialized, enabling effective cross-modal information exchange. At first, we align both feature vectors to form the multimodal embedding $M \in \mathbb{R}^{2k}$ and leverage a sliding window of size $2d$ to explore inter-modality dynamics. Through the sliding window, each feature vector can be segmented into multiple portions, each termed a local portion. Feature vectors of each modality are split into n segments of size $2d_k$. Each segment is equivalent to a local portion, and the segmentation procedure for a feature vector of a modality is defined as follows:

$$\begin{aligned} t_i &= [ft_{s\cdot(i-1)+1}; ft_{s\cdot(i-1)+2}; \ldots; ft_{s\cdot(i-1)+d}] \\ v_i &= [fv_{s\cdot(i-1)+1}; fv_{s\cdot(i-1)+2}; \ldots; fv_{s\cdot(i-1)+d}] \end{aligned} \tag{2}$$

where, ft_i and fv_i denote the textual and visual modalities, d is the window size, s is the stride, and n_i represents the ith local portion of both textual and visual modalities with $i \in [1, n]$ and n is the number of local portions for each modality. Each modality has $n = \lfloor (k-d)/s \rfloor + 1$ local portions in total.

(ii) Multi-head Cross-Modal Alignment Enhancer (MCAE) Sub-module: MCAE involves several crucial steps to effectively enhance the capture of cross-modal relationships between textual and visual chunks segmented earlier. We denote these chunks as t_i for textual and v_i for visual, where i represents the index of the local portion. At first, we perform linear transformations on these chunks to obtain *query* ($Q_{t_i} = t_i \cdot W_{tq}, Q_{v_i} = v_i \cdot W_{vq}$), *key* ($K_{t_i} = t_i \cdot W_{tk}, K_{v_i} = v_i \cdot W_{vk}$), and *value* ($V_{t_i} = t_i \cdot W_{tv}$, $V_{v_i} = v_i \cdot W_{vv}$) vectors for both textual and visual modalities using learned weight matrices ($W_{tq}, W_{tk}, W_{tv}, W_{vq}, W_{vk}, W_{vv}$). Next, multi-head attention is employed to capture cross-modal correlations. Each attention head focuses on distinct aspects of the cross-modal correlation between textual (t_i) and visual (v_i) chunks. For each pair of textual and visual chunks (t_i, v_i), we calculate attention scores (α_{t_i}, α_{v_i}) independently using scaled dot-product attention:

$$\alpha_{t_i} = softmax\left(Q_{t_i} \cdot (K_{v_i})^T / \sqrt{d_k}\right) \; ; \; \alpha_{v_i} = softmax\left(K_{v_i} \cdot (Q_{t_i})^T / \sqrt{d_k}\right) \tag{3}$$

where, d_k is the dimension of the key vectors, $\mathcal{X}^T$ represents the transpose of matrix $\mathcal{X}$.

Using the computed attention scores, we generate the weighted sums of the value vectors (V_{t_i} and V_{v_i}) associated with each chunk. The attended representation Z_{t_i} for textual chunk t_i, and Z_{v_i} for visual chunk v_i are calculated with cross-modality, as follows:

$$Z_{t_i} = \sum_j \alpha_{t_i,j} \cdot V_{v_j} \;\; ; \;\; Z_{v_i} = \sum_j \alpha_{v_i,j} \cdot V_{t_j} \tag{4}$$

where, $\alpha_{t_i,j}$ and $\alpha_{v_i,j}$ are the attention scores for textual chunk t_i and visual chunk v_j, respectively. Finally, we concatenate the attended representations from all attention heads for each chunk. This concatenation aggregates information from multiple aspects of the cross-modal correlation.

$$Z_{t_i}^{\text{concat}} = [Z_{t_i}^{(1)}; Z_{t_i}^{(2)}; ..., Z_{t_i}^{(h)}], Z_{v_i}^{\text{concat}} = [Z_{v_i}^{(1)}; Z_{v_i}^{(2)}; ..., Z_{v_i}^{(h)}] \tag{5}$$

Here, h represents the number of attention heads. $Z_{t_i}^{\text{concat}}$ and $Z_{v_i}^{\text{concat}}$ are the concatenated representations for textual and visual chunks, respectively.

(iii) Multimodal Bi-linear Pooling Representation Synthesizer (MBPR) Submodule: Although $Z_{t_i}^{\text{concat}}$ and $Z_{t_i}^{\text{concat}}$ in Sect. 4.3, independently conduct image-text multimodal recognition but to enhance the model's performance further a final multimodal representation of the meme S_i is obtained by passing these multi-head attention chunks ($Z_{t_i}^{\text{concat}}$ and $Z_{v_i}^{\text{concat}}$) through a bilinear pooling. This pooling operation aggregates information from various aspects of the cross-modal correlation:

$$M_i = Z_{t_i}^{\text{concat}} \otimes Z_{v_i}^{\text{concat}} = Z_{t_i}^{\text{concat}} \cdot W \cdot Z_{v_i}^{\text{concat}^T} \tag{6}$$

where, M_i represents the multimodal representation for meme S_i, W is a learnable weight matrix used in the bilinear pooling operation; $\otimes$ represents the element-wise multiplication. The final multimodal representation is obtained by passing M_i through a layer-normalization layer [3] and then adding it with t_i through a residual connection.

$$o_i^{tv} = \text{LayerNorm}(M_i) + t_i \tag{7}$$

4.4 Humor-Aware Sarcasm Synced Teacher Training (HuS-STT) Module

The Teacher Model comprises two classifiers in a multi-task learning manner: the sarcasm classifier ($\mathbf{T}_{\text{sarcasm}}$) and the humor classifier ($\mathbf{T}_{\text{humor}}$). We use two task-specific singular feed-forward neural nets (FFN) with $softmax$ activation, which takes the shared concatenated multimodal representation (o_i^{tv}) of a meme sample S_i, calculated in Eqn. 7 as input and output classes for both the classifiers, shown in the following Eqn. 8:

$$\hat{y_t} = P(Y_i|o_i^{tv}, W_i, b_i) = softmax\left(o_i^{tv} W_i + b_i\right) \tag{8}$$

where, $\hat{y_t}$ is the prediction probability of selecting the ith class (Y_i), bias b_i, and weight matrix W_i for t $\in$ {sarcasm, humor}. We use the categorical cross entropy as a loss function:

$$\mathcal{L}_{\text{t}}^{T} = -\frac{1}{N}\sum_{i=1}^{N}\sum_{c=1}^{C} y_i^{t,c} \log p_i^{t,c} \tag{9}$$

where, N is the number of samples, C is the number of classes in task $t \in \{sar, hum\}$, $y_i^{(t),c}$ is 1 if sample i belongs to class c and 0 otherwise. $p_i^{(t),c}$ is the predicted probability that sample i belongs to class c.

In addition to task-specific loss, our multi-task learning method incorporates contrastive loss to improve the model's multimodal representation understanding. The context-aware multimodal representations, denoted as ($Z_{t_i}^{\text{concat}}$, $Z_{v_i}^{\text{concat}}$ in Eqn. 5), are presumed to encapsulate similar contextual information related to a particular meme S_i. These representations are aligned within the same semantic space to harness cross-modal information through contrastive learning during training effectively. More precisely, the proposed model's training involves contrasting the multimodal representation (i.e., $Z_{t_i}^{\text{concat}}$) with another multimodal representation ($Z_{v_i}^{\text{concat}}$) for a specific meme sample S_i, against representations from other memes within the sampled batch.

$$\mathcal{L}_{MM}^{T} = -\log \frac{\exp\left(\text{sim}\left(\boldsymbol{Z}_{t_i}^{\text{concat}}, \boldsymbol{Z}_{v_i}^{\text{concat}}\right)/\tau\right)}{\sum_{k=1[k\neq i]}^{2N} \exp\left(\text{sim}\left(\boldsymbol{Z}_{t_k}^{\text{concat}}, \boldsymbol{Z}_{v_k}^{\text{concat}}\right)/\tau\right)} \tag{10}$$

where, sim is the cosine-similarity, N is the batch size, and τ is the temperature to scale the logits. Therefore, the overall loss is a weighted sum of the multitask losses for sarcasm and humor tasks ($\mathcal{L}_{\text{sar}}^{T}$ and $\mathcal{L}_{\text{hum}}^{T}$) and contrastive loss ($\mathcal{L}_{MM}^{T}$).

By minimizing the task-specific losses for each task, as defined in Eqn. 9, the teacher model strives to make accurate predictions for sarcasm and humor. The weights (α, β, and γ) control the relative importance of each loss.

$$\mathcal{L}_{overall}^{T} = \alpha \cdot \mathcal{L}_{\text{sar}}^{T} + \beta \cdot \mathcal{L}_{\text{hum}}^{T} + \gamma \cdot \mathcal{L}_{MM}^{T} \tag{11}$$

4.5 Sarcasm-Illuminator Student Model Training (SaI-SMT) Module

To train the student model ($\mathbf{S}_{\text{sarcasm}}$), we first compute the difference between the teacher model's predictions for sarcasm and humor tasks. This difference in feature distributions (Δper), denoted as a perturbation, suggests characteristics unique to sarcasm. We add this feature to the student model's input, encouraging it to learn the features distinguishing sarcasm from humor, even when the sarcasm is masked by humor.

$$\Delta_{per}^{i} = \hat{y}_{hum}^{i} - \hat{y}_{sar}^{i} \tag{12}$$

The resulting feature representation is then fed through a $softmax$ layer in the student model for the final sarcasm prediction where $\hat{y}_{sar}^{s}$ represents the predicted probability of selecting the ith class (Y_i), with b_i as the bias and W_i as the weight matrix specifically for the sarcasm class.

$$\hat{y}_{sar}^{s} = softmax((o_i^{tv} + \Delta_{per}^{i})W_i + b_i) \tag{13}$$

where, o_i^{tv} is the multimodal representation derived from its text T_i and image I_i using the M3F-CmU fusion (f_{MC}) technique; $o_i^{tv} = f_{MC}(T_i, V_i)$. Notably, we used the same M3F-CmU technique (refer to Sect. 4.3) to fuse the textual and visual features in the student model due to its effective performance. To train $\mathbf{S}_{\text{sarcasm}}$, we compute the total loss ($\mathcal{L}_{\text{total}}$) by combining the cross-entropy loss ($\mathcal{L}_{\text{sar}}^{CE}$) and knowledge-distillation loss ($\mathcal{L}_{\text{sar}}^{KD}$).

$$\mathcal{L}_{\text{sar}}^{CE} = -\sum[y_{sar} \log \hat{y}_{sar}^{s} + (1 - y_{sar}) \log(1 - \hat{y}_{sar}^{s})] \quad (14)$$

where, $\hat{y}_{sar}^{s}$ represents the Student Model's sarcasm predictions and y_{sar} denotes the ground truth labels for sarcasm classification.

The $\mathcal{L}_{\text{sar}}^{KD}$ is calculated using Kullback-Leibler (KL) divergence [19] between the soft targets $\hat{y}_{sar}$ from the Teacher Model ($\mathbf{T}_{\text{sarcasm}}$) and probability ($\hat{y}_{sar}^{s}$)) from student model ($\mathbf{S}_{\text{sarcasm}}$) with temperature scaling to soften the logits before applying softmax.

$$\mathcal{L}_{\text{sar}}^{KD} = \frac{\tau^2}{|D|} \sum_{i \in D} KL\left(\, ((\hat{y}_{sar})_i; \tau) \, , ((\hat{y}_{sar}^{s})_i; \tau) \, \right) \quad (15)$$

where, the temperature parameter τ controls the entropy of the output distribution; higher temperature τ means higher entropy in the soft labels. We scale the loss by τ^2 to keep gradient magnitudes approximately constant when changing the temperature [19]. We omit τ for brevity.

Finally, the total loss $\mathcal{L}_{\text{total}}$ for the Student Model is obtained by combining $\mathcal{L}_{\text{sar}}^{CE}$ and $\mathcal{L}_{\text{sar}}^{KD}$ with appropriate weights, as below:

$$\mathcal{L}_{\text{total}} = \beta \cdot \mathcal{L}_{\text{sar}}^{CE} + (1 - \beta) \cdot \mathcal{L}_{\text{sar}}^{KD} \quad (16)$$

where β is a hyperparameter determining the trade-off between the two loss terms.

5 Results and Analysis

This section presents the experimental details and model results that outline a comparison between the baseline models, our proposed model, and different variants of our proposed models for the sarcastic meme identification task. To evaluate the proposed model, we use macro-level Precision (P), Recall (R), F1 score (F1), and accuracy (Acc).

5.1 Experimental Details

We use Pytorch Lightning (*lightning.ai*) framework with mCLIP and XLM-R [14] tokenizer, covering 100 languages. The model is trained for 60 epochs with a batch size of 32 and Adam optimizer [29] with learning-rate $= 3 \times 10^{-5}$, $\beta1 = 0.9$, $\beta2 = 0.999$, $\epsilon = 10^{-8}$. We used 7416 meme samples with a train: test ratio of $80 : 20$, where 15% of the train set was used for model validation. During the training of the HuS-STT module, through grid search, we fixed the α, β, and γ hyperparameters (Eqn. 11) in the $\mathcal{L}_{overall}^{T}$ as 0.5, 0.4, and 0.4, respectively.

5.2 Comparison with Baselines

In Table 2, we present the results of baseline models and our proposed model for the sarcastic meme identification task, considering both single-task learning (STL) and multi-task learning (MTL) scenarios, using SHMH dataset. Notably, the mCLIP-based model outperformed other baselines in STL and MTL scenarios, forming the foundation for our proposed method. Furthermore, it is noteworthy that the multimodal baselines consistently outperformed their unimodal counterparts for both tasks, achieving a substantial 15%–17% increase in F1 score. The proposed model, $Mu2STS$, maintained its superior performance across both tasks compared to the developed baselines.

Table 2. Comparative results for sarcasm and humor tasks of baseline models and proposed model in STL and MTL frameworks on SHMH dataset. We observe that the performance gains are statistically significant using t-test [56] with p-values (<0.05), which signifies a 95% confidence interval. * is representing the baseline model.

Models			Modality		STL								MTL							
					Sarcasm				Humor				Sarcasm				Humor			
			T	I	Acc	F1	P	R	Acc	F1	P	R	Acc	F1	P	R	Acc	F1	P	R
Unimodal Baselines		CNN with FastText [10]	✓		36.59	34.18	32.25	31.62	52.92	26.23	28.84	24.06	40.42	35.30	36.56	34.13	56.94	29.32	30.92	27.88
		LSTM with FastText	✓		46.80	45.67	43.78	44.54	59.81	29.07	38.93	23.20	50.63	47.01	48.09	45.99	63.37	32.30	41.81	26.32
	Pre-trained	m-BERT [40]	✓		54.34	51.71	51.23	52.41	56.89	31.83	40.13	26.38	58.17	55.12	54.72	55.52	59.11	35.58	42.11	30.81
		LaBSE [17]	✓		56.79	53.64	52.01	54.00	59.93	39.27	43.92	35.51	61.62	58.77	59.04	58.42	64.91	42.66	47.81	38.52
		Muril [26]	✓		59.63	56.89	56.71	57.61	62.74	40.32	41.44	39.26	63.07	60.79	59.37	62.27	67.11	43.14	45.47	41.04
		Indic BERT [25]	✓		53.30	51.89	51.72	51.00	63.37	37.07	34.95	39.47	58.75	56.17	57.93	54.51	66.30	41.17	39.93	42.49
		VGG-19 [53]		✓	55.27	44.98	43.04	41.36	66.88	38.54	36.84	40.41	58.16	45.53	47.31	43.87	68.72	42.01	40.82	43.28
		ResNet [18]		✓	51.30	46.74	46.47	45.78	55.14	33.32	33.05	33.60	57.49	49.27	49.13	49.41	61.02	36.29	35.17	37.49
		ViT [16]		✓	53.81	47.31	45.59	45.16	59.89	40.53	40.08	40.98	56.02	47.66	48.23	47.11	63.37	42.97	42.73	43.21
Multimodal Baselines	Early Fusion	CNN+VGG	✓	✓	48.58	46.02	47.26	47.71	63.77	42.8	42.73	42.87	53.41	52.14	51.57	52.73	65.75	45.85	45.67	46.04
		LSTM+VGG	✓	✓	39.38	43.06	41.84	43.99	61.14	40.16	38.69	41.74	42.15	47.71	47.07	48.37	64.76	44.69	43.66	45.76
		mBERT+ViT	✓	✓	57.02	61.60	57.76	57.22	57.05	39.80	41.50	38.24	62.33	61.38	62.38	60.41	61.03	43.43	44.83	42.11
		LaBSE++ViT	✓	✓	58.98	54.51	52.71	54.81	60.29	42.47	44.15	40.91	63.51	58.15	57.75	58.55	64.52	46.11	46.39	45.83
		Muril+ViT	✓	✓	64.86	60.49	56.45	54.57	56.59	44.16	46.61	41.95	67.12	59.06	62.56	55.93	60.04	46.65	48.79	44.69
		Indic BERT+ViT	✓	✓	60.79	56.86	55.88	52.86	53.93	40.35	38.99	41.80	65.22	57.53	58.73	56.38	59.76	44.46	43.82	45.11
	Pre-trained	LXMERT [58]	✓	✓	65.09	53.27	49.88	52.76	57.67	44.08	45.88	42.42	68.45	55.10	54.92	55.27	59.43	47.09	47.75	46.44
		VisualBERT [34]	✓	✓	62.90	**61.93**	59.47	**58.56**	54.66	39.35	41.94	37.06	67.32	61.47	62.43	**60.55**	58.75	44.39	46.02	42.88
		BLIP [32]	✓	✓	65.83	57.73	58.47	57.02	53.91	44.85	45.81	43.93	59.05	58.67	58.33	59.03	61.08	48.85	49.08	48.63
		ALBEF [33]	✓	✓	63.47	54.82	56.88	52.92	55.21	47.24	**48.03**	46.48	60.82	55.74	55.49	56.01	59.44	47.89	48.32	47.47
		$mCLIP^*$ [46]	✓	✓	**67.27**	61.87	**59.95**	57.37	**58.83**	**47.52**	47.16	**47.89**	**72.12**	**61.53**	**63.28**	59.88	**65.39**	**51.31**	**52.48**	**50.21**
Proposed		$Mu2STS$ (Ours)	✓	✓	**69.52**	**64.04**	**63.28**	**64.81**	**69.23**	**52.08**	**50.13**	**54.19**	**76.03**	**67.79**	**67.92**	**67.67**	**71.32**	**57.41**	**55.43**	**59.53**

5.3 Ablation Study

In Table 3, we present the results of various model variants to assess the impact of each module in our architecture:

Role of Modalities: Our study reveals that the unimodal versions of our proposed models (i.e., $Mu2STS^{I^-}$ and $Mu2STS^{T^-}$) exhibit notably inferior performance compared to the multimodal setup across both tasks. Notably, $Mu2STS$ consistently outperforms the unimodal variations by a substantial margin of 3%–5% F1 scores. This significant improvement can be attributed to effectively utilizing both textual and visual modalities, enhancing sarcasm comprehension in meme-related tasks.

Role of M3F-CmU Module: Employing simple concatenation instead of the M3F-CmU module led to decreased model performance. This drop in performance can be attributed to the limitations of simple concatenation, where not every element of the textual modality effectively communicates with the visual modalities. The M3F-CmU module is crucial in enhancing cross-modal understanding and preserving the intricate relationships between the different modalities, resulting in improved performance.

Role of Contrastive Learning Loss ($\mathcal{L}_{MM}$): The absence of the contrastive loss $\mathcal{L}MM$ during training led to a significant drop in the proposed model's performance. This highlights the crucial role played by the contrastive loss in promoting effective discrimination between multimodal representations of different memes while preserving similarity among those with similar contexts.

Table 3. Ablation Study: Role of different modalities or modules in our proposed model for the sarcasm and humor tasks. Note that each MTL model is trained by following multitasking of both tasks.

Variations	Model Name	Modality		Sarcasm				Humor			
		Text	Image	Acc	F1	P	R	Acc	F1	P	R
Proposed	$Mu2STS$	✓	✓	76.03	67.79	67.92	67.67	71.32	57.41	55.43	59.53
Unimodal	$Mu2STS^{I-}$	✓	—	63.76	63.63	64.82	62.47	67.24	54.21	54.87	53.56
	$Mu2STS^{T-}$	—	✓	62.09	61.47	60.94	62.01	69.88	55.36	55.69	55.04
Proposed w/o M3F-CmU	$Mu2STS^{M3F-CmU-}$	✓	✓	63.27	62.63	61.82	63.47	67.98	52.66	52.06	53.28
Proposed w/o Student Model	$Mu2STS^{stu-}$	✓	✓	61.78	60.16	64.33	60.97	71.32	57.41	55.43	59.53
Proposed w/o added noise	$Mu2STS^{noise-}$	✓	✓	65.64	64.8	64.28	65.33	71.32	57.41	55.43	59.53
Proposed w/o L_{MM}	$Mu2STS^{LMM-}$	✓	✓	66.98	65.35	66.31	64.43	71.32	57.41	55.43	59.53

Role of Teacher-Student Model: The Teacher-Student model plays a crucial role in improving the performance of sarcasm detection. When comparing the model's performance with and without the teacher-student framework, it becomes evident that the latter results in a decrease of 5%–10% in every metric. This performance drop highlights the effectiveness of knowledge distillation in enhancing sarcasm detection tasks.

Role of Added Perturbation to Student Model: Removing the perturbation, designed to distinguish between humor and sarcasm, from the student model results in a significant 2.99% decrease in the F1-score of the proposed model. This underscores the perturbation's crucial role in enhancing the model's capacity to differentiate between sarcasm and humor.

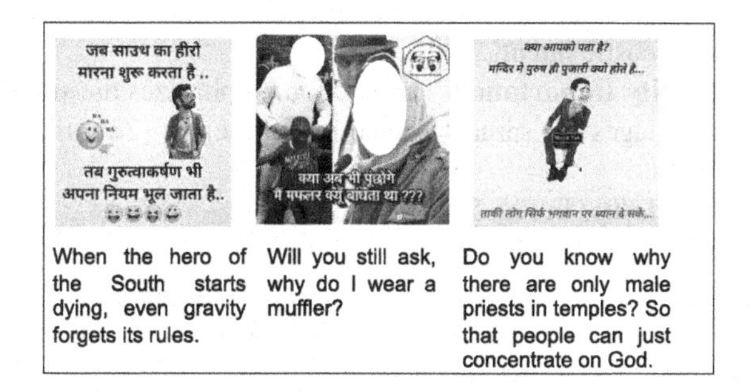

Fig. 3. Case studies of $Mu2STS$ for test sets.

5.4 Detailed Analysis

Qualitative Analysis with Case Study. Using Fig. 3, we qualitatively analyze our proposed framework through the predictions obtained from the *mclip* baseline model and the proposed $Mu2STS$ model. All the samples of Fig. 3 have the gold label 'highly sarcastic.' While the baseline model *mCLIP* without the perturbation predicted these memes as non-sarcastic and humorous, primarily fixating on literal text meanings due to the emoticons and funny images, missing the underlying sarcasm, $Mu2STS$ correctly identifies these as 'highly sarcastic' memes with its ability to differentiate humor from sarcasm. It removes the humor layer to unveil the underlying, frequently sarcastic message. It demonstrates $Mu2STS$ strength in meme context comprehension by effectively differentiating humor and sarcasm.

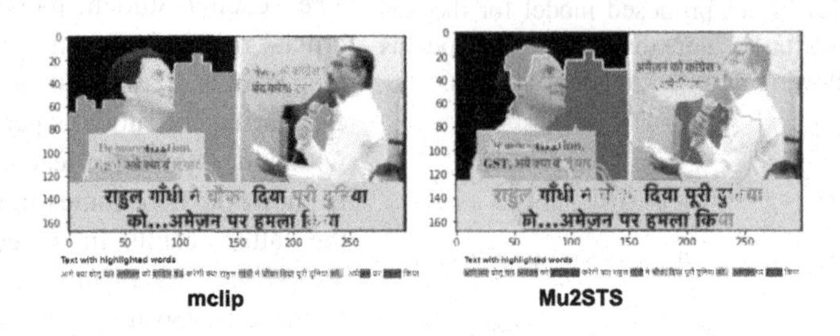

Fig. 4. Visualization by LIME [47] for baseline model $mclip$ and proposed $Mu2STS$.

Explainability and Diagnostics. Once our $Mu2STS$ model is trained, it relies on specific sarcasm-related features within memes to make predictions. To explain how our model makes predictions, we use LIME (Locally Interpretable Model-Agnostic Explanations), a widely used interpretability method. LIME helps us clarify our model's reasoning in an easily understandable way. In Fig. 4, we can see that for the given test sample, certain image regions (e.g., a person's face) and specific words in the text significantly influence $Mu2STS$'s accurate predictions. In contrast, the baseline *mclip* model struggles to differentiate between sarcasm and humor effectively, lacking the ability to recognize sarcastic intent.

Diagnostics of Modality Importance. This section emphasizes the significance of utilizing both text and images for sarcasm detection in memes [28], as shown in Fig. 3.

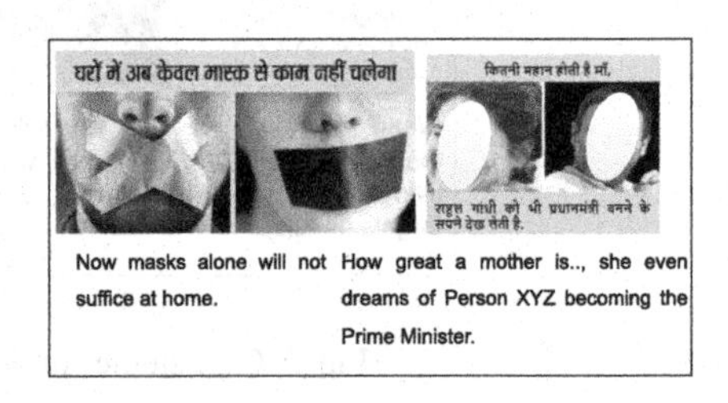

Fig. 5. Example samples where unimodal systems fail whereas proposed multimodal model $Mu2STS$ effectively predicted the sarcasm class.

Failure of Textual Modality: In Fig. 5, the left-most meme is initially labeled non-sarcastic by $Mu2STS^{I^-}$ when visual information is absent. However, when visual content (showing men with tape on their mouths) is included in the model ($Mu2STS$), it correctly identifies the meme as slightly sarcastic. This demonstrates how visuals can provide context and aid sarcastic meme detection, complementing textual information. ***Failure of Visual Modality:*** Similarly, in the right-most meme in Fig. 3 (b), $Mu2STS^{T^-}$, relying only on the image (showing two politicians are laughing), fails to identify the meme's sarcasm, while $Mu2STS$ correctly identifies it as highly sarcastic. It shows that the text modality provides information on intent and meaning through keywords, phrases, sarcasm, irony, and language identification.

5.5 Cross-Lingual Generalization

Table 4 shows the results of **zero-shot learning** using our proposed model and its variations, evaluated on the *Memotion* [50] and *Memotion 2.0* [39] dataset to gauge cross-lingual adaptability for the sarcasm detection task. Incorporating a perturbation-based teacher-student model $Mu2STS$ enhances the baseline model's ($mclip$) performance by up to 5%-6% F1 score, highlighting the significance of domain knowledge in cross-lingual meme analysis. The outperformance of $Mu2STS$ over $mclip$ demonstrates its effectiveness, especially when handling slang and jargon uncommon in English and Hindi data.

Table 4. Cross-lingual generalization over publicly available Memotion dataset

Models	Memotion				Memotion 2.0			
	Acc	F1	P	R	Acc	F1	P	R
$mclip$	50.85	34.43	31.67	37.73	80.76	64.15	66.84	61.67
$Mu2STS^{I-}$	47.28	31.08	31.62	30.57	49.28	41.09	42.79	39.53
$Mu2STS^{T-}$	53.73	33.06	32.53	33.62	53.89	44.86	48.94	41.42
$Mu2STS^{M3F-CmU}$	51.49	35.21	33.84	36.71	69.44	68.11	69.73	66.57
$Mu2STS^{stu-}$	52.81	37.86	38.72	37.05	68.41	61.67	60.57	62.82
$Mu2STS$	**55.74**	**39.11**	**39.69**	**38.56**	**72.55**	**69.64**	**70.96**	**68.37**

5.6 Comparison with the State-of-the-Art (SOTA) Models

Table 5 compares our proposed model's performance with several SOTA models. Compared to the other models, our proposed model has the highest value for all the metrics. These results signify the strong capability of our model to identify sarcasm by utilizing humor within the context of multimodal memes. It also shows the robustness of the *M3F-CmU* fusion module as compared to existing SOTA methods. Additionally, the increase in precision and recall scores further demonstrates the model's ability to identify sarcastic memes while correctly minimizing false positives and negatives.

Table 5. Performance of SOTA models for sarcasm detection on SHMH dataset. Our proposed model is statistically significant to all the baselines ($p < 0.04$)

Models	Acc	F1	P	R
Zhou et al. [66]	56.58	55.63	53.18	58.32
Hossain et al. [21]	53.93	47.98	48.04	47.93
Chauhan et al. (i) [12]	61.12	56.49	59.12	54.09
Chauhan et al. (ii) [13]	59.83	59.92	58.94	60.93
Sharma et al. (i) [52]	59.23	55.78	55.74	55.82
Sharma et al. (ii) [52]	63.94	56.74	58.82	54.81
Pramanick et al. [43]	65.08	60.63	61.47	59.81
Qin et al. [44]	60.68	59.99	59.96	60.04
Tian et al. [60]	66.61	63.91	62.27	65.64
Bandyopadhyay et al. [7]	66.64	63.37	65.01	63.11
$Mu2STS$ (Ours)	**76.03**	**67.79**	**67.92**	**67.67**

6 Error Analysis and Limitations

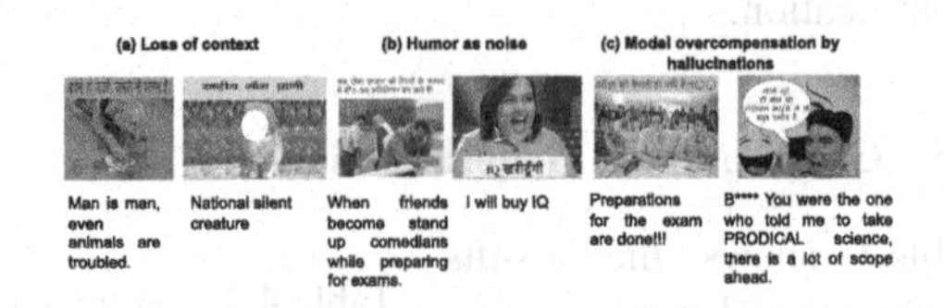

Fig. 6. Examples of miss-classification by the proposed model $Mu2STS$.

Despite its high performance, our proposed model $Mu2STS$ still incurred misclassifications in certain instances. To gain insight into these errors, we identify key reasons for misclassifications by our $Mu2STS$: ***(i) Loss of Context:*** contextual nuances that sarcasm often relies on, leading to misclassification when such context is lacking (Fig. 6.(a)), ***(ii) Humor as Noise:*** instances where humor elements in memes distract the model from sarcasm, causing misclassification (Fig. 6.(b)), and ***(iii) Model Overcompensation by Hallucinations:*** situations where the model overcompensates and misclassifies highly sarcastic memes as slightly sarcastic due to cautious predictions (Fig. 6.(c)).

7 Conclusion and Future Work

In this paper, we explored the challenges of sarcasm detection in memes, highlighting its unique nature in multimodal data. We explored the fine line that differentiates sarcasm from humor, emphasizing that while they often intertwine, humor is not always the primary intent in sarcasm, and introduced the *Mu2STS* model, demonstrating its effectiveness in Hindi memes and cross-lingual contexts. We emphasized the roles of knowledge distillation, multimodal fusion, and perturbation in enhancing the model. Comprehensive evaluations showed its superiority over existing methods. In the future, we will focus on improving the model's understanding of memes by incorporating robust common-sense knowledge to identify hidden cues in sarcasm.

Acknowledgements. The research reported in this paper is an outcome of the project "**HELIOS: Hate, Hyperpartisan, and Hyperpluralism Elicitation and Observer System,**" sponsored by Wipro AI Labs, India.

References

1. Shrout, P.E., Spitzer, R.L., Fleiss, J.L.: Quantification of agreement in psychiatric diagnosis revisited. Arch. Gen. Psychiatry **44**, 2 (1987)
2. Agrawal, A., An, A., Papagelis, M.: Leveraging transitions of emotions for sarcasm detection. In: Huang, J., et al. (eds.) Proceedings of the 43rd International ACM SIGIR Conference on Research and Development in Information Retrieval, SIGIR 2020, Virtual Event, China, 25–30 July 2020, pp. 1505–1508. ACM (2020). https://doi.org/10.1145/3397271.3401183
3. Ba, J.L., Kiros, J.R., Hinton, G.E.: Layer normalization (2016)
4. Babanejad, N., Davoudi, H., An, A., Papagelis, M.: Affective and contextual embedding for sarcasm detection. In: Proceedings of the 28th International Conference on Computational Linguistics, pp. 225–243. International Committee on Computational Linguistics, Barcelona, Spain, December 2020. https://doi.org/10.18653/v1/2020.coling-main.20. https://aclanthology.org/2020.coling-main.20

5. Baishya, A.K.: The conquest of the world as meme: memetic visuality and political humor in critiques of the Hindu right wing in India. Media Cult. Soc. **43**, 1113–1135 (2021). https://api.semanticscholar.org/CorpusID:234224857
6. Bamman, D., Smith, N.A.: Contextualized sarcasm detection on Twitter. In: International Conference on Web and Social Media (2015). https://api.semanticscholar.org/CorpusID:15054136
7. Bandyopadhyay, D., Kumari, G., Ekbal, A., Pal, S., Chatterjee, A., Bn, V.: A knowledge infusion based multitasking system for sarcasm detection in meme. In: Kamps, J., et al. (eds.) Advances in Information Retrieval, pp. 101–117. Springer, Cham (2023). https://doi.org/10.1007/978-3-031-28244-7_7
8. Bansal, S., Garimella, V., Suhane, A., Patro, J., Mukherjee, A.: Code-switching patterns can be an effective route to improve performance of downstream NLP applications: a case study of humour, sarcasm and hate speech detection. In: Proceedings of the 58th Annual Meeting of the Association for Computational Linguistics, pp. 1018–1023. Association for Computational Linguistics, Online, July 2020. https://doi.org/10.18653/v1/2020.acl-main.96. https://aclanthology.org/2020.acl-main.96
9. Bedi, M., Kumar, S., Akhtar, M.S., Chakraborty, T.: Multi-modal sarcasm detection and humor classification in code-mixed conversations. IEEE Trans. Affect. Comput. **14**(2), 1363–1375 (2023). https://doi.org/10.1109/TAFFC.2021.3083522
10. Bojanowski, P., Grave, E., Joulin, A., Mikolov, T.: Enriching word vectors with subword information. arXiv preprint arXiv:1607.04606 (2016)
11. Castro, S., Cubero, M., Garat, D., Moncecchi, G.: Is this a joke? Detecting humor in Spanish tweets. In: Montes-y-Gómez, M., Escalante, H.J., Segura, A., Murillo, J.D. (eds.) IBERAMIA 2016. LNCS (LNAI), vol. 10022, pp. 139–150. Springer, Cham (2016). https://doi.org/10.1007/978-3-319-47955-2_12
12. Chauhan, D.S., Dhanush, S.R., Ekbal, A., Bhattacharyya, P.: All-in-one: a deep attentive multi-task learning framework for humour, sarcasm, offensive, motivation, and sentiment on memes. In: Proceedings of the 1st Conference of the Asia-Pacific Chapter of the Association for Computational Linguistics and the 10th International Joint Conference on Natural Language Processing, pp. 281–290. Association for Computational Linguistics, Suzhou, China, December 2020. https://aclanthology.org/2020.aacl-main.31
13. Chauhan, D.S., Dhanush, S.R., Ekbal, A., Bhattacharyya, P.: Sentiment and emotion help sarcasm? A multi-task learning framework for multi-modal sarcasm, sentiment and emotion analysis. In: Proceedings of the 58th Annual Meeting of the Association for Computational Linguistics, pp. 4351–4360. Association for Computational Linguistics, July 2020. https://doi.org/10.18653/v1/2020.acl-main.401. https://aclanthology.org/2020.acl-main.401
14. Conneau, A., et al.: Unsupervised cross-lingual representation learning at scale (2019). https://doi.org/10.48550/ARXIV.1911.02116. https://arxiv.org/abs/1911.02116
15. Dimitrov, D., et al.: Detecting propaganda techniques in memes. In: Proceedings of the 59th Annual Meeting of the Association for Computational Linguistics and the 11th International Joint Conference on Natural Language Processing (Volume 1: Long Papers), pp. 6603–6617. Association for Computational Linguistics, August 2021. https://doi.org/10.18653/v1/2021.acl-long.516. https://aclanthology.org/2021.acl-long.516
16. Dosovitskiy, A., et al.: An image is worth 16x16 words: transformers for image recognition at scale. CoRR abs/2010.11929 (2020). https://arxiv.org/abs/2010.11929
17. Feng, F., Yang, Y., Cer, D., Arivazhagan, N., Wang, W.: Language-agnostic BERT sentence embedding. CoRR abs/2007.01852 (2020). https://arxiv.org/abs/2007.01852
18. He, K., Zhang, X., Ren, S., Sun, J.: Deep residual learning for image recognition. CoRR abs/1512.03385 (2015). http://arxiv.org/abs/1512.03385
19. Hinton, G., Vinyals, O., Dean, J.: Distilling the knowledge in a neural network (2015)

20. Hossain, E., Sharif, O., Hoque, M.M.: MUTE: a multimodal dataset for detecting hateful memes. In: Proceedings of the 2nd Conference of the Asia-Pacific Chapter of the Association for Computational Linguistics and the 12th International Joint Conference on Natural Language Processing: Student Research Workshop, pp. 32–39. Association for Computational Linguistics, November 2022. https://aclanthology.org/2022.aacl-srw.5
21. Hossain, E., Sharif, O., Hoque, M.M., Akber Dewan, M.A., Siddique, N., Hossain, M.A.: Identification of multilingual offense and troll from social media memes using weighted ensemble of multimodal features. J. King Saud Univ. Comput. Inf. Sci. **34**(9), 6605–6623 (2022). https://doi.org/10.1016/j.jksuci.2022.06.010. https://www.sciencedirect.com/science/article/pii/S1319157822002166
22. Hossain, N., Krumm, J., Vanderwende, L., Horvitz, E., Kautz, H.: Filling the blanks (hint: plural noun) for mad libs humor. In: Proceedings of the 2017 Conference on Empirical Methods in Natural Language Processing, pp. 638–647. Association for Computational Linguistics, Copenhagen, Denmark, September 2017. https://doi.org/10.18653/v1/2020.acl-main.96. https://aclanthology.org/2020.acl-main.96
23. Jing, L., Song, X., Ouyang, K., Jia, M., Nie, L.: Multi-source semantic graph-based multimodal sarcasm explanation generation. In: Proceedings of the 61st Annual Meeting of the Association for Computational Linguistics (Volume 1: Long Papers), pp. 11349–11361. Association for Computational Linguistics, Toronto, Canada, July 2023. https://doi.org/10.18653/v1/2023.acl-long.635. https://aclanthology.org/2023.acl-long.635
24. Joshi, A., Sharma, V., Bhattacharyya, P.: Harnessing context incongruity for sarcasm detection. In: Proceedings of the 53rd Annual Meeting of the Association for Computational Linguistics and the 7th International Joint Conference on Natural Language Processing (Volume 2: Short Papers), pp. 757–762. Association for Computational Linguistics, Beijing, China, July 2015. https://doi.org/10.3115/v1/P15-2124. https://aclanthology.org/P15-2124
25. Kakwani, D., et al.: IndicNLPSuite: monolingual corpora, evaluation benchmarks and pre-trained multilingual language models for Indian languages. In: Findings of EMNLP (2020)
26. Khanuja, S., et al.: MuRIL: multilingual representations for Indian languages. CoRR abs/2103.10730 (2021). https://arxiv.org/abs/2103.10730
27. Kiela, D., et al.: The hateful memes challenge: detecting hate speech in multimodal memes. In: Larochelle, H., Ranzato, M., Hadsell, R., Balcan, M., Lin, H. (eds.) Advances in Neural Information Processing Systems, vol. 33, pp. 2611–2624. Curran Associates, Inc. (2020). https://proceedings.neurips.cc/paper/2020/file/1b84c4cee2b8b3d823b30e2d604b1878-Paper.pdf
28. Kiela, D., et al.: The hateful memes challenge: detecting hate speech in multimodal memes (2021)
29. Kingma, D.P., Ba, J.: Adam: a method for stochastic optimization. CoRR abs/1412.6980 (2015)
30. Kirk, H., et al.: Memes in the wild: assessing the generalizability of the hateful memes challenge dataset. In: Proceedings of the 5th Workshop on Online Abuse and Harms (WOAH 2021), pp. 26–35. Association for Computational Linguistics, August 2021. https://doi.org/10.18653/v1/2021.woah-1.4. https://aclanthology.org/2021.woah-1.4
31. Kumari, G., Das, A., Ekbal, A.: Co-attention based multimodal factorized bilinear pooling for internet memes analysis. In: Proceedings of the 18th International Conference on Natural Language Processing (ICON), pp. 261–270. NLP Association of India (NLPAI), National Institute of Technology Silchar, Silchar, India, December 2021. https://aclanthology.org/2021.icon-main.31
32. Li, J., Li, D., Xiong, C., Hoi, S.: BLIP: bootstrapping language-image pre-training for unified vision-language understanding and generation (2022). https://doi.org/10.48550/ARXIV.2201.12086. https://arxiv.org/abs/2201.12086

33. Li, J., Selvaraju, R.R., Gotmare, A.D., Joty, S., Xiong, C., Hoi, S.: Align before fuse: vision and language representation learning with momentum distillation (2021). https://doi.org/10.48550/ARXIV.2107.07651. https://arxiv.org/abs/2107.07651
34. Li, L.H., Yatskar, M., Yin, D., Hsieh, C.J., Chang, K.W.: VisualBERT: a simple and performant baseline for vision and language. Arxiv (2019)
35. Liu, H., Wang, W., Li, H.: Towards multi-modal sarcasm detection via hierarchical congruity modeling with knowledge enhancement. In: Proceedings of the 2022 Conference on Empirical Methods in Natural Language Processing, pp. 4995–5006. Association for Computational Linguistics, Abu Dhabi, United Arab Emirates, December 2022. https://doi.org/10.18653/v1/2022.emnlp-main.333. https://aclanthology.org/2022.emnlp-main.333
36. Lou, C., Liang, B., Gui, L., He, Y., Dang, Y., Xu, R.: Affective dependency graph for sarcasm detection. In: Proceedings of the 44th International ACM SIGIR Conference on Research and Development in Information Retrieval, SIGIR 2021, pp. 1844–1849. Association for Computing Machinery, New York, NY, USA (2021). https://doi.org/10.1145/3404835.3463061
37. Lunando, E., Purwarianti, A.: Indonesian social media sentiment analysis with sarcasm detection. In: 2013 International Conference on Advanced Computer Science and Information Systems (ICACSIS), pp. 195–198 (2013). https://doi.org/10.1109/ICACSIS.2013.6761575
38. Min, C., Li, X., Yang, L., Wang, Z., Xu, B., Lin, H.: Just like a human would, direct access to sarcasm augmented with potential result and reaction. In: Proceedings of the 61st Annual Meeting of the Association for Computational Linguistics (Volume 1: Long Papers), pp. 10172–10183. Association for Computational Linguistics, Toronto, Canada, July 2023. https://doi.org/10.18653/v1/2023.acl-long.566. https://aclanthology.org/2023.acl-long.566
39. Mishra, S., et al.: Memotion 3: dataset on sentiment and emotion analysis of codemixed Hindi-English memes. CoRR abs/2303.09892 (2023). https://doi.org/10.48550/arXiv.2303.09892
40. Pires, T., Schlinger, E., Garrette, D.: How multilingual is multilingual BERT? In: Proceedings of the 57th Annual Meeting of the Association for Computational Linguistics, pp. 4996–5001. Association for Computational Linguistics, Florence, Italy, July 2019. https://doi.org/10.18653/v1/P19-1493. https://aclanthology.org/P19-1493
41. Poria, S., Cambria, E., Hazarika, D., Vij, P.: A deeper look into sarcastic tweets using deep convolutional neural networks. In: Proceedings of COLING 2016, the 26th International Conference on Computational Linguistics: Technical Papers, pp. 1601–1612. The COLING 2016 Organizing Committee, Osaka, Japan, December 2016. https://aclanthology.org/C16-1151
42. Pramanick, S., Roy, A., Patel, V.M.: Multimodal learning using optimal transport for sarcasm and humor detection. In: Proceedings of the IEEE/CVF Winter Conference on Applications of Computer Vision (WACV), pp. 3930–3940, January 2022
43. Pramanick, S., Sharma, S., Dimitrov, D., Akhtar, M.S., Nakov, P., Chakraborty, T.: MOMENTA: a multimodal framework for detecting harmful memes and their targets. CoRR abs/2109.05184 (2021). https://arxiv.org/abs/2109.05184
44. Qin, L., et al.: MMSD2.0: towards a reliable multi-modal sarcasm detection system. In: Findings of the Association for Computational Linguistics: ACL 2023, pp. 10834–10845. Association for Computational Linguistics, Toronto, Canada, July 2023. https://doi.org/10.18653/v1/2023.findings-acl.689. https://aclanthology.org/2023.findings-acl.689
45. Radford, A., et al.: Learning transferable visual models from natural language supervision. In: Meila, M., Zhang, T. (eds.) Proceedings of the 38th International Conference on Machine Learning. Proceedings of Machine Learning Research, vol. 139, pp. 8748–8763. PMLR, 18–24 July 2021. https://proceedings.mlr.press/v139/radford21a.html
46. Radford, A., et al.: Learning transferable visual models from natural language supervision (2021). https://doi.org/10.48550/ARXIV.2103.00020. https://arxiv.org/abs/2103.00020

47. Ribeiro, M.T., Singh, S., Guestrin, C.: "why should I trust you?": explaining the predictions of any classifier. In: Proceedings of the 22nd ACM SIGKDD International Conference on Knowledge Discovery and Data Mining, San Francisco, CA, USA, 13–17 August 2016, pp. 1135–1144 (2016)
48. Rijhwani, S., Sequiera, R., Choudhury, M., Bali, K., Maddila, C.S.: Estimating code-switching on Twitter with a novel generalized word-level language detection technique. In: Proceedings of the 55th Annual Meeting of the Association for Computational Linguistics (Volume 1: Long Papers), pp. 1971–1982. Association for Computational Linguistics, Vancouver, Canada, July 2017. https://doi.org/10.18653/v1/P17-1180. https://aclanthology.org/P17-1180
49. Schifanella, R., de Juan, P., Tetreault, J., Cao, L.: Detecting sarcasm in multimodal social platforms. In: Proceedings of the 24th ACM International Conference on Multimedia, MM 2016, pp. 1136–1145. Association for Computing Machinery, New York, NY, USA (2016). https://doi.org/10.1145/2964284.2964321
50. Sharma, C., et al.: SemEval-2020 task 8: memotion analysis- the visuo-lingual metaphor!, pp. 759–773, December 2020. https://doi.org/10.18653/v1/2020.semeval-1.99. https://aclanthology.org/2020.semeval-1.99
51. Sharma, S., Akhtar, M.S., Nakov, P., Chakraborty, T.: DISARM: detecting the victims targeted by harmful memes. In: Findings of the Association for Computational Linguistics: NAACL 2022, pp. 1572–1588. Association for Computational Linguistics, Seattle, United States, July 2022. https://doi.org/10.18653/v1/2022.findings-naacl.118. https://aclanthology.org/2022.findings-naacl.118
52. Sharma, S., Siddiqui, M.K., Akhtar, M.S., Chakraborty, T.: Domain-aware self-supervised pre-training for label-efficient meme analysis. In: Proceedings of the 2nd Conference of the Asia-Pacific Chapter of the Association for Computational Linguistics and the 12th International Joint Conference on Natural Language Processing (Volume 1: Long Papers), pp. 792–805. Association for Computational Linguistics, Online Only, November 2022. https://aclanthology.org/2022.aacl-main.60
53. Simonyan, K., Zisserman, A.: Very deep convolutional networks for large-scale image recognition. In: International Conference on Learning Representations (2015)
54. Srivastava, H., Varshney, V., Kumari, S., Srivastava, S.: A novel hierarchical BERT architecture for sarcasm detection. In: Proceedings of the Second Workshop on Figurative Language Processing, pp. 93–97. Association for Computational Linguistics, July 2020. https://doi.org/10.18653/v1/2020.figlang-1.14. https://aclanthology.org/2020.figlang-1.14
55. Stieger, S., Formann, A.K., Burger, C.: Humor styles and their relationship to explicit and implicit self-esteem. Personality Individ. Differ. **50**(5), 747–750 (2011). https://doi.org/10.1016/j.paid.2010.11.025. https://www.sciencedirect.com/science/article/pii/S0191886910005751
56. Student: the probable error of a mean. Biometrika, pp. 1–25 (1908)
57. Suryawanshi, S., Chakravarthi, B.R., Arcan, M., Buitelaar, P.: Multimodal meme dataset (MultiOFF) for identifying offensive content in image and text. In: Proceedings of the Second Workshop on Trolling, Aggression and Cyberbullying, pp. 32–41. European Language Resources Association (ELRA), Marseille, France, May 2020. https://aclanthology.org/2020.trac-1.6
58. Tan, H., Bansal, M.: LXMERT: learning cross-modality encoder representations from transformers. In: Proceedings of the 2019 Conference on Empirical Methods in Natural Language Processing and the 9th International Joint Conference on Natural Language Processing (EMNLP-IJCNLP), pp. 5100–5111. Association for Computational Linguistics, Hong Kong, China, November 2019. https://doi.org/10.18653/v1/D19-1514. https://aclanthology.org/D19-1514

59. Tay, Y., Luu, A.T., Hui, S.C., Su, J.: Reasoning with sarcasm by reading in-between. In: Proceedings of the 56th Annual Meeting of the Association for Computational Linguistics (Volume 1: Long Papers), pp. 1010–1020. Association for Computational Linguistics, Melbourne, Australia, July 2018. https://doi.org/10.18653/v1/P18-1093. https://aclanthology.org/P18-1093
60. Tian, Y., Xu, N., Zhang, R., Mao, W.: Dynamic routing transformer network for multimodal sarcasm detection. In: Proceedings of the 61st Annual Meeting of the Association for Computational Linguistics (Volume 1: Long Papers), pp. 2468–2480. Association for Computational Linguistics, Toronto, Canada, July 2023. https://doi.org/10.18653/v1/2023.acl-long.139. https://aclanthology.org/2023.acl-long.139
61. Tsur, O., Davidov, D., Rappoport, A.: ICWSM - a great catchy name: semi-supervised recognition of sarcastic sentences in online product reviews. In: Proceedings of the International AAAI Conference on Web and Social Media, vol. 4, no. 1, pp. 162–169 (2010). https://doi.org/10.1609/icwsm.v4i1.14018. https://ojs.aaai.org/index.php/ICWSM/article/view/14018
62. Xiong, T., Zhang, P., Zhu, H., Yang, Y.: Sarcasm detection with self-matching networks and low-rank bilinear pooling, pp. 2115–2124 (2019). https://doi.org/10.1145/3308558.3313735
63. Xiong, T., Zhang, P., Zhu, H., Yang, Y.: Sarcasm detection with self-matching networks and low-rank bilinear pooling. In: The World Wide Web Conference (2019). https://api.semanticscholar.org/CorpusID:86385192
64. Zhang, M., Zhang, Y., Fu, G.: Tweet sarcasm detection using deep neural network. In: Proceedings of COLING 2016, the 26th International Conference on Computational Linguistics: Technical Papers, pp. 2449–2460. The COLING 2016 Organizing Committee, Osaka, Japan, December 2016. https://aclanthology.org/C16-1231
65. Zhao, W., Huang, Q., Xu, D., Zhao, P.: Multi-modal sarcasm generation: dataset and solution. In: Findings of the Association for Computational Linguistics: ACL 2023, pp. 5601–5613. Association for Computational Linguistics, Toronto, Canada, July 2023. https://doi.org/10.18653/v1/2023.findings-acl.346. https://aclanthology.org/2023.findings-acl.346
66. Zhou, Y., Chen, Z.: Multimodal learning for hateful memes detection (2020)

Robustness in Fairness Against Edge-Level Perturbations in GNN-Based Recommendation

Ludovico Boratto[1], Francesco Fabbri[2], Gianni Fenu[1], Mirko Marras[1](✉), and Giacomo Medda[1]

[1] University of Cagliari, Cagliari, Italy
{ludovico.boratto,mirko.marras}@acm.org, {fenu,giacomo.medda}@unica.it
[2] Spotify, Barcelona, Spain
francescof@spotify.com

Abstract. Efforts in the recommendation community are shifting from the sole emphasis on utility to considering beyond-utility factors, such as fairness and robustness. Robustness of recommendation models is typically linked to their ability to maintain the original utility when subjected to attacks. Limited research has explored the robustness of a recommendation model in terms of fairness, e.g., the parity in performance across groups, under attack scenarios. In this paper, we aim to assess the robustness of graph-based recommender systems concerning fairness, when exposed to attacks based on edge-level perturbations. To this end, we considered four different fairness operationalizations, including both consumer and provider perspectives. Experiments on three datasets shed light on the impact of perturbations on the targeted fairness notion, uncovering key shortcomings in existing evaluation protocols for robustness. As an example, we observed perturbations affect consumer fairness on a higher extent than provider fairness, with alarming unfairness for the former. Source code: https://github.com/jackmedda/CPFairRobust.

Keywords: Robustness · Fairness · Recommendation · GNN · Perturbation · Multi-Stakeholder · Provider · Consumer

1 Introduction

Individuals are increasingly interacting with recommender systems, enjoying the benefits of personalized services provided by e-commerce and streaming platforms. These services are designed to adapt to the preferences and interests of consumers about the content they discover, while also meeting the expectations of content providers, who seek visibility and engagement. However, the experiences of these stakeholders can be compromised by specialized attacks targeting recommender systems. These attacks aim to manipulate the recommendations generated by the systems according to the attacker's objectives [1,42,63].

The effectiveness of attacks against recommender systems has been demonstrated across a diverse range of recommendation models, including those based

N. Goharian et al. (Eds.): ECIR 2024, LNCS 14610, pp. 38–55, 2024.
https://doi.org/10.1007/978-3-031-56063-7_3

on k-nearest neighborhood [42], matrix factorization [31], association rules [57], recurrent neural networks [49], and graphs [19,39]. In this context, 'poisoning' attacks have become particularly prevalent [51,63]. These attacks primarily involve data perturbation during the training stage [1,51], often through the introduction of fake users, also known as 'shilling' attacks [1,42,49]. Concerted efforts have been increasingly devoted to enhancing the robustness of recommender systems against various attacks [12,29,63–65]. Fraudster detection [10,65] and adversarial learning [12,29,55] have emerged as the primary defensive strategies. The former seeks to identify and mitigate the influence of fake users, while the latter introduces perturbations to strengthen models against adversarial samples.

Differently from other research fields (e.g., computer sensing [3,15,16,59], code generation [36], and program understanding [62]), attacks against recommendation have prioritized the maximization of model disruption, often at the expense of constructive objectives, without considering their impact on model robustness. The existing methods for evaluating robustness merely compare recommendation utility before/after attacks [63]. This practice is unfortunately limited, given that the overall utility can remain stable even when recommendations are significantly altered [41]. This limitation prevents from detecting the impact of attacks on beyond-accuracy objectives, such as trustworthiness [52], fairness [54], and explainability [66]. Few prior works consider robustness in beyond-accuracy properties, such as bias [46] and sparseness [68], but do not cover fairness [63]. In other domains, this interplay has already been addressed [35,43].

In this paper, we provide a novel comprehensive analysis on the robustness of graph-based recommender systems in terms of fairness, referred to as *robustness in fairness*. Specifically, we investigate the extent to which the system fairness remains stable, from both the consumer [7,13,54] and the provider [25,26,47] sides, under attack scenarios. We address this issue on systems based on graphs due to their state-of-the-art performance and the extensive range of attacks on graph data [14,19,39,51,64,65]. Adding and deleting edges is a popular technique for attacks in graph data [51]. To this end, we extended an approach that perturbs a graph at the edge-level to explain the predictions in several downstream tasks [30,34]. This approach iteratively performs poisoning-like attacks against recommender systems based on Graph Neural Networks (GNNs) and monitors fairness as the user-item interaction graph gets gradually perturbed, encompassing different types of perturbations and fairness operationalizations. Although our experimental evaluation is driven by the employed attack, it is important to note that the attack itself does not constitute the main contribution. Rather, this paper specifically aims to study the robustness in fairness in recommendation and explore the nuances in the GNN models' outcomes after attacks.

Our study operates within a white-box scenario, simulating the role of an attacker aiming to compromise the group fairness of a recommender system. In this scenario, the perturbation process involves modifying the input graph that feeds into the GNN. Such attacks may have real-world consequences, including compromising a company's reputation in the public eye [38], both through media coverage and legal implications that may result in sanctions and other repercussions. Concerning the recent regulations in terms of robustness and fairness of automated systems [21,40], such consequences could illustrate a worrying scenario. Our experimental study showcases an extensive characterization of robustness in fairness against poisoning-like attacks, by employing three datasets and three GNN-based recommender systems. The tested models exhibit a higher sensitivity to attacks tailored for group consumer fairness compared with provider one. Specifically, the unfairness levels across consumer groups can be increased by a restrained amount of perturbations, whereas the impact on provider fairness is limited by the prior unfairness level exhibited in the original recommendations.

2 Related Work

2.1 Attacks and Robustness in Recommendation

The researchers addressing attacks and robustness in recommendation do not necessarily see these properties as interconnected, although most of the literature in robustness regards attacks [63]. In fact, several papers solely focused on identifying attacks and treating them as strategies for achieving maximum disruptions [14,19,31,42,49,57] by injecting fake users to increase the recommendation of specific items [19,42], or adversarially generate unnoticeable fake profiles [14]. Conversely, other works focused on improving robustness without necessarily viewing the adopted attack as the actual contribution [12,41,60,61,65]. For instance, [65] detected suspicious users as fraudsters using neural random forests. Despite these advancements, a comprehensive analysis of the impact of attacks on the models' robustness, particularly in terms of accuracy and other critical properties, is notably absent. This gap in research is remarkable, especially when compared to analogous studies conducted in other fields [3,15,16,36,43,59,62].

2.2 Fairness in Recommendation

Due to recently issued regulations [21,40], researchers are increasingly prioritizing beyond-accuracy aspects in recommendation, as explainability [66] and fairness [54]. The relevant amount of recent works studying consumer and provider (un)fairness addressed their assessment [7,8,25,26,47], mitigation [5,6,9,22,32], and explanation [17,23,24,37]. Despite calls for unifying the goals of robustness and fairness in recommendation [63], to the best of our knowledge, [58] is the only work that focused on both properties. Specifically, [58] proposed a fair and distributionally robust method to solve the distribution shift problem between the training and testing sets. Unfortunately, no study addressed the assessment of robustness in group consumer/provider fairness against specialized attacks.

2.3 Robustness and Beyond-Accuracy Aspects

Some studies on robustness in recommendation considered beyond-accuracy properties, e.g., bias [46] and sparseness [68], as pertaining to a kind of robustness [63]. However, their scope does not cover the fairness property envisioned in our study. On the other hand, the literature in other fields has witnessed the introduction of novel techniques of certified robustness for text classification [43], and novel attacks that target the fairness of classifiers [38,48]. Nevertheless, their works regard classification tasks, where attacks and robustness methods differ from the ones employed in recommendation, as those targeted by this paper.

3 Methodology

3.1 Perturbation Task in Graph-Based Recommendation

Our perturbation task tailored for GNN-based recommender systems aims to perturb the adjacency matrix through edge perturbations to alter the predicted recommendation lists, and test the systems' robustness in fairness. We then distinguish between the recommendation task and the proper perturbation task.

Recommendation Task. In a typical recommendation scenario, a model learns the preferences of a set of users U from their past interactions with a catalog of items I. The network of user-item interactions can be represented by means of an undirected bipartite graph $G = (V, E)$, where $V = U \cup I$ is the set of nodes, E is the set of edges between user and item nodes. G can be encoded in a $n \times n$ adjacency matrix A, where $A_{u,i} > 0$ denotes an edge links the user u with the item i, otherwise $A_{u,i} = 0$. We can feed A to a GNN f to predict the probability of missing user-item links. Specifically, f can be parameterized by a weight matrix W and optimized to recommend to each user a list of the top-k items sorted by the predicted linking probability in descending order. Let $q_u@k$ be the top-k list recommended to user u and $Q@k$ the set of all $q_u@k, \forall u \in U$.

Perturbation Task. Following [63], robustness can be estimated by the disparity between the performance measured with the original (non-perturbed) data and the perturbed data. A model reporting a disparity lower than a threshold ϵ and a perturbation bounded by a constant γ would be denoted as (γ, ϵ)-robust. In our graph-based scenario, the original data is the adjacency matrix A, and we denote its perturbed version as $\tilde{A}$. Given our fairness-related task, we define the performance by means of a fairness metric M. We can then formally define the (γ, ϵ)-robustness in fairness of our recommender system f as follows:

$$\Delta = M(f(\tilde{A}, W), A) - M(f(A, W), A), \quad \|\Delta\|_2^2 \leq \epsilon, \quad |\tilde{E}| \leq \gamma \tag{1}$$

where M estimates the fairness level based on the outcome of f and A, and $\tilde{E}$ denotes the set of candidate edges for perturbation. Although [63] does not guide the selection of ϵ and γ, a small ϵ guarantees a greater level of robustness, while a small γ reflects an attack that is harder to detect. Focused on analyzing the robustness in fairness, we do not set a fixed bound on the extent of edge perturbations, i.e. $\gamma = +\infty$. Addressing edge perturbations as deletion and addition, the range of $|\tilde{E}|$ is then $[1, |E|]$ for deletion, and $[1, |U| \times |I| - |E|]$ for addition.

We seek to test the robustness in fairness in recommendation by identifying the edge perturbations that maximize the disparity in (1), prioritizing fewer perturbations. In other words, we aim to optimize the following objective function:

$$\min_{\hat{p}} \quad - \left\| M(f(\tilde{A}, W; \hat{p}), A) - M(f(A, W), A) \right\|_2^2 + \lambda \left\| \Gamma(\tilde{A}, A) \right\|_2^2 \tag{2}$$

where Γ is a distance function [34,51], $\hat{p}$ is a trainable weight used to identify the edges to be perturbed, $\lambda \in \mathbb{R}$ is a hyper-parameter that is used to control the weight between the two terms. The minus sign applied to the first term optimizes the objective function towards maximizing the disparity Δ, resulting in unfair recommendations. However, the original fairness level estimated by M is not affected by the perturbation process. Hence, we can simplify (2) as follows:

$$\min_{\hat{p}} \quad -M(f(\tilde{A}, W; \hat{p}), A) + \lambda \left\| \Gamma(\tilde{A}, A) \right\|_2^2 \tag{3}$$

3.2 Graph Perturbation Mechanism

Following works of explainability in GNNs [30,34], we use a sparsification method to obtain a binary perturbation tensor from a trainable real-valued weight (first described by [50]) and extend it to the recommendation scenario. First, we enlarge the space of the candidate edges to the entire graph. Second, we replace the perturbation matrix P used in [34] with a binary perturbation vector $p \in \{0, 1\}^{|\tilde{E}|}$, such that solely the relevant user-item connections are perturbed instead of affecting also self-loops, user-user and item-item links. p can be derived from the trainable weight $\hat{p}$ (2) by applying a sigmoid transformation and a binarization to $\hat{p}$, so as to map values lower than 0.5 to 0, otherwise to 1.

The perturbation mechanism can be thought as a substitution process that updates the entries of the adjacency matrix A with the entries of p, resulting in $\tilde{A}$. A fixed relation between the 2D index of A and the 1D index of p establishes which candidate edge $(u, i) \in \tilde{E}$ will be perturbed by the j-th entry of p. The entries update process resulting in $\tilde{A}$ is formally defined as:

$$\tilde{A} = A \dotplus p, \quad \tilde{A}_{u,i} = \begin{cases} p_j & \text{if } (u, i) \in \tilde{E} \\ A_{u,i} & \text{otherwise} \end{cases} \tag{4}$$

$\dotplus$ denotes the perturbation operator for edge deletion $\dotplus$ *Del* or addition $\dotplus$ *Add*.

This perturbation mechanism is performed iteratively by gradually modifying A to generate $\tilde{A}$, until the perturbed edges optimize the targeted task. Specifically, we initialize $\hat{p}$ based on the perturbation type (deletion or addition), such that no edge is affected in A, i.e. $A + p = A$. At each iteration, we generate $\tilde{A}$ with $\hat{p}$, and feed $\tilde{A}$ to the GNN f to produce the corresponding recommendations. The latter are processed by (3) to estimate the fairness level, the distance between $\tilde{A}$ and A, and to update the weight $\hat{p}$ accordingly. Finally, the process stops based on a predefined criterion, e.g., the impact of the last perturbation.

3.3 Fairness Notion and Operationalization

Fairness Notion. We proceed to define the fairness metric M. We follow recent works [2,7,20,22,23,32,47,56] that emphasized the relevance of the group fairness notion of *demographic parity* from both the consumer [7,32,56] and provider side [22,23,47]. For the former side, demographic parity is satisfied if consumer groups experience the same level of recommendation utility. For the latter side, the notion is satisfied if the probability of being recommended is equal across provider groups, proportionally to their representation in the catalog.

We ground our work on a binary setting as previous studies [7,22,23,32], where each stakeholder set Z ($Z \subseteq U$ for consumers, $Z \subseteq I$ for providers[1]) can be partitioned in two groups, Z_1 and Z_2. Multiple attributes, e.g., gender and age, of each stakeholder could produce a distinct partition of Z in two groups. Demographic parity (DP) can then be operationalized as the following disparity:

$$DP = \left\| S(f(A, W), A^{Z_1}) - S(f(A, W), A^{Z_2}) \right\|_2^2 \quad (5)$$

where S represents the metric used to estimate the performance w.r.t. the corresponding stakeholder, e.g., exposure for a provider, A^{Z_1} and A^{Z_2} respectively denote the adjacency sub-matrices with regard to the two partitions Z_1 and Z_2.

Consumer and Provider Fairness Operationalization. Depending on the adopted metric S, we can define specific operationalizations of DP, which reflect distinct perspectives of the unfairness issue. For each stakeholder, we contemplate two types of operationalization, a rank-aware and a rank-agnostic one.

We underline that DP represents M in (1)-(2)-(3), given that it estimates the fairness performance of a recommender system. Therefore, we define each operationalization of DP, the corresponding metric S used for evaluation, and the differentiable approximation of S to be used in the objective function in (3):

- **Consumer Preference** (CP): it estimates the consumer fairness as the disparity across consumers groups in rank-aware top-k recommendation utility, which can be measured by the Normalized Discounted Cumulative Gain (NDCG@k). Following [44,56], NDCG (N@k) is approximated as the differentiable function $\widehat{NDCG}$, where the rank of an item is defined in terms of the pairwise preference with respect to any other item in the catalog.

[1] We do not consider other features and associate each item with a distinct provider.

- **Consumer Satisfaction** (CS): it estimates the consumer fairness as the disparity across consumers groups in rank-agnostic top-k recommendation utility, which can be measured by the Precision (P@k). We optimize P@k by treating the recommendation task as a binary classification task, by using a sigmoid function followed by a binary cross entropy loss. This task aims to include relevant items in the top-k list, regardless of their position.
- **Provider Exposure** (PE): it estimates the provider fairness as the disparity in exposure across providers groups. Following [22,23,25,26], we define the exposure of a generic provider group I_* as the average number of exposures in $Q@k$ (estimated by an indicator function $\mathbb{1}[\cdot]$ as [22,23]), discounted by the importance of their position [25,26,47] (as for DCG), and normalized by the ideal exposure [25,26] (as for NDCG). Formally:

$$\text{Exposure}(I_* \mid Q@k) = \frac{|I|}{|I_*|}\frac{1}{|U|}\sum_{u\in U}\frac{\sum_{j=1}^{k}\frac{\mathbb{1}[i_j \in I_*]}{log_2(j+1)}}{\sum_{j=1}^{k}\frac{\mathbb{1}[i_j \in I]}{log_2(j+1)}} \tag{6}$$

  We leverage the approximation proposed by [23], where the indicator function is replaced by the predicted linking probability (item relevance).
- **Provider Visibility** (PV): it estimates the provider fairness as the disparity in visibility across providers groups. Following [18,25,26], we define the visibility of a generic provider group I_* as the average number of exposures in $Q@k$ (estimated by $\mathbb{1}[\cdot]$ as for PE). Formally:

$$\text{Visibility}(I_* \mid Q@k) = \frac{|I|}{|I_*|}\frac{1}{|U|k}\sum_{u\in U}\sum_{i\in q_u@k}\mathbb{1}[i \in I_*] \tag{7}$$

  We approximate PV by replacing Visibility$(\cdot,\cdot)$ with the loss function used for CS, which would aim to include the items of I_* in the top-k list.

CP and CS $\in [0,1]$, PE and PV $\in [0, \frac{|I|}{|I_1|}]$, for which 0 denotes fairness.

The approximations of PE and PV are inspired by [23], but the authors measure the disparity only according to the top-k items, limiting the exposure/visibility information, given that k is usually small. It causes the gradient to be computed only for the top-k items, which will not necessarily be included in the final recommendations, due, for instance, to items already enjoyed by some users. Instead, we set k to be 10% of the items catalog size $|I|$ to expand the operational scope of PE and PV. This choice also helps when just one of the groups is represented in $Q@k$, and enables our perturbation task to better optimize the presence of one of the groups in positions closer to the top-k ones.

4 Experimental Evaluation

The following experiments aim to answer the following research questions:

RQ1: What is the extent to which edge perturbations impact the robustness in fairness of recommender systems?

RQ2: Are the adopted models similarly affected in terms of alterations in robustness in fairness as edge perturbations gradually increase?
RQ3: Which consumer or provider group should be more affected by the edge perturbation to nuke the models' robustness in fairness?

Table 1. Left: datasets' statistics (*Repr*: *Representation*, *O*: *Older*, *Y*: *Younger*, *F*: *Female*, *M*: *Males*). **Right**: original models' performance in N@10 (%) and P@10 (%).

		ML1M [27]	LF1K [11]	INS [32]
# Users		6,040	268	346
# Items		3,706	51,609	20
# Interactions		1,000,209	200,586	1,879
Domain		Movie	Music	Insurance
Repr.	Age	O : 43.4%	O : 42.2%	O : 49.4%
		Y : 56.6%	Y : 57.8%	Y : 50.6%
	Gender	F : 28.3%	F : 42.2%	F : 23.4%
		M : 71.7%	M : 57.8%	M : 76.6%

	INS		LF1K		ML1M	
	N@10	P@10	N@10	P@10	N@10	P@10
GCMC	76.40	9.60	39.66	38.47	12.62	11.35
LGCN	78.07	9.74	39.81	38.36	12.68	11.35
NGCF	78.39	9.74	39.71	38.43	12.94	11.65

4.1 Evaluation Setting

Evaluation Protocol. The perturbation process is run for 200 epochs, but early stopped if the increment in Δ after 15 consecutive epochs is lower than 0.001.

Given our objective of testing the robustness in fairness, we do not perform a classic poisoning attack as defined in [1,51,63]. Specifically, the combination of datasets, models, perturbation types, and fairness operationalizations sums up to 108 attacks, but we aim to also address intermediary perturbation stages, which would result in an impracticable amount of models re-training processes. To this end, we estimate the first term of Δ by substituting the original adjacency matrix A with the perturbed one $\tilde{A}$ at the inference stage, and maintaining the models' parameters constant. If $\tilde{A}$ was generated by external tools or approaches, such poisoning-like attack would reflect a white/grey-box setting, with the unique requirement of having access to the saved representation of A.

Models. We rely on Recbole [67] and select GCMC [4], LightGCN (LGCN) [28], and NGCF [53] as the GNN-based recommender systems for our study. Though the set of employed models is limited, they cover different architectures to learn the users' preferences: GCMC leverages an auto-encoder structure, LGCN learns from linear relationships between users and items, NGCF adopts features transformation and nonlinear activation on the message-passing step. While the last two models generate the recommendations by the learnt user and item embeddings, GCMC performs a complete forward process during inference. Thus, GCMC represents a more suitable candidate for the attacker, who does not need to force the embeddings re-generation after the graph perturbation.

Datasets. We rely on [7], which includes MovieLens-1M [27] (ML1M) and LFM-1K [11] (LF1K)[2]. We also consider Insurance [33] (INS) and discard consumers with less than 5 interactions. Table 1 reports the datasets' statistics and the original models' performance. We use the items' popularity to partition the provider set in short-head I_1 and long-tail I_2 items such that $\frac{|I_1|}{|I_2|} = \frac{1}{4}$ as in [22,23]. For each user, the interactions are sorted in ascending order of recency, split by a ratio 7:1:2, and each split respectively assigned to the training, validation, and testing set. We use the validation set to select the best original model, and the ground truth of the testing set to optimize the fairness operationalizations (Sect. 3.3). Using the test relevance judgements would result in a powerful attack and support our main concern, i.e. the analysis of robustness in fairness[3].

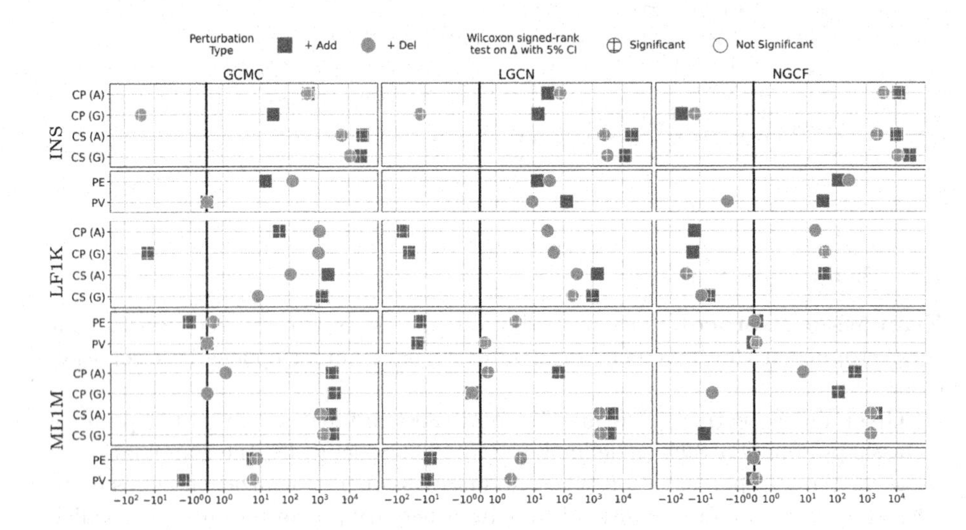

Fig. 1. Impact of edges addition ($\dotplus$ *Add*) and deletion ($\dotplus$ *Del*) on the robustness in fairness, reported as the relative difference in M between the non-perturbed and perturbed model, i.e. $\Delta/M(f(A,W),A)$ ((1)). A stands for *Age*, G for *Gender*.

4.2 RQ1: Impact on Robustness in Fairness

We first assess the impact of edge-level perturbations on the robustness in fairness, i.e. Δ, where M corresponds to the formulation of DP in (5) and its specialized operationalizations, e.g., PE for provider exposure fairness.

Figure 1 reports the impact on robustness in fairness as the relative difference in M between the original fairness level and after each perturbation type, i.e. Δ divided by the second term in (1). The reported values pertain to the iteration where at least one edge was perturbed and $\tilde{A}$ affected the most Δ for each model.

[2] The timestamp of each (u, i) refers to the last interaction between u and a i's song.
[3] This design choice does not constitute the issue highlighted in [45].

The formulation of Δ in (1) enables us to easily denote positive values as increment in fairness level disparity across groups, i.e. increment in DP, otherwise a decrement of the latter. Therefore, positive values reflect a successful outcome for the attacker, whose goal is to make the system generate unfair recommendations. Negative values derive from instances of $\tilde{A}$ that reduce the unfairness level even at the first iterations. The x-axis has been symmetrically log-normalized to highlight the remarkable impact on Δ on the systems. For instance, both $\dotplus$ *Add* and $\dotplus$ *Del* significantly caused an impact for CS on LGCN under ML1M, precisely increasing DP by more than 1,000%. Also the fairness levels under INS were remarkably affected, reaching values higher than 10,000%.

Successful attacks are especially reported in terms of edge deletions, given that most of the points labeled as $\dotplus$ *Del* are depicted at the right of the zero line. Conversely, $\dotplus$ *Add* caused varied results, ranging from reductions of fairness level disparity, e.g., on LGCN under LF1K, to disruptions in robustness in fairness, e.g., on GCMC under ML1M for the consumer side. This observation could be related to the addition of information in the graph due to $\dotplus$ *Add*, but also to the larger edge sample space for the latter compared with $\dotplus$ *Del*. The experiments on provider fairness confirm $\dotplus$ *Del* as a more effective perturbation attack than $\dotplus$ *Add*. However, several settings report an optimal robustness, e.g., GCMC and NGCF under LF1K, and other ones a negative orientation of Δ, e.g., LGCN under LF1K and ML1M. Across the models, NGCF exhibits the least sensitivity to perturbations, especially on provider fairness and under LF1K, given that most of the points are close to the zero line. This may be attributed to the feature transformation and nonlinear activation applied in the NGCF message-passing scheme, which diminish the impact of the perturbed graph on the predictions.

4.3 RQ2: Robustness in Fairness Under Incremental Perturbations

The previous research question aimed to highlight if the edge perturbations could be able to affect the robustness in fairness of the considered models, accounting only for the iteration with the highest impact on Δ. It is then unclear to what extent the robustness is affected by the gradual increment of edge perturbations.

To this end, we report the DP for each iteration of the perturbation process through the points in Fig. 2. For each setting, a horizontal dashed line denotes the original DP to highlight the gap caused by the increment of perturbed edges. A robust model would then be described by points close to the dashed line, while far points conceive the perturbed edges significantly affected the fairness level. Even the size of such points is important: big points close to the dashed line denote the model is robust despite a relevant amount of perturbed edges.

Consumer-Side. Across the datasets, NGCF is the most robust model, given that in most of the setting the points are distributed closer to the original DP compared with the other two systems. This likely stems from the observation highlighted in the previous section. GCMC and LGCN behave in a similar way, especially in terms of CS, for which the addition of edges ($\dotplus$ *Add*) has a more significant influence on DP compared with the deletion ($\dotplus$ *Del*). However, LGCN

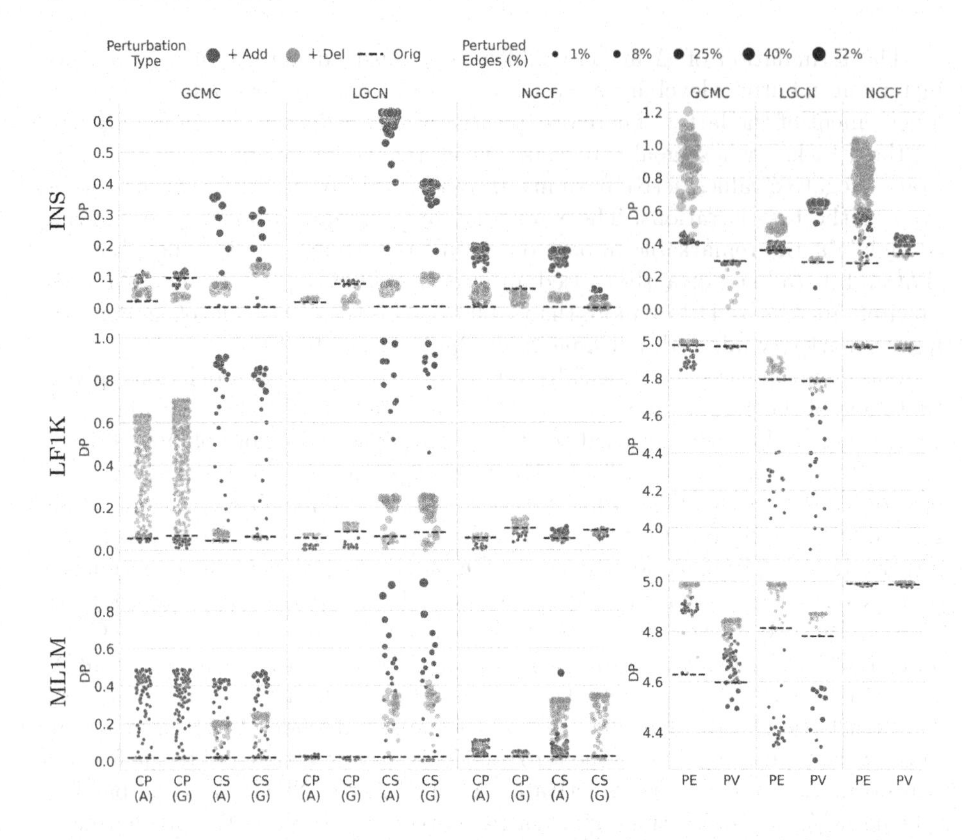

Fig. 2. Trend of the operationalizations of DP across different stages of the perturbation process. Each stage reflects a fraction of perturbed edges w.r.t. $\tilde{E}$, depicted as points gradually larger as the amount of perturbed edges increases. The horizontal dashed line labeled as *Orig* denotes the DP of the recommendations generated by the non-perturbed system. In the consumer-side results, *A* stands for *Age*, *G* for *Gender*.

reports a systematic robustness for the rank-aware operationalization CP with only a few small points reported in the proximity of the original DP.

Some experiments underline the attack could be influenced by the dataset and sensitive attribute. The former influence factor can be observed on GCMC, for which + *Del* significantly affected CP under LF1K; a similar result was reported under ML1M, but due to the other perturbation type + *Add*. The latter influence factor can be observed under INS on NGCF, where the impact of + *Add* on DP is reported only across age groups, regardless of the operationalization.

Provider-Side. Differently from the consumer-side, the original systems report a high degree of unfairness, close to the maximum value $\frac{|I|}{|I_1|} = 5$. Hence, the edge perturbations adopted by a potential attacker cannot significantly affect the robustness in fairness, as highlighted under LF1K on all models, and under ML1M on NGCF, where the original level of PE and PV is remarkable.

In the other settings, deleting edges is more reliable in influencing the robustness in provider fairness compared with $+$ *Add*. This behavior is especially reported under INS, but also under ML1M on GCMC and LGCN.

Other observations contrast with the consumer-side evaluation. First, NGCF exhibits a high degree of sensitivity to the perturbations under INS, as emphasized by the gradual increment in PE as more edges are deleted. This is possibly due to the limited size of INS, which prevents NGCF from robustly learning the users' preferences in a generalized way. Second, the differences in trend between GCMC and LGCN are more remarked, with $+$ *Add* affecting more the robustness of GCMC under ML1M w.r.t. the LGCN's one. This can be explained by the GCMC encoder architecture, which possibly interprets the added edges as noise, which is instead captured as new information by the linear step in LGCN.

4.4 RQ3: Edge Perturbations Influence on Groups

Group unfairness denotes one of the groups is favored by the system, e.g., with a higher level of exposure. We denote such group as *advantaged* and the other group as *disadvantaged*. We seek to discover whether the impact on Δ is due to a greater amount of edge perturbations applied to the advantaged or the disadvantaged group. To this end, we define *Edge Impact* (EI) as the ratio between the

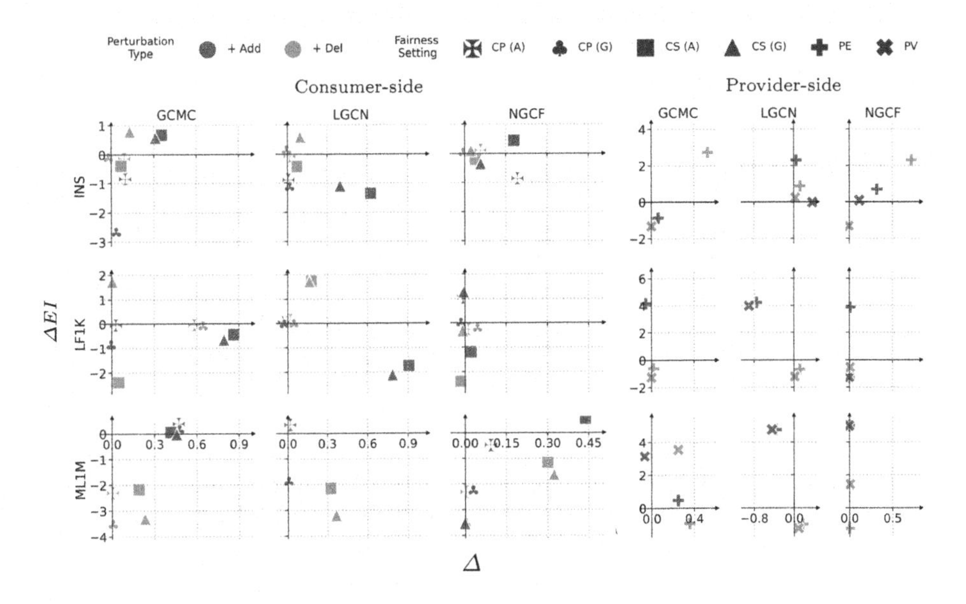

Fig. 3. Relationship between ΔEI (y-axis), i.e. the disparity in edge perturbations distribution between the advantaged and the disadvantaged group, and Δ (x-axis), i.e. the disparity in fairness level before and after the attack. Successful attacks ($\Delta > 0$) are caused by targeting more the advantaged group if $\Delta EI > 0$, otherwise the disadvantaged group was targeted. *A* stands for *Age*, *G* for *Gender*. Some points overlap.

distribution of edge perturbations on the nodes of a group and the representation of the latter. Let $\neg\tilde{E} \subseteq \tilde{E}$ be the subset of edges actually perturbed, formally $EI_* = \frac{|\neg\tilde{E}_*|/|\neg\tilde{E}|}{|Z_*|/|Z|}$, where Z_* is a group of consumers or providers, and $\neg\tilde{E}_*$ is the subset of edges perturbed w.r.t. the nodes in Z_*. We also define the difference $\Delta EI = EI_{Adv} - EI_{Disadv}$ between the advantaged and the disadvantaged group.

In Fig. 3, we depict ΔEI on the y-axis and Δ on the x-axis to study their relationship. The values of Δ pertain to our first experiment in Sect. 4.2.

In the consumer side, most of the settings reporting an impact on Δ were caused by edge perturbations prioritizing the disadvantaged group, i.e. $\Delta EI < 0$. In particular, this behavior regards $\dotplus$ *Del*, which removed interactions of the disadvantaged group to, intuitively, reduce its recommendation utility. Conversely, we expect $\dotplus$ *Add* to report $\Delta EI > 0$. However, this is actually true only under ML1M on LGCN and NGCF, and slightly under INS on GCMC and NGCF. Other settings, e.g., under LF1K, illustrate Δ was affected by adding more edges to the disadvantaged group, highlighting that a higher number of interactions does not systematically correspond to a higher recommendation utility.

In the provider side, the disparity in edge perturbations distribution is more remarked in comparison with the consumer-side results. However, such disparity does not reflect a relevant impact on Δ in most settings, as already highlighted in other experiments. Successful attacks ($\Delta > 0$) are mostly reported under INS for PE, where ΔEI is systematically above the origin. Hence, the impact on the robustness in provider fairness was caused by deleting or adding more edges to the advantaged group. Other scenarios where Δ was affected, e.g., on GCMC under ML-1M, confirm such observation or report close points to the x-axis.

5 Conclusions

In this paper, we present the concept of robustness in fairness and raised attention towards the issues caused by related attacks in recommendation. Compared to prior work, our analysis aimed to assess the robustness in fairness of GNN-based recommender systems against poisoning-like attacks based on edge-level perturbations, focusing on the models' robustness and not on the attack itself.

Even though the range of considered models is limited, they represent consistent baselines in the literature, and cover varied GNN architectures. It follows that the adopted models represent perfect candidates for an analysis of robustness in fairness, a topic that is still unexplored in recommendation compared with other fields. Nevertheless, future works will cover a wider set of models.

Moreover, although the adopted attack is focused to test the robustness in fairness, the same approach could be extended to targeted attacks, e.g., against specific consumers or providers. In other words, based on how each stakeholder partitions are established, our attack could be used to modify the model performance to specifically favor one of the tailored groups. We plan to investigate other types of attack, such as those relying on perturbations based on re-wiring.

Additionally, our white-box setting could be simplified to a grey-box one by removing the assumption of having access to the model parameters. Given that the attack solely requires the substitution of the adjacency matrix with its perturbed version, the latter could be generated through a different process. Future work will be focused on exploring grey- or black-box settings.

Acknowledgement. We acknowledge financial support under the National Recovery and Resilience Plan (NRRP), Miss. 4 Comp. 2 Inv. 1.5 - Call for tender No. 3277 published on Dec 30, 2021 by the Italian Ministry of University and Research (MUR) funded by the European Union - NextGenerationEU. Prj. Code ECS0000038 eINS Ecosystem of Innovation for Next Generation Sardinia, CUP F53C22000430001, Grant Assignment Decree N. 1056, Jun 23, 2022 by the MUR.

References

1. Anelli, V.W., Deldjoo, Y., Noia, T.D., Merra, F.A.: Adversarial recommender systems: attack, defense, and advances. In: Ricci, F., Rokach, L., Shapira, B. (eds.) Recommender Systems Handbook, pp. 335–379. Springer, New York (2022). https://doi.org/10.1007/978-1-0716-2197-4_9
2. Atzori, A., Fenu, G., Marras, M.: Explaining bias in deep face recognition via image characteristics. In: Proceedings of the IEEE International Joint Conference on Biometrics, IJCB, pp. 1–10. IEEE (2022)
3. Barbu, A., et al.: ObjectNet: a large-scale bias-controlled dataset for pushing the limits of object recognition models. In: Proceedings of the Annual Conference on Neural Information Processing Systems, NeurIPS, pp. 9448–9458 (2019)
4. den Berg, R.V., Kipf, T.N., Welling, M.: Graph convolutional matrix completion. CoRR abs/1706.02263 (2017)
5. Boratto, L., Fabbri, F., Fenu, G., Marras, M., Medda, G.: Counterfactual graph augmentation for consumer unfairness mitigation in recommender systems. In: Proceedings of the 32nd ACM International Conference on Information and Knowledge Management, CIKM, pp. 3753–3757. ACM (2023)
6. Boratto, L., Fenu, G., Marras, M.: Interplay between upsampling and regularization for provider fairness in recommender systems. User Model. User Adapt. Interact. **31**(3), 421–455 (2021)
7. Boratto, L., Fenu, G., Marras, M., Medda, G.: Consumer fairness in recommender systems: contextualizing definitions and mitigations. In: Hagen, M., et al. (eds.) ECIR 2022. LNCS, vol. 13185, pp. 552–566. Springer, Cham (2022). https://doi.org/10.1007/978-3-030-99736-6_37
8. Boratto, L., Fenu, G., Marras, M., Medda, G.: Practical perspectives of consumer fairness in recommendation. Inf. Process. Manag. **60**(2), 103208 (2023)
9. Burke, R., Sonboli, N., Ordonez-Gauger, A.: Balanced neighborhoods for multi-sided fairness in recommendation. In: Proceedings of the Conference on Fairness, Accountability and Transparency, FAT, vol. 81, pp. 202–214. PMLR (2018)
10. Cao, Y., Chen, X., Yao, L., Wang, X., Zhang, W.E.: Adversarial attacks and detection on reinforcement learning-based interactive recommender systems. In: Proceedings of the 43rd International ACM SIGIR Conference on Research and Development in Information Retrieval, SIGIR, pp. 1669–1672. ACM (2020)
11. Celma, Ò.: Music Recommendation and Discovery - The Long Tail, Long Fail, and Long Play in the Digital Music Space. Springer, Heidelberg (2010). https://doi.org/10.1007/978-3-642-13287-2

12. Chen, H., Zhou, K., Lai, K., Hu, X., Wang, F., Yang, H.: Adversarial graph perturbations for recommendations at scale. In: Proceedings of the 45th International ACM SIGIR Conference on Research and Development in Information Retrieval, SIGIR, pp. 1854–1858. ACM (2022)
13. Chen, J., Dong, H., Wang, X., Feng, F., Wang, M., He, X.: Bias and debias in recommender system: a survey and future directions. ACM Trans. Inf. Syst. **41**(3), 67:1–67:39 (2023)
14. Christakopoulou, K., Banerjee, A.: Adversarial attacks on an oblivious recommender. In: Bogers, T., Said, A., Brusilovsky, P., Tikk, D. (eds.) Proceedings of the 13th ACM Conference on Recommender Systems, RecSys, pp. 322–330. ACM (2019)
15. Croce, F., Gowal, S., Brunner, T., Shelhamer, E., Hein, M., Cemgil, A.T.: Evaluating the adversarial robustness of adaptive test-time defenses. In: Proceedings of the International Conference on Machine Learning, ICML, vol. 162, pp. 4421–4435. PMLR (2022)
16. Croce, F., Hein, M.: Reliable evaluation of adversarial robustness with an ensemble of diverse parameter-free attacks. In: Proceedings of the 37th International Conference on Machine Learning, ICML, vol. 119, pp. 2206–2216. PMLR (2020)
17. Deldjoo, Y., Bellogín, A., Noia, T.D.: Explaining recommender systems fairness and accuracy through the lens of data characteristics. Inf. Process. Manag. **58**(5), 102662 (2021)
18. Fabbri, F., Croci, M.L., Bonchi, F., Castillo, C.: Exposure inequality in people recommender systems: the long-term effects. In: Proceedings of the Sixteenth International AAAI Conference on Web and Social Media, ICWSM, pp. 194–204. AAAI Press (2022)
19. Fang, M., Yang, G., Gong, N.Z., Liu, J.: Poisoning attacks to graph-based recommender systems. In: Proceedings of the 34th Annual Computer Security Applications Conference, ACSAC, pp. 381–392. ACM (2018)
20. Fenu, G., Marras, M., Medda, G., Meloni, G.: Fair voice biometrics: impact of demographic imbalance on group fairness in speaker recognition. In: Proceedings of the 22nd Annual Conference of the International Speech Communication Association, Interspeech, pp. 1892–1896. ISCA (2021)
21. Floridi, L., Holweg, M., Taddeo, M., Silva, J., Mokander, J., Wen, Y.: capAI - a procedure for conducting conformity assessment of AI systems in line with the EU artificial intelligence act. SSRN Electron. J. (2022)
22. Ge, Y., et al.: Towards long-term fairness in recommendation. In: Proceedings of the Fourteenth ACM International Conference on Web Search and Data Mining, WSDM, pp. 445–453. ACM (2021)
23. Ge, Y., et al.: Explainable fairness in recommendation. In: Proceedings of the 45th International ACM SIGIR Conference on Research and Development in Information Retrieval, SIGIR, pp. 681–691. ACM (2022)
24. Ghazimatin, A., Balalau, O., Roy, R.S., Weikum, G.: PRINCE: provider-side interpretability with counterfactual explanations in recommender systems. In: WSDM 2020: The Thirteenth ACM International Conference on Web Search and Data Mining, pp. 196–204. ACM (2020)
25. Gómez, E., Zhang, C.S., Boratto, L., Salamó, M., Marras, M.: The winner takes it all: geographic imbalance and provider (un)fairness in educational recommender systems. In: Proceedings of the 44th International ACM SIGIR Conference on Research and Development in Information Retrieval, SIGIR, pp. 1808–1812. ACM (2021)

26. Gómez, E., Zhang, C.S., Boratto, L., Salamó, M., Ramos, G.: Enabling cross-continent provider fairness in educational recommender systems. Future Gener. Comput. Syst. **127**, 435–447 (2022)
27. Harper, F.M., Konstan, J.A.: The MovieLens datasets: history and context. ACM Trans. Interact. Intell. Syst. **5**(4), 19:1–19:19 (2016)
28. He, X., Deng, K., Wang, X., Li, Y., Zhang, Y., Wang, M.: LightGCN: simplifying and powering graph convolution network for recommendation. In: Proceedings of the 43rd International ACM SIGIR Conference on Research and Development in Information Retrieval, SIGIR, pp. 639–648. ACM (2020)
29. He, X., He, Z., Du, X., Chua, T.: Adversarial personalized ranking for recommendation. In: Proceedings of the 41st International ACM SIGIR Conference on Research & Development in Information Retrieval, SIGIR, pp. 355–364. ACM (2018)
30. Kang, B., Lijffijt, J., Bie, T.D.: Explanations for network embedding-based link predictions. In: Kamp, M., et al. (eds.) Proceedings of the International Workshops of the European Conference on Machine Learning and Principles and Practice of Knowledge Discovery in Databases, ECML PKDD, vol. 1524, pp. 473–488. Springer, Cham (2021). https://doi.org/10.1007/978-3-030-93736-2_36
31. Li, B., Wang, Y., Singh, A., Vorobeychik, Y.: Data poisoning attacks on factorization-based collaborative filtering. In: Proceedings of the Annual Conference on Neural Information Processing Systems, NeurIPS, pp. 1885–1893 (2016)
32. Li, Y., Chen, H., Fu, Z., Ge, Y., Zhang, Y.: User-oriented fairness in recommendation. In: Proceedings of the Web Conference, TheWebConf, pp. 624–632. ACM/IW3C2 (2021)
33. Li, Y., Chen, H., Xu, S., Ge, Y., Zhang, Y.: Towards personalized fairness based on causal notion. In: Proceedings of the 44th International ACM SIGIR Conference on Research and Development in Information Retrieval, SIGIR, pp. 1054–1063. ACM (2021)
34. Lucic, A., ter Hoeve, M.A., Tolomei, G., de Rijke, M., Silvestri, F.: CF-GNNExplainer: counterfactual explanations for graph neural networks. In: Proceedings of the International Conference on Artificial Intelligence and Statistics, AISTATS, vol. 151, pp. 4499–4511. PMLR (2022)
35. Marras, M., Korus, P., Jain, A., Memon, N.D.: Dictionary attacks on speaker verification. IEEE Trans. Inf. Forensics Secur. **18**, 773–788 (2023)
36. Mastropaolo, A., et al.: On the robustness of code generation techniques: an empirical study on GitHub Copilot. In: Proceedings of the 45th IEEE/ACM International Conference on Software Engineering, ICSE, pp. 2149–2160. IEEE (2023)
37. Medda, G., Fabbri, F., Marras, M., Boratto, L., Fenu, G.: GNNUERS: fairness explanation in GNNs for recommendation via counterfactual reasoning. CoRR abs/2304.06182 (2023)
38. Mehrabi, N., Naveed, M., Morstatter, F., Galstyan, A.: Exacerbating algorithmic bias through fairness attacks. In: Proceedings of the Thirty-Fifth AAAI Conference on Artificial Intelligence, AAAI, Thirty-Third Conference on Innovative Applications of Artificial Intelligence, IAAI, The Eleventh Symposium on Educational Advances in Artificial Intelligence, EAAI, pp. 8930–8938. AAAI Press (2021)
39. Nguyen, T.T., et al.: Poisoning GNN-based recommender systems with generative surrogate-based attacks. ACM Trans. Inf. Syst. **41**(3), 58:1–58:24 (2023)
40. Noia, T.D., Tintarev, N., Fatourou, P., Schedl, M.: Recommender systems under European AI regulations. Commun. ACM **65**(4), 69–73 (2022)
41. Oh, S., Ustun, B., McAuley, J.J., Kumar, S.: Rank list sensitivity of recommender systems to interaction perturbations. In: Proceedings of the 31st ACM Interna-

tional Conference on Information & Knowledge Management, CIKM, pp. 1584–1594. ACM (2022)
42. O'Mahony, M.P., Hurley, N.J., Silvestre, G.C.M.: Recommender systems: attack types and strategies. In: Proceedings of the Twentieth National Conference on Artificial Intelligence and the Seventeenth Innovative Applications of Artificial Intelligence Conference, AAAI, pp. 334–339. AAAI Press/The MIT Press (2005)
43. Pruksachatkun, Y., Krishna, S., Dhamala, J., Gupta, R., Chang, K.: Does robustness improve fairness? Approaching fairness with word substitution robustness methods for text classification. In: Proceedings of the Findings of the Association for Computational Linguistics: ACL/IJCNLP. Findings of ACL, ACL/IJCNLP 2021, pp. 3320–3331. ACL (2021)
44. Qin, T., Liu, T., Li, H.: A general approximation framework for direct optimization of information retrieval measures. Inf. Retr. **13**(4), 375–397 (2010)
45. Rahmani, H.A., Naghiaei, M., Dehghan, M., Aliannejadi, M.: Experiments on generalizability of user-oriented fairness in recommender systems. In: Proceedings of the 45th International ACM SIGIR Conference on Research and Development in Information Retrieval, SIGIR, pp. 2755–2764. ACM (2022)
46. Sato, M., Takemori, S., Singh, J., Ohkuma, T.: Unbiased learning for the causal effect of recommendation. In: Proceedings of the Fourteenth ACM Conference on Recommender Systems, RecSys, pp. 378–387. ACM (2020)
47. Singh, A., Joachims, T.: Fairness of exposure in rankings. In: Proceedings of the 24th ACM SIGKDD International Conference on Knowledge Discovery & Data Mining, KDD, pp. 2219–2228. ACM (2018)
48. Solans, D., Biggio, B., Castillo, C.: Poisoning attacks on algorithmic fairness. In: Hutter, F., Kersting, K., Lijffijt, J., Valera, I. (eds.) ECML PKDD 2020. LNCS (LNAI), vol. 12457, pp. 162–177. Springer, Cham (2021). https://doi.org/10.1007/978-3-030-67658-2_10
49. Song, J., et al.: PoisonRec: an adaptive data poisoning framework for attacking black-box recommender systems. In: Proceedings of the 36th IEEE International Conference on Data Engineering, ICDE, pp. 157–168. IEEE (2020)
50. Srinivas, S., Subramanya, A., Babu, R.V.: Training sparse neural networks. In: Proceedings of the IEEE Conference on Computer Vision and Pattern Recognition Workshops, CVPR Workshops, pp. 455–462. IEEE (2017)
51. Sun, L., et al.: Adversarial attack and defense on graph data: a survey. IEEE Trans. Knowl. Data Eng. **35**(8), 7693–7711 (2023)
52. Wang, S., Zhang, X., Wang, Y., Liu, H., Ricci, F.: Trustworthy recommender systems. CoRR abs/2208.06265 (2022)
53. Wang, X., He, X., Wang, M., Feng, F., Chua, T.: Neural graph collaborative filtering. In: Proceedings of the 42nd International ACM SIGIR Conference on Research and Development in Information Retrieval, SIGIR 2019, Paris, France, 21–25 July 2019. ACM (2019)
54. Wang, Y., Ma, W., Zhang, M., Liu, Y., Ma, S.: A survey on the fairness of recommender systems. ACM Trans. Inf. Syst. (2022)
55. Wu, C., Wu, F., Wang, X., Huang, Y., Xie, X.: Fairness-aware news recommendation with decomposed adversarial learning. In: Proceedings of the Thirty-Fifth AAAI Conference on Artificial Intelligence, AAAI, pp. 4462–4469. AAAI Press (2021)
56. Wu, H., Ma, C., Mitra, B., Diaz, F., Liu, X.: A multi-objective optimization framework for multi-stakeholder fairness-aware recommendation. ACM Trans. Inf. Syst. (2022). Just Accepted

57. Yang, G., Gong, N.Z., Cai, Y.: Fake co-visitation injection attacks to recommender systems. In: Proceedings of the 24th Annual Network and Distributed System Security Symposium, NDSS. The Internet Society (2017)
58. Yang, H., Liu, Z., Zhang, Z., Zhuang, C., Chen, X.: Towards robust fairness-aware recommendation. In: Proceedings of the 17th ACM Conference on Recommender Systems, RecSys, pp. 211–222. ACM (2023)
59. Yin, D., Lopes, R.G., Shlens, J., Cubuk, E.D., Gilmer, J.: A Fourier perspective on model robustness in computer vision. In: Proceedings of the Annual Conference on Neural Information Processing Systems, NeurIPS, pp. 13255–13265 (2019)
60. Yuan, F., Yao, L., Benatallah, B.: Exploring missing interactions: a convolutional generative adversarial network for collaborative filtering. In: Proceedings of the 29th ACM International Conference on Information and Knowledge Management, CIKM, pp. 1773–1782. ACM (2020)
61. Yue, Z., Zeng, H., Kou, Z., Shang, L., Wang, D.: Defending substitution-based profile pollution attacks on sequential recommenders. In: Proceedings of the Sixteenth ACM Conference on Recommender Systems, RecSys, pp. 59–70. ACM (2022)
62. Zeng, Z., Tan, H., Zhang, H., Li, J., Zhang, Y., Zhang, L.: An extensive study on pre-trained models for program understanding and generation. In: Proceedings of the 31st ACM SIGSOFT International Symposium on Software Testing and Analysis, ISSTA, pp. 39–51. ACM (2022)
63. Zhang, K., et al.: Robust recommender system: a survey and future directions. CoRR abs/2309.02057 (2023)
64. Zhang, S., Yin, H., Chen, T., Huang, Z., Cui, L., Zhang, X.: Graph embedding for recommendation against attribute inference attacks. In: Proceedings of The Web Conference 2021, TheWebConf, pp. 3002–3014. ACM/IW3C2 (2021)
65. Zhang, S., Yin, H., Chen, T., Nguyen, Q.V.H., Huang, Z., Cui, L.: GCN-based user representation learning for unifying robust recommendation and fraudster detection. In: Proceedings of the 43rd International ACM SIGIR Conference on Research and Development in Information Retrieval, SIGIR, pp. 689–698. ACM (2020)
66. Zhang, Y., Chen, X.: Explainable recommendation: a survey and new perspectives. Found. Trends Inf. Retr. **14**(1), 1–101 (2020)
67. Zhao, W.X., et al.: RecBole: towards a unified, comprehensive and efficient framework for recommendation algorithms. In: Proceedings of the 30th ACM International Conference on Information and Knowledge Management, CIKM, pp. 4653–4664. ACM (2021)
68. Zheng, J., Ma, Q., Gu, H., Zheng, Z.: Multi-view denoising graph auto-encoders on heterogeneous information networks for cold-start recommendation. In: Proceedings of the 27th ACM SIGKDD Conference on Knowledge Discovery and Data Mining, KDD, pp. 2338–2348. ACM (2021)

Is Google Getting Worse? A Longitudinal Investigation of SEO Spam in Search Engines

Janek Bevendorff[1](✉), Matti Wiegmann[2], Martin Potthast[2,3], and Benno Stein[2]

[1] Leipzig University, Leipzig, Germany
janek.bevendorff@uni-weimar.de
[2] Bauhaus-Universität Weimar, Weimar, Germany
[3] ScaDS.AI, Leipzig, Germany

Abstract. Many users of web search engines have been complaining in recent years about the supposedly decreasing quality of search results. This is often attributed to an increasing amount of search-engine-optimized but low-quality content. Evidence for this has always been anecdotal, yet it's not unreasonable to think that popular online marketing strategies such as affiliate marketing incentivize the mass production of such content to maximize clicks. Since neither this complaint nor affiliate marketing as such have received much attention from the IR community, we hereby lay the groundwork by conducting an in-depth exploratory study of how affiliate content affects today's search engines. We monitored Google, Bing and DuckDuckGo for a year on 7,392 product review queries. Our findings suggest that all search engines have significant problems with highly optimized (affiliate) content—more than is representative for the entire web according to a baseline retrieval system on the ClueWeb22. Focussing on the product review genre, we find that only a small portion of product reviews on the web uses affiliate marketing, but the majority of all search results do. Of all affiliate networks, Amazon Associates is by far the most popular. We further observe an inverse relationship between affiliate marketing use and content complexity, and that all search engines fall victim to large-scale affiliate link spam campaigns. However, we also notice that the line between benign content and spam in the form of content and link farms becomes increasingly blurry—a situation that will surely worsen in the wake of generative AI. We conclude that dynamic adversarial spam in the form of low-quality, mass-produced commercial content deserves more attention. (Code and data: https://github.com/webis-de/ECIR-24).

Keywords: Web Search Quality · Search Engine Optimization · Web Spam

J. Bevendorff and M. Wiegmann—Equal contribution.

N. Goharian et al. (Eds.): ECIR 2024, LNCS 14610, pp. 56–71, 2024.
https://doi.org/10.1007/978-3-031-56063-7_4

1 Introduction

Web search engines are possibly the most important information access technologies today. It may therefore be a troubling sign that a noticeable number of social media users are sharing their observation that search engines are becoming less and less capable of finding genuine and useful content satisfying their information needs. Reportedly, a torrent of low-quality content, especially for product search, keeps drowning any kind of useful information in search results.

Previous research has shown that most pages returned by web search engines have some degree of search engine optimization (SEO) [22], with conflicting effects on users' perception of page quality [33]. The dynamics of search engine optimization and the web in general have always been a problem for search providers. It's not far-fetched to assume a connection between SEO and a perceived degradation of quality and to ask whether search providers are losing this battle. Unfortunately, search providers offer little insight into their efforts to curb SEO and SEO's dynamic nature is difficult to capture in a static and standardized test collection, which may explain why SEO has received relatively little attention from the research community in terms of retrieval effectiveness studies. Zobel [37] argues in this context that retrieval research is therefore susceptible to Goodhart's Law, owing to the difficulty of quantifying the qualitative goal of user satisfaction. Hence, measuring relevance may yield only initially convincing but ultimately impractical results.

In this paper, we systematically investigate for the first time whether and to which degree "Google is getting worse." We focus on comparative product reviews that offer tests and purchase recommendations as a key indicator of search quality. Such reviews often contain affiliate product links, which refer customers to a seller. The referring entity (the "affiliate") then receives a commission for clicks or purchases resulting from the referral. Affiliate marketing is essentially built on the trust of customers in the affiliate [15]. However, since users often trust their search engines already [20,31], the affiliate inherits this trust as a byproduct of a high ranking. This creates a conflict of interest between affiliates, search providers, and users. With "relevance" being an imperfect metric, affiliates then turn to optimizing rankings instead of investing in high-quality reviews.

Our first contribution is an investigation of the SEO properties of comparative review pages found on the result pages of Google (by proxy of Startpage), Bing, and DuckDuckGo for 7,392 product review queries (Sect. 3). We compare these findings with the results of the BM25 baseline search engine ChatNoir [5] and the raw ClueWeb22 [30] (Sect. 4). We find that the majority of high-ranking product reviews in the result pages of commercial search engines (SERPs) use affiliate marketing, and significant amounts are outright SEO product review spam. The baseline system retrieves both at much lower rates, more consistent with the overall low base rate of affiliate marketing in the ClueWeb22 as a whole. We also find strong correlations between search engine rankings and affiliate marketing, as well as a trend toward simplified, repetitive, and potentially AI-generated content. Our second contribution is a longitudinal analysis of the ongoing competition between SEO and the major search engines over the period of one year (Sect. 5). We find that search engines do intervene and that ranking updates, especially from Google, have a temporary positive effect, though search engines seem to lose the cat-and-mouse game that is SEO spam.

2 Related Work

SEO is an integral part of today's web. Lewandowski et al. [22] estimate that at least 80% of all web pages use it in some form. To show this, they employ 41 measures to assess the degree of optimization, among which are checks for SEO plugins, lists of URLs, page-level HTML features, or load speed. A recent study by Schultheiß et al. [33] investigates the compatibility between SEO and content quality on medical websites with a user study. The study finds an inverse relationship between a page's optimization level and its perceived expertise, indicating that SEO may hurt at least subjective page quality. Other studies and expert interviews by the same authors conclude that despite its prevalence, (German) lay users are largely oblivious of SEO and its effects [21,34], and that less knowledgeable users tend to trust Google in particular more than others [33].

SEO is a double-edged sword. On the one hand, it makes high-quality pages easier to find, but is on the other hand also a sharp tool for pushing up low-quality results in the search rankings. This necessarily begs the question whether topical relevance is a good proxy for utility and user satisfaction. User-based effectiveness measures are typically modeled after the notion of *gain*, where a user interacts with the result list, accumulating utility from encountered documents until they decide to stop [6,25,26]. Costs incurred by interacting with complex search result pages (SERPs) can also be included to improve stopping rank prediction [3]. However, the framework is mostly descriptive and still relies on a good definition of utility, which in practice comes down to user click data and topical relevance judgments by third-party annotators. This leaves a gap [37] between the quantitative detached proxy measures of relevance and the qualitative goal of utility that can be widened by adversarial optimization.

Epstein and Robertson [12] demonstrate the power of SEO to influence the outcome of elections, but to our knowledge, little research has been published on how to combat it. Recent retrieval systems consider ranking fairness [28,32,36] to avoid biasing the results towards individual providers. Although motivated differently, this can potentially avoid over-ranking individual highly optimized pages, but De Jonge and Hiemstra [10] already demonstrate that fairness measures are insufficient to prevent SEO in general. Kurland and Tennenholtz [19] also find that besides generic spam detection, little research on adversarial content optimization exists. They further seem to agree with us in that this may be due to the difficulty of modeling competitive processes with static test collections and thus call for a rigorous game-theoretical search modeling framework.

In our study, we also investigate the role of affiliate marketing in product reviews and its relationship with SEO. Previous work on affiliate marketing often focuses on fraud rather than SEO abuse. Most affiliate fraud falls into one of four categories [1,11]: (1) conversion hijacking via adware or loyalty software (i.e., adding affiliate tags to links on the fly through a malicious client-side software), (2) cookie-stuffing [8,35] (i.e., planting malicious cookies in users' browsers),

(3) typo squatting (i.e., buying domain names with typos and redirecting to the target domain with an added affiliate tag), and (4) user tracking with affiliate cookies. Affiliate link spam is often a part of "long-tail" SEO spam [16,23], where low-frequency queries are targeted with spam gateway pages to dominate niche queries. Unfortunately, spam mitigation research [2,7,9,24] rarely considers affiliate marketing at all.

Since our work focuses on product reviews, it is also related to research on fake reviews [27], review spam in general [17], as well as review quality and helpfulness assessment [29]. However, these studies focus more on user-contributed reviews on retail websites and less on editorial content in blogs or dedicated product test and review portals.

3 Data and Feature Extraction

To analyze prevalence and impact of SEO spam in product reviews on search engines, we created a large collection of top 20 SERPs for 7,392 product review queries. The SERPs were scraped repeatedly over the course of a year from Startpage (a privacy frontend to Google), Bing, and DuckDuckGo. The linked pages were archived as Web ARChive (WARC[1]) files. From those, we extracted a set of page-level features inspired by Google's SEO [14] and affiliate marketing guidelines [13] that indicate text complexity and quality, HTML page structure, the use of affiliate marketing, and whether a page looks like a product review.

Product Review Queries. To find review pages for a wide range of products, we curated a large list of search queries of the form "best *product category*," where *product category* is a placeholder for a category taken from one of two publicly available product category taxonomies: (1) the GS1 Global Product Classification (GPC, November 2021) and (2) the Google Product Taxonomy (GPT, Version 2021-09-21). We combined the leaf nodes of both taxonomies and cleaned them up by excluding categories we expected would produce atypical results or that don't lend themselves for actual product reviews, such as live animals, crops, fresh produce, and large vehicles like airplanes or boats. We manually reviewed the resulting queries and discarded near-duplicates and queries with artifacts or poor wording, resulting in a final list of 7,392 unique search queries. The lists contains several typical product search queries like *best headphones*, but also many long-tail queries like *best anvils* or *best alphabet toys*.

Commercial Search Engines. We retrieved results from Startpage, Bing, and DuckDuckGo as representative commercial search engines. Although DuckDuckGo claims to utilize many different data sources, we found the results to be extremely similar to Bing. For all queries, we retrieved the top 20 English-language (organic, non-ad) SERPs, resulting in about 148,000 hits and 128,000 unique URLs per scrape. The first collection was done after Google's July product reviews update on August 24th, 2022. The scrape was repeated around every two weeks starting October 26th, 2022, until September 19th, 2023. The period

[1] https://iipc.github.io/warc-specifications/specifications/warc-format/warc-1.1/.

spans eleven substantial updates to Google's ranker (among which are three helpful content updates and three product review updates). From the retrieved result lists, we crawled the corresponding web pages, and archived them into WARC files. For the Startpage results, we also included page assets and rendered screenshots for later use.

Baseline Retrieval. We employed the research search engine ChatNoir [5] as a baseline, which offers access to the ClueWeb22B [30], a recent collection of the 200 million most popular web pages, via a web-based API. ChatNoir at its core uses an Okapi BM25 retrieval model and hence serves as a basic and purely document-based whitebox retrieval baseline. The querying and archiving process is analogous to the commercial search engines.

SEO Page Features. For our analysis of the page SEO properties, we compiled a set of page features inspired by Google's SEO [14] and affiliate marketing guidelines [13]. We adapted the guidelines that could be operationalized most easily at page level without rendering the page or executing dynamic content.

The features resulting from this process measure (1) the length and lexical diversity of the main content (extracted using the Resiliparse library [4]), `<a>` anchor and `<img>` alt texts, `<meta>` descriptions, and `<h1>` headings by extracting word and character counts, type-token ratios (TTR; the ratio of unique words to total word count), function word ratios (FWR; the ratio of function words to total word count), and Flesch reading ease scores [18]; (2) the structuredness of a page by counting `<h1>`, `<h2>`, `<img>`, and `<a>` tags, the ratio of `<p>` and `<h[1-6]>` elements to main content words, the use of Open Graph or JSON linked data (JSON-LD), and the existence of breadcrumb navigations; (3) the length and depth of the page URL, i.e., how many files and directories follow after the domain; (4) the number and ratio of affiliate links, the ratio between affiliate links and main content, the number and ratio of site-internal links, as well as the use of `nofollow` link relations; (5) the reuse of topic keywords from headings in the remaining content to measure keyword stuffing.

The features were calculated on a total of 6.6 million results from Startpage (Google), Bing, and DuckDuckGo, and 122,000 results from ChatNoir. We used English as the search language, though we also received a few German results due to unavoidable geo-personalization. Pages without detectable main content were discarded. We extracted the same features also from another 79 million English-language pages from the raw ClueWeb22B dataset as a representation of the long-tail web behind the retrieval frontends.

Review Classification. To test whether a page actually is a review, we performed a simple regular-expression-based keyword classification of the `<h[1-6]>` page headings. Phrases such as "best ...," "top picks," "our favorite ...," "how to use ...," "... we've tested," "what is the best ...," "... review," or various combinations thereof with numbers or other keywords are indicative of review content. We evaluated this approach by drawing a balanced random sample of 100 positively and 100 negatively classified pages from all Startpage and Bing scrapes. These were then annotated manually as either *review* or *non-review* by the two main authors of this paper with almost perfect agreement (Cohen's $\kappa = .96$). Based on this ground truth, the review classification accuracy was 79% and the precision 85%, which we find decent for such a simple classifier.

Affiliate Link Analysis. We counted the number of affiliate links placed on a page by comparing all anchor URLs to a list of typical patterns that we compiled for the nine largest and most influential online affiliate networks. These are in alphabetical order: *Ali Express*, *Amazon Associates*, *Awin*, *CJ*, *ClickBank*, *eBay*, *FlexOffers*, *Refersion*, and *ShareASale*. The list is based on publicly available web information about affiliate network market shares and participating seller counts. We consider a web page to use affiliate marketing if at least one anchor URL in the HTML source matches one of the patterns. To increase the recall of this method, we resolved short links from `bit.ly`, `amzn.to`, `ebay.us`, and `fxo.co` with a single HEAD request prior to matching them against the list of patterns.

Website Content Categorization. For a qualitative analysis of the contents of the SERPs, we manually annotated the top-30 domains of each scrape (cf. Table 1) with the following seven classes: (1) *Authentic Review Sites* serving high-quality comparative reviews and real product tests (e.g., `nytimes.com/wirecutter`, `consumerreports.org`); (2) *Magazines*, news papers, or other editorial pages which also discuss and review products as a (less serious) side hustle—often as a separate division or on a dedicated subdomain (e.g., `nymag.com`); (3) *Review Content Farms* producing low-effort product listicles, pseudo-reviews, and buyer's guides in large quantities, but with (superficial) editorial content on the side—these are sometimes also found on separate subdomains of otherwise more reputable sites (e.g., `reviewed.com`); (4) *Review Spam* consisting of seemingly generated product listicles without any genuine editorial content (e.g., `blinkx.tv`); (5) *Web Shops* like `amazon.com`; (6) *Social Media* or other community sites with user-generated content. (7) *Other* sites like product manufacturer websites or anything else that doesn't fall into the other classes.

The annotation was done independently by the two main authors with substantial agreement (Cohen's $\kappa = .70$, Accuracy $= .76$). The raters disagreed in 7 cases between review farms and magazines and in 5 cases between review farms and spam. This only speaks to the overall low quality standards for affiliate web content, making it increasingly difficult to distinguish benign content from low-grade or spam content. We resolved these disagreements in favor of the site.

4 SEO Spam in Product Reviews

Our first concern in this work is the general prevalence of low-quality SEO content and spam in search engine results and its driving motivation. We therefore first analyze which of the content features are predictive of rank and thus indicate potential SEO engineering on a page. We then analyze quantitatively and qualitatively the use of affiliate marketing as a measure of monetization level.

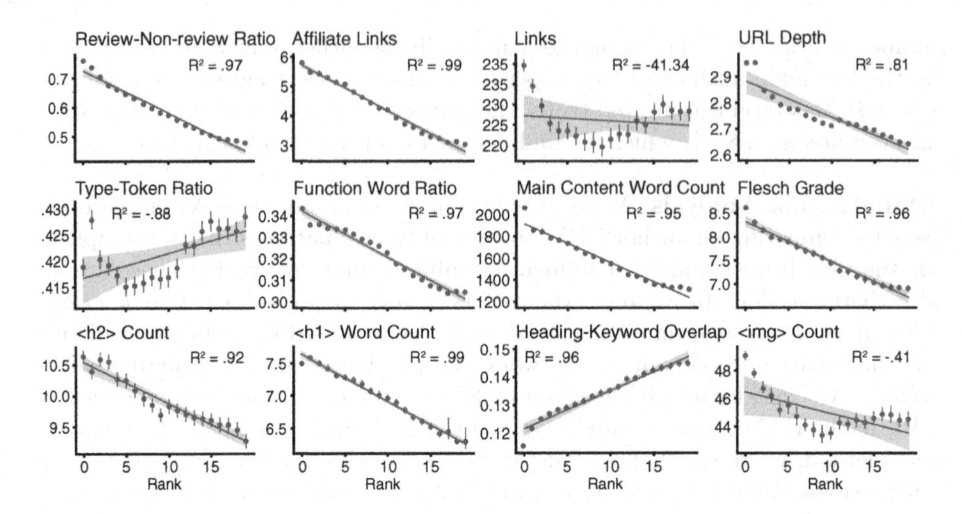

Fig. 1. Selected correlations between rank (independent variable) and the average number of affiliate links, review status, etc. across all Startpage and Bing scrapes. Most averages (not all shown here) correlate either perfectly with rank or have at least a non-linear, non-monotonic relationship. Error bars indicate the bootstrapped 95% confidence intervals of the rank bins. Shaded areas indicate the 95% confidence intervals of the regression line. Note: This plot shows global trends and allows no conclusions about individual pages, since each point is only the mean over all pages at that rank.

Measuring SEO in Review Pages. Across the SERPs of all Startpage and Bing scrapes, we find that rank is indeed a very good predictor of most of our page features. The inverse is not necessarily true, meaning that our page features are not effective SEO exploits, but are nonetheless able to measure SEO at a global population level with highly-correlated sample means. Figure 1 shows a selection of SEO features and their correlation with rank, which we discuss in the following.[2] Cleaned of extreme pages with more than 40 affiliate links per page, these features point to SEO engineering in that pages with better (i.e., numerically lower) ranks are more repetitive (FWR: $r = -.99$; TTR: $r = .59$, $p = .006$) but also more readable (Flesch: $r = .94$). They are also indicators of lower-quality, possibly mass-produced, or even AI-generated content. The FWR relationship is stronger than the TTR relationship, which breaks down for the top-5 ranks. Highly ranked pages also have shallower URLs ($r = -.92$) and longer main content text ($r = -.98$), and are more structured, i.e., they have a lower ratio of text paragraphs to headings ($r = -.99$). Contrary to our initial expectations, pages with better ranking also have a lower overlap between heading keywords and body text ($r = .98$), which indicates that headings become—although more keyword-heavy–also more generic and less specific to the content on the page. Based on manual inspection, a possible explanation could be that "Review" pages become increasingly "thinner" with more affiliate links, i.e., they become more list-like with only a bit of filler text for every featured product that

[2] r is Pearson's correlation coefficient with $p \ll .001$, unless stated otherwise.

provides little value for a user. Some features are more weakly correlated, but still show remarkable non-monotonic and non-linear or piecewise linear relationships (such as number of links or images on a page, which are negatively and only approximately linearly correlated until rank 10 and then reverse direction).

The inverse correlations where the page features themselves are used as independent variables (binned and cleaned of outliers beyond the 95th percentile) to predict the rank also hold, although with smaller effect sizes. Given the nature of global averages, the effect sizes are large enough to detect a trend, but too small to predict the actual ranking. Thus, we can say that longer pages are on average ranked higher ($r = -.91$) and have more affiliate links ($r = -.81$), but the absolute regression coefficient is quite shallow and close to a horizontal line around an average rank 10 (of 20) with medium to high determination ($R^2 = .81$). This means our features can measure certain global effects of SEO, but are not sufficient ranking factors in and of themselves (which would have been quite surprising anyway).

Affiliate Marketing on the Web. Our analysis of the relationship between SEO and the use of affiliate marketing in product reviews in particular reveals a strong positive relationship. First, pages with affiliate links are much more common on Startpage (29%), Bing (42%), and DuckDuckGo (41%) SERPs than on ChatNoir SERPs (18%) and vastly more common than in the ClueWeb22 overall (2.35%). The largest affiliate network across all result pages is *Amazon Associates* by an order of magnitude, followed by *Awin*, *ShareASale*, *CJ*, and *eBay*. There are major differences in the overall numbers of pages with affiliate links returned by the different search engines. Startpage retrieves on average ca. 12,000 pages with 1–10 affiliate links for all product queries. Bing and DuckDuckGo return almost 20,000 in the earlier scrapes and ca. 16,000 in later scrapes (more on this in Sect. 5). ChatNoir retrieves the fewest affiliate pages with only 9,400. For the range of 10–20 links, the search engines return 9,000 (Startpage), 13,000–18,000 (Bing/DuckDuckGo), and 8,200 pages (ChatNoir).

Second, higher-ranked pages have clearly more affiliate links ($r = -.99$, see Fig. 1). Comparing the mean of all pages with the median and 95th percentile of affiliate pages, this is best explained by a mix of both individual pages with high affiliate counts and more affiliate pages in general among the top ranks. We find no conclusive relationship between rank and normal (non-affiliate) links on a page across the whole top-20 range ($r = -.15$, $p = .523$). This confirms that highly ranked pages have indeed more affiliate links and not only more links in general, though non-affiliate links are indeed correlated for first 10 ranks ($r = -.92$, $p = .0002$). The inverse relationship is not nearly as strong ($r = -.62$, $p = .001$) with weak linear determination ($R^2 = -.59$), so, thankfully and unsurprisingly, affiliate links alone cannot predict the rank.

Third, our qualitative site content classification (see Table 1) shows that several spam domains are frequently among the top ranks, some with hundreds of links per page (see also Fig. 3). All inspected pages with more than 100 affiliate links were from spam sites and pages with more than 20 links were at least increasingly likely to be from spam or low-quality affiliate review farm sites

Table 1. Number of websites per review content category for all search engine scrapes (top 20 websites for Startpage, Bing, DuckDuckGo, top 30 for ChatNoir).

Class	Startpage		DDG		Bing		ChatNoir	
Authentic Review Site	1	(2%)	3	(5%)	2	(3%)	0	(0%)
Magazine	14	(31%)	17	(27%)	14	(23%)	2	(7%)
Review Farm	7	(16%)	9	(15%)	9	(15%)	3	(10%)
Spam	4	(9%)	19	(31%)	14	(23%)	1	(3%)
Web Shop	14	(31%)	10	(16%)	9	(15%)	15	(50%)
Social Media	5	(11%)	4	(6%)	3	(5%)	5	(17%)
Other	0	(0%)	0	(0%)	11	(18%)	4	(13%)

designed primarily to harvest clicks. Of all search engines, ChatNoir returns the fewest pages with excessive amounts of affiliate links and the fewest sites in the spam and review farm categories. Bing and DuckDuckGo are especially vulnerable and frequently return up to 2–5 times as many spam pages as Startpage or ChatNoir. As a result, we will base most further analyses only on pages with fewer than 40 affiliate links, which corresponds to the 95th percentile of Startpage results (90th for Bing/DuckDuckGo, 96th for ChatNoir, 99.9th for the ClueWeb22). We find this amount of blatant yet well-ranked affiliate spam peculiar and concerning and it goes to show how important thorough spam filtering is. It is unclear whether on- or off-page SEO (such as link networks) helped in making these spam domains visible. We found through `archive.org` that some of the identified spam sites were quite likely sold or hijacked (such as `socialmoms.com` or `distrotest.net`), while others (like `pulptastic.com`) intentionally mix spam reviews with (possibly generated or scraped) editorial content.

Reviews Vs. Non-reviews. Of the retrieved pages, more than half are identified as review pages by our keyword classification (Startpage: 54%, Bing/DuckDuckGo: 59%). Again, we see that the global mean likelihood of being a review ($r = -.98$) is almost perfectly predicted by a page's rank. We take this as strong evidence that the search engines fundamentally understood our information need. ChatNoir retrieves fewer review pages in total: only 39% of the result pages are reviews (ClueWeb22 base rate: 7%). There are several valid explanations for this behavior: (1) the retrieval algorithm is worse and does not retrieve review pages well, (2) the ClueWeb is a smaller dataset containing fewer review pages in total; (3) SEO on the review pages targets live search engines and ChatNoir simplistic BM25 model ignores many features that identify those pages as relevant to more advanced retrieval systems, such as Google's; (4) Google and Bing as the two major search engines and thus primary SEO targets are particularly vulnerable, which is also a consequence of (3).

Reviews being mostly monetized with affiliate marketing—which relies on trust and visibility—makes at least some amount of SEO quite likely, which

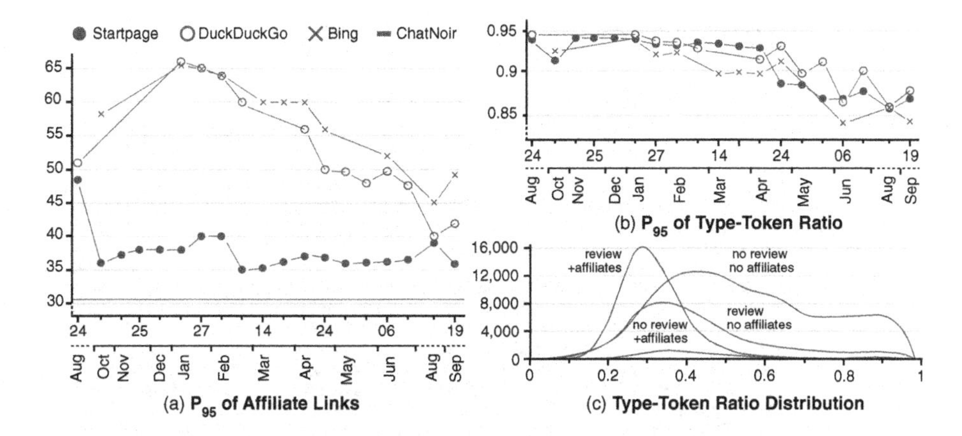

Fig. 2. (a) 95th percentile of per-page affiliate links in the SERPs for all search engines over time (including pages with more than 40 affiliate links). (b) 95th percentile of TTR over time. (c) Count and TTR distribution split by review vs. non-review and with vs. without affiliate links.

conflicts with users' needs for accurate and unbiased information. This is a compelling argument, because a page's likelihood of both being a review but also spam share the same predictor: affiliate links. Figure 2c shows the distribution of type-token ratio (TTR) values over all pages. Review pages that use affiliate marketing have the overall lowest TTR. Review pages without affiliate links and non-review pages with affiliate links have a slightly higher TTR and non-review pages without affiliate links have the highest. This shows that highly commercialized pages are on average simpler and use more repetitive vocabulary, which is a strong indicator of lower-quality content.

5 Temporal Analysis of Product Reviews

Our second major contribution in this work is the temporal analysis of the search quality in terms of (1) the prevalence of pages from the "Review Content Farm" or "Spam" categories (Sect. 3) and (2) SEO-indicating content features.

A temporal analysis of the product review search results should reveal one of three trends: (1) Search engines are truly getting worse, i.e., they are losing the battle against SEO content. In this case we should see a long-term increase in spam and a decline in overall quality. (2) Search engines are winning the upper hand and we see the inverse of this. (3) SEO is a constant battle and we see repeated breathing patterns of review spam entering and leaving the results as search engines and SEO engineers take turns adjusting their parameters. Our evidence suggests that all search engines have some success to show for. Particularly Bing and DuckDuckGo have substantially improved their results, albeit on an overall lower level than Startpage (Google). Yet despite these gains, it seems like (3) is the most likely scenario.

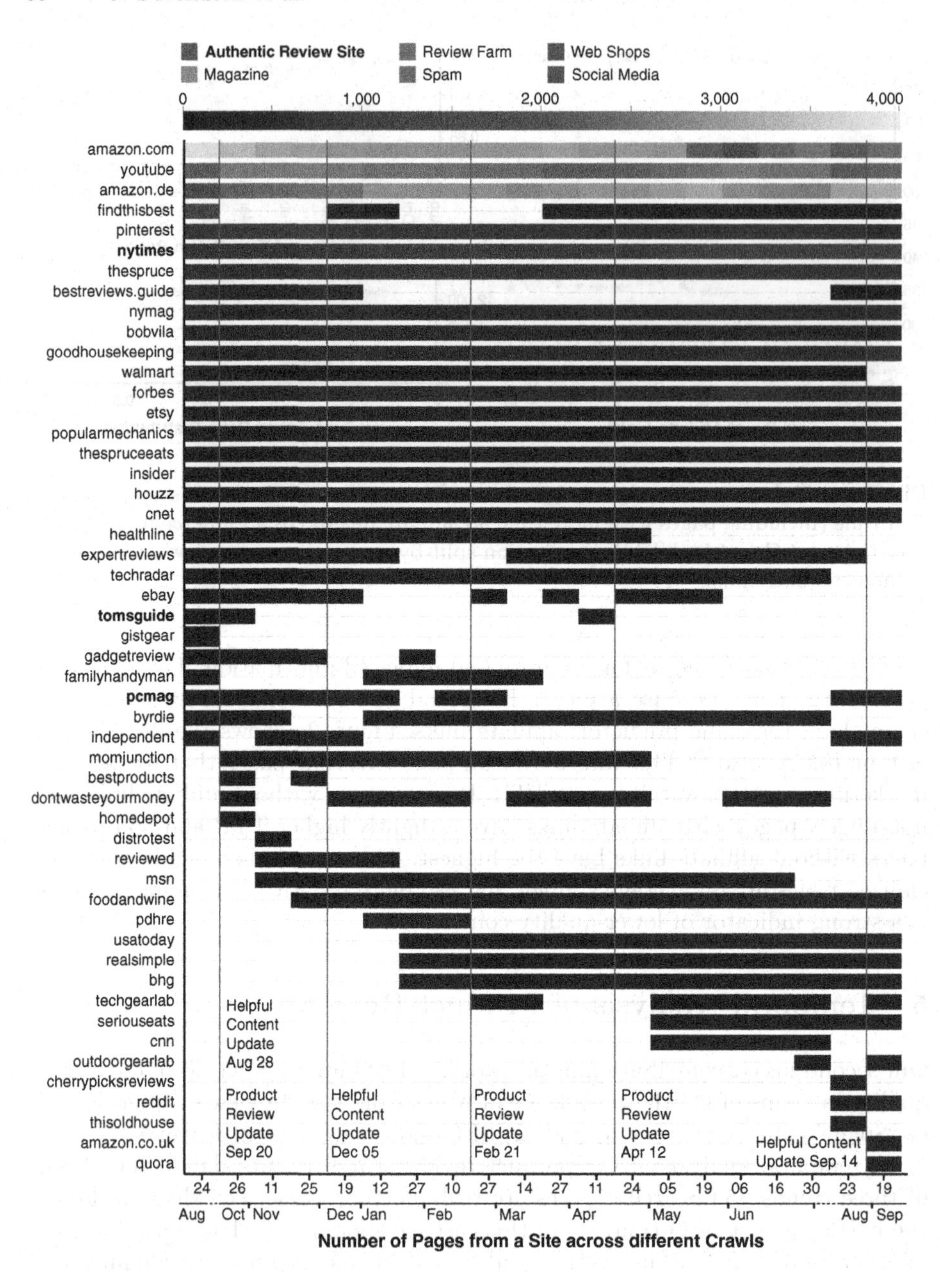

Fig. 3. The 30 most common sites from the Startpage SERPs and their frequencies over time (as page counts per scrape). Blue lines indicate Google's ranker updates. Sites were manually categorized based on the content they served at their first retrieval time. (Color figure online)

Frequent Websites Over Time. Figure 3 shows the union page counts of the 30 most common domains from each of our Startpage scrapes on a time axis, as well as notable ranker updates that happened between the scrapes. Each domain was manually classified into one of the categories from Sect. 3.

We see that *review spam* sites are usually short-lived and de-indexed or penalized quickly, especially after ranker updates. *Review farm* sites and some more frequent *spam* sites such as `findthisbest.com` and `bestreviews.guide` are more persistent and remain throughout multiple consecutive scrapes. However, they often vanish (at least for a while) after Google's ranker updates, either immediately or up to two weeks later, which is the usual update rollout time. The "August 2022 Helpful Content Update" and the "September 2022 Product Reviews Update" combined had the most profound effect, shaving off ranks from all major sites and causing several spam sites to disappear entirely.

Reputable sites, (*web shops*, *social media*, *authentic review sites*) often appear in many consecutive scrapes at similar frequencies. *Magazines* and pure *product review farms* are somewhat, but not as persistent. The most common category of new pages that enter the top 30 for the first time seems to be *magazines*, which hints that having a separate low-quality review section to support a site's primary content is a successful and lucrative business model.

Table 1 shows that *magazines* are also the dominant category besides *web shops*. Bing and DuckDuckGo are also notably less robust against *review spam* than Startpage. Some of the *review spam* sites we identified still existed at the time of the last scrape. Most of them, however, were already defunct at that point. A common pattern we observed is that these start off as seemingly legitimate review pages, but at some point flip to delivering mass spam in the form of scraped or directly embedded Amazon search results. It is unclear what exactly triggers the sudden disappearance of such a spam page, but a significant update to Google's ranking seems like a likely explanation.

SEO Content Over Time. Our SEO content features analyzed over time reveal positive effects of the ranker updates, particularly in the average number of affiliate links per page. Figure 2a shows the 95th percentile of average per-page affiliate link counts by search engine. After September 2022, a large drop can be observed in the Startpage results, while Bing and DuckDuckGo continue to climb until around January. The February update produces another dent in Startpage's curve and the April update yet another albeit smaller one. On the other hand, we also see affiliate pages slowly regaining their lost momentum between updates, indicating a constant struggle. Yet, even after all updates, the commercial search engines still have significantly higher percentile scores than the ChatNoir baseline and although the baseline does return spam domains, it usually does so with much lower frequency.

Interestingly, Google started downranking at least some affiliate pages in the last two of our scrapes, starting end of August, 2023 ($r = -.92$; previously: $r \approx -.99$), resulting in a significantly flatter regression coefficient ($\beta = -.07$, $R^2 = .81$; previously: $\beta \approx -.20$, $R^2 \approx .98$). Whether this is a short-lived change or a lasting trend remains to be seen.

The most profound change during our measurement period, however, is in Bing's (and thus DuckDuckGo's) affiliate link count per page. Between February and August 2023, Bing reduced their 95th percentile by almost 70%, yet still hovered at a rather high level at the end of the measurement period. With Bing being less transparent about their ranker updates, we cannot tell if this is due to a massive crackdown on affiliate spam on their behalf or just windfall gains from Google's updates causing certain pages to disappear. The effects are softened, though still visible if we exclude pages with more than 40 affiliate links.

We see further in Fig. 2b that the average page type-token ratio has decreased consistently over time across all search engines, which shows that while mass affiliate spam may have been contained to some degree, the overall content quality may not have improved.

6 Conclusion

In this paper, we investigate the common observation that "Google is getting worse" by examining its search results for its susceptibility to SEO-driven low-quality content along with those of other major search engines, and in comparison to baselines. We focus on product review search, which we consider particularly vulnerable to affiliate marketing due to its inherent conflict of interest between users, search providers, and content providers.

We conduct two main analyses. First, we investigate what kind of content is retrieved by product review queries and how much SEO influences rankings in this web genre. We correlate page-level quality attributes with search engine rank and find strong relationships between them. Although we cannot predict the rank of individual pages, at the population level, we can conclude that higher-ranked pages are on average more optimized, more monetized with affiliate marketing, and they show signs of lower text quality.

Second, we examine how search results change over time and whether the changes made by search engine operators improve the overall quality of the results. We find that search engines measurably target SEO and affiliate spam with their ranker updates. Google's updates in particular are having a noticeable, yet mostly short-lived, effect. In fact, the Google results seem to have improved to some extent since the start of our experiment in terms of the amount of affiliate spam. Yet, we can still find several spam domains and also see an overall downwards trend in text quality in all three search engines, so there is still quite a lot of room for improvement.

The constant struggle of billion-dollar search engine companies with targeted SEO affiliate spam should serve as an example that web search is a dynamic game with many players, some with bad intentions. Addressing this kind of dynamic, fast-changing, and monetization-driven adversarial SEO content is difficult to do with static evaluation. Going forward, we plan to evaluate how we can better build and evaluate truly robust web IR systems in competitive environments.

Acknowledgments. This publication has received funding from the European Commission under grant agreement № 101070014 (OpenWebSearch.eu).

References

1. Amarasekara, B., Mathrani, A., Scogings, C.: Stuffing, sniffing, squatting, and stalking: sham activities in affiliate marketing. Libr. Trends **68**(4), 659–678 (2020)
2. Asdaghi, F., Soleimani, A.: An effective feature selection method for web spam detection. Knowl.-Based Syst. **166**, 198–206 (2019)
3. Azzopardi, L., Thomas, P., Craswell, N.: Measuring the utility of search engine result pages: an information foraging based measure. In: The 41st International ACM SIGIR Conference on Research & Development in Information Retrieval, SIGIR 2018, pp. 605–614. Association for Computing Machinery, New York, NY, USA, 27 June 2018. https://doi.org/10.1145/3209978.3210027
4. Bevendorff, J., Potthast, M., Stein, B.: FastWARC: optimizing large-scale web archive analytics. In: Wagner, A., Guetl, C., Granitzer, M., Voigt, S. (eds.) 3rd International Symposium on Open Search Technology (OSSYM 2021). International Open Search Symposium, October 2021
5. Bevendorff, J., Stein, B., Hagen, M., Potthast, M.: Elastic ChatNoir: search engine for the ClueWeb and the common crawl. In: Pasi, G., Piwowarski, B., Azzopardi, L., Hanbury, A. (eds.) ECIR 2018. LNCS, vol. 10772, pp. 820–824. Springer, Cham (2018). https://doi.org/10.1007/978-3-319-76941-7_83
6. Carterette, B.: System effectiveness, user models, and user utility: a conceptual framework for investigation. In: Proceedings of the 34th International ACM SIGIR Conference on Research and Development in Information Retrieval, SIGIR 2011, pp. 903–912. Association for Computing Machinery, New York, NY, USA, 24 July 2011. https://doi.org/10.1145/2009916.2010037
7. Castillo, C., Donato, D., Gionis, A., Murdock, V., Silvestri, F.: Know your neighbors: web spam detection using the web topology. In: Proceedings of the 30th Annual International ACM SIGIR Conference on Research and Development in Information Retrieval, SIGIR 2007, pp. 423–430. Association for Computing Machinery, New York, NY, USA, July 2007
8. Chachra, N., Savage, S., Voelker, G.M.: Affiliate crookies: characterizing affiliate marketing abuse. In: Proceedings of the 2015 Internet Measurement Conference, IMC 2015, pp. 41–47. Association for Computing Machinery, New York, NY, USA, October 2015. https://doi.org/10.1145/2815675.2815720
9. Chandra, A., Suaib, M., Beg, R.: Google search algorithm updates against web spam. Inform. Eng. Int. J. **3**(1), 1–10 (2015)
10. De Jonge, T., Hiemstra, D.: UNFair: search engine manipulation, undetectable by amortized inequity. In: Proceedings of the 2023 ACM Conference on Fairness, Accountability, and Transparency, FAccT 2023, pp. 830–839. Association for Computing Machinery, New York, NY, USA, 12 June 2023. https://doi.org/10.1145/3593013.3594046
11. Edelman, B., Brandi, W.: Information and incentives in online affiliate marketing. Citeseer (2013)
12. Epstein, R., Robertson, R.E.: The search engine manipulation effect (SEME) and its possible impact on the outcomes of elections. Proc. Nat. Acad. Sci. U.S.A. **112**(33), E4512–21 (2015). https://doi.org/10.1073/pnas.1419828112
13. Google Search Central: Affiliate programs (2022). https://developers.google.com/search/docs/advanced/guidelines/affiliate-programs. Accessed 17 June 2022
14. Google Search Central: Write high quality product reviews (2022). https://developers.google.com/search/docs/advanced/ecommerce/write-high-quality-product-reviews. Accessed 17 June 2022

15. Gregori, N., Daniele, R., Altinay, L.: Affiliate marketing in tourism: determinants of consumer trust. J. Travel Res. **53**(2), 196–210 (2014). https://doi.org/10.1177/0047287513491333
16. Gyongyi, Z., Garcia-Molina, H.: Spam: it's not just for inboxes anymore. Computer **38**(10), 28–34 (2005)
17. Heydari, A., Tavakoli, M.A., Salim, N., Heydari, Z.: Detection of review spam: a survey. Expert Syst. Appl. **42**(7), 3634–3642 (2015)
18. Kincaid, J.P., Fishburne, R.P. Jr., Rogers, R.L., Chissom, B.S.: Derivation of new readability formulas (automated readability index, fog count and Flesch reading ease formula) for navy enlisted personnel (1975)
19. Kurland, O., Tennenholtz, M.: Competitive search. In: Proceedings of the 45th International ACM SIGIR Conference on Research and Development in Information Retrieval, SIGIR 2022, pp. 2838–2849. Association for Computing Machinery, New York, NY, USA, 7 July 2022. https://doi.org/10.1145/3477495.3532771
20. Lewandowski, D., Kerkmann, F., Rümmele, S., Sünkler, S.: An empirical investigation on search engine ad disclosure. J. Am. Soc. Inf. Sci. **69**(3), 420–437 (2018)
21. Lewandowski, D., Schultheiß, S.: Public awareness and attitudes towards search engine optimization. Behav. Inf. Technol. **42**(8), 1025–1044 (2023). https://doi.org/10.1080/0144929X.2022.2056507
22. Lewandowski, D., Sünkler, S., Yagci, N.: The influence of search engine optimization on Google's results: a multi-dimensional approach for detecting SEO. In: WebSci, pp. 12–20. ACM (2021)
23. Liao, X., Liu, C., McCoy, D., Shi, E., Hao, S., Beyah, R.A.: Characterizing long-tail SEO spam on cloud web hosting services. In: Bourdeau, J., Hendler, J., Nkambou, R., Horrocks, I., Zhao, B.Y. (eds.) Proceedings of the 25th International Conference on World Wide Web, WWW 2016, Montreal, Canada, 11–15 April 2016, pp. 321–332. ACM (2016). https://doi.org/10.1145/2872427.2883008
24. Liu, J., Su, Y., Lv, S., Huang, C.: Detecting web spam based on novel features from web page source code. Secur. Commun. Netw. **2020** (2020)
25. Moffat, A., Thomas, P., Scholer, F.: Users versus models: what observation tells us about effectiveness metrics. In: Proceedings of the 22nd ACM International Conference on Information & Knowledge Management, CIKM 2013, pp. 659–668. Association for Computing Machinery, New York, NY, USA, 27 October 2013. https://doi.org/10.1145/2505515.2507665
26. Moffat, A., Zobel, J.: Rank-biased precision for measurement of retrieval effectiveness. ACM Trans. Inf. Syst. Secur. **27**(1), 1–27 (2008). https://doi.org/10.1145/1416950.1416952
27. Mohawesh, R., et al.: Fake reviews detection: a survey. IEEE Access **9**, 65771–65802 (2021)
28. Morik, M., Singh, A., Hong, J., Joachims, T.: Controlling fairness and bias in dynamic learning-to-rank. In: Proceedings of the 43rd International ACM SIGIR Conference on Research and Development in Information Retrieval, SIGIR 2020, pp. 429–438. Association for Computing Machinery, New York, NY, USA, 25 July 2020. https://doi.org/10.1145/3397271.3401100
29. Ocampo Diaz, G., Ng, V.: Modeling and prediction of online product review helpfulness: a survey. In: Proceedings of the 56th Annual Meeting of the Association for Computational Linguistics (Volume 1: Long Papers), pp. 698–708. Association for Computational Linguistics, Melbourne, Australia, July 2018
30. Overwijk, A., Xiong, C., Liu, X., VandenBerg, C., Callan, J.: ClueWeb 22: 10 billion web documents with visual and semantic information. arXiv (2022). https://doi.org/10.48550/ARXIV.2211.15848. https://arxiv.org/abs/2211.15848

31. Purcell, K., Rainie, L., Brenner, J.: Search engine use 2012 (2012)
32. Raj, A., Ekstrand, M.D.: Measuring fairness in ranked results: an analytical and empirical comparison. In: Proceedings of the 45th International ACM SIGIR Conference on Research and Development in Information Retrieval, SIGIR 2022, pp. 726–736. Association for Computing Machinery, New York, NY, USA, 7 July 2022. https://doi.org/10.1145/3477495.3532018
33. Schultheiß, S., Häußler, H., Lewandowski, D.: Does search engine optimization come along with high-quality content?: A comparison between optimized and non-optimized health-related web pages. In: CHIIR, pp. 123–134. ACM (2022)
34. Schultheiß, S., Lewandowski, D.: "Outside the industry, nobody knows what we do" SEO as seen by search engine optimizers and content providers. J. Doc. **77**(2), 542–557 (2020). https://doi.org/10.1108/JD-07-2020-0127
35. Snyder, P., Kanich, C.: Characterizing fraud and its ramifications in affiliate marketing networks. J. Cybersecur. **2**(1), 71–81 (2016)
36. Zehlike, M., Yang, K., Stoyanovich, J.: Fairness in ranking: a survey, 25 March 2021
37. Zobel, J.: When measurement misleads: the limits of batch assessment of retrieval systems. SIGIR Forum **56**(1), 1–20 (2023). https://doi.org/10.1145/3582524.3582540

GLAD: Graph-Based Long-Term Attentive Dynamic Memory for Sequential Recommendation

Deepanshu Pandey(✉), Arindam Sarkar(✉), and Prakash Mandayam Comar(✉)

Amazon Development Center, Bengaluru, India
{deepnsp,arindsar,prakasc}@amazon.com

Abstract. Recommender systems play a crucial role in the e-commerce stores, enabling customers to explore products and facilitating the discovery of relevant items. Typical recommender systems are built using n most recent user interactions, where value of n is chosen based on trade-off between incremental gains in performance and compute/memory costs associated with processing long sequences. State-of-the-art recommendation models like Transformers, based on attention mechanism, have quadratic computation complexity with respect to sequence length, thus limiting the length of past customer interactions to be considered for recommendations. Even with the availability of compute resources, it is crucial to design an algorithm that strikes delicate balance between long term and short term information in identifying relevant products for personalised recommendation. Towards this, we propose a novel extension of Memory Networks, a neural network architecture that harnesses external memory to encapsulate information present in lengthy sequential data. The use of memory networks in recommendation use-cases remains limited in practice owing to their high memory cost, large compute requirements and relatively large inference latency, which makes them prohibitively expensive for online stores with millions of users and products. To address these limitations, we propose a *novel* transformer-based sequential recommendation model GLAD, with external graph-based memory that *dynamically* scales user memory by adjusting the memory size according to the user's history, while facilitating the flow of information between users with similar interactions. We establish the efficacy of the proposed model by benchmarking on multiple public datasets as well as an industry dataset against state-of-the-art sequential recommendation baselines.

1 Introduction

On e-commerce stores with large product catalogs, personalized product recommendations are essential to improve the customers' journey by helping them easily find relevant products from millions of available items. Customer preferences are influenced by various factors such as seasonality, trends, personal experiences, social media etc. Thus, it is essential to effectively process the dynamic

N. Goharian et al. (Eds.): ECIR 2024, LNCS 14610, pp. 72–88, 2024.
https://doi.org/10.1007/978-3-031-56063-7_5

nature of product preferences, by clearly differentiating their immediate needs from their long term interests. Balancing the long-term and short-term considerations can be challenging because short-term interactions can be sporadic that change rapidly and may not always be indicative of the user's true long term preferences. On the other hand, understanding long-term preferences require significant storage and processing resources to identify trends and patterns in the user's historical behavior. This is further complicated in situations where users have changing preferences/interests over time. Developing a recommendation system that can effectively balance both short-term and long-term considerations is an ongoing challenge in e-commerce recommendation systems [19,25,34].

State-of-the-art transformer based sequential recommendation models like Bert4Rec [32], CORE [12] and SASRec [13] are expensive to scale to long sequence lengths at inference time due to the quadratic nature of self-attention. Since the interaction history can be extremely long for some users, a common practice is to make the predictions based on the last n items in the user history, where n is a hyperparameter which depends on the resource availability and inference time constraints. However, this results in discarding of rich interaction data, especially for the highly active users with long histories [19,21,33]. To efficiently represent changing user preferences over time, and effectively combine long and short term preferences, we propose a *novel* sequential recommendation model, GLAD (Graph-based Long-Term Attentive Dynamic memory), based on Memory Networks [31,37], by introducing the concept of discretionary memory cells based on length of users' interaction history. Not only does the proposed model enhance quality of recommendations, but being modular, it provides a way to add dynamic memory as a component to any of the existing sequential models. Where efficiently processing and remembering long term dependencies in sequential data is essential. The major contributions of our work are as follows:

- We propose a *novel* deep neural network model for sequential recommendation which leverages a *dynamic* graph-based external memory. The proposed approach has significantly less memory complexity compared conventional memory networks while maintaining low inference latency.
- We propose an *attentive* reading mechanism over graph memory conditioned on the target/query item to be scored by the model, which further enhances the quality of the recommendations.
- For real world applications, which are often memory constrained, we propose a *clustered* graph memory variant of our model which reduces memory consumption and also enables information flow between similar users.

2 Methodology

In this work, we focus on the task of next item prediction, where the goal is to predict the next item the user is likely to interact with based on the past history. Specifically, given a list of N_i items which a user U_i has interacted with, we want to rank a candidate set of target items $\{T_1, T_2, .., T_K\}$ on the basis of relevance to the user. The sequence of items interacted with by a given user

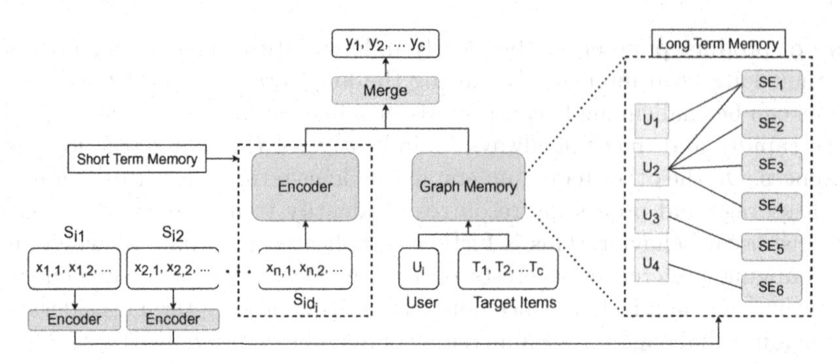

Fig. 1. The proposed GLAD model architecture.

is sorted based on the interaction time, and is then split into $d_i = \lceil \frac{N_i}{c} \rceil$ subsequences of maximum size c. We denote the mth subsequence corresponding to the ith user U_i as S_{im}. Intuitively, the sequence of items S_{im} captures short-term behaviour in a local temporal window, forming an abstract short term memory of product preferences. We then create a bi-partite graph structure between users and their subsequences. Corresponding to each subsequence S_{im}, we create a subsequence node in the graph. Since it is possible for multiple users to share exactly the same subsequence (especially when c is small), corresponding to each unique subsequence, we create a node in the graph SE_j independent of user index i. Note that the users with rich past interaction history will be connected to more subsequence nodes compared to users with sparse interactions. We use a self-attention based sequence encoder to generate embedding h_{SE_j} for each subsequence. These are used to initialize the embeddings of the corresponding sub-sequence nodes in the graph, which forms the *dynamic* long-term memory. Finally, the short-term (h_{SE_j}) and the long-term representations captured by a Graph Neural Network (GNN) over the graph-structured memory are fused via a gating mechanism to generate the final representation for the user. Proposed model architecture is presented in Fig. 1.

2.1 Interaction Sequence Encoder

We use Bert4Rec [32] as the backbone transformer model to process user interaction sequences, and refer to it as the Interaction Sequence Encoder. Since the interaction sequences for users can be quite large, we divide each sequence into smaller chunks (S_{im}) of equal lengths c (last subsequence is padded if required), and pass them independently through the encoder, which is pretrained on the Masked Item Prediction (*Cloze*) [32] task. The representation of the final chunk S_{id_i} for the user U_i from the trained model is denoted by z_{U_i} and can be thought of as a representation of short-term interaction preferences for the user, as it considers the last L items in the user's interaction sequence. Finally, the encoder is trained end-to-end with the graph-based memory to make the final recommendations (Sect. 2.5). Note that, it is possible to replace Bert4Rec with similar models that can produce a sequence representation from interaction sequences.

2.2 Dynamic Graph Based Memory

To capture and retain long-term user preferences, we create an external memory component in the form of a user interaction graph. The memory is in the form of a bipartite graph with two type of nodes: 1) user nodes U_i, and 2) sequence nodes SE_j. Each of the unique sub-sequence chunks S_{im} is given a distinct sequence node in the graph and connected to the corresponding user node, as shown in Fig. 1. Each of the nodes is represented by a high-dimensional node embedding h_{SE_j}, and are initialised using the Interaction Sequence Encoder. Once the graph is initialised, a Graph Neural Network (GNN) is used to aggregate information into the user node embeddings from the neighbouring sequence nodes. This results in aggregation of all the historical interaction information stored for the user in their user node U_i, and the resulting user node embedding is then used as the user's long-term memory representation. For our experiments, GraphSAGE [6] architecture was used as GNN for the graph memory, where the representation of a graph layer is as follows:

$$h^{l+1}_{\mathcal{N}(U_i)} = agg(\{h^{l}_{SE_j}, \forall SE_j \in \mathcal{N}(U_i)\});\ h^{l+1}_{U_i} = \sigma(W^{l}_{U}\,(h^{l}_{U_i} || h^{l+1}_{\mathcal{N}(U_i)})) \quad (1)$$

Here, $h^{l+1}_{U_i}$ is the node embedding for user node U_i at $l+1^{th}$ layer, $\mathcal{N}(U_i)$ represents the set of neighbouring nodes of U_i and W^{l}_{U} is projection matrix for user nodes at layer l. $||$ represents the concatenation operator. The aggregate (agg) function used here is sum. Similarly for the sequence nodes:

$$h^{l+1}_{\mathcal{N}(SE_j)} = agg(\{h^{l}_{U_i}, \forall U_i \in \mathcal{N}(SE_j)\});\ h^{l+1}_{SE_j} = \sigma(W^{l}_{SE}\,(h^{l}_{SE_j} || h^{l+1}_{\mathcal{N}(SE_j)})) \quad (2)$$

Note that unlike conventional memory networks that allocate a fixed-size memory for each user [1,33], the graph-based memory model allocates memory based on the size of user interaction histories, which results in efficient use of physical memory. The choice of GraphSAGE for neighborhood aggregation is motivated by the fact that it takes into account both the neighborhood embeddings and the node embedding of the current node to generate node embeddings for the next layer. We believe that this is helpful in propagating the information from initial node embeddings to subsequent layers, since the initial representations for the nodes are rich and contain compressed information about user sequences. It is possible to replace GraphSage with other GNN architectures in the architecture, and we leave this as an area for future research and exploration (Fig. 2).

Clustered Sequence Representations: Although the graph-based memory model is able to aggregate historical information for the user, there is little to no information exchange between users since they are connected only to their own sub-sequence nodes. Further, the GNN captures less information for users with

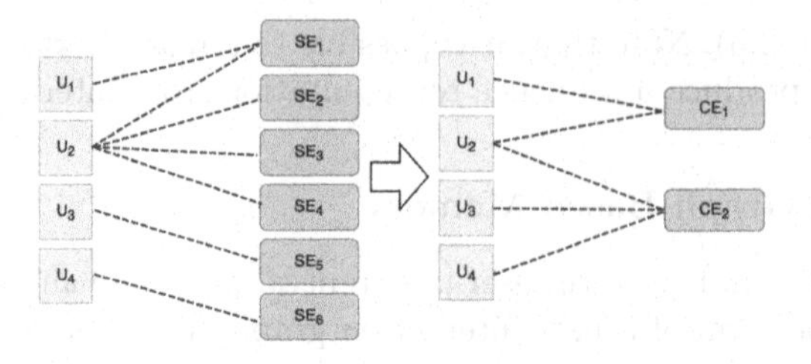

Fig. 2. Visualization of the compression of user-item interaction graph. Here CE_j are Cluster Embeddings. Semantically similar nodes (1,2,3), (4,5,6) are clustered.

lesser interactions, since these users are connected to only a few sequence nodes. To solve these issues, we cluster similar sequence nodes together into cluster nodes. In the clustered memory model, the outgoing connections from user nodes to sequence nodes are now grouped to form connections to cluster nodes CE_j. Cluster node embeddings are computed by mean-pooling over sequence node embeddings. This connects the users who have interacted with similar sequences as *2*-hop neighbors, enabling the model to capture information from similar users via multi-hop message-passing. Further, it reduces the memory footprint of the model as now only the cluster embeddings need to be stored, rather than storing all the sequence embeddings for user interaction sequences. To accommodate the new user interactions and tackle the gradual drift in user preference and purchase patterns, several sequential clustering mechanisms [4,22] can be leveraged to update the clustered graph memory with new user interaction sequences.

2.3 Output Gating

To appropriately weigh the recent and historical interactions, a gating function is learned from data, and is used to combine the short-term sequence embedding from the encoder and the long-term memory representation from the graph memory to generate the final representation for user U_i:

$$g = \sigma(W_T(z_{U_i}) + W_G(h^L_{U_i})) \tag{3}$$

where σ is the sigmoid function, $z_{U_i} \in R^e$ is the output of the encoder, and $(h^L_{U_i}) \in R^e$ is the final (L^{th}) GNN layer node representation for user node U_i. $W_T \in R^{e\times e}$ and $W_G \in R^{e\times e}$ are projections for the transformer and graph outputs respectively and e is the hidden dimension of the model. The final output of the model $h_{U_i} \in R^e$ is a weighted combination of the two outputs, aggregated via elementwise product ($\odot$) with the gating weights $g \in R^e$. h_{U_i} gathers all the relevant information for user U_i from both the transformer encoder and the graph memory. For each target item T_i in the candidate set, the relevance to the user U_i is computed by taking dot (.) product of the final representation h_{U_i} and the target item representation h_{T_k}, resulting in similarity score for item, s_k:

$$s_k = h_{U_i} \cdot h_{T_k}; \text{ where } h_{U_i} = g \odot z_{U_i} + (1-g) \odot h^L_{U_i} \tag{4}$$

During inference, the target items are re-ranked based on these similarity scores and top items are chosen as the final recommendations for user U_i. Note that the candidate set of target items $\{T_1, T_2, ...\}$ can vary for different users, depending on the candidate generation [5,43] algorithm.

2.4 Attentive Reading of Graph Memory

The graph memory model aggregates the long term user preferences. However, for a given target item, some of the memory nodes in the user memory graph might be more relevant than others. To selectively capture useful information, the graph memory is further enhanced by replacing the last layer of the GNN with an attention based aggregation of neighbourhood information. The aggregation of neighbourhood information for user node U_i in the final L^{th} layer is determined by the equation (for clustered graph, SE will be replaced CE):

$$\alpha_{SE_j} = (h_{T_k} \cdot h_{SE_j}^{L-1}); \; h_{U_i T_k}^{L} = \sum_{SE_j \in N(U_i)} \alpha_{SE_j} h_{SE_j}^{L-1} \tag{5}$$

which replaces Eq. 1 for final layer computation for user node U_i. Here, $N(U_i)$ represents the set of neighbouring nodes of node U_i and T_k is the kth target item. The target conditioned attentive read operation is similar to the target attention proposed by [44], and enables our model to selectively read over the most relevant parts of the user history conditioned on the target item. The final output $h_{U_i T_k}$ gathers all the relevant information from the graph memory for user U_i with respect to target item T_k. Once the user node representation $h_{U_i T_k}^{L}$ is computed, it replaces $h_{U_i}^{L}$ in Eq. 3, while Eq. 4 is modified to:

$$s_k = h_{U_i T_k} \cdot h_{T_k}, \text{ where, } h_{U_i T_k} = g \odot z_{U_i} + (1 - g) \odot h_{U_i T_k}^{L} \tag{6}$$

2.5 End-to-End Training

We train our model in two phases. In *Phase 1*, we pretrain the interaction sequence encoder on Masked Item Prediction task, with the sequence chunks S_{im}'s. We then create the graph-based external memory with user and sequence nodes. We pass each of the training chunks, with a mask token concatenated at the end through the trained interaction sequence encoder and use the final layer transformer output for the mask token as the sequence representation. We initialise the sequence node embeddings in the graph with their corresponding sequence representations while the user node embeddings are initialised with random vectors. For clustered graph, the sequence nodes are first clustered together using k-means clustering, and the cluster nodes are stored in the graph instead of sequence nodes.

In *Phase 2*, we combine representations from the transformer encoder and the graph memory using the gating layer and train the entire model end-to-end. For each sequence chunk for a user U_i, we capture the transformer sequence representation and consider this to be the short term user preference. We then sample a sub-graph containing the user node for U_i from the graph memory, along with all its 2-hop neighbors and compute the user node representation for U_i using the GNN, which is used as the long-term memory component for the user. The model is trained using the cross-entropy loss given by:

$$\mathcal{L}_{CE} = -\sum_{k=0}^{K} y_k \log p_k, \text{ where, } p_k = \frac{e^{s_k}}{\sum_{m=0}^{K} e^{s_m}} \tag{7}$$

where T is the set of candidate items to be ranked, and $|T| = K$. p_k is the probability score generated from the similarity scores s_k given by the model for the kth target item, and $y_k \in \{0, 1\}$ denotes the actual label for the item.

3 Related Work

The early years of recommendation systems saw extensive use of Collaborative Filtering (CF) [16,30] methods. Matrix Factorization [17], used inner product between user and item representations to model the user-item interactions. This was followed by neural approaches like Neural Collaborative Filtering (NCF) [7]. However, these models fail to capture the sequential nature of user interactions.

Sequential Recommendation: Early Markov Chain based models like FPMC [26] showed that a user interaction is closely related to the last interactions made by the user, but lacked the capability to utilise long user histories. Following the success of RNNs [3,11,28] in NLP tasks, models like GRU4Rec [9,10] leveraged RNNs for next-item prediction, but suffered from non-parallelizable operations and catastrophic forgetting over long sequences. The use of attention mechanism in later models like NARM [18], STAMP [19] etc., managed to successfully alleviate some of these issues. Models like SRGNN [38] and GCSAN [39] additionally leveraged information from graph based structures constructed from user sessions, but are difficult to scale to large datasets due to item graph construction for each user. Further improvements followed with the introduction of Transformer [35] based models. Engrained with the powerful self-attention mechanism to process sequential data, models like SASRec [13], Bert4Rec [32], CORE [12] have become the go-to approach for building sequential recommendation systems.

Memory Networks: While sequential models effectively capture the short-term user preferences, they often lack in capturing the long-term user interest due to their inability to process very long user histories. To tackle this problem, the idea of having a dedicated memory has been proposed. Memory Networks

[31,37] with dedicated external storage are capable of explicitly storing historical information, and have read, write and update operations defined for the memory store. User-Memory Networks (RUM) [1] proposed a dedicated external memory per user for more personalised recommendations. MA-GNN [21] uses a graph-based structure as short-term memory, and uses a shared memory network for long term memory. HPMN [25] proposes hierarchical memory units per user to capture multiscale user behavior for lifelong learning. In DMAN [33], the authors build on the idea of memory networks and couple a sequential recommendation model with conventional user memory, thus leveraging both long and short-term user preference information. While the user-specific memory cells intuitively capture more personalized patterns per user, a major drawback of memory networks is that the memory requirement can be prohibitively large, and poor scalability with increasing number of users, as they allocate fixed size memory matrices per user. Naturally, this either results in excessive compression of memory for users with long histories, and/or wastage of memory space for users having short histories. In contrast to the existing approaches, the proposed model GLAD leverages state-of-art transformer encoder network to capture short term user preferences, and efficiently maintains long term user interest representation by allocating differential memory space to the users based on their interaction history size via a *novel* graph-based *dynamic memory*.

4 Experiments

We establish the efficacy of the proposed model by performing extensive quantitative and qualitative analyses. We benchmark our model against representative baseline models on both public and industry datasets to answer multiple research questions (RQ) on predictive performance and incremental benefit of each component, and address them by performing extensive experimentation.

Datasets: We use the publicly available Amazon Reviews datasets, which have been a popular benchmark for recommendation system models [1,14,32,44]. Specifically, we use the Electronics and Home & Kitchen categories [23] from the Amazon Reviews dataset for our experiments. We also experiment with a private dataset (hereby referred to as E-COM dataset) created by sampling anonymized user interaction logs over a fixed period of time for an e-commerce site. Summary statistics for the datasets are: (1) **E-COM**: 1.2M users, 2.2M items, 39.6M interactions, (2) **Electronics**: 204K users, 247k items, 20M interactions, and (3) **Home & Kitchen**: 207k users, 310k items, 22M interactions. The maximum number of interactions for E-COM, Electronics and Home & Kitchen are 8919, 592 and 446, while the median values are 20, 13 and 13 respectively.

Baselines: We compare our model against 8 competitive baselines (Table 1). We have included representative models with memory component: **FRUM** and **STAMP**, RNN/Transformer based models: **GRU4Rec**, **Bert4Rec**, **CORE**,

CL4SRec (the latter two additionally utilise contrastive learning), and graph based models: **SRGNN** and **GCSAN**. Among the baselines, only FRUM explicitly captures the long-term user context by using external memory, but it adds to the memory footprint due to a large fixed size memory per user, unlike our model which is endowed with a dynamic graph-based memory. Also, due to the sequential nature of memory updates in FRUM, the training and inference latency is high (Table 5). In contrast, for GLAD, memory updates are simply edge updates in the memory graph. Models like Bert4Rec, CORE and CL4SRec, lack an efficient mechanism to represent the long term user preferences. Among graph models, both SRGNN and GCSAN create a separate graph per user, based on the session data available for them, which is expensive over long sequences. We propose an efficient long term preference representation via a subsequence graph instead. We have omitted early works like Collaborative Filtering [8,27] and models like SASRec [14] and Caser [34] since the newer models like Bert4Rec, GCSAN etc. outperform them, and some of the larger memory networks like DMAN due to their extremely large memory and compute requirements.

Experimental Setup: We compare the performance of the models on the popular task of next-item prediction [8,14,34]. The per-user interactions are sorted by the timestamp and converted into sequences. The last item from each user sequence is retained for the test data, second last item for validation and the rest of the sequence is used for training the models. Since the training sequences can be very large for some of the users, we further divide these into smaller, equal-sized chunks. For training, the last item in the sequence chunks is considered to be ground truth and is paired with a set of 100 negatively sampled items to be ranked and the top-k items with highest scores are taken as recommendations from a model. For the recommendation task, we only use item ID as feature. We report Hit-Rate (hit@k) which measures the proportion of users for whom the item they interacted with was in the top-k predictions by the model, and Mean Reciprocal Rank (mrr@k), which additionally quantifies if the model is able to rank the interacted item towards the top of the list. For all results on the ECOM dataset, metrics are reported as relative gains with respect to the FRUM baseline to maintain dataset anonymity.

Sequence Masking: We pre-train the Bert4Rec backbone on *Cloze* task. For GLAD, the last item in each of the chunks is masked and the chunks are passed sequentially through the model, where the model tries to predict the masked item correctly. During validation and testing, for both the models, only the last chunk for each of the users is passed after adding a mask token at the end of the sequence. The sequence masking process is as follows:

Training sequence: [$x_1, x_2, ..., x_k, x_{k+1}, ..., x_{2k}, x_{2k+1}, x_{2k+2}$]

Pretraining (Bert4Rec):
Sub-sequence 1 = [$x_1, \{mask\}, x3..., \{mask\}, x_k$]
Target 1 = [x_2, x_{k-1}]
Sub-sequence 2 = [$x_{k+1}, x_{k+2}, \{mask\}, ..., x_{2k-1}, \{mask\}$]
Target 2 = [x_{k+3}, x_{2k}]

Training (GLAD):
Sub-sequence 1 = [$x_1, x_2, ..., x_{k-1}, \{mask\}$]
Target 1 = [x_k]
Sub-sequence 2 = [$x_{k+1}, x_{k+2}, , x_{2k-1}, \{mask\}$]
Target 2 = [x_{2k}]

Validation: [$x_{k+2}, x_{k+3}, ..., x_{2k}, \{mask\}$]
Targets: [x_{2k+1}]

Test: [$x_{k+3}, x_{k+4}, ..., x_{2k+1}, \{mask\}$]
Targets: [x_{2k+2}]

Implementation Details: We use a 2-layer Bert4Rec model as the backbone sequence encoder along with a 2-layer GraphSAGE network as GNN for the graph memory. The embedding dimension size e is chosen as 256. The chunk size is set to 50 for E-COM and 20 for Electronics and Home & Kitchen. We use k-means clustering to create the set of clusters for the clustered graph memory. We choose a clustering factor of 100 for E-COM dataset, which means approximately 100 nodes are clustered together to form one cluster node. Since both Electronics and Home & Kitchen datasets are smaller in size, we do not cluster the graph memory for these datasets. The model is trained with a Cross Entropy loss with Adam [15] optimiser, Cosine Annealing Scheduler [20] and a negative sampling strategy. For hyperparameter tuning of all the models (including our model), we start with the optimal hyperparameters suggested by the authors, and leveraged a mix of manual tuning and grid search. Except for the FRUM model, we use Recbole [40–42] implementations of the baselines. Note that for some of the models, like Bert4Rec, the choice of hyperparameters and the number of training epochs is critical (as was also observed in [24]). Accordingly, to address these shortcomings of Bert4Rec implementation in Recbole, we use a custom Transformer implementation on top of the RecBole training framework. We perform all our experiments on a distributed setup of 4 Nvidia V100 GPUs. We use Deep Graph Library (DGL) [36] for implementation of the graph component in our model. In our graph implementation, each of the operations are batched to ensure maximum training efficiency. Since processing entire graph at once is expensive, while creating user node representations from the graph memory, we sample all the users in the batch along with their 1-hop and 2-hop neighbors to create a sub-graph and apply GNN over the subgraph to get user embeddings.

Table 1. hit@k and mrr@k metrics on next-item prediction task. For E-COM, we report relative gains with respect to FRUM baseline for dataset anonymity. Abbreviations used: E-COM (EC), Electronics (EL), Home & Kitchen (H&C).

		FRUM	GRU4Rec	STAMP	SRGNN	Bert4Rec	CL4SRec	GCSAN	CORE	**GLAD**
EC	hit@1	—	0.0693	0.1277	0.1325	0.1523	0.1413	0.1441	0.1548	**0.1609**
	hit@5	—	0.1335	0.1694	0.1740	0.1767	0.1824	0.1847	0.1780	**0.1889**
	hit@10	—	0.1491	0.1705	0.1770	0.1737	0.1844	0.1856	0.1857	**0.1965**
	mrr@10	—	0.0939	0.1426	0.1473	0.1555	0.1561	0.1586	0.1631	**0.1727**
EL	hit@1	0.1728	0.2054	0.2485	0.2264	0.2524	0.2350	0.2347	0.2073	**0.3639**
	hit@5	0.3438	0.4559	0.4912	0.4783	0.5020	0.4840	0.4862	0.4347	**0.5714**
	hit@10	0.4207	0.5786	0.5949	0.5874	0.6142	0.5939	0.5977	0.5471	**0.6665**
	mrr@10	0.1719	0.3123	0.3509	0.3335	0.3594	0.3410	0.3416	0.3047	**0.4525**
H&C	hit@1	0.1569	0.1845	0.2439	0.2018	0.2056	0.2013	0.2029	0.2098	**0.3426**
	hit@5	0.3098	0.4270	0.4337	0.4362	0.4376	0.4369	0.4387	0.4348	**0.5118**
	hit@10	0.3981	0.5236	0.5435	0.5467	0.5446	0.5476	0.5491	0.5493	**0.5888**
	mrr@10	0.1590	0.2886	0.3014	0.3020	0.3049	0.3017	0.3036	0.3051	**0.4145**

4.1 Discussion

The performance metrics for our proposed model along with popular sequential recommendation baselines are summarized in Table 1. In this section, we discuss the results in detail, and examine various qualitative aspects of the model.

RQ1. *How does the proposed model (GLAD) perform in comparison to the representative sequential recommendation models?*

It can be observed from Table 1, that our model outperforms the state of the art baselines by a significant margin. We observe clear gains across all metrics for the proposed GLAD model over the existing baselines. As expected, among the other models, it can be seen that transformer based architectures perform better in general compared to the memory/attention only models. The observed gains of the proposed GLAD over other baselines can be attributed to: 1) incorporating the graph structure, which helps our model learn from similar users, 2) target based attention mechanism (Eq. 5) that selectively attends to node-representations from long-term memory optimally, and 3) the gating mechanism (Eq. 6) that strikes a delicate balance between the long term and short term component. On the ECOM dataset, our model outperforms the closest baseline by 0.32%–1.18% on hit-rate@k. For large industry scale datasets, this is a significant improvement, given the millions of impressions on daily basis. For instance, with an average purchase value of 100 USD [2], a conversion-rate of ∼0.2%, and attribution of conversion to recommendation of ∼25% [29], 1.18%

improvement in hit@10 roughly translates to incremental gain of USD 600 per million impressions.

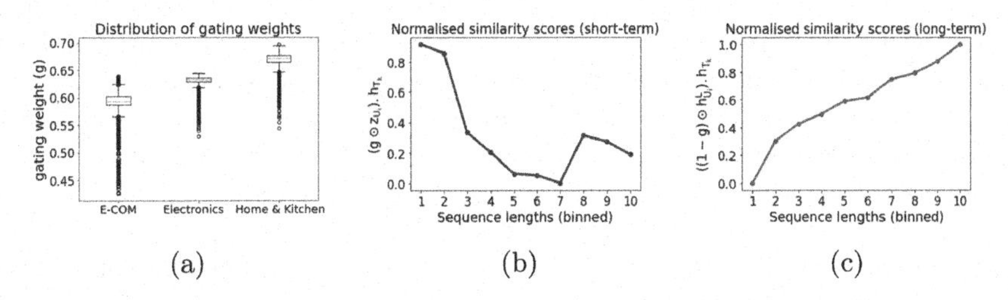

(a) (b) (c)

Fig. 3. Visualisation of gating weights. Here, (a) Distribution of gating weights, (b) Normalised similarity scores of encoder representation, and (c) Normalised similarity scores of graph memory with increasing sequence lengths.

RQ2. *Does the proposed dynamic graph-based memory effectively leverage long-term dependencies in user interactions?*

We analyse the performance of GLAD model with graph memory on different segments of users (top x% in terms of sequence lengths) based on the overall number of interactions (Table 2). It can be observed that not only does the addition of graph based memory gives consistent improvements over the backbone model (Bert4Rec), the performance gains are greater for users with larger number of interactions. This reinforces the argument that the graph based memory component is able to efficiently retrieve long term user preferences to make relevant recommendations.

Table 2. Relative gain of the GLAD model over Bert4Rec for users with varying length of historical interactions on the ECOM dataset. Sequence length is given as top % of the population for dataset anonymity.

Seq. len	Δhit@10	Δmrr@10
top 0.05%	**+3.38%**	**+4.51%**
top 1%	**+3.87%**	**+2.73%**
top 5%	**+4.47%**	**+3.46%**
top 15%	+2.31%	+1.17%
top 50%	+1.03%	−0.26%
all users	+1.73%	+0.39%

Table 3. Impact of different model components of the proposed model. Results on ECOM dataset are relative to FRUM baseline.

Dataset	Model	hit@10	mrr@10
E-COM	Bert4Rec	0.1737	0.1555
	GLAD(G)	0.1749	0.1427
	GLAD(G+C)	0.1878	0.1594
	GLAD(G+C+A)	**0.1965**	**0.1727**
Electronics	Bert4Rec	0.6142	0.3594
	GLAD(G)	0.6107	0.3608
	GLAD(G+A)	**0.6665**	**0.4525**
Home & Kitchen	Bert4Rec	0.5446	0.3049
	GLAD(G)	0.5653	0.3230
	GLAD(G+A)	**0.5888**	**0.4145**

In order to understand the model's behavior with respect to users with different sequence lengths, we plot and analyse the gating weights g. About 35%–50% weight is given to long term memory highlighting the need for using memory network to gain additional performance improvements (Fig. 3a). To gain further insights into contribution of short term vs. long term memory component, we calculate the dot-product similarity scores for these components with the target items, and plot the similarity scores. To maintain dataset anonymity, we do not disclose the actual sequence lengths. Sequences are binned into equidistant ordinal bins based on sequence lengths, and we report the average similarity scores (normalised) for each bin. These are calculated as $(g \odot z_{U_i}).h_{T_k}$ for interaction encoder output (Fig. 3b) and $((1-g) \odot h^L_{U_i}).h_{T_k}$ for the graph memory component (Fig. 3c). It can be observed that moving from users with smaller interactions histories to users with longer histories, the impact of graph memory increases, while the contribution of short-term representation from the encoder keeps decreasing, which shows that the proposed graph structure effectively leverages long-term user preferences for users with long histories.

RQ3. *What is the incremental benefit of various model components in improving the quality of the final recommendations?*

We experiment with four different model variants and report the incremental gains in performance in Table 3. All the models use Bert4Rec as the interaction sequence encoder. GLAD(G) represents the proposed model with graph memory without clustering, GLAD(C) represents the clustered version of this model and GLAD(C+A) represents the final clustered model with target conditional attention. It can be observed that the variant with clustering and target attention model performs the best highlighting the importance of higher order information sharing between different users, and selective memory read operations.

RQ4. *Is the proposed graph-memory a generic memory module that can be used with other sequential recommendation models?*

To establish the proposed memory component as a general extension which can be used with any of the sequential recommendation models, we experiment with different backbone models and tabulate the results in Table 4. It can be observed that addition of graph based memory results in a significant boost in the performance of the backbone models in both the cases.

Table 4. Performance of **GLAD** with different backbones (G+) on the E-COM dataset.

Model	hit@5	hit@10	mrr@10
STAMP	0.1694	0.1705	0.1425
G+STAMP	**0.1852**	**0.1871**	**0.1559**
Bert4Rec	0.1767	0.1737	0.1555
G+Bert4Rec	**0.1889**	**0.1965**	**0.1727**

Table 5. Comparison of inference latency of representative models. Results correspond to a set of ∼1MM customers.

Time	FRUM	GRU4Rec	Bert4Rec	GCSAN	GLAD
Total	7500 s	5520 s	1260 s	1500 s	2025 s
[7pt] Avg.	7.14 ms	5.25 ms	1.19 ms	1.42 ms	1.92 ms

RQ5. *Is the model suitable for deployment for low latency applications?*

While complex models bring in higher capacity, it is important for a model to be within acceptable latency bounds for good user experience in real world recommendation systems. Table 5 shows the inference latency comparison of our model with the other baselines on sampled E-COM dataset. It can be observed that unlike conventional memory networks, the increase in inference time due to proposed memory component is insignificant relative to the baseline models.

RQ6. *How does the memory footprint of GLAD compare against traditional memory networks?*

We estimate and compare the memory requirements of our proposed model for the memory component with that of two representative memory networks.

$$Mem(GLAD) = N_{seq} \times e/c_f + N_U \times e \tag{8}$$

Here, c_f is the clustering factor (100 here), and N_{seq} is the number of sequences. N_{seq} can be re-written as $l_{seq} \times N_U$, where l_{seq} is the average number of sequence nodes connected to a user, which is approximately 1.2 for the E-COM dataset (chunk-size = 50), and N_U represents the number of users. The effective number of parameters thus roughly reduces to:

$$Mem(GLAD) = l_{seq} \times N_U \times e/c_f + N_U \times e = 1.01 \times N_U \times e \tag{9}$$

For FRUM model with f memory slots, the relative memory requirement is:

$$Mem(FRUM) = f \times N_U \times e \approx f \times Mem(GLAD) \tag{10}$$

whereas, the number of parameters in the memory component for an L layered DMAN model is:

$$Mem(DMAN) = L \times f \times N_U \times e \approx f \times L \times Mem(GLAD) \tag{11}$$

5 Conclusion

In this work, we present a novel sequential model for personalized recommendations. The proposed architecture encodes the most recent interactions of the user to form a highly predictive short-term memory representation, and at the same time has the ability to query a larger set of past interactions through a shared graph-based memory formed using past user interaction subsequences. The graph component allocates memory for the user based on the size of the interaction history, thus making efficient use of the available memory, without having noticeable increase in the inference time. We improve the scalability of the model via clustering the graph nodes, resulting in further reduction in the number of sequence nodes to be stored for the graphs while enabling information flow between similar users. Finally, we propose a target conditioned read operation on the clustered graph memory for more accurate memory retrieval. The proposed model outperforms representative sequential recommendation baselines. Moreover, the graph-based memory is modular by design, and can be added to any sequential recommendation model to act as a store for historical interactions.

References

1. Chen, X., et al.: Sequential recommendation with user memory networks. In: Proceedings of the Eleventh ACM International Conference on Web Search and Data Mining (2018)
2. Chevalier, S.: U.S. online shopping order value by device 2022|Statista – statista.com (2022). https://www.statista.com/statistics/439516/us-online-shopping-order-values-by-device/
3. Chung, J., Gulcehre, C., Cho, K., Bengio, Y.: Empirical evaluation of gated recurrent neural networks on sequence modeling (2014)
4. Cook, P.: Sequential k-means (2008). https://www.cs.princeton.edu/courses/archive/fall08/cos436/Duda/C/sk_means.htm
5. Covington, P., Adams, J.K., Sargin, E.: Deep neural networks for YouTube recommendations. In: Proceedings of the 10th ACM Conference on Recommender Systems (2016)
6. Hamilton, W.L., Ying, Z., Leskovec, J.: Inductive representation learning on large graphs. In: NIPS (2017)
7. He, X., Liao, L., Zhang, H., Nie, L., Hu, X., Chua, T.S.: Neural collaborative filtering. In: Proceedings of the 26th International Conference on World Wide Web, WWW 2017, pp. 173–182. International World Wide Web Conferences Steering Committee, Republic and Canton of Geneva, CHE (2017). https://doi.org/10.1145/3038912.3052569
8. He, X., Liao, L., Zhang, H., Nie, L., Hu, X., Chua, T.S.: Neural collaborative filtering. In: Proceedings of the 26th International Conference on World Wide Web (2017)
9. Hidasi, B., Karatzoglou, A.: Recurrent neural networks with top-k gains for session-based recommendations. In: Proceedings of the 27th ACM International Conference on Information and Knowledge Management, CIKM 2018, pp. 843–852. Association for Computing Machinery, New York, NY, USA (2018). https://doi.org/10.1145/3269206.3271761
10. Hidasi, B., Karatzoglou, A., Baltrunas, L., Tikk, D.: Session-based recommendations with recurrent neural networks. CoRR abs/1511.06939 (2016)
11. Hochreiter, S., Schmidhuber, J.: Long short-term memory. Neural Comput. **9**, 1735–1780 (1997)
12. Hou, Y., Hu, B., Zhang, Z., Zhao, W.X.: CORE: simple and effective session-based recommendation within consistent representation space. In: Proceedings of the 45th International ACM SIGIR Conference on Research and Development in Information Retrieval (2022)
13. Kang, W.C., McAuley, J.: Self-attentive sequential recommendation. In: 2018 IEEE International Conference on Data Mining (ICDM), pp. 197–206 (2018). https://doi.org/10.1109/ICDM.2018.00035
14. Kang, W., McAuley, J.J.: Self-attentive sequential recommendation. CoRR abs/1808.09781 (2018). http://arxiv.org/abs/1808.09781
15. Kingma, D.P., Ba, J.: Adam: a method for stochastic optimization. CoRR abs/1412.6980 (2015)
16. Koren, Y., Bell, R.: Advances in Collaborative Filtering, pp. 77–118. Springer, Heidelberg (2015). https://doi.org/10.1007/978-1-4899-7637-6_3
17. Koren, Y., Bell, R., Volinsky, C.: Matrix factorization techniques for recommender systems. Computer **42**(8), 30–37 (2009). https://doi.org/10.1109/MC.2009.263

18. Li, J., Ren, P., Chen, Z., Ren, Z., Lian, T., Ma, J.: Neural attentive session-based recommendation. In: Proceedings of the 2017 ACM on Conference on Information and Knowledge Management, CIKM 2017, pp. 1419–1428. Association for Computing Machinery, New York, NY, USA (2017). https://doi.org/10.1145/3132847.3132926
19. Liu, Q., Zeng, Y., Mokhosi, R., Zhang, H.: STAMP: short-term attention/memory priority model for session-based recommendation. In: Proceedings of the 24th ACM SIGKDD International Conference on Knowledge Discovery and Data Mining, KDD 2018, pp. 1831–1839. Association for Computing Machinery, New York, NY, USA (2018). https://doi.org/10.1145/3219819.3219950
20. Loshchilov, I., Hutter, F.: SGDR: stochastic gradient descent with warm restarts. In: 5th International Conference on Learning Representations, ICLR 2017, Toulon, France, 24–26 April 2017, Conference Track Proceedings. OpenReview.net (2017). https://openreview.net/forum?id=Skq89Scxx
21. Ma, C., Ma, L., Zhang, Y., Sun, J., Liu, X., Coates, M.: Memory augmented graph neural networks for sequential recommendation. Proc. AAAI Conf. Artif. Intell. **34**(04), 5045–5052 (2020). https://doi.org/10.1609/aaai.v34i04.5945, https://ojs.aaai.org/index.php/AAAI/article/view/5945
22. MacQueen, J.: Some methods for classification and analysis of multivariate observations (1967)
23. Ni, J., Li, J., McAuley, J.: Justifying recommendations using distantly-labeled reviews and fine-grained aspects. In: Proceedings of the 2019 Conference on Empirical Methods in Natural Language Processing and the 9th International Joint Conference on Natural Language Processing (EMNLP-IJCNLP), pp. 188–197. Association for Computational Linguistics, Hong Kong, China, November 2019. https://doi.org/10.18653/v1/D19-1018, https://aclanthology.org/D19-1018
24. Petrov, A., Macdonald, C.: A systematic review and replicability study of BERT4Rec for sequential recommendation. In: Proceedings of the 16th ACM Conference on Recommender Systems, RecSys 2022, pp. 436–447. Association for Computing Machinery, New York, NY, USA (2022). https://doi.org/10.1145/3523227.3548487
25. Ren, K., et al.: Lifelong sequential modeling with personalized memorization for user response prediction. In: Proceedings of the 42nd International ACM SIGIR Conference on Research and Development in Information Retrieval, SIGIR 2019, pp. 565–574. Association for Computing Machinery, New York, NY, USA (2019). https://doi.org/10.1145/3331184.3331230
26. Rendle, S., Freudenthaler, C., Schmidt-Thieme, L.: Factorizing personalized Markov chains for next-basket recommendation, pp. 811–820 (2010). https://doi.org/10.1145/1772690.1772773
27. Sarwar, B.M., Karypis, G., Konstan, J.A., Riedl, J.: Item-based collaborative filtering recommendation algorithms. In: The Web Conference (2001)
28. Schuster, M., Paliwal, K.K.: Bidirectional recurrent neural networks. IEEE Trans. Signal Process. **45**, 2673–2681 (1997)
29. Sharma, A., Hofman, J.M., Watts, D.J.: Estimating the causal impact of recommendation systems from observational data. In: Roughgarden, T., Feldman, M., Schwarz, M. (eds.) Proceedings of the Sixteenth ACM Conference on Economics and Computation, EC 2015, Portland, OR, USA, 15–19 June 2015, pp. 453–470. ACM (2015). https://doi.org/10.1145/2764468.2764488
30. Su, X., Khoshgoftaar, T.M.: A survey of collaborative filtering techniques. Adv. Artif. Intell. **2009**, 2 (2009). https://doi.org/10.1155/2009/421425

31. Sukhbaatar, S., Szlam, A.D., Weston, J., Fergus, R.: End-to-end memory networks. In: NIPS (2015)
32. Sun, F., et al.: BERT4Rec: sequential recommendation with bidirectional encoder representations from transformer. In: Proceedings of the 28th ACM International Conference on Information and Knowledge Management, CIKM 2019, pp. 1441–1450. Association for Computing Machinery, New York, NY, USA (2019). https://doi.org/10.1145/3357384.3357895
33. Tan, Q., et al.: Dynamic memory based attention network for sequential recommendation. In: AAAI (2021)
34. Tang, J., Wang, K.: Personalized top-n sequential recommendation via convolutional sequence embedding. In: Proceedings of the Eleventh ACM International Conference on Web Search and Data Mining (2018)
35. Vaswani, A., et al.: Attention is all you need. In: Guyon, I., et al. (eds.) Advances in Neural Information Processing Systems, vol. 30. Curran Associates, Inc. (2017). https://proceedings.neurips.cc/paper/2017/file/3f5ee243547dee91fbd053c1c4a845aa-Paper.pdf
36. Wang, M., et al.: Deep graph library: a graph-centric, highly-performant package for graph neural networks. arXiv: Learning (2019)
37. Weston, J., Chopra, S., Bordes, A.: Memory networks. In: Bengio, Y., LeCun, Y. (eds.) 3rd International Conference on Learning Representations, ICLR 2015, San Diego, CA, USA, 7–9 May 2015, Conference Track Proceedings (2015). http://arxiv.org/abs/1410.3916
38. Wu, S., Tang, Y., Zhu, Y., Wang, L., Xie, X., Tan, T.: Session-based recommendation with graph neural networks. Proc. AAAI Conf. Artif. Intell. **33**(01), 346–353 (2019). https://doi.org/10.1609/aaai.v33i01.3301346, https://ojs.aaai.org/index.php/AAAI/article/view/3804
39. Xu, C., et al.: Graph contextualized self-attention network for session-based recommendation. In: Proceedings of the 28th International Joint Conference on Artificial Intelligence, IJCAI 2019, pp. 3940–3946. AAAI Press (2019)
40. Xu, L., et al.: Recent advances in RecBole: extensions with more practical considerations (2022)
41. Zhao, W.X., et al.: RecBole 2.0: towards a more up-to-date recommendation library. In: Proceedings of the 31st ACM International Conference on Information & Knowledge Management, pp. 4722–4726 (2022)
42. Zhao, W.X., et al.: RecBole: towards a unified, comprehensive and efficient framework for recommendation algorithms. In: CIKM, pp. 4653–4664. ACM (2021)
43. Zhou, C., Ma, J., Zhang, J., Zhou, J., Yang, H.: Contrastive learning for debiased candidate generation in large-scale recommender systems. In: Proceedings of the 27th ACM SIGKDD Conference on Knowledge Discovery & Data Mining (2020)
44. Zhou, G., et al.: Deep interest network for click-through rate prediction. In: Proceedings of the 24th ACM SIGKDD International Conference on Knowledge Discovery & Data Mining (2017)

Effective Adhoc Retrieval Through Traversal of a Query-Document Graph

Erlend Frayling(✉), Sean MacAvaney, Craig Macdonald, and Iadh Ounis

University of Glasgow, Glasgow, UK
{erlend.frayling,sean.macavaney,craig.macdonald,iadh.ounis}@glasgow.ac.uk

Abstract. Adhoc retrieval is the task of effectively retrieving information for an end-user's information need, usually expressed as a textual query. One of the most well-established retrieval frameworks is the two-stage retrieval pipeline, whereby an inexpensive retrieval algorithm retrieves a subset of candidate documents from a corpus, and a more sophisticated (but costly) model re-ranks these candidates. A notable limitation of this two-stage framework is that the second stage re-ranking model can only re-order documents, and any relevant documents not retrieved from the corpus in the first stage are entirely lost to the second stage. A recently-proposed Adaptive Re-Ranking technique has shown that extending the candidate pool by traversing a document similarity graph can overcome this recall problem. However, this traversal technique is agnostic of the user's query, which has the potential to waste compute resources by scoring documents that are not related to the query. In this work, we propose an alternative formulation of the document similarity graph. Rather than using document similarities, we propose a weighted bipartite graph that consists of both document nodes and query nodes. This overcomes the limitations of prior Adaptive Re-Ranking approaches because the bipartite graph can be navigated in a manner that explicitly acknowledges the original user query issued to the search pipeline. We evaluate the effectiveness of our proposed framework by experimenting with the TREC Deep Learning track in a standard adhoc retrieval setting. We find that our approach outperforms state-of-the-art two-stage re-ranking pipelines, improving the nDCG@10 metric by 5.8% on the DL19 test collection.

1 Introduction

Adhoc retrieval tasks aim to retrieve and rank information relevant to queries. Queries are typically provided by an end-user and represent some information need the end-user has. The most effective retrieval methods tend to use a multi-stage cascading approach, wherein an inexpensive first-stage retrieval algorithm (e.g., a lexical retriever like BM25 [29] or a dense retriever such as TAS-B [11]) returns a set of candidate documents, which are subsequently re-scored using a more costly and sophisticated model (such as a BERT-based cross-encoder) [18]. Although the second stage "re-ranker" can improve the precision of the top

N. Goharian et al. (Eds.): ECIR 2024, LNCS 14610, pp. 89–104, 2024.
https://doi.org/10.1007/978-3-031-56063-7_6

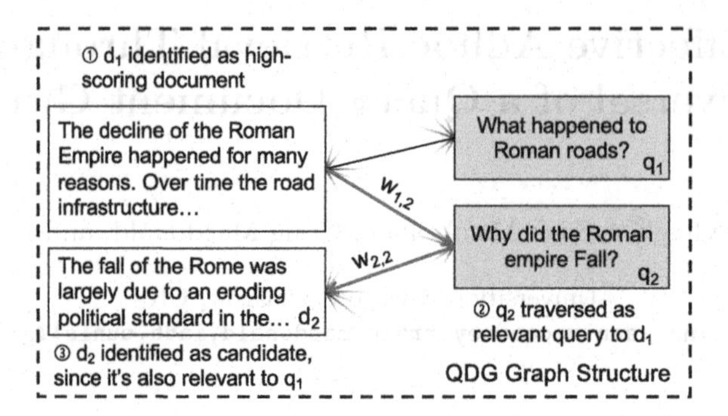

Fig. 1. Query-Document Graph (QDG): A candidate document (d_1) is likely to contain similar content to another document (d_2) if they both contain information relevant to a specific query (q_2).

results, it is ultimately limited by the recall of the first stage. No matter how well a second-stage model re-ranks the retrieved documents, the model can only score documents that were retrieved by the first stage, regardless of the retriever used. Adaptive Re-Ranking (ARR) [21], presents a solution to this second-stage recall limitation by progressively adding additional documents to the candidate pool, which are similar to the documents that were re-ranked highest. Intuitively, a document estimated to be highly relevant to a specific query is likely to be more similar to other highly ranked documents than those not relevant to the query. This intuition is formally known as the Cluster Hypothesis [13]: similar documents are likely to be relevant to the same query.

In this work, we propose an alternative mechanism for identifying related documents that is based on the *converse* of the Cluster Hypothesis. Simply put, we hypothesis that: if two documents are relevant to the same query, they are likely to be similar. Following this hypothesis, we propose that similar documents can be linked to one another through queries that the documents are both relevant to. To test this hypothesis, we propose building a bipartite graph structure that maps between documents and queries that are relevant to the documents, which we call a Query-Document Graph (QDG). The nodes of the QDG represent either queries or documents, and the edges are weighted by the estimated relevance between the document and the query. Figure 1 shows an example of a QDG. Further, we propose methods of exploring a QDG that (1) enables the estimation of document-document similarity, and (2) can be used in an ARR setting to help identify additional relevant documents. Specifically, we propose two ARR methods to use the QDG during retrieval, called (i) reverted adaptive retrieval (RAR) and (ii) resource selection.

Through experiments, we find that using QDG in the ARR framework can be used in retrieval tasks to find additional relevant documents as effectively as traditional nearest neighbour lexical graphs, and obtains similar performance in MS MARCO benchmark evaluation tests. In addition, we find that our approaches outperform dense retrieval approaches in terms of the nDCG metric.

Our contributions can be summarised as follows: (1) We present a new document-similarity method based on the hypothesis that documents relevant to the same query are similar; (2) We demonstrate how this new document-similarity method can be utilised with the Cluster Hypothesis in the context of information retrieval and how it can be constructed as a bipartite graph using query generation models; (3) We demonstrate how we can use the weighted document-query bipartite graph in an Adaptive Re-ranking setting to perform retrieval in a variety of ways; (4) We demonstrate that we can use several methods of graph exploration with Adaptive Re-ranking to significantly improve upon typical re-ranking pipelines, improving the nDCG@10 metric by 5.8% compared to a state-of the art two-stage re-ranking pipeline.

2 Related Work

In adhoc retrieval tasks, neural models such as BERT, T5 and XLNET [7,28,37] can be fine-tuned to determine the relevance between queries and documents by using learned semantic embedding representation of text [18], typically outperforming retrieval models that rely on lexical term matches, such as BM25. Neural models are typically too costly to run on entire document collections due to their large size [10,14]. Therefore, neural ranking models are commonly deployed as second-stage re-rankers - they reorder a candidate subset of the whole corpus retrieved by an inexpensive (but less effective) first-stage model. An emergent limitation of such a two-stage pipeline design is that recall cannot be improved after the first stage ranking. Ind eed, the neural re-ranker can only reorder the documents retrieved by the first-stage retriever, and any relevant documents that were not retrieved in the first stage are 'lost'. This results in an efficiency-effectiveness trade-off, whereby retrieving a higher number of documents in the first stage of retrieval improves recall, but increases the computational cost of re-ranking. As a whole, the two-stage pipeline tries to approximate the performance of an end-to-end neural ranker. Indeed, there have been several attempts to reduce the computational costs using a standalone neural ranker, i.e. dense retrieval [15,19,20,36]. However, it is still generally considered an expensive procedure and comes with its own trade-offs, especially in terms of the computational requirements at indexing time and the storage requirements of the pre-computed representations.

Another technique that is typically used to improve recall after re-ranking is Pseudo-Relevance Feedback (PRF) [31]. Traditional PRF reformulates the original query with additional terms taken from the top ranked documents assumed to be relevant to the query. The first-stage ranker then uses this enhanced query to retrieve more candidate documents [1,2,12]. PRF has also been applied to improve dense retrieval [17,35,38], where feedback of the top ranked documents is expressed in the form of the embeddings [16]. Recent work in transformer-based solutions like Doc2Query [25] and DocT5Query [24] have been used to enrich the initial document representation by applying sequence-to-sequence text generation on documents to generate queries that may be relevant to the document's content. Further pre-processing steps - in particular, relevance filtering - can improve the quality of the document expansion terms [8]. Other learned sparse retrieval systems combine both document and query expansion, alongside term re-weighting [23].

An earlier idea for improving the two-stage pipeline uses generated queries specifically for PRF. Pickens *et al.* [26] proposed Reverted Indexing (RI), which builds an index of which terms or queries are most indicative of each document in a corpus. After a first pass retrieval, the terms associated to the retrieved documents can then be used as an expanded query. One of our proposed Adaptive Retrieval methods is inspired by Reverted Indexing.

In ARR [21], documents from the first-stage retrieval are re-ranked in batches. Each time a candidate batch is re-ranked, an additional discovery task is executed to find similar documents to the most relevant, highest re-ranked documents. These similar documents are obtained using a document-similarity graph, and any identified similar documents not already present in the first-stage retrieval document collection are queued to be re-ranked alongside the documents from the first-stage retrieved documents. This process is repeated, alternating between re-ranking batches from the first stage retrieval and re-ranking batches of documents similar to those already re-ranked highly, which were discovered with the similarity graph until a re-ranking budget, in terms of number of documents, has been consumed. This process is based on the principle of the Cluster Hypothesis [9,34], which states that similar documents will be relevant to the same queries. Using ARR, it is possible to identify additional documents similar to those that the re-ranker has already assessed as relevant to the user's query. This process, therefore, presents an effective method of overcoming the first stage-recall limitation as additional similar documents may not have been present in the first-stage retrieval candidate collection and shows significant improvements in recall over typical two-stage pipelines [21].

One notable limitation of the ARR framework is that the document-similarity graph is agnostic to the user's original query. The graph identifies documents similar to those already predicted as highly relevant to the query by the re-ranker. However, it does not explicitly account for the user's query in any way. In our work, we instead propose to use a bipartite graph of generated queries and documents to discover similar documents that also consider document relevance to the actual user's query. To our knowledge, only one other work [32] has attempted to directly map documents to a space of their generated queries, however they do not consider a graph-based approach. In our approach, neighbouring documents linked by generated queries can be prioritised for adaptive retrieval based on how similar their linking query in the bipartite graph is to the original user query. We present our work as a reformulation of ARR, accounting for the user query with this new graph mechanism.

3 Query-Document Graph

In this section, we introduce our idea as to how similar documents can be found through queries that documents are relevant to. We also propose that such query-document links can be developed into a bipartite graph structure where the nearest neighbour documents can be identified through connecting query nodes.

The Cluster Hypothesis can be stated using notation as follows:

$$Rel(d_1, q) \wedge Sim(d_1, d_2) \implies Rel(d_2, q) \quad (1)$$

In other words, if a document d_1 is relevant to a query q, and d_1 is similar to another document d_2, then d_2 is also likely to be relevant to q. Central to our idea of QDG, we believe that the converse of the Cluster Hypothesis statement also indicates the similarity between two documents. Specifically, if two documents are relevant to the same query, then those documents are likely to be similar:

$$Rel(d_1, q) \wedge Rel(d_2, q) \implies Sim(d_1, d_2) \quad (2)$$

More broadly, we envisage these relations between documents and queries as a bipartite graph structure, where multiple documents may be relevant to the same queries - forming multi-hop connections in the graph. Our proposed logic for this graph connecting documents and queries, and for how similar documents can be identified using the graph, which we call a Query-Document Graph (QDG), is illustrated in Fig. 1.

The sets of unique documents and unique queries of a corpus can represent two disjointed sets of vertices in a bipartite graph such that no two documents are directly connected, instead always connecting through a shared query vertex. When building such a structure, two key challenges arise. Firstly, obtaining a large set of associated relevant queries for each document in a corpus may be challenging. Secondly, we require a method of using the weighted edges of the graph so that the order of nearest neighbour documents can be calculated when a document is linked to multiple other documents through query vertices. We describe in detail how these two challenges are approached in the following two sections to build a weighted bipartite graph and how this can be used to identify nearest neighbour documents. Finally, we note that other works have also modeled query-document relationships as a bipartite graph [3,6]. However, to the best of our knowledge, we are the first to make use of this graph in an adaptive re-ranking setting.

3.1 Query Population

To build QDG for a corpus, it is necessary to first obtain a set of queries for each document and an associated relevance score is necessary to represent edge weights. For our work, we generate queries for each document using a Doc2Query generative model - generating T queries for each document in the collection [24]. We use a neural re-ranking model to provide a relevance estimate of how well-suited each generated query is to its source document. We remove duplicate queries per document and, inspired by work on Reverted Indexing [26], invert that list, such that we obtain a two-way look-up structure, i.e. document-to-query and query-to-document, with associated relevance scores that map traversals across the bipartite graph from any document or query vertex.

3.2 Ranking Neighbours

A mapping of documents and queries with associated relevance scores for each document makes it possible to explore documents connected through queries as a bipartite graph. The weights of the edges between documents and queries are associated with estimated relevance scores between the documents and queries. Indeed, between a tuple of documents $\langle d_1, d_2 \rangle$, there exists two aspects of relevance connecting them via a common generated query, q'. Hence, we can estimate the similarity between d_1 and d_2 as the product of the relevance scores along each edge, as follows:

$$Sim(d_1, d_2) = Rel(q', d_1) * Rel(q', d_2) \tag{3}$$

where $Rel(\cdot)$ is a relevance estimation function, such as a cross-encoding neural re-ranker. The intuition is that for two documents to be similar to one another, they should be similar to the same query. By scoring all documents that are two hops from a source document (d_1), we can compute a weighted ranking of the most similar documents to the source document d_1.

Equation (3) is sufficient to determine the similarity between two documents in isolation. However, in practice, we often have an additional important signal available: the user's original query q_u. In this case, it becomes more important to emphasise edges that are similar to the actual user's information need, rather than relevance to our source document. We, therefore, can further condition the similarity on the user's query, and replace the $Rel(q', d)$ component with a query similarity measure as follows:

$$Sim(d_1, d_2 \mid q_u) = Sim(q_u, q') * Rel(q', d_2) \tag{4}$$

Note that in this formulation, the source document (d_1) is ignored entirely; instead, it is only used insofar as to identify potentially-relevant queries in the graph to traverse.

4 QDG in an Adaptive Retrieval Framework

We now describe different ways a QDG can be utilised in a retrieval pipeline. To begin, we identify the Adaptive Re-ranking pipeline as a natural fit for the QDG due to its use of a corpus graph component. In the original ARR work, this pipeline retrieves a candidate set of documents with a first stage retriever, and then re-ranks batches of documents from this candidate pool. Using a corpus graph component (originally a lexical nearest neighbour graph) the pipeline identifies documents that similar to the documents in the batch to be re-ranked are added to a special pool called the frontier. The re-ranking process then continues, but batches of documents are chosen alternatively from the original candidate pool or the frontier pool, adding new documents to the frontier for every batch that is re-ranked, until a re-ranking budget is reached. This frontier helps to find documents not retrieved by the first-stage retriever and helps overcome the first-stage retriever.

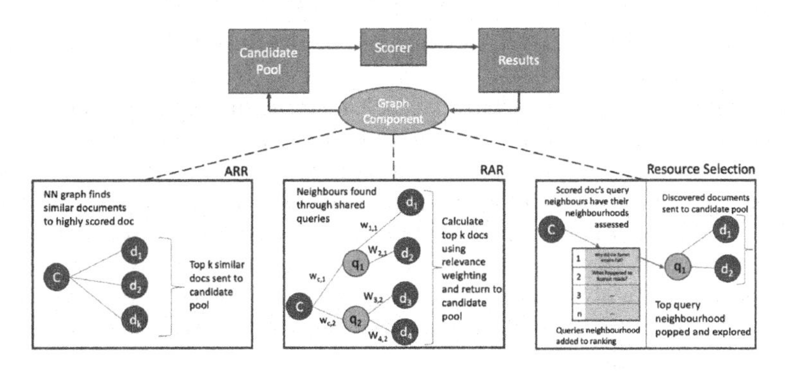

Fig. 2. The fundamental Adaptive Re-Ranking pipeline remains while the graph-based component is altered. (Left) Original Adaptive Re-Ranking approach. (Centre) Reverted Adaptive Retrieval (Right) The Resource Selection approach.

As a first idea, we propose to directly use a QDG as the corpus graph component in the Adaptive Re-ranking framework. Using the logic described in Sect. 3.2, similar documents to those already re-ranked in the adaptive re-ranking pipeline can be found with the QDG to add to the frontier, with the added benefit that the similar documents will also be identified using the original user query (as portrayed in Eq. (4)).

Secondly, we again replace the corpus graph component in the adaptive re-ranking pipeline with the QDG graph. However in this implementation, we treat each query vertex as a neighborhood of documents, i.e. all document vertices one hop away from a specific query vertex make up a neighbourhood. For a re-ranked document in the Adaptive Re-ranking pipeline, we then score, using the neural re-ranker, a small number of documents from each query vertex neighbourhood to prioritise the full query vertex neighbourhood for re-ranking in the Adaptive Re-ranking system. In this way, we can estimate the relevance of the neighbourhood of a retrieved document directly using the pipeline's re-ranker and can prioritise specific neighbourhoods of documents.

Figure 2 provides a graphical summary of our contributions. It first shows the original ARR approach using a typical nearest neighbour graph in the Adaptive Re-ranking pipeline, alongside our two proposed adaptations. As seen in Fig. 2, we essentially change the graph-based part of the ARR framework - the component that provides similar documents to those already re-ranked highly to the frontier - the rest of the Adaptive Re-ranking framework remains the same. In the remainder of this section, we describe in detail these two adaptations, Reverted Adaptive Retrieval (Sect. 4.1) and Resource Selection (Sect. 4.2).

4.1 Reverted Adaptive Retrieval

In incorporating QDG directly as the graph component of ARR, a new logic emerges for finding additional relevant documents - we call this new process

Reverted Adaptive Retrieval (RAR), inspired by the Reverted Indexing approach of Pickens *et al.* [26].

To explain RAR, we resort back to the Cluster Hypothesis (Eq. (1)) and the equation for the converse of the Cluster Hypothesis (Eq. (2)). In the case of RAR, the QDG is responsible for suggesting additional documents to score based on their similarity to the top-scoring documents seen so far. By instantiating Eq. (2) into Eq. (1) for the original Cluster Hypothesis, we can now discern the central intuition behind RAR. In particular, our intuition is that, given a document d_1 that is scored highly to the query q, both d_1 and another document d_2 are relevant to another query q', then there is a high probability that d_2 may also be relevant to the original query q, i.e.:

$$Rel(d_1, q) \land Rel(d_1, q') \land Rel(d_2, q') \implies Rel(d_2, q) \tag{5}$$

We envisage several advantages that are associated with RAR. Firstly, RAR aims to search the graph for documents that are closely related to the user's original query and not just those documents similar to the candidate document which is the case with the original ARR. Secondly, documents found using a reverted approach are directly connected to the candidate document through a related query. This allows to inspect the path in the graph that has been used to identify a document as potentially relevant, thereby providing a potential explanation for why a document was retrieved.

4.2 Resource Selection

In our first proposed retrieval framework, RAR, there is no guarantee that the neighbours added to the frontier will be relevant to the query - there is no process to identify which neighbourhoods should be explored and which neighbourhoods should be ignored, as a whole. We hypothesise that some neighbourhoods may contain no relevant documents at all and thus, a potential efficiency problem can occur where resources are used to process non-relevant documents that have been added to the frontier from low-quality neighbourhoods.

To tackle this problem, we propose a resource selection approach that prioritises entire neighbourhoods in the frontier based on an estimate of whether or not they are likely to contain relevant documents. This is schematically visualised in the right-hand box of Fig. 2. The new frontier consists of prioritised *neighbourhoods*, rather than individual documents. The process remains similar to ARR. A batch of documents is re-ranked, and then for each re-ranked document, a subset of neighbour documents for each neighbour query will also be scored. The average score of this subset of documents, per neighbour query, will be used to give each query neighbourhood a priority ranking. The central idea of this framework is that a neighbourhood with very few documents predicted to be relevant will receive a low average score from the subset of nearest neighbours. In contrast, we predict that a neighbourhood with many documents that are estimated to be relevant will score high for the subset. By prioritising the neighbourhoods based on a small scored subset, we can prioritise whole neighbourhoods that likely have many additional relevant documents. We argue this leads to a more informed selection of neighbourhoods and neighbouring documents to re-rank, and should reduce the resources used to re-rank documents that are not really relevant to the original query.

5 Experiments

We examine the effectiveness of our RAR and Resource Selection methods using the QDG by performing experiments to answer three research questions:

- **RQ1:** Does RAR bring additional relevant documents during adaptive re-ranking?
- **RQ2:** Does Resource Selection bring additional relevant documents during adaptive ranking?
- **RQ3:** Do our approaches enable earlier discovery of additional documents? i.e. can more relevant documents be found with a smaller re-ranking budget?

5.1 Experimental Setup

We address our research questions by performing retrieval experiments using the TREC Deep Learning 2019 (DL19) [4] and 2020 (DL20) [5] test collections. Our methods were developed on DL19, with DL20 serving as a held-out evaluation set. Both of these datasets use the MSMARCO passage ranking corpus consisting of 8.8 million passages. In the evaluation of our experiments, we are concerned with both precision and recall metrics, so we use nDCG, reporting the official task measure of nDCG with a rank cutoff of 10 (nDCG@10) to provide meaningful comparisons with other works, and also with cutoff 100 (nDCG@100). We also compute Mean Average Precision (MAP), and Recall at cutoff 100 (R@100).

To build a QDG for the MSMARCO corpus, we obtain a collection of 80 generated queries for each passage in MSMARCO, which were originally generated by a Doc2Query generative model [24]. We then follow the process of high-quality query generation described in Sect. 3, by using the ELECTRA cross-encoder re-ranker model [27] to estimate the relevance between a document and each generated query - inspired by the work of Gospodinov *et al.* [8] who use the same cross-encoder model for relevance assessment to identify poor quality queries generated by the Doc2query model.

We compare our methods to the original ARR approach, and with a two-stage retrieval pipeline that uses BM25 [30] – a classical lexical (sparse) retrieval model – for the first stage retrieval, and an ELECTRA cross-encoder as the second stage re-ranker, due to its superior performance to BERT-based re-rankers [27]. We also compare to two end-to-end dense retrieval approaches, ColBERT [15] & TAS-B [11]. In total, we apply six retrieval pipeline configurations:

- **MonoELECTRA**: A two-stage retrieval pipeline, where a BM25 first-stage retrieval is re-ranked by the MonoELECTRA cross-encoder, with no adaptive component.

- **ColBERT-E2E**: A state-of-the-art end-to-end dense neural ranking model that uses multiple representations per query and document.
- **TAS-B**: A state-of-the-art end-to-end dense neural ranking model that uses a single representation per query and document.
- **ARR**: This applies the original ARR framework that re-ranks documents from a BM25 initial first-stage ranking, uses a BM25 lexical graph, and MonoELECTRA as the re-ranker.
- **RAR**: This applies RAR (Sect. 4.1), using the QDG as the graph component in the ARR framework and using the BM25 retrieval model for first stage retrieval.
- **Resource Selection**: As above, but applying the Resource Selection (Sect. 4.2).

All retrieval pipeline configurations are implemented using PyTerrier [22], which provides pre-indexed retrieval models for both BM25 on the MSMARCO collection and a framework for building neural re-ranking pipelines. We obtain the original BM25 corpus graph structure used by the original ARR paper from their Github repository [21]. To measure query-query similarity, we use the `neeva/query2query` model obtained from the Huggingface model repository.

Finally, we follow the original ARR work when setting operational parameters. We use a re-ranking batch size of 16 for the re-ranker in all pipelines. We therefore set an overall re-ranking budget of 100, and evaluate performance with metrics at a rank cutoff of 100 or below. In line with the ARR work, we set a limit of 8 on the number of k-nearest neighbours to be returned in the both ARR and RAR pipelines. In the case of the Resource Selection pipeline, we assess the top 4 closest neighbour documents to each query neighbour documents to rank each query vertex, and retrieve the next top 8, after discounting the top 4 that were already scored in the assessment phase. The source code of our Adaptive Re-ranking pipelines and the generated queries are available at https://github.com/terrierteam/ecir2024_rar.

5.2 Results

Table 1 presents the results of different retrieval pipeline configurations evaluated on the DL19 and DL20 datasets. We first analyse the results compared to the MonoELECTRA baseline and the original ARR approach, and then compare the performance of our methods to the dense retrieval methods.

From the DL19 results, we see that the Resource Selection method performs significantly better than the MonoELECTRA baseline in terms of nDCG@10 - the only approach to do so on DL19 - achieving a 5.8% improvement. RAR and ARR also show improvements over the baseline for nDCG@10. In terms of nDCG@100 and MAP, the Resource Selection algorithm is also the best-performing pipeline. R@100 is the only metric where the Resource Selection approach is not the highest-performing for DL19. There, the ARR pipeline is the best model, achieving the best score of 0.559, a 14.5% improvement over the baseline. It is noteworthy, however, that although ARR retrieves more relevant documents, it does not necessarily identify documents it ranks highly, as

Table 1. Results for the DL19 and DL20 displaying RAR alongside ARR and other retrieval pipelines. **Bold** text indicates highest performing score in a given metric. * indicates statistical significance in the corresponding metric w.r.t. the baseline Mono-ELECTRA approach (paired t-test, $p < 0.05$).

Pipeline	TREC DL 2019				TREC DL 2020			
	NDCG@10	NDCG@100	MAP	R@100	NDCG@10	NDCG@100	MAP	R@100
MonoELECTRA	0.706	0.565	0.365	0.488	0.698	0.588	0.417	0.584
w/ ARR	0.724	0.607*	0.405*	**0.559***	0.724*	0.620*	0.440*	0.615
w/ RAR	0.740	0.608*	0.400	0.543*	**0.744***	**0.642***	**0.481***	0.665*
w/ Resource Selection	**0.747***	0.623*	0.417*	0.551*	0.726*	0.625*	0.446*	0.622*
Dense Retrieval Baselines								
ColBERT E2E	0.693	0.602	**0.431**	0.578	0.687	0.626	0.465	0.710
TAS-B	0.716	**0.636**	0.405	0.609	0.684	0.632	0.449	**0.713**

evidenced by its lower nDCG scores. Both RAR and Resource selection also improve recall significantly compared to the baseline, improving by 11.3% and 12.9%, respectively.

For DL20, the RAR pipeline achieves the best performance in all metrics: for the nDCG@10 metric, it achieves a 6.5% improvement over the baseline; for nDCG@100 a 9.2% improvement; for MAP a 15.3%; and finally for R@100, RAR achieves a 13.8% improvement - all of which are statistically significant. Likewise, our Resource Selection pipeline also shows statistical significance over the MonoELECTRA baseline on all metrics. The ARR pipeline shows a significant improvement in the nDCG@10, nDCG@100 and MAP metrics, however, it is not statistically significant in the R@100 metric w.r.t. the baseline - though it does offer improved performance in this metric.

Across both tasks, DL19 and DL20, the RAR and Resource selection pipelines consistently achieve higher recall than the MonoELECTRA baseline method - though only our two methods are significantly better than the baseline for R@100 on the DL20 queryset. This indicates that the Adaptive Re-Ranking approaches are capable of overcoming the recall problem associated with the first stage retrieval of the baseline pipeline.

Comparing our methods to the dense retrieval methods, we observe that for DL19, both Resource selection and RAR outperform both of the dense retrieval methods in terms of nDCG@10. Our best performing approach, Resource Selection, improves on the best performing dense approach, TAS-B, by 4.3%. Both our approaches also beat ColBERT-E2E for nDCG@100 (but TAS-B performs best overall for that measure). In terms of the recall, the dense retrieval approaches outperform all others. This is also the case for the recall metric on the DL20 dataset. However, in DL20, for every other metric, we observe that RAR outperforms the dense retrieval methods. Here, for the nDCG@10 metric, Resource Selection also outperforms the dense retrieval methods. To summarise, while the dense retrieval approaches recall more relevant documents, they do not necessarily rank these documents highly. Our approaches provide a better trade-off in this regard, as indicated by the observed nDCG values.

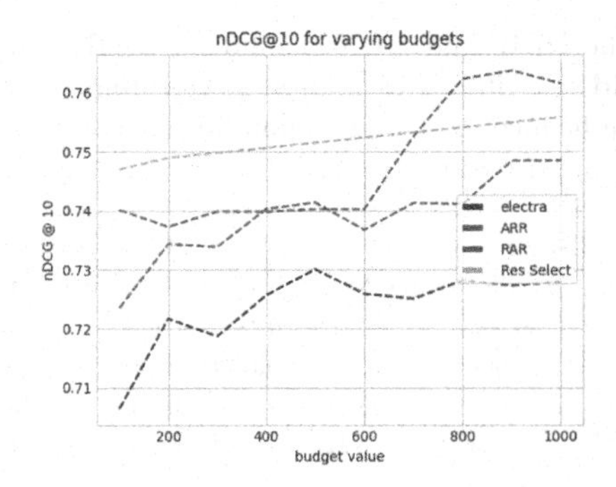

Fig. 3. nDCG@10 scores of each retrieval pipeline performing the DL19 search task across at different re-ranking budget values ranging from 100 to 1000.

We can now answer RQ1 and RQ2; both RAR and Resource Selection find additional relevant documents during retrieval, compared to the baseline MonoELECTRA two-stage re-ranker pipeline. We find that our methods perform better than ARR and the MonoELECTRA baseline for nDCG@10 in DL19, and better in all metrics than ARR in DL20.

Figure 3 shows a line graph for four of the pipelines we experimented with. It shows how nDCG@10 varies (y-axis) as the re-ranking budget (x-axis) varies from 100 to 1000 in increments of 100 for each pipeline. The values to the far left of each plot are those already displayed in Table 1. From Fig. 3, we see that both of our approaches, RAR and Resource Selection, achieve the highest nDCG@10 scores at a re-ranking budget of 100. This indicates that more relevant documents are found earlier in the retrieval process when using our approaches. Notably, our Resource Selection approach (denoted Res Select in Fig. 3) achieves the highest nDCG@10 for most re-ranking budgets below 1000, up to a budget of 700. From this point, RAR achieves the best performance from budgets of 700 to 1000. In answering RQ3, our approaches showed improved nDCG@10 scores at low re-ranking budgets compared to the MonoELECTRA re-ranking baseline and the original ARR approach. RAR and Resource selection can indeed find more relevant documents than other approaches sooner, providing the re-ranker component with more relevant documents to score.

6 Qualitative Analysis

As mentioned in Sect. 4.1, since the QDG connects document nodes through related queries, it is possible to trace back the path to determine why a document was retrieved. In this section, we demonstrate this through a qualitative example. Fig. 4 proves an example where RAR improves retrieval performance over the baseline approach in the DL19 task. One relevant document (4489656) was retrieved by both the RAR and baseline retrieval pipelines, while a second document (1901879) was only retrieved by the RAR pipeline (i.e., it was not retrieved by BM25). Instead, it was retrieved using the QDG, since it is

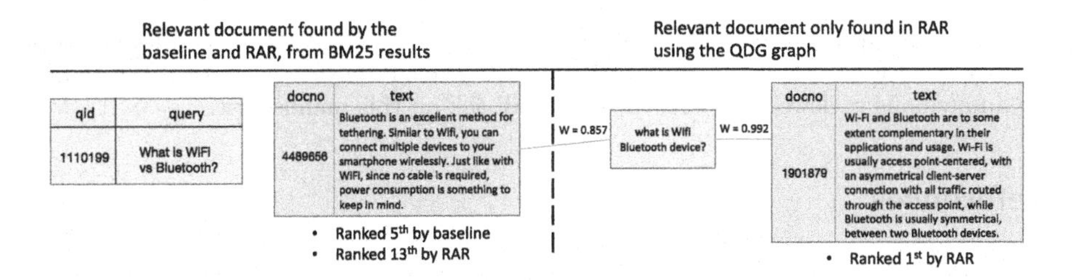

Fig. 4. An example RAR graph traversal that resulted in improved effectiveness.

closely connected to the first document through a query ("what is Wifi Bluetooth device?") which both documents are estimated to be highly relevant to. This trace could help explain to a user or researcher why the document was retrieved, for instance, by stating that the document is relevant to a query that is similar to that entered by the user.

7 Conclusions

Overcoming the recall problem is a central issue for re-ranking systems. Although the recently-proposed Adaptive Re-Ranking approach helps address the issue, its document similarity measure is agnostic of the query, which can lead to related-yet-non-relevant documents being scored in the re-ranking process. Drawing on intuitions based on the converse of the Cluster Hypothesis, we proposed two alternative approaches for identifying potentially similar documents that is conditioned on the user's original query by leveraging a new index structure – a query-document graph – that links documents and their generated queries. Through experimentation, we successfully used these approaches in two Adaptive Re-Ranking settings, which ultimately yielded improved precision of the top results and improved overall recall. We also found that the process identifies relevant documents earlier in the re-ranking process, potentially reducing the number of documents that need to be re-ranked to achieve high effectiveness.

Our approach has several limitations to address in future work, however. First, it relies on a query generation process during indexing time, which is known to be computationally expensive [33]. Future work could explore how to reduce this burden through techniques like relevance filtering [8] or by using latent (rather than text-based) query representations. The approach also adds computational overhead to the neighbourhood lookup process that cannot be pre-computed, since it relies on the user's current query. We expect that clustering or quantisation approaches could reduce this computational burden. Nevertheless, this work fills an important research gap in both index-time modelling of query-document relationships and overcoming limitations of Adaptive Re-Ranking systems.

Acknowledgement. We acknowledge EPSRC grant EP/R018634/1: Closed-Loop Data Science for Complex, Computationally- & Data-Intensive Analytics. We thank the anonymous reviewers for their helpful feedback on this manuscript.

References

1. Amati, G., Carpineto, C., Romano, G.: Query difficulty, robustness, and selective application of query expansion. In: Advances in Information Retrieval - 26th European Conference on Information Retrieval, pp. 127–137 (2004)
2. Amati, G., Van Rijsbergen, C.J.: Probabilistic models of information retrieval based on measuring the divergence from randomness. ACM Trans. Inf. Syst. (TOIS) **20**(4), 357–389 (2002)
3. Boldi, P., Bonchi, F., Castillo, C., Donato, D., Gionis, A., Vigna, S.: The query-flow graph: model and applications. In: Proceedings of the 17th ACM Conference on Information and Knowledge Management, pp. 609–618 (2008)
4. Craswell, N., Mitra, B., Yilmaz, E., Campos, D., Voorhees, E.M.: Overview of the TREC 2019 deep learning track. In: Proceedings of the Twenty-Eighth Text REtrieval Conference (2019)
5. Craswell, N., Mitra, B., Yilmaz, E., Campos, D., Voorhees, E.M., Soboroff, I.: TREC deep learning track: Reusable test collections in the large data regime. In: Proceedings of the 44th International ACM SIGIR Conference on Research and Development in Information Retrieval, pp. 2369–2375 (2021)
6. Craswell, N., Szummer, M.: Random walks on the click graph. In: Proceedings of the 30th International ACM SIGIR Conference on Research and Development in Information Retrieval, pp. 239–246 (2007)
7. Devlin, J., Chang, M., Lee, K., Toutanova, K.: BERT: pre-training of deep bidirectional transformers for language understanding. In: Proceedings of the 2019 Conference of the North American Chapter of the Association for Computational Linguistics: Human Language Technologies, NAACL-HLT, pp. 4171–4186 (2019)
8. Gospodinov, M., MacAvaney, S., Macdonald, C.: Doc2Query–: when less is more. In: Advances in Information Retrieval - 45th European Conference on Information Retrieval, pp. 414–422 (2023)
9. Hearst, M.A., Pedersen, J.O.: Reexamining the cluster hypothesis: scatter/gather on retrieval results. In: Proceedings of the 19th International ACM SIGIR Conference on Research and Development in Information Retrieval, pp. 76–84 (1996)
10. Hofstätter, S., Hanbury, A.: Let's measure run time! extending the IR replicability infrastructure to include performance aspects. In: Proceedings of the Open-Source IR Replicability Challenge co-located with 42nd International ACM SIGIR Conference on Research and Development in Information Retrieval, pp. 12–16 (2019)
11. Hofstätter, S., Lin, S., Yang, J., Lin, J., Hanbury, A.: Efficiently teaching an effective dense retriever with balanced topic aware sampling. In: Proceedings of the 44th International ACM SIGIR Conference on Research and Development in Information Retrieval, pp. 113–122 (2021)
12. Jaleel, N.A., et al.: UMass at TREC 2004: novelty and HARD. In: Proceedings of the Thirteenth Text REtrieval Conference (2004)
13. Jardine, N., van Rijsbergen, C.J.: The use of hierarchic clustering in information retrieval. Inf. Storage Retr. **7**(5), 217–240 (1971)
14. Johnson, J., Douze, M., Jégou, H.: Billion-scale similarity search with GPUs. IEEE Trans. Big Data **7**(3), 535–547 (2019)

15. Khattab, O., Zaharia, M.: ColBERT: efficient and effective passage search via contextualized late interaction over BERT. In: Proceedings of the 43rd International ACM SIGIR Conference on Research and Development in Information Retrieval, pp. 39–48 (2020)
16. Li, C., et al.: NPRF: a neural pseudo relevance feedback framework for ad-hoc information retrieval. In: Proceedings of the 2018 Conference on Empirical Methods in Natural Language Processing, pp. 4482–4491 (2018)
17. Li, H., Zhuang, S., Mourad, A., Ma, X., Lin, J., Zuccon, G.: Improving query representations for dense retrieval with pseudo relevance feedback: a reproducibility study. In: Advances in Information Retrieval - 44th European Conference on Information Retrieval, pp. 599–612 (2022)
18. Lin, J., Nogueira, R.F., Yates, A.: Pretrained Transformers for Text Ranking: BERT and Beyond. Morgan & Claypool Publishers, San Rafael (2021)
19. MacAvaney, S., Nardini, F.M., Perego, R., Tonellotto, N., Goharian, N., Frieder, O.: Efficient document re-ranking for transformers by precomputing term representations. In: Proceedings of the 43rd International ACM SIGIR Conference on Research and Development in Information Retrieval, pp. 49–58 (2020)
20. MacAvaney, S., Nardini, F.M., Perego, R., Tonellotto, N., Goharian, N., Frieder, O.: Expansion via prediction of importance with contextualization. In: Proceedings of the 43rd International ACM SIGIR Conference on Research and Development in Information Retrieval, pp. 1573–1576 (2020)
21. MacAvaney, S., Tonellotto, N., Macdonald, C.: Adaptive re-ranking with a corpus graph. In: Proceedings of the 31st ACM International Conference on Information & Knowledge Management, pp. 1491–1500 (2022)
22. Macdonald, C., Tonellotto, N.: Declarative experimentation in information retrieval using PyTerrier. In: Proceedings of the 2020 ACM SIGIR International Conference on the Theory of Information Retrieval, pp. 161–168 (2020)
23. Nguyen, T., MacAvaney, S., Yates, A.: A unified framework for learned sparse retrieval. In: Advances in Information Retrieval - 45th European Conference on Information Retrieval, pp. 101–116 (2023)
24. Nogueira, R., Lin, J.: From doc2query to docTTTTTquery (2019). https://cs.uwaterloo.ca/~jimmylin/publications/Nogueira_Lin_2019_docTTTTTquery-v2.pdf
25. Nogueira, R.F., Yang, W., Lin, J., Cho, K.: Document expansion by query prediction. CoRR abs/1904.08375 (2019)
26. Pickens, J., Cooper, M., Golovchinsky, G.: Reverted indexing for feedback and expansion. In: Proceedings of the 19th ACM Conference on Information and Knowledge Management, pp. 1049–1058 (2010)
27. Pradeep, R., Liu, Y., Zhang, X., Li, Y., Yates, A., Lin, J.: Squeezing water from a stone: a bag of tricks for further improving cross-encoder effectiveness for reranking. In: Advances in Information Retrieval - 44th European Conference on Information Retrieval, pp. 655–670 (2022)
28. Raffel, C.: Exploring the limits of transfer learning with a unified text-to-text transformer. J. Mach. Learn. Res. **21**(1), 5485–5551 (2020)
29. Robertson, S.E., Walker, S., Jones, S., Hancock-Beaulieu, M., Gatford, M.: Okapi at TREC-3. In: Proceedings of the Third Text REtrieval Conference, pp. 109–126 (1994)
30. Robertson, S.E., Zaragoza, H.: The probabilistic relevance framework: BM25 and beyond. Found. Trends Inf. Retr. **3**(4), 333–389 (2009)
31. Rocchio Jr, J.J.: Relevance feedback in information retrieval. The SMART retrieval system: experiments in automatic document processing (1971)

32. Salamat, S., Arabzadeh, N., Zarrinkalam, F., Zihayat, M., Bagheri, E.: Learning query-space document representations for high-recall retrieval. In: Advances in Information Retrieval - 45th European Conference on Information Retrieval, pp. 599–607 (2023)
33. Scells, H., Zhuang, S., Zuccon, G.: Reduce, reuse, recycle: green information retrieval research. In: Proceedings of the 45th International ACM SIGIR Conference on Research and Development in Information Retrieval, pp. 2825–2837 (2022)
34. Voorhees, E.M.: The cluster hypothesis revisited. In: Proceedings of the 8th International ACM SIGIR Conference on Research and Development in Information Retrieval, pp. 188–196 (1985)
35. Wang, X., Macdonald, C., Tonellotto, N., Ounis, I.: Pseudo-relevance feedback for multiple representation dense retrieval. In: Proceedings of the 2021 ACM SIGIR International Conference on the Theory of Information Retrieval, pp. 297–306 (2021)
36. Xiong, L., et al.: Approximate nearest neighbor negative contrastive learning for dense text retrieval. In: 9th International Conference on Learning Representations (2021)
37. Yang, Z., Dai, Z., Yang, Y., Carbonell, J.G., Salakhutdinov, R., Le, Q.V.: XLNet: generalized autoregressive pretraining for language understanding. In: Advances in Neural Information Processing Systems 32: Annual Conference on Neural Information Processing Systems 2019, pp. 5754–5764 (2019)
38. Yu, H., Xiong, C., Callan, J.: Improving query representations for dense retrieval with pseudo relevance feedback. In: Proceedings of the 30th ACM International Conference on Information & Knowledge Management, pp. 3592–3596 (2021)

How to Forget Clients in Federated Online Learning to Rank?

Shuyi Wang[1(✉)], Bing Liu[2], and Guido Zuccon[1]

[1] The University of Queensland, Brisbane, Australia
{shuyi.wang,g.zuccon}@uq.edu.au
[2] CSIRO, Brisbane, Australia

Abstract. Data protection legislation like the European Union's General Data Protection Regulation (GDPR) establishes the *right to be forgotten*: a user (client) can request contributions made using their data to be removed from learned models. In this paper, we study how to remove the contributions made by a client participating in a Federated Online Learning to Rank (FOLTR) system. In a FOLTR system, a ranker is learned by aggregating local updates to the global ranking model. Local updates are learned in an online manner at a client-level using queries and implicit interactions that have occurred within that specific client. By doing so, each client's local data is not shared with other clients or with a centralised search service, while at the same time clients can benefit from an effective global ranking model learned from contributions of each client in the federation.

In this paper, we study an effective and efficient unlearning method that can remove a client's contribution without compromising the overall ranker effectiveness and without needing to retrain the global ranker from scratch. A key challenge is how to measure whether the model has unlearned the contributions from the client c^* that has requested removal. For this, we instruct c^* to perform a poisoning attack (add noise to this client updates) and then we measure whether the impact of the attack is lessened when the unlearning process has taken place. Through experiments on four datasets, we demonstrate the effectiveness and efficiency of the unlearning strategy under different combinations of parameter settings.

Keywords: Online Learning to Rank · Federated Learning · Federated Online Learning to Rank · Machine Unlearning

1 Introduction

In Online Learning to Rank (OLTR), ranking models keep evolving by being updated using users' implicit feedback (e.g. click data) on the relevance between queries and documents in an online manner [1,31,38,47,50]. Though OLTR provides a mechanism to learn effective ranking models, it also raises privacy concerns as it requires to collect users' interaction data to the server for centralised

N. Goharian et al. (Eds.): ECIR 2024, LNCS 14610, pp. 105–121, 2024.
https://doi.org/10.1007/978-3-031-56063-7_7

training. This paradigm is thus not suitable to privacy-sensitive situations where users do not want to share their data. To support privacy protection, a new OLTR setting which characterizes no data sharing – Federated Online Learning to Rank (FOLTR) – has been explored [22,42]. In this setting, each client (i.e. user) exploits its own data to update the ranker locally; it then sends the ranker update, instead of its data, to a central server. The server aggregates the received updates to derive an updated global ranker, which is subsequently broadcast to each client. This federated paradigm is suitable to both web-scale training of rankers, where many clients are involved (cross-device FOLTR[1]), and the institutional training of rankers, where only a few institutions or organisations are involved (cross-silo FOLTR (See Footnote 1)).

A considerate federated learning system should consider the possibility for clients to leave the federation and request the contributions of their data erased [23–25]. This possibility – dubbed the *right to be forgotten* – is contemplated in modern data protection legislation, such as in the General Data Protection Regulation (GDPR) emanated by the European Union [27]. However, the design of existing FOLTR systems is defective as lacking an *unlearning* mechanism to forget certain users' contributions. An effective and efficient unlearning mechanism is not straightforward to design. A naive way is to ask all remaining clients to re-execute the training of ranker from scratch. This carries implications in terms of disruption of service and comes with large computational costs, even if the update was done in an offline manner (counterfactually on log data stored in each client), rather than in an online manner (which in turns is impractical as it needs the users to interact again with the search results). Therefore, a more reasonable unlearning mechanism for FOLTR is necessary, but has not been studied. In this paper, we aim to fill this gap and provide an initial investigation of unlearning mechanisms for FOLTR.

In particular, we strive to overcome two main challenges. The first challenge is how to *efficiently* unlearn without requiring an unreasonable amount of additional computation. Also, the obtained new ranker is expected to have comparable effectiveness to the one retrained from scratch. The second challenge is how to evaluate the *effectiveness* of an unlearning method. In a FOLTR system with many clients, the effect on ranker's effectiveness of removing a client can be marginal, or even unnoticeable. A natural question from a user leaving the federation is: how can it be proven that the impact of my data on the ranker has been erased? An adequate evaluation method is then required to verify whether an unlearning process is effective in forgetting.

To address these challenges and facilitate future studies, we build the first benchmark for unlearning in FOLTR. We adapt an effective unlearning technique emerging from the general federated machine unlearning field (FedEraser [23]) to the context of FOLTR with adaptation into online training. In our method, some historical updates are stored in the local devices and re-used to help retrain a new ranker with much less additional computation cost. In addition, for evaluation, we adopt a poisoning attack method [44] to magnify and control the effect of

[1] Similar concepts as cross-device and cross-silo Federated Learning [21].

the client leaving the federation. Through extensive empirical experimentation across four learning-to-rank datasets, we study the effectiveness and efficiency of the unlearning method and the factors influencing its performance. The utility of our evaluation method is also verified. This paper is the first study that investigates unlearning in FOLTR systems, where it is not clear that advances in general federated learning translate to similar improvements[2]. This is because of the significant difference between FOLTR from general classification tasks, e.g. ranking vs. classification, online learning vs. offline learning[3], implicit user feedback vs. ground-truth labels, etc. In addition to the impact on FOLTR, our benchmark can enrich the task-level diversity for the evaluation of general federated unlearning methods.

2 Related Work

Federated Online Learning to Rank. FOLTR systems consider a decentralized OLTR [18,20,31,34,39,48–50] scenario where data owners (clients) collaboratively train a ranker in an online manner under the coordination of a central server without the need of sharing their data. Though FOLTR is still largely unexplored, few existing works have established an initial landscape of this research area. The Federated OLTR with Evolutionary Strategies (FOLtR-ES) [22] method was the first FOLTR system proposed in the literature. FOLtR-ES extends the centralised OLTR scenario into federated setting [29] and uses Evolution Strategies as optimization method [33]. While FOLtR-ES performs well on small-scale datasets under specific evaluation metrics, its effectiveness does not generalize to large-scale datasets and standard OLTR metrics [42]. To improve the effectiveness of FOLTR, Wang et al. [41] proposed an alternative method named FPDGD [41], which leverages the state-of-the-art OLTR method, the Pairwise Differentiable Gradient Descent (PDGD) [31], and adapts it to the Federated Averaging (FedAvg) framework [29]. FPDGD's effectiveness is comparable to that of centralized OLTR methods and is currently the state-of-the-art FOLTR method. Though, we noticed that Wang et al. [43] pointed out that FPDGD's effectiveness has been shown to deteriorate if data is not distributed identically and independently (non-IID) across clients.

This paper extends the landscape of existing works by adding an unlearning mechanism to a FOLTR system. In our experiments we use FPDGD [41] as the base FOLTR system. To avoid entangling different sources of challenges, we do not consider non-IID settings [43] and privacy-preserving mechanisms [41] within unlearning in our experiments. The study of unlearning across these more complex experimental settings is left for future work.

[2] For example, many methods that are successful for dealing with non identical and independently distributed data (non-IID) in general federated learning do not work in FOLTR [43].

[3] In federated learning, the local model can be trained on the local data repeatedly across several epochs, while in FOLTR, training data is acquired in real time as user interactions occur and it cannot be repeated and reused (e.g., a user cannot be asked to submit the same query they did in the past, and perform the same interactions).

Machine Unlearning. Our work is related to machine unlearning [5,30], which pertains to the removal of any evidence of a chosen data point from the model, a process commonly known as selective amnesia. Except ensuring the removal of certain data points from the model being the primary objective, the unlearning procedure should also do not affect the model's effectiveness. Machine unlearning has been explored in both centralised [5,7] and federated settings [9,17,23,40], but never for FOLTR – this is a novel contribution of our work.

Methods in machine unlearning can be broadly classified into two families: exact unlearning and approximate unlearning. Exact unlearning methods are designed to provide a theoretical guarantee that the methods can completely remove the influence of the data to be forgotten [4,6,12,15,28]; but a limitation of these methods is that they can only be applied to simple machine learning models. Approximate unlearning methods, on the other hand, are characterized by higher efficiency, which is achieved by relying on specific assumptions regarding the accessibility of training information, and by permitting a certain amount of reduction in the model's effectiveness [2,11,13]. These methods can be used on more complex machine learning models (e.g., deep neural networks). The existing unlearning methods for federated learning belong to the approximate unlearning family [23–25]: our method builds upon one such methods, FedEraser [23]. However, most of these methods are applied to classification tasks and no previous work considers either ranking or FOLTR systems.

Evaluation of Machine Unlearning. A key challenge posed by the task of unlearning in the federated learning context is how to evaluate whether an unlearning method has successfully removed the contributions from the client that requested to leave the federation. This evaluation is at times termed as unlearning verification [30], which specifically aims to certify that the unlearned model is unrelated to the data that needed to be removed. Due to the stochastic nature of the training for many machine learning models, it is difficult to distinguish the individual clients' models and their unlearned counterparts after a certain group of data is removed.

To evaluate the effectiveness of unlearning, a group of methods leverage membership inference attacks [10,19,23], i.e. a kind of attack method that predicts whether a data item belongs to the training data of a certain model [35]: it is then straightforward to adapt this type of attacks to verify if the unlearned samples participated in the unlearning process. However, conducting successful membership inference attacks needs subtle design and training of the inference model. The effectiveness of such attacks on FOLTR is unknown as it is an unexplored area. Thus, we do not consider this type of verification in our study.

Another type of methods is inspired from the idea of poisoning attacks; in our experiments, we embrace this direction for evaluating unlearning. These methods add arbitrary noise [46] or backdoor triggers [14,17,37,45] to the data that is needed to be deleted, with the aim of manipulating the effectiveness of the trained model. After the unlearning of poisoned datasets has taken place, the impact from the arbitrary poisoning or backdoor triggers should be reduced in the unlearned model: the extent of this reduction (and thus model gains) determines the extent of the success of unlearning.

Table 1. Notation used in this paper.

Symbol	Description
c_i	client i in the FOLTR system
c^*	the unlearned client that requested removal from the FOLTR system
n_i	number of local updates for client i, before unlearning takes place
n'_i	number of local updates for client i, during the unlearning process
T	global update rounds before the unlearning request
$T_{unlearn}$	global update rounds for federated unlearning,
	also equals to number of stored local updates for unlearning
Δt	interval of time between stored local updates
M_i^{local}	local ranking model of client i before unlearning
ΔM_i	local update of client i in federated learning
$\Delta M_i^{unlearn}$	local update of client i in federated unlearning
ΔM_i^{mal}	compromised local update of the client to be unlearned (client c^*)
z	parameter for poisoning attack (Sect. 3.4)

A significant difference between our work and existing unlearning works is that we consider the context of FOLTR, which has specific challenges not present in common classification tasks (e.g., ranking vs. classification, online learning vs. offline learning, etc.). Thus, methods proposed in the general machine learning community may not work in FOLTR. Instead of diving into a specific problem, we establish the first unlearning benchmark – including both unlearning method and evaluation setting – for FOLTR. By doing so, we facilitate future works which focus on the specific challenges of unlearning in FOLTR.

3 Methodology

3.1 Preliminary

Table 1 summarises the main notations used in this paper. In OLTR, for a certain query q and its candidate documents $\mathbf{D}$, a ranking model M is used to compute a relevance score for each candidate document $d \in \mathbf{D}$. The search engine displays the documents according to their scores in descending order, and collects user's interactions with the result page. In a centralised setting, M is iteratively trained on the server based on the features of each query-document pair and collected interaction data.

On the contrary, in FOLTR, a global ranking model M^{global} is initialised in a central server and distributed to each client. At training step t, each client c_i holds a local ranking model M_i^{local} (received from the central server) and trains the local model with local data (queries, documents, interactions) for n_i local updating times. After the local training phase[4], each client sends its local update

[4] In our empirical study, we adapt FPDGD in which PDGD algorithm is used in the local training phase. Detailed method is specified in the original paper [41].

$\Delta M_{i,t} = M_i^{local} - M_t^{global}$ to a central server, that aggregates the updates from all clients to generate an updated global model. The most common aggregation rule is FedAvg [29] which updates the global model using the weighted average of all local updates:

$$M_{t+1}^{global} = M_t^{global} + \sum_i \frac{n_i}{\sum n_i} \Delta M_{i,t} \tag{1}$$

The newly-updated global model will be broadcast to each client and replace each local M_i^{local}. The whole process is repeated continuously.

3.2 Unlearning in FOLTR

We next illustrate the unlearning process of FedEraser [23], including how we adapted to doing unlearning for a FOLTR system.

Suppose that, during the FOLTR process and after T global update rounds, a client c^* requests to leave the federation and remove all the contributed local updates. Every Δt global update rounds (i.e. at rounds $\{1, 1+\Delta t, 1+2\Delta t, ...\}$ etc.), we instruct each client to store their local updates $\Delta M_{i,t}$. The number of local updates stored by each client should then be $T_{unlearn} = [\frac{T}{\Delta t}]$.

The unlearning process takes place as outlined below, in which steps (2)–(4) are performed iteratively and the iterations correspond to the global update rounds $\{1, 1+\Delta t, 1+2\Delta t, ...\}$ in the original FOLTR process:

1. The global ranking model is initialized in the same way as in the original FOLTR process, and is passed to the clients.
2. Then, each client c_i but c^* (which left the federation), updates the local model by n_i' steps ($n_i' < n_i$, the local step in unlearning n_i' is by design smaller than that before unlearning, i.e. n_i).
3. Each client calibrates the local update $\Delta M_{i,t}'$ using the stored historical local update $\Delta M_{i,t}$ according to:

$$\Delta M_{i,t}^{unlearn} = ||\Delta M_{i,t}|| \frac{\Delta M_{i,t}'}{||\Delta M_{i,t}'||} \tag{2}$$

   where $||\Delta M_{i,t}||$ indicates the step size of the global update and $\frac{\Delta M_{i,t}'}{||\Delta M_{i,t}'||}$ indicates the direction of the update ($||\cdot||$ is L_2-norm).
4. Each client c_i sends the calibrated updates $\Delta M_{i,t}^{unlearn}$ to the central server and the global model is updated using Eq. 1.

After $T_{unlearn}$ times of global aggregation, the unlearned global model is obtained. In theory, the impact of client c^* is still imposed through $\Delta M_{i,t}$ in Eq. 2. But this impact is weakened since $\Delta M_{i,t}$ is used in fewer global updating times and the way of using $\Delta M_{i,t}$ is different from the original FOLTR process (Eq. 1). The actual impact will be evaluated as in Sect. 3.4.

3.3 Efficiency Analysis

In total, the above unlearning approach only requires $n'_i \cdot T_{unlearn}$ local updates for each client c_i. Retraining the federated ranker from scratch instead requires $n_i \cdot T$ updates (n'_i is set to be less than n_i to reduce the local computation times). Thus, in terms of training efficiency compared to the baseline condition of retraining from scratch, the unlearning approach provides a reduction of $(n_i \cdot T)/(n'_i \cdot T_{unlearn}) = \frac{n_i}{n'_i} \cdot \Delta t$ local updates for each client c_i. Along with a reduction in local training, the unlearning process also reduces the communications required between each client and the central server by $\frac{T}{T_{unlearn}} = \Delta t$ times. This reduced number of updates comes at the expense of some extra space required to store the $T_{unlearn}$ updates[5]. For each client, the storage space cost is $\lceil \frac{T}{\Delta t} \rceil \cdot ||\Delta M_{i,t}||$, where $||\Delta M_{i,t}||$ indicates the L_2-norm of each local updates and it is determined by ranking model structure and the number of features.

3.4 Evaluating Unlearning

The impact of a typical client on the whole FOLTR system can be marginal. For the purpose of evaluating the unlearning approach, we need to magnify the effect of the client that is leaving the FOLTR system and make this effect relatively controllable. In previous research on machine unlearning and federated unlearning, the effectiveness of unlearning is verified by comparing the effectiveness of the model before and after unlearning [13]. This evaluation method comes with a drawback: effective unlearning is associated with a loss in model effectiveness – a situation that is undesirable and that would penalise methods that can unlearn and do not hurt model effectiveness. An alternative direction has been recently proposed: leverage poisoning or backdoor attacks to evaluate unlearning [17,19,36,37,45,46].

Inspired by poisoning attacks methods for federated learning [3] and FOLTR systems [44] where malicious clients compromise the effectiveness of the trained models by poisoning the local training data or the model updates, we add noise to the local updates of the client to be unlearned. By doing so, we expect the unlearned clients to be clearly distinguishable from others. Due to the noise being injected, the original (i.e., before unlearning) global model performs worse to some extent compared to the model trained without the poisoned client (i.e., retrained from scratch). However, after unlearning, if the contribution of the poisoned client has been successfully removed, the overall effectiveness of the unlearned global model should improve, achieving similar effectiveness as if it was retrained from scratch. We instantiate this intuition by compromising the unlearned client's (c^*) local model after each local updating phase using:

$$M_{c^*}^{mal} = -z \cdot M_{c^*}^{local} \tag{3}$$

[5] These local updates would ideally be stored within each client, but they could be stored instead in the central server: this though would required extra communication cost to provide the local updates back to the clients when needed.

where $z > 0$ represents how much we compromise the local model, while the negative coefficient is added to change the local model to its opposite direction. Thus, the compromised local update for the client to be unlearned is:

$$\Delta M_{c^*}^{mal} = -z \cdot \Delta M_{c^*}^{local} - (z+1) \cdot M_t^{global} \tag{4}$$

In the global updating phase before unlearning (Eq. 1), we replace $\Delta M_{c^*}^{local}$ with $\Delta M_{c^*}^{mal}$ for c^*. In our experiments, we set $z = 2$ as we found it sufficient in degrading the effectiveness of the global model. Note here our poisoning method does not have the burden of hiding from detection as in real poisoning attack scenarios and thus it is simple and its parameters can be tuned as per need.

A key tenet of this evaluation is that the ability to remove such a distinguishable client is equivalent to the ability to remove a much less unique client: we are unsure whether this assumption holds true, and we are not aware of relevant literature that clearly support this.

4 Experimental Setup

Datasets. We evaluate using 4 common learning-to-rank (LTR) datasets: MQ2007 [32], MSLR-WEB10k [32], Yahoo [8], and Istella-S [26]. Each dataset contains query ids and candidate document lists for each query, which is formalized as exclusive query-document pairs. Each query-document pair is represented by a multi-dimensional feature vector and annotated relevance label of the corresponding document. Among the selected datasets, MQ2007 [32] is the smallest with 1,700 queries, 46-dimensional feature vectors, and 3-level relevance assessments (from *not relevant*: 0 to *very relevant*: 2). Provided by commercial search engines, the other three datasets are larger and more recent. MSLR-WEB10k has 10,000 queries and each query is associated with 125 documents on average, each represented with 136 features. Yahoo has 29,900 queries and each query-document pair has 700 features. Istella-S is the largest, with 33,018 queries, 220 features, and an average of 103 documents per query. These three commercial datasets are all annotated for relevance on a five-grade-scale: from *not relevant* (0) to *perfectly relevant* (4). Both MQ2007 and MSLR-WEB10k datasets have five data splits (stored separately in five data folders) while Yahoo and Istella-S contain only one. Our experimental results are averaged across all data splits.

User Simulations. We follow the standard setup for user simulations in OLTR [31,41,43] by randomly selecting queries for a user and relying on the *Cascade Click Model* (CCM) click model [16] for simulating user's clicks based the relevance label and ranking position of the candidate documents. Specifically, for each query, we limit the search engine result page (SERP) to 10 documents. User clicks on the displayed ranking list are generated based on the SDBN click model. Each user is assumed to inspect every document displayed in a SERP from top to bottom while clicks the document with click probability $P(click = 1|rel(d))$ conditioned with the actual relevance label. After a click happens, the user will stop the browsing session with stop probability $P(stop = 1|click = 1, rel(d))$, or continue otherwise. In our experiments, same as the aforementioned previous works, we consider three widely-used instantiations of SDBN click model: *perfect, navigational, informational.* A *perfect* user inspects every document in a SREP with high chance to click high relevant documents thus provides very reliable feedback. The *navigational* user also searches for reasonably relevant document but has a higher chance of stopping browsing after one click. The *informational* user provides the most noisy click feedback as they do not have a specific preference on what to click and when to stop. The implementation of these click models is detailed in Table 2.

Federated Setup. We consider 10 clients participating in the FOLTR process; among these 10, one client requests to be unlearned. The original federated setup (before unlearning) involves 5 local updating steps ($n_i = 5$) among all participants and 10,000 global updating steps (T). We assume the client requests to leave the federation at global step $T = 10,000$. During the original training in our FOLTR experiments (i.e., before the unlearning request is proposed), each client holds a copy of the current ranker and updates the local ranker by issuing $n_i = 5$ queries along with the respective click responses. After the local updating finishes, the central server will receive the updated ranker from each client and aggregate all local messages to update the global one. At each Δt global steps (i.e. at rounds $\{1, 1 + \Delta t, 1 + 2\Delta t, ...\}$ etc.), each client will also keep a copy of their local ranker update to the local device for the calibration purpose in the unlearning process. In our evaluation of unlearning (results in Fig. 3), we set $\Delta t = 10$ thus the total number of stored local updates is $\lfloor \frac{T}{\Delta t} \rfloor = 1000$, equaling to the global steps in the unlearning process ($T_{unlearn}$). For further hyper-parameter analysis (results in Table 3), we set a wider value ranges of $\Delta t \in \{5, 10, 20\}$. The unlearning process follows the same federated setup, with $T_{unlearn}$ global steps, $n'_i \in \{1, 2, 3, 4\}$ local steps for the remaining 9 clients. We experiment with a linear model as the ranker with learning rate $\eta = 0.1$ and zero initialization, which we train with and rely on the configuration of the state-of-the-art FOLTR method, FPDGD [41].

Evaluation Metric. As we limit each SERP to 10 documents, nDCG@10 is used for evaluating the overall offline effectiveness of the global ranker both before and after unlearning. Effectiveness is measured by averaging the nDCG scores of the global ranker on the queries in the held-out test dataset. This is in line with previous work on OLTR and FOLTR. Unlike previous work [20, 31, 41], online evaluation is not considered in this work, as we do not need to monitor user experience during model update. Instead, we focus on measuring the overall impact of the unlearned client (the "attacker" in Fig. 2) comparing to other baselines, and the dynamic or final performance gain during the unlearning process (results in Fig. 3 and Table 3) following our evaluation process specifically for unlearning (specified in Sect. 3.4).

Table 2. Instantiations of the CCM click model used to simulate user behaviour. *rel*(d): relevance label for document d. In MQ2007, only three-levels of relevance are used: we report values for this in brackets.

rel(*d*)	$P(click = 1 \mid rel(d))$					$P(stop = 1 \mid click = 1, rel(d))$				
	0	1	2	3	4	0	1	2	3	4
per.	0.0 (0.0)	0.2 (0.5)	0.4 (1.0)	0.8 (–)	1.0 (–)	0.0 (0.0)	0.0 (0.0)	0.0 (0.0)	0.0 (–)	0.0 (–)
nav.	0.05 (0.05)	0.3 (0.5)	0.5 (0.95)	0.7 (–)	0.95 (–)	0.2 (0.2)	0.3 (0.5)	0.5 (0.9)	0.7 (–)	0.9 (–)
inf.	0.4 (0.4)	0.6 (0.7)	0.7 (0.9)	0.8 (–)	0.9 (–)	0.1 (0.1)	0.2 (0.3)	0.3 (0.5)	0.4 (–)	0.5 (–)

5 Results and Analysis

5.1 Validation of Evaluation Methodology

The unlearning evaluation methodology is based on instantiating the unlearned client as a malicious actor that injects noise in the learning process. We start by investigating if our evaluation methodology could be effective in identifying whether the unlearning has happened. For this, we consider three configurations (visualized in Fig. 1):

1. 9H-1M (green line): effectiveness obtained when one of the 10 clients is set to produce noisy updates (i.e. one client behaves maliciously).
2. 10H-0M (black line): effectiveness obtained when all 10 clients behave in an honest way, (i.e. no client is acting maliciously).
3. 9H-0M (pink line): effectiveness obtained when considering only the 9 honest clients from above, without the client that produces noisy updates (i.e. no malicious client).

Figure 1 clarifies the relationship between these rankers – all rankers share the same common set of 9 honest clients, but they differ in the 10th client considered: a malicious client for 9H-1M, an honest client for 10H-0M, and no 10th client for 9H-0M.

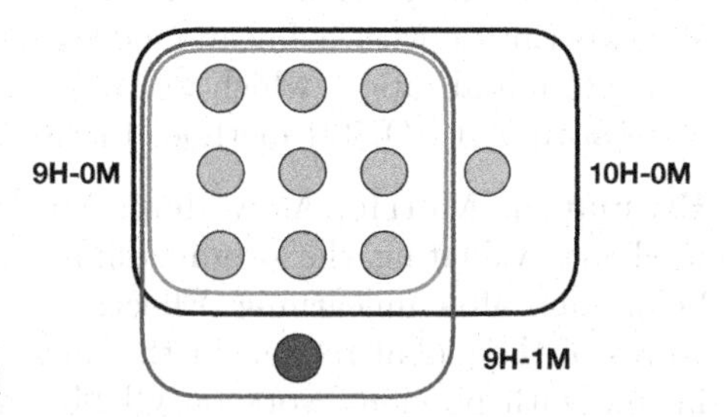

Fig. 1. Relationships between FOLTR configurations: 9H-1M (green line), 10H-0M (black), 9H-0M (pink). Circles are clients. (Color figure online)

Figure 2 reports the results obtained by these three conditions across three click modes on the MSLR-WEB10k dataset. Other datasets display similar trends; figures are omitted here for space constraints but are made available for completeness at https://github.com/ielab/2024-ECIR-foltr-unlearning. The results highlight that it is the addition of the malicious client (i.e. client c^* that will be the target of the unlearning) that sensibly reduces the effectiveness of the ranker. Note that the 9H-0M is the ranker one would obtain if the unlearning process was implemented as a federated re-training of the ranker from scratch by only considering the 9 clients remaining after the removal of client c^*. Comparing this ranker with the 10H-0M, we further highlight the need for evaluation based on the malicious client. In fact, 10H-0M and 9H-0M only consider honest clients, but in the 9H-0M one of these clients has been removed – but the effectiveness of two rankers is indistinguishable.

5.2 Effectiveness of Unlearning

We now investigate the effectiveness of the unlearning method. For this, we consider ranker 9H-1M and we perform unlearning to remove client c^*, which is the malicious client; this leads to ranker $\mathcal{U}$(9H-1M). Table 3 reports the effectiveness of the global model after unlearning has taken place and under different settings of hyper-parameters n_i' and Δt. A more detailed analysis of the impact of n_i' and Δt is presented in Sect. 5.3.

In Fig. 3, we report the offline effectiveness on four datasets obtained by the unlearning mechanism during the $[\frac{T}{\Delta t}] = 1,000$ global update times, where we set $\Delta t = 10$ and $n_i' = 3$. Compared to the original model 9H-1M[6], the unlearned model $\mathcal{U}$(9H-1M) achieves better effectiveness and it gradually converges towards the effectiveness of the 9H-0M model, showing that the unlearned model is able to successfully remove the impact of the unlearned client c^*.

5.3 Hyper-parameters Analysis

Next, we study the sensitivity of the unlearning method to its two hyper-parameters: the number of local updates for unlearning $n_i' \in \{1, 2, 3, 4\}$ and interval of time between stored updates $\Delta t \in \{5, 10, 20\}$. We report the results of this analysis in Table 3 with the final nDCG@10 score after the unlearning (or

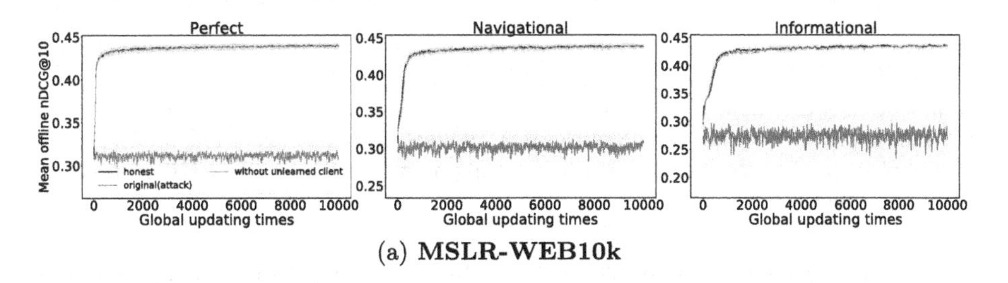

(a) **MSLR-WEB10k**

Fig. 2. Offline effectiveness (nDCG@10) obtained under the 9H-1M (green line), 10H-0M (black line), 9H-0M (pink line) FOLTR configurations with three click modes (*Perfect, Navigational, Informational*). Results are averaged across all dataset splits and experimental runs. These results motivate the use of the evaluation methodology based on the malicious client to evaluate the effectiveness of unlearning. (Color figure online)

[6] The effectiveness of 9H-1M is not shown in Fig. 3 for clarity. The reader can cross reference Fig. 3 with Fig. 2, which instead contains the effectiveness of 9H-1M.

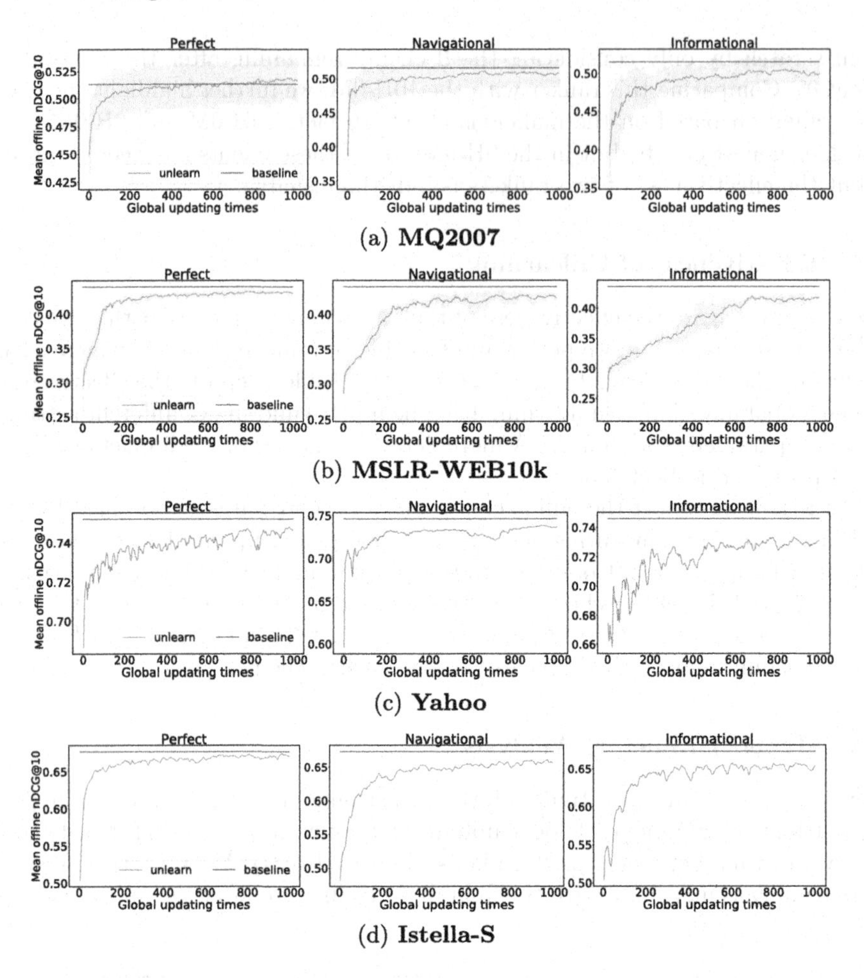

Fig. 3. Comparison between the offline effectiveness (nDCG@10) after the unlearning method is applied (ranker $\mathcal{U}$(9H-1M) denoted as "unlearn") and the ranker is retrained from scratch after client c^* is removed (ranker 9H-0M denoted as "baseline"). For the unlearning process, we set $n_i' = 3$ with $\Delta t = 10$ and show the evaluation values across all global steps. For the baseline setup, we only show the final nDCG@10 score after retraining finishes.

retraining). The effectiveness of ranker 9H-0H, i.e. the global model re-trained from scratch for $T = 10,000$ iterations and without client c^*, represents the baseline condition.

Impact of $\mathbf{n_i'}$. By design, lower values of n_i' lead to higher time savings as the time required by each local update is similar. However, lower values of n_i' may mean there is insufficient training data in each iteration, and this can lead to lower ranking effectiveness. Our experimental findings display this trade-off with relative higher effectiveness obtained when more local updates (n_i') are used during the unlearning process.

Table 3. Offline effectiveness (nDCG@10) of (1) the ranker trained from scratch after client c^* has been removed (9H-0M), and (2) the ranker for which unlearning is performed with our method ($\mathcal{U}$(9H-1M)). Effectiveness is analysed with respect to the hyper-parameters $n'_i \in \{1, 2, 3, 4\}$ and $\Delta t \in \{5, 10, 20\}$, under three different click models, and averaged across dataset splits. The best results are highlighted in boldface with superscripts denoting results for statistical significance study (paired Student's t-test with $p \leq 0.05$ with Bonferroni correction).

Dataset	#	Ranker	Click Model: Perfect			Click Model: Navigational			Click Model: Informational		
			$\Delta t = 5$	$\Delta t = 10$	$\Delta t = 20$	$\Delta t = 5$	$\Delta t = 10$	$\Delta t = 20$	$\Delta t = 5$	$\Delta t = 10$	$\Delta t = 20$
MQ2007		9H-0M	0.513	**0.513**	**0.513**	**0.516**	**0.516**	**0.516**	**0.515**	**0.515**	**0.515**
	a	$\mathcal{U}$(9H-1M), $n'_i = 4$	0.514	0.513	0.511	0.508	0.507	0.501	0.506	0.502	0.492
	b	$\mathcal{U}$(9H-1M), $n'_i = 3$	0.514	0.513	0.507	0.510	0.504	0.499	0.505	0.498	0.491
	c	$\mathcal{U}$(9H-1M), $n'_i = 2$	**0.515**	0.510	0.507	0.505	0.499	0.499	0.502	0.497	0.483
	d	$\mathcal{U}$(9H-1M), $n'_i = 1$	0.511	0.506	0.501	0.504	0.498	0.496	0.489	0.491	0.484
MSLR-10k		9H-0M	**0.439**	**0.439**^{d}	**0.439**^{cd}	**0.439**^{d}	**0.439**^{cd}	**0.439**^{abcd}	**0.436**^{cd}	**0.436**^{abcd}	**0.436**^{abcd}
	a	$\mathcal{U}$(9H-1M), $n'_i = 4$	0.436	0.433	0.429	0.433	0.428	0.422	0.429	0.413	0.389
	b	$\mathcal{U}$(9H-1M), $n'_i = 3$	0.435	0.432	0.427	0.429	0.428	0.414	0.424	0.412	0.361
	c	$\mathcal{U}$(9H-1M), $n'_i = 2$	0.432	0.431	0.423	0.426	0.421	0.389	0.420	0.386	0.348
	d	$\mathcal{U}$(9H-1M), $n'_i = 1$	0.430	0.422	0.414	0.418	0.382	0.344	0.393	0.343	0.332
Yahoo		9H-0M	**0.752**	**0.752**^{bcd}	**0.752**^{abcd}	**0.746**^{bd}	**0.746**^{bcd}	**0.746**^{abcd}	**0.746**^{abcd}	**0.746**^{abcd}	**0.746**^{abcd}
	a	$\mathcal{U}$(9H-1M), $n'_i = 4$	0.748	0.746	0.741	0.743	0.743	0.734	0.738	0.725	0.726
	b	$\mathcal{U}$(9H-1M), $n'_i = 3$	0.748	0.740	0.742	0.738	0.739	0.735	0.736	0.725	0.721
	c	$\mathcal{U}$(9H-1M), $n'_i = 2$	0.746	0.746	0.742	0.738	0.735	0.733	0.732	0.725	0.715
	d	$\mathcal{U}$(9H-1M), $n'_i = 1$	0.746	0.739	0.732	0.734	0.726	0.725	0.730	0.720	0.713
Istella-s		9H-0M	**0.677**^{cd}	**0.677**^{cd}	**0.677**^{abcd}	**0.673**^{bcd}	**0.673**^{abcd}	**0.673**^{abcd}	**0.673**^{abcd}	**0.673**^{abcd}	**0.673**^{abcd}
	a	$\mathcal{U}$(9H-1M), $n'_i = 4$	0.673	0.672	0.664	0.665	0.659	0.650	0.663	0.655	0.654
	b	$\mathcal{U}$(9H-1M), $n'_i = 3$	0.673	0.670	0.663	0.664	0.662	0.643	0.664	0.658	0.639
	c	$\mathcal{U}$(9H-1M), $n'_i = 2$	0.669	0.670	0.663	0.661	0.652	0.637	0.659	0.653	0.635
	d	$\mathcal{U}$(9H-1M), $n'_i = 1$	0.669	0.662	0.657	0.653	0.637	0.604	0.649	0.641	0.600

Impact of Δt. Larger values of Δt correspond to less global updates needed for unlearning. In fact, the required global updating time is $\lceil \frac{T}{\Delta t} \rceil$, where in our experiments $T = 10{,}000$. Table 3 shows that, in most cases, the model for which unlearning has taken place delivers higher ranking effectiveness for small values of Δt. This higher effectiveness, however, comes at the cost of extra time required by the unlearning process.

6 Conclusion

This paper is the first study that investigates unlearning in Federated Online Learning to Rank. For this, we adapt the FedEraser method [23], developed for general federated learning problems, to the unique context of federated online learning to rank, where rankers are learned in an online manner on implicit interactions (e.g. clicks). We further modify the method to leverage stored historic local updates to guide and accelerate the process of unlearning. To evaluate the effectiveness of the unlearning method, we adapt the idea of poisoning attacks to the context of determining whether the contributions of a client to be unlearned made to the ranker are effectively erased from the ranker itself. Experimental results on four popular LTR datasets show both the effectiveness and the efficiency of the unlearning method. In particular: (1) the search effectiveness of the global model once unlearning takes place converges to the effectiveness of the ranker if the removed client did not take part in the FOLTR system from the start, (2) the contributions made to the ranker by the unlearned client are effectively removed, and (3) less local and global steps are required by unlearning compared to retraining the model from scratch.

References

1. Ai, Q., Yang, T., Wang, H., Mao, J.: Unbiased learning to rank: online or offline? ACM Trans. Inf. Syst. (TOIS) **39**(2), 1–29 (2021)
2. Aldaghri, N., Mahdavifar, H., Beirami, A.: Coded machine unlearning. IEEE Access **9**, 88137–88150 (2021)
3. Baruch, G., Baruch, M., Goldberg, Y.: A little is enough: circumventing defenses for distributed learning. In: Advances in Neural Information Processing Systems, vol. 32 (2019)
4. Baumhauer, T., Schöttle, P., Zeppelzauer, M.: Machine unlearning: linear filtration for logit-based classifiers. Mach. Learn. **111**(9), 3203–3226 (2022)
5. Bourtoule, L., et al.: Machine unlearning. In: 2021 IEEE Symposium on Security and Privacy (SP), pp. 141–159. IEEE (2021)
6. Brophy, J., Lowd, D.: Machine unlearning for random forests. In: Meila, M., Zhang, T. (eds.) Proceedings of the 38th International Conference on Machine Learning, ICML 2021, 18–24 July 2021, Virtual Event. Proceedings of Machine Learning Research, vol. 139, pp. 1092–1104. PMLR (2021). https://proceedings.mlr.press/v139/brophy21a.html
7. Cao, Y., Yang, J.: Towards making systems forget with machine unlearning. In: 2015 IEEE Symposium on Security and Privacy, SP 2015, San Jose, CA, USA, 17–21 May 2015, pp. 463–480. IEEE Computer Society (2015). https://doi.org/10.1109/SP.2015.35
8. Chapelle, O., Chang, Y.: Yahoo! Learning to rank challenge overview. In: Proceedings of the Learning to Rank Challenge, pp. 1–24. PMLR (2011)
9. Che, T., et al.: Fast federated machine unlearning with nonlinear functional theory. In: International Conference on Machine Learning, pp. 4241–4268. PMLR (2023)
10. Chen, M., Zhang, Z., Wang, T., Backes, M., Humbert, M., Zhang, Y.: When machine unlearning jeopardizes privacy. In: Proceedings of the 2021 ACM SIGSAC Conference on Computer and Communications Security, pp. 896–911 (2021)
11. Chen, M., Zhang, Z., Wang, T., Backes, M., Humbert, M., Zhang, Y.: Graph unlearning. In: Yin, H., Stavrou, A., Cremers, C., Shi, E. (eds.) Proceedings of the 2022 ACM SIGSAC Conference on Computer and Communications Security, CCS 2022, Los Angeles, CA, USA, 7–11 November 2022, pp. 499–513. ACM (2022). https://doi.org/10.1145/3548606.3559352
12. Chen, Y., Xiong, J., Xu, W., Zuo, J.: A novel online incremental and decremental learning algorithm based on variable support vector machine. Clust. Comput. **22**(Supplement), 7435–7445 (2019)
13. Chundawat, V.S., Tarun, A.K., Mandal, M., Kankanhalli, M.: Zero-shot machine unlearning. IEEE Trans. Inf. Forensics Secur. **18**, 2345–2354 (2023)
14. Gao, X., et al.: VeriFi: towards verifiable federated unlearning. arXiv preprint arXiv:2205.12709 (2022)

15. Ginart, A., Guan, M.Y., Valiant, G., Zou, J.: Making AI forget you: data deletion in machine learning. In: Wallach, H.M., Larochelle, H., Beygelzimer, A., d'Alché-Buc, F., Fox, E.B., Garnett, R. (eds.) Advances in Neural Information Processing Systems 32: Annual Conference on Neural Information Processing Systems 2019, NeurIPS 2019, Vancouver, BC, Canada, 8–14 December 2019, pp. 3513–3526 (2019). https://proceedings.neurips.cc/paper/2019/hash/cb79f8fa58b91d3af6c9c991f63962d3-Abstract.html
16. Guo, F., Liu, C., Wang, Y.M.: Efficient multiple-click models in web search. In: Proceedings of the second ACM International Conference on Web Search and Data Mining, pp. 124–131 (2009)
17. Halimi, A., Kadhe, S., Rawat, A., Baracaldo, N.: Federated unlearning: how to efficiently erase a client in FL? arXiv preprint arXiv:2207.05521 (2022)
18. Hofmann, K., Schuth, A., Whiteson, S., De Rijke, M.: Reusing historical interaction data for faster online learning to rank for IR. In: Proceedings of the Sixth ACM International Conference on Web Search and Data Mining, pp. 183–192 (2013)
19. Hu, H., Salcic, Z., Dobbie, G., Chen, J., Sun, L., Zhang, X.: Membership inference via backdooring. In: The 31st International Joint Conference on Artificial Intelligence (IJCAI 2022) (2022)
20. Jia, Y., Wang, H.: Learning neural ranking models online from implicit user feedback. In: Proceedings of the ACM Web Conference 2022, pp. 431–441 (2022)
21. Kairouz, P., et al.: Advances and open problems in federated learning. Found. Trends® Mach. Learn. **14**(1–2), 1–210 (2021)
22. Kharitonov, E.: Federated online learning to rank with evolution strategies. In: Proceedings of the Twelfth ACM International Conference on Web Search and Data Mining, pp. 249–257 (2019)
23. Liu, G., Ma, X., Yang, Y., Wang, C., Liu, J.: FedEraser: enabling efficient client-level data removal from federated learning models. In: 2021 IEEE/ACM 29th International Symposium on Quality of Service (IWQOS), pp. 1–10. IEEE (2021)
24. Liu, Y., Ma, Z., Liu, X., Ma, J.: Learn to forget: user-level memorization elimination in federated learning. CoRR abs/2003.10933 (2020). https://arxiv.org/abs/2003.10933
25. Liu, Y., Xu, L., Yuan, X., Wang, C., Li, B.: The right to be forgotten in federated learning: an efficient realization with rapid retraining. In: IEEE INFOCOM 2022 - IEEE Conference on Computer Communications, London, United Kingdom, 2–5 May 2022, pp. 1749–1758. IEEE (2022). https://doi.org/10.1109/INFOCOM48880.2022.9796721
26. Lucchese, C., Nardini, F.M., Orlando, S., Perego, R., Silvestri, F., Trani, S.: Post-learning optimization of tree ensembles for efficient ranking. In: Proceedings of the 39th International ACM SIGIR conference on Research and Development in Information Retrieval, pp. 949–952 (2016)
27. de Magalhães, S.T.: The European union's general data protection regulation (GDPR). In: Cyber Security Practitioner's Guide, pp. 529–558. World Scientific (2020)
28. Mahadevan, A., Mathioudakis, M.: Certifiable machine unlearning for linear models. arXiv preprint arXiv:2106.15093 (2021)
29. McMahan, B., Moore, E., Ramage, D., Hampson, S., Arcas, B.A.: Communication-efficient learning of deep networks from decentralized data. In: Artificial Intelligence and Statistics, pp. 1273–1282. PMLR (2017)
30. Nguyen, T.T., Huynh, T.T., Nguyen, P.L., Liew, A.W.C., Yin, H., Nguyen, Q.V.H.: A survey of machine unlearning. arXiv preprint arXiv:2209.02299 (2022)

31. Oosterhuis, H., de Rijke, M.: Differentiable unbiased online learning to rank. In: Proceedings of the 27th ACM International Conference on Information and Knowledge Management, pp. 1293–1302 (2018)
32. Qin, T., Liu, T.Y.: Introducing LETOR 4.0 datasets. arXiv preprint arXiv:1306.2597 (2013)
33. Salimans, T., Ho, J., Chen, X., Sidor, S., Sutskever, I.: Evolution strategies as a scalable alternative to reinforcement learning. arXiv preprint arXiv:1703.03864 (2017)
34. Schuth, A., Oosterhuis, H., Whiteson, S., de Rijke, M.: Multileave gradient descent for fast online learning to rank. In: Proceedings of the Ninth ACM International Conference on Web Search and Data Mining, pp. 457–466 (2016)
35. Shokri, R., Stronati, M., Song, C., Shmatikov, V.: Membership inference attacks against machine learning models. In: 2017 IEEE Symposium on Security and Privacy (SP), pp. 3–18. IEEE (2017)
36. Sommer, D.M., Song, L., Wagh, S., Mittal, P.: Towards probabilistic verification of machine unlearning. arXiv preprint arXiv:2003.04247 (2020)
37. Sommer, D.M., Song, L., Wagh, S., Mittal, P.: Athena: probabilistic verification of machine unlearning. Proc. Priv. Enhancing Technol. **3**, 268–290 (2022)
38. Wang, H., Kim, S., McCord-Snook, E., Wu, Q., Wang, H.: Variance reduction in gradient exploration for online learning to rank. In: Proceedings of the 42nd International ACM SIGIR Conference on Research and Development in Information Retrieval, pp. 835–844 (2019)
39. Wang, H., Langley, R., Kim, S., McCord-Snook, E., Wang, H.: Efficient exploration of gradient space for online learning to rank. In: The 41st International ACM SIGIR Conference on Research & Development in Information Retrieval, pp. 145–154 (2018)
40. Wang, J., Guo, S., Xie, X., Qi, H.: Federated unlearning via class-discriminative pruning. In: Proceedings of the ACM Web Conference 2022, pp. 622–632 (2022)
41. Wang, S., Liu, B., Zhuang, S., Zuccon, G.: Effective and privacy-preserving federated online learning to rank. In: Proceedings of the 2021 ACM SIGIR International Conference on Theory of Information Retrieval, pp. 3–12 (2021)
42. Wang, S., Zhuang, S., Zuccon, G.: Federated online learning to rank with evolution strategies: a reproducibility study. In: European Conference on Information Retrieval (2021)
43. Wang, S., Zuccon, G.: Is non-IID data a threat in federated online learning to rank? In: Proceedings of the 45th International ACM SIGIR Conference on Research and Development in Information Retrieval, pp. 2801–2813 (2022)
44. Wang, S., Zuccon, G.: An analysis of untargeted poisoning attack and defense methods for federated online learning to rank systems. In: Proceedings of the 2023 ACM SIGIR International Conference on Theory of Information Retrieval, pp. 215–224 (2023)
45. Wu, C., Zhu, S., Mitra, P.: Federated unlearning with knowledge distillation. arXiv preprint arXiv:2201.09441 (2022)
46. Yuan, W., Yin, H., Wu, F., Zhang, S., He, T., Wang, H.: Federated unlearning for on-device recommendation. In: Proceedings of the Sixteenth ACM International Conference on Web Search and Data Mining, pp. 393–401 (2023)
47. Yue, Y., Joachims, T.: Interactively optimizing information retrieval systems as a dueling bandits problem. In: Proceedings of the 26th Annual International Conference on Machine Learning, pp. 1201–1208 (2009)

48. Zhao, T., King, I.: Constructing reliable gradient exploration for online learning to rank. In: Proceedings of the 25th ACM International on Conference on Information and Knowledge Management, pp. 1643–1652 (2016)
49. Zhuang, S., Qiao, Z., Zuccon, G.: Reinforcement online learning to rank with unbiased reward shaping. Inf. Retr. J. **25**(4), 386–413 (2022)
50. Zhuang, S., Zuccon, G.: Counterfactual online learning to rank. In: Jose, J.M., et al. (eds.) ECIR 2020, Part I. LNCS, vol. 12035, pp. 415–430. Springer, Cham (2020). https://doi.org/10.1007/978-3-030-45439-5_28

An Adaptive Feature Selection Method for Learning-to-Enumerate Problem

Satoshi Horikawa[1], Chiyonosuke Nemoto[1], Keishi Tajima[2(✉)], Masaki Matsubara[1], and Atsuyuki Morishima[1]

[1] University of Tsukuba, 1-2 Kasuga, Tsukuba, Ibaraki 305-8550, Japan
{satoshi.horikawa.2017b,chiyonosuke.nemoto.2014b}@mlab.info,
{masaki,mori}@slis.tsukuba.ac.jp

[2] Kyoto University, Yoshida-Honmachi, Kyoto 606-8501, Japan
tajima@i.kyoto-u.ac.jp

Abstract. In this paper, we propose a method for quickly finding a given number of instances of a target class from a fixed data set. We assume that we have a noisy query consisting of both useful and useless features (e.g., keywords). Our method finds target instances and trains a classifier simultaneously in a greedy strategy: it selects an instance most likely to be of the target class, manually label it, and add it to the training set to retrain the classifier, which is used for selecting the next item. In order to quickly inactivate useless query features, our method compares discriminative power of features, and if a feature is inferior to any other feature, the weight 0 is assigned to the inferior one. The weight is 1 otherwise. The greedy strategy explained above has a problem of bias: the classifier is biased toward target instances found earlier, and deteriorates after running out of similar target instances. To avoid it, when we run out of items that have the superior features, we re-activate the inactivated inferior features. By this mechanism, our method adaptively shifts to new regions in the data space. Our experiment shows that our binary and adaptive feature weighting method outperforms existing methods.

Keywords: data extraction · ranking method · relevance feedback

1 Introduction

Suppose we want to quickly find a given number of instances of some target class from a fixed large data set. If we have enough sample data, we can train a classifier for the class, but sometimes we have no such training data. For example, when we want to find news articles related to some new topic from a large news corpus, we usually have no labeled training data for such a new topic.

In such a case, a possible strategy is to find target instances and train a classifier simultaneously. We select an item most likely to be of the target class

This work was supported by JSPS KAKENHI Grant Number 22H00508, 23H03405, and JST CREST Grant Number JPMJCR22M2, Japan.

N. Goharian et al. (Eds.): ECIR 2024, LNCS 14610, pp. 122–136, 2024.
https://doi.org/10.1007/978-3-031-56063-7_8

by using the current classifier (initially random), and manually label it. If it is of the target class, we add it to the set of found target instances. No matter what its label is, we also add it to the training data set, and re-train the classifier. We repeat this process until we obtain a given number of target instances. Our goal is to minimize the number of "misses", i.e., the number of non-target instances we manually label before obtaining a given number of target instances.

This problem is sometimes called *the learning-to-enumerate* problem [11], and is related to relevance feedback. Note that it is different from the active learning problem [1,6,13,23,30], where we want to obtain a good classifier with the minimum number of items we label regardless of their classes. In active learning, we choose an item to label that would best improve the classifier regardless of its class, but in our problem, we prefer to label instances of the target class.

In the learning-to-enumerate problem, therefore, there exists a trade-off between exploitation and exploration. We need to choose either an item that is more likely to be of the target class (exploitation), or an item that would better improve the classifier (exploration). Jörger et al. [11] conducted an experiment with 19 public data sets, and reported that an exploitation-only strategy with a random forest classifier achieved the best performance in most cases without any pattern regarding class bias, number of features, or total number of instances, that influenced their results in any consistent way.

One problem in the exploitation-only strategy for data extraction from a fixed data set is that the classifier is biased toward target instances found earlier. If the target instances are distributed across several clusters, the classifier is biased toward the clusters found earlier, and after we have extracted all the instances in those clusters, the performance of the classifier suddenly degrades.

Another issue in our problem setting is formulation of useful queries. Because we assume a new target class without existing training data, we do not know what features (e.g., keywords) we should use in the query. Therefore, we assume that we only have a noisy query including both useful and useless features.

We propose *AdaFeaSE* (Adaptive Feature Selection for Enumeration), a method for data extraction from a fixed data set. It consists of the following two mechanisms for solving the two issues above. First, to quickly discard useless features, we compare discriminative power of features in a pairwise manner, and if a feature is inferior to any other with statistically significant difference, we inactivate it by giving weight 0. Otherwise a feature is active and given weight 1. Second, when some active feature runs out of matching items, we remove the feature from the set of the candidate features, re-activate inactive features, and compare their discriminative power again to re-select the features to use. By this mechanism, we can switch from an exhausted cluster to other clusters.

For example, suppose we want to find news articles on some topic from a news corpus by using keywords. There is, however, no single keyword that can find all the articles on the topic. Instead, the articles on the topic form several clusters, each of which corresponds to some keywords. We do not know those keywords in advance, and we can only come up with a set of keywords including both useful and useless ones. We run AdaFeaSE with these keywords as the features,

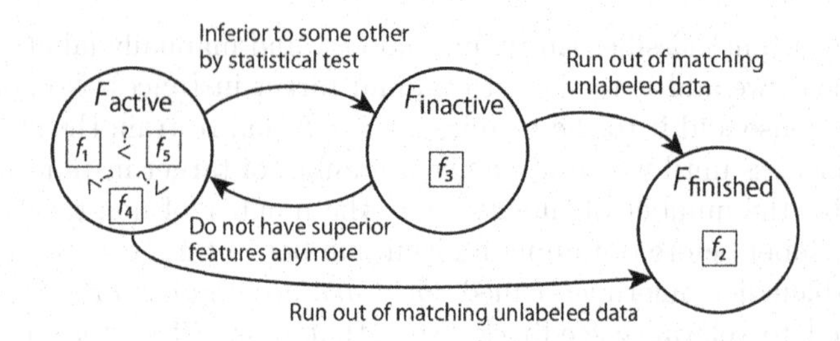

Fig. 1. State transition of features in our method. An active feature is inactivated if it is inferior to some other active feature by statistical test. On the contrary, an inactive feature is reactivated if no active feature is superior to it anymore. When an active or inactive feature runs out of matching unlabeled data, it is moved to the finished state.

and after several rounds of selecting and labeling items, suppose AdaFeaSE has found a useful keyword f that can find articles in one of the clusters with high precision. AdaFeaSE then inactivates the other keywords by assigning weight 0, and find target instances with relying on f. Because we extract articles from a fixed data set, we run out of articles including the keyword f at some point. AdaFeaSE then removes f, reactivate the other keywords that were given the weight 0, and compare their discriminative power to re-select the keywords to use. By this mechanism, AdaFeaSE focuses on each cluster in turn.

Figure 1 summarizes the state transition of features among three states in AdaFeaSE: $\mathcal{F}_{active}$, $\mathcal{F}_{inactive}$, $\mathcal{F}_{finished}$. $\mathcal{F}_{finished}$ is the state representing that the feature has run out of the matching items.

We conducted experiments where we find articles related to given new topics from a fixed news corpus. To simulate a situation where we do not know what words are useful to find relevant articles, and have to come up with candidate words resulting in a set of words including both useful and useless ones, we collected candidate words through crowdsourcing. The results of our experiments show the superiority of our binary and adaptive feature selection method over the standard learning techniques.

2 Related Work

Feature selection has been an important research topic in machine learning [2,3,9,12,15,20]. In particular, filter methods [21] use statistical tests such as Chi square tests for choosing prospective features. The filter methods test the correlation between features and the target class on the training data set, and either choose some fixed number of features with high correlation, or choose those with correlation higher than some threshold. On the other hand, AdaFeaSE tests whether each feature is significantly inferior to some other feature in the discriminative power, inactivate inferior ones temporarily, and reactivate them when the superior ones have run out of matching data items. By this mechanism, we adaptively change features to use. When the target class is distributed over multiple clusters, useful features tend to be disjunctive, that is, each feature covers only one (or some) of the clusters. Our method chooses a small number of features effective for one (or some) of the clusters, and dynamically changes the features in order to focus on each cluster one by one.

Dimensionality reduction [19,26,27] is also a popular approach for reducing the number of features for improving the learning speed when we have many features. It is, however, effective when there are many redundant features that have strong correlation with each other. The problem we focus on is not a redundant feature set but a noisy feature set including many useless ones, e.g., a feature set that non-expert users come up with. Useless features in such a feature set do not necessarily have strong correlation with each other.

As explained before, we have a trade-off between exploitation and exploration in our problem setting. It is similar to the multi-armed bandit problem [16]. There have been studies that model feature selections as a kind of multi-armed bandit problem. Durand et al. [5] applied combinational bandits to online feature selection problem. In their setting, items arrive sequentially and the learner is allowed to access a small number of features for each item. Therefore, the problem is how to choose a subset of features to observe in order to maximize the classification rate. On the other hand, in our problem setting, we choose an item to label, and if we pay a constant cost for labeling it, we can observe all features of the item. In other words, in their setting, we can try one feature in one attempt, and we cannot choose an item to try, while in our setting, we can try all features in one attempt, and we need to choose one item to try.

As explained in the previous section, Jörger et al. [11] compared various methods of balancing exploitation and exploration in the learning-to-enumerate problem setting, and their conclusion is that a simple exploitation-only method is the best. This means that the methods for active learning, e.g., uncertainly sampling, which focus on exploration, do not perform well in our setting.

In our experiments, we collect candidate features (key phrases) through crowdsourcing, which results in a noisy feature set including many useless ones. There have been attempts to obtain features for classifications through crowdsourcing [4,25,32]. In those attempts, they show positive and negative samples to help workers to find useful features. By contrast, we assume that we do not have labeled sample data for training, thus cannot show such samples to workers.

A system proposed in [22] also uses human power to select useful features. When their system asks a user to label a document, it also picks words by using a standard feature selection method, and asks the user to determine if they are relevant to specific classes. This method can be combined with ours.

Binary weighting is not necessarily good for classification tasks [10,17]. Our experimental result also show that our binary weighting method is inferior to some standard methods when they have been trained with enough data. However, our result also show that binary weighting can be a good quick approximation in the early stage of the training when we have a feature set including a few useful ones and many useless ones. In addition, our experimental result shows that feature sets human workers come up with really have such a characteristic.

Algorithm 1. The basic procedure of the exploitation-only strategy

Input: items $\mathcal{X}$, features $\mathcal{F}$, target class C, workers W, number of needed instances n
Output: set of found target instances $\mathcal{X}_{found}$
1: $\mathcal{X}_{unlabeled} \leftarrow \mathcal{X}$, $\mathcal{X}_{found} \leftarrow \emptyset$, $\mathcal{X}_{train} \leftarrow \emptyset$, initialize the model $\mathcal{M}$
2: **while** $|\mathcal{X}_{found}| < n$ **do**
3: **for all** $x_i \in \mathcal{X}_{unlabeled}$ **do** compute target class score p_i of x_i by using $\mathcal{M}$
4: $x \leftarrow x_i$ with the highest p_i
5: Query the workers W for the label y of x
6: **if** $y = C$ **then** add x to $\mathcal{X}_{found}$
7: Remove x from $\mathcal{X}_{unlabeled}$
8: Add (x, y) to $\mathcal{X}_{train}$
9: Re-train $\mathcal{M}$ on $\mathcal{X}_{train}$ for $\mathcal{X}_{unlabeled}$
10: **end while**

The exploitation-only strategy in the learning-to-enumerate problem is similar to relevance feedback. We query the labels of top-ranked items for improving rankings. Our work is particularly related to query reduction problem [7,8,14] where we remove words from verbose queries. Our method is different from theirs in that our method does not aim to find a set of features that best describes the whole target class. Instead, we find features with high precision even if they work only for a small subset of the target class, and we dynamically change the features to use in order to cover the whole target class.

3 Proposed Method

We first show the basic procedure of the exploitation-only strategy in the learning-to-enumerate problem in Algorithm 1. Let W be a set of human workers that always give the correct label to a queried item, $\mathcal{X}$ be the data set, and C be the target class. We repeatedly choose the most likely item $x \in \mathcal{X}$ and query W for its label y until we find n instances of C. To minimize the number of misses, i.e., labeling of non-target instances, we train a model $\mathcal{M}$, and choose the next item to label based on the target class score given by the current $\mathcal{M}$ (Lines 3–4). Every time we label a new item, we re-train the model $\mathcal{M}$ (Lines 7–9).

This procedure always exploits the current knowledge, and never explore new regions in the data space. It has been reported in [11] that the exploitation-only strategy is the best for the learning-to-enumerate problem. Based on their results, we adopt the exploitation-only strategy as the basis of our method.

In our framework, human workers play two roles. First, we ask workers W' to propose prospective features $\mathcal{F}$. We also ask workers W to label chosen items at Line 5 in Algorithm 1. We assume that labels given by W (by using majority voting) are always correct, while features proposed by W' are noisy because suggesting a candidate feature is a difficult open question, while determining whether an item belong to a specific class is a closed question.

Next we explain our method AdaFeaSE. Let $\mathcal{F} = \{f_1, \dots, f_m\}$ be the set of features proposed by the workers W'. Let $f_j(x_i)$ denote the value of the feature

f_j of the item x_i. In AdaFeaSE, $\mathcal{M}$ is represented by $(\mathcal{F}_{active}, \mathcal{F}_{inactive}, \mathcal{F}_{finished})$, where the three components are the following subsets of $\mathcal{F}$:

- $\mathcal{F}_{active}$: currently active features,
- $\mathcal{F}_{inactive}$: currently inactive features,
- $\mathcal{F}_{finished}$: features that no remaining unlabeled item has.

They are disjoint decomposition of $\mathcal{F}$, that is:

$\mathcal{F}_{active} \cap \mathcal{F}_{inactive} = \mathcal{F}_{inactive} \cap \mathcal{F}_{finished} = \mathcal{F}_{active} \cap \mathcal{F}_{finished} = \emptyset$, and
$\mathcal{F}_{active} \cup \mathcal{F}_{inactive} \cup \mathcal{F}_{finished} = \mathcal{F}$.

When we initialize $\mathcal{M}$, all features in $\mathcal{F}$ are in $\mathcal{F}_{active}$. When updating $\mathcal{M}$, if no remaining unlabeled item has f_i (when the feature is not binary, we use a threshold), we move f_i to $\mathcal{F}_{finished}$. In addition, if we determine that $f_i \in \mathcal{F}_{active}$ is inferior to some $f_j \in \mathcal{F}_{active}$, we move f_i to $\mathcal{F}_{inactive}$. On the contrary, if no $f_j \in \mathcal{F}_{active}$ is superior to $f_i \in \mathcal{F}_{inactive}$ anymore (because of the update of the statistics or state transition of f_j to $\mathcal{F}_{finished}$), we move f_i back to $\mathcal{F}_{active}$.

By this initialize and update procedure, each feature moves among three states in the following way. First, all features are in the active state. Each feature then may go back and forth between the active and inactive state. However, once it moves to the finished sate, it never moves back to the other states (Fig. 1).

In order to compare features f_i and f_j in terms of the discriminating power to the target class, we maintain a table $\mathcal{S}$ where i-th (and j-th) row stores the number of target and non-target instances in $\mathcal{X}_{train}$ that has the feature f_i (and f_j). To determine whether one is inferior to the other with statistically significant difference, we use Fischer's exact test when at least one of the four parameters in the contingency table is smaller than 5, and we use Chi-squared test otherwise.

Algorithm 2 shows the sub-routines used by AdaFeaSE for initializing and updating $\mathcal{M}$, and for computing the target score p_i for x_i at the corresponding parts in Algorithm 1. In the procedure for re-training $\mathcal{M}$ in Algorithm 2, we first update $\mathcal{S}$ based on the current $\mathcal{X}_{train}$. After that we first revoke features in $\mathcal{F}_{active}$ and $\mathcal{F}_{inactive}$ that no remaining data items in $\mathcal{X}_{unlabeled}$ has (Line 7–8). Next, we reactivate all f in $\mathcal{F}_{inactive}$ that have no active superior f' anymore (Line 9–11). We compare f and f' based on the current $\mathcal{S}$. Some f may be reactivated because of the updated $\mathcal{S}$, and some f may be reactivated because some f' that was superior to f has moved to $\mathcal{F}_{finished}$. Finally, we temporarily inactivate all f that are inferior to any f' that is currently active (Line 12–14).

$\mathcal{M}$ computes the target class score p_i of x_i by the formula $p_i = \sum_{f \in \mathcal{F}_{active}} f(x_i)$. In other words, we compute p_i by a linear combination of the features with the binary weight 1 for active features and 0 for other features. We use p_i for ranking the remaining data items based on their likeliness of being of the target class.

The theorem below shows the time complexity of AdaFeaSE.

Algorithm 2. AdaFeaSE (sub-routines invoked from Algorithm 1)

1: **procedure** Initialize $\mathcal{M}$
2: $\quad\mathcal{F}_{active} \leftarrow \mathcal{F}$, $\mathcal{F}_{inactive} \leftarrow \emptyset$, $\mathcal{F}_{finished} \leftarrow \emptyset$
3: **end procedure**
4:
5: **procedure** Re-train $\mathcal{M}$ on $\mathcal{X}_{train}$ for $\mathcal{X}_{unlabeled}$
6: $\quad$Update the statistics table $\mathcal{S}$ based on $\mathcal{X}_{train}$
7: $\quad$**for all** $f \in \mathcal{F}_{active} \cup \mathcal{F}_{inactive}$ **do**
8: $\quad\quad$**if** no $x \in \mathcal{X}_{unlabeled}$ has the feature f **then** move f to $\mathcal{F}_{finished}$
9: $\quad$**for all** $f \in \mathcal{F}_{inactive}$ **do**
10: $\quad\quad$**if** no $f' \in \mathcal{F}_{active}$ is superior to f statistically significantly based on $\mathcal{S}$ **then**
11: $\quad\quad\quad$move f to $\mathcal{F}_{active}$
12: $\quad$**for all** $f \in \mathcal{F}_{active}$ **do**
13: $\quad\quad$**if** f is inferior to any $f' \in \mathcal{F}_{active}$ statistically significantly based on $\mathcal{S}$ **then**
14: $\quad\quad\quad$move f to $\mathcal{F}_{inactive}$
15: **end procedure**
16:
17: **procedure** Compute target class score p_i of x_i by $\mathcal{M}$
18: $\quad p_i \leftarrow \sum_{f \in \mathcal{F}_{active}} f(x_i)$
19: **end procedure**

Theorem 1. *The worst case time complexity of AdaFeaSE is* $O(|\mathcal{F}|^2|\mathcal{X}|)$. □

It is proportional to $|\mathcal{F}|^2$ in the worst case because of the pairwise comparison of the features. However, the worst case occurs only when $O(|\mathcal{F}|)$ rows of the table $\mathcal{S}$ are updated (thus we need to re-evaluate all pairs of them), which rarely happens in practice. The results of our experiments with the data from a real application show that only a few rows are updated in the most cases. Thus we expect that AdaFeaSE scales well in practice with the number of features in $\mathcal{F}$.

4 Experiments

We conducted experiments to evaluate our method. We used 6,684 news articles published by Yahoo! Japan [29] in 2016. We set two target classes. Class Scandal is the class of articles on scandals of celebrities. We define celebrities as people who sometimes appear on TV as their jobs. Class Toyota is the class of articles on Toyota's business performance or on events that can affect it.

To create the ground truth, we crowdsourced the tasks of giving two binary labels corresponding to the two classes to each article through Yahoo! Crowdsourcing [28]. We assigned three workers to each article, and adopted their results if they all agree on the label. If they do not agree, we hired another three workers to give labels, and adopted the results of majority voting. Table 1 summarizes the results. Only a small portion of the 6,684 articles belong to the target classes.

We also submitted 200 microtasks to Yahoo! Crowdsourcing to collect phrases that workers think are good clues for distinguishing articles of each target class. We obtained 386 and 656 phrases for the Scandal and Toyota classes, respectively.

Table 1. Statistics of data sets

Target Class	Articles			Phrases	
	Positive	Negative	Total	Nominated	Used
Scandal	252	6,432	6,684	386	286
Toyota	79	6,605	6,684	656	272

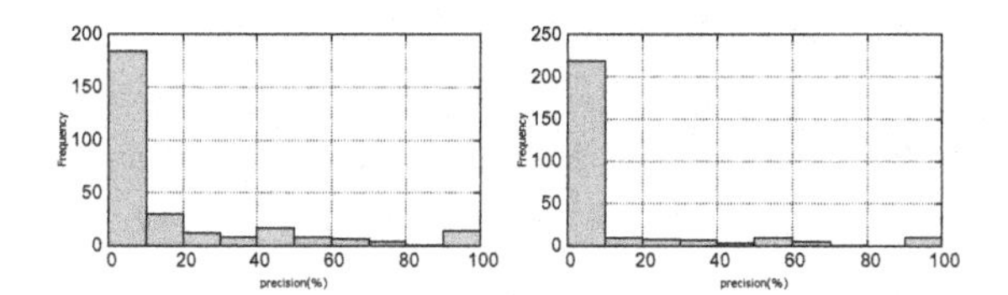

Fig. 2. Histogram over the precision of features (left: Scandal, right: Toyota).

For example, the phrases for Toyota class include: automatic driving and crude oil. We regard each obtained phrase as a binary feature of the articles; it has the value 1 if the phrase appears in the article and 0 otherwise.

The phrases nominated by crowds include many phrases that appear in no article in our data set. Out of 386 and 656 phrases nominated for Scandal and Toyota class, 286 and 272 phrases, respectively, appear in at least one article, as summarized in Table 1. The other phrases were excluded from our experiment.

Figure 2 shows the histograms of the number of phrases nominated for Scandal (left) and Toyota (right) distributed over the precision, where the precision represents the ratio of the number of true positive articles to the number of articles including the phrase. As shown in these histograms, most phrases have very low precision while there are also some phrases with high precision.

Those high-precision phrases, however, include phrases that appear only in a couple of articles. In our adaptive feature selection method, a high-precision feature can be useful even if its recall is low, but a feature that only a couple of instances have may not be useful because we may run out of the data with the feature before the classifier learns that the feature is useful. Because of that, there are a very small number of features that are really useful in our data sets.

Even if the precision is low, if a feature has strong negative correlation, it is useful in discriminating target instances. Figure 3 shows the pointwise mutual information $\log_2 (P(q|p)/P(q))$ between p meaning that the phrase appears in the article, and q meaning that the article is the target instance. It shows that few features have negative mutual information, and they only have small quantity of information compared with those with positive one. Similar results are expected when features are nominated by human users. Our binary weighting method cannot give negative weights to features while the standard learning methods can, but it is not a significant disadvantage of our method when we use features nominated by human users because of these properties of such features. When we allow users to nominate a feature as a negative feature, e.g., a keyword whose

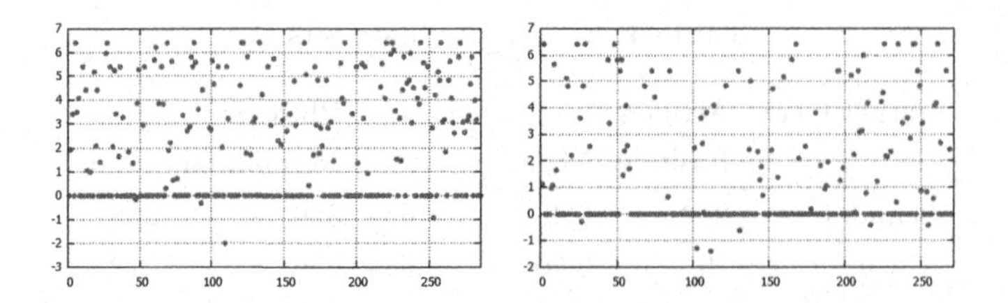

Fig. 3. Pointwise mutual information of features and class (left: Scandal, right: Toyota)

Table 2. Precision@k of AdaFeaSE with different α values

Method (α)	Scandal			Toyota		
	@100	@500	@1K	@100	@500	@1K
AdaFeaSE (0.05)	.468	.251	.179	.219	.079	.055
AdaFeaSE (0.10)	.448	.275	.180	.213	.073	.056
AdaFeaSE (0.15)	.444	.257	.180	.226	.076	.056

absence suggests that the document is a target instance, we can flip its value, i.e., we define the feature takes the value 1 when the instance does not have it.

4.1 Experimental Result

We ran our method and six baseline methods on our data set to compare their performance. We ran each method for two tasks, one for Scandal class and one for Toyota class. The six baselines include logistic regression (LR), random forest (RF), which [11] concluded is the best in the learning-to-enumerate problem, and Lasso. We also included the combination of PCA (Principal Component Analysis) and these three baseline methods. We first run PCA (cumulative proportion threshold is 0.8) on the data set with the nominated features to obtain the data set in the reduced dimension, and run one of the baselines on that data set.

Note that our goal is to quickly find target instances when starting with no training data set, and the neural network-based methods [6,18,24,31], which require a large training data set, are not effective in our problem setting.

In Algorithm 1, we ask crowds to label a data (Line 5). In this experiment, we assume that they always return the correct answers, i.e., the ground truth we created by hiring three or six workers as explained before. We leave an experiment with crowd answers that are not always correct for future work.

AdaFeaSE has a parameter α, which is the level of significance for statistical test. First, we run AdaFeaSE with three α values: $\alpha = 0.05, 0.10, 0.15$. Table 2 shows Precision@k of the three methods for $k = 100, 500, 1000$ for the two tasks. Presision@k here means that the ratio of the true positives in the k data items chosen by the first k rounds of the while loop in Algorithm 1. Table 2 shows that these α values do not largely affect the performance of AdaFeaSE. In other words, AdaFeaSE is very robust and is not sensitive to the parameter α.

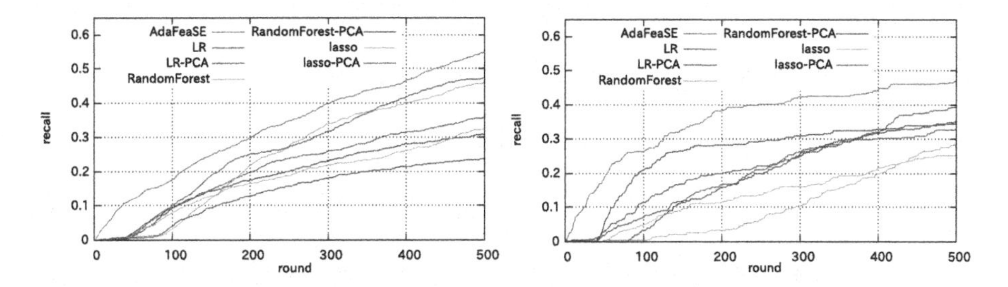

Fig. 4. Recall of AdaFeaSE and baselines at early rounds (left: Scandal, right: Toyota)

Table 3. Performance of AdaFeaSE and the second best methods (@50, @100, @300)

Class	Round	Recall (AdaFeaSE)	Second Best	Recall (Second)	Diff	Ratio
Scandal	@50	0.119	Lasso + PCA	0.021	0.098	567%
	@100	0.178	Lasso + PCA	0.099	0.079	180%
	@300	0.398	RF	0.340	0.058	117%
Toyota	@50	0.172	LR + PCA	0.053	0.119	324%
	@100	0.273	LR + PCA	0.210	0.063	130%
	@300	0.425	LR + PCA	0.309	0.116	138%

Figure 4 shows the recall of AdaFeaSE and the baselines after each round in the two tasks (left: Scandal, right: Toyota). Because these methods include randomness (they randomly choose an instance when there are ties), we run each method five times and take the average of the recall values. For AdaFeaSE, we chose the best α in each case. Because our goal is to quickly find some number of target instances, the performance at the early rounds is of our interest. Therefore, Fig. 4 only shows the early rounds upto the 500th round. These graphs show that AdaFeaSE outperforms the baselines at those early rounds.

Table 3 shows the comparison of AdaFeaSE and the second best methods in the early rounds. It shows that the difference is significant. AdaFeaSE is outperformed by some others (not necessarily the second best one in Table 3) when they are trained enough. However, when we want to quickly find a small number of target instances, such as 30 instances, AdaFeaSE can find them earlier.

4.2 Analysis on Why AdaFeaSE Works

In our experiment, AdaFeaSE outperformed the baselines in the earlier rounds. The next question is why it works. We made the following hypotheses: First, by inactivating features that are inferior to any others, AdaFeaSE can avoid overestimation of useless features at the early rounds. Second, because AdaFeaSE detects features running out of matching items, and adaptively changes the features to use, it works better when the target class is distributed across many clusters. We conducted experiments for validating these two hypotheses.

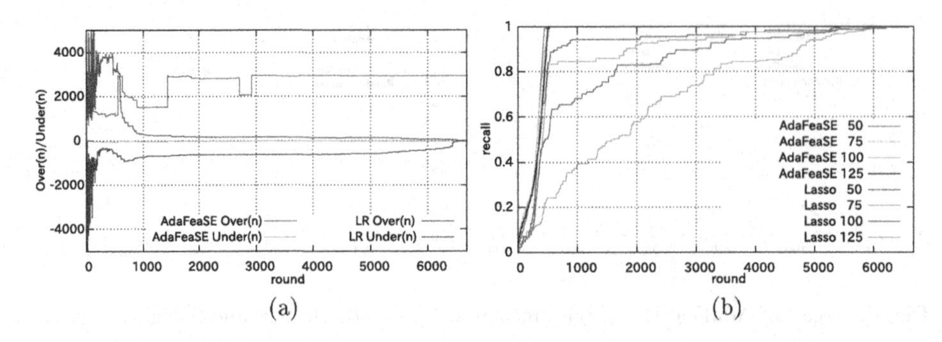

Fig. 5. (a) Over/under-estimation of weights at each round, (b) Recall by AdaFeaSE and Lasso at each round for generated data sets with different numbers of clusters.

To validate the first hypothesis, we measured how over/under-estimation of the weights change as we proceed through the rounds. We choose logistic regression for the comparison, and use the Scandal data set. We do not choose methods with PCA for the comparison because their feature sets are different from that of AdaFeaSE, and we cannot directly compare over/under-estimation of the features. We choose logistic regression because it was the method that retrieved all the target instances at the earliest round, and we use its final weights as the reference value for defining over/under-estimation as we explain below.

We first define two measures for evaluating the ratio of overestimation and underestimation. Let $W_t = \langle w_{t,1}, \ldots, w_{t,m} \rangle$ be the vector of weights for the features $f_1, \ldots, f_m$ at round t. Let also $W = \langle w_1, \ldots, w_m \rangle$ be the final weights output by logistic regression for the features $f_1, \ldots, f_m$ after all 6,684 articles are added to the training set. Because logistic regression could retrieve all the target instances earlier than AdaFeaSE, we regard the final weights output by logistic regression as the best available approximation of the ideal weights.

We then normalize the vector W_t by dividing the coefficients by $\sum_j |w_{t,j}|$. Let $\bar{W}_t = \langle \bar{w}_{t,1}, \ldots, \bar{w}_{t,m} \rangle$ denote the normalized vector. That is, $\bar{w}_{t,i} = w_{t,i} / \sum_j |w_{t,j}|$. The normalized final weights, $\bar{w}_i$, are similarly defined by $\bar{w}_i = w_i / \sum_j |w_j|$.

We then define the overestimation ratio and underestimation ratio of weights at the round t as follows:

$$Over(t) = \sum_{i=1}^{m} \begin{cases} \frac{\bar{w}_{t,i} - \bar{w}_i}{|\bar{w}_i|} & (\bar{w}_{t,i} > \bar{w}_i) \\ 0 & (otherwise) \end{cases} \tag{1}$$

$$Under(t) = \sum_{i=1}^{m} \begin{cases} \frac{\bar{w}_{t,i} - \bar{w}_i}{|\bar{w}_i|} & (\bar{w}_{t,i} < \bar{w}_i) \\ 0 & (otherwise) \end{cases} \tag{2}$$

They represent how feature weights at round t deviate from the final weights output by logistic regression. Figure 5(a) shows $Over(t)$ and $Under(t)$ for AdaFeaSE (red line) and logistic regression (blue line). The x-axis is the round $t = 1$ to $6,684$, and y-axis is $Over(t)/Under(t)$. Values closer to 0 is better. Figure 5(a)

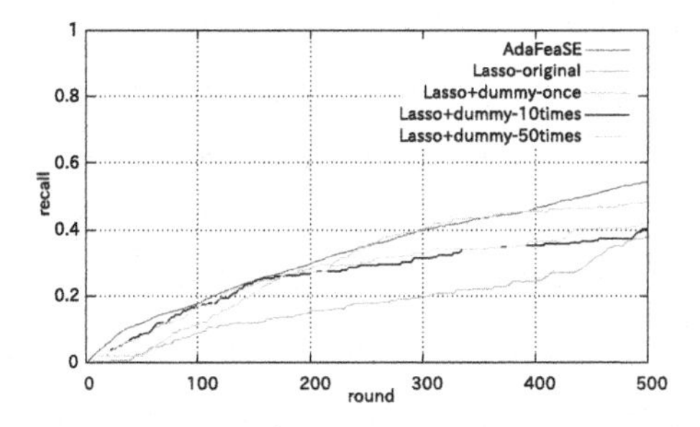

Fig. 6. Recall by pre-trained Lasso simulating initial weight values 1 and AdaFeaSE.

shows that AdaFeaSE has lower $Over(t)$ in the early rounds. We believe this is one of the reasons why AdaFeaSE could find more target instances at the early rounds. On the other hand, logistic regression has lower $Over(t)$ at $t > 700$. $Over(t)$ of AdaFeaSE largely increases at several t where some useful features are moved to $\mathcal{F}_{finished}$, and less useful features are re-activated.

Next, to validate the second hypothesis, we conducted an experiment with generated data sets consisting of different number of clusters. We generated four data sets, those with 50, 75, 100, and 125 clusters. Each of them consists of 6,684 documents, among which 252 are target instances. We also assume that 250 phrases are used as the features. These numbers are the same as those of our Scandal data set. The four data sets are different from each other only in the number of features that only target instances have. Each data set includes 50, 75, 100, and 125 phrases that appear only in the target instances, respectively. We call these features positive features. Because they appear only in the target instances, precision of these features is 1. They are also disjoint with each other, i.e., no document includes more than one positive feature. They are also complete, i.e., every target instance has one of the positive features. Therefore, the recall totally achieved by all the positive features is 1. Because of this assignment of features, each positive feature forms a cluster. Therefore, each data set includes 50, 75, 100, and 125 clusters, respectively. Because we have 250 features, each data set has 200, 175, 150, and 125 features that are not positive features, respectively. Precision of these features are set to 0.04, which is the precision of features that randomly appear in both target and non-target instances.

Figure 5(b) shows the performance of AdaFeaSE and Lasso on these four data sets. We chose Lasso for this comparison because it was the second best method at the early rounds for the scandal data set. The performance of Lasso deteriorates as the number of clusters increases. On the other hand, the performance of AdaFeaSE does not largely change when the number of clusters increases. This result supports our hypothesis that one of the reason of the superiority of AdaFeaSE is its ability to switch from a cluster to a cluster.

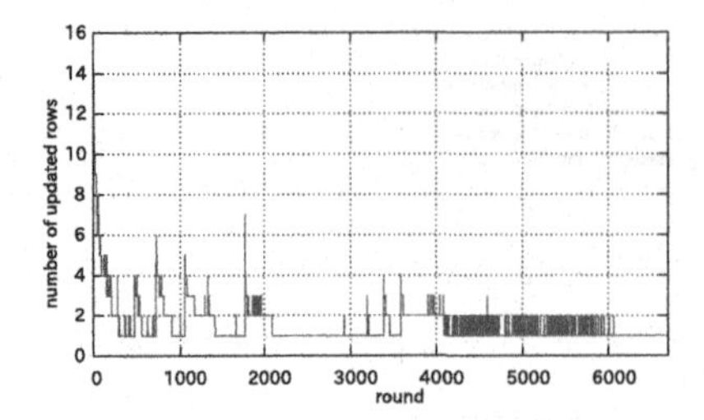

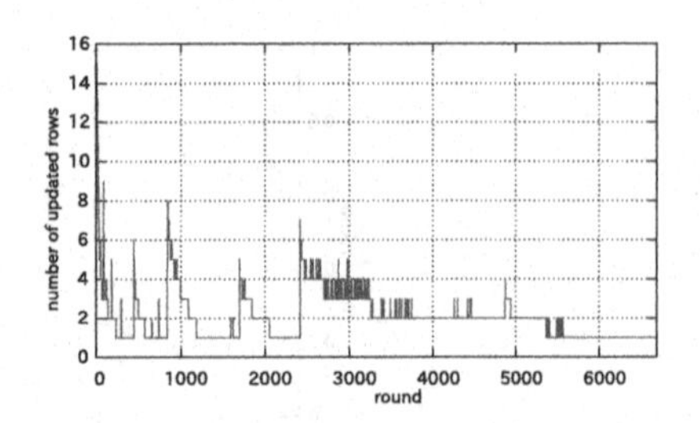

Fig. 7. The number of updated rows in $\mathcal{S}$ at each round (left: Scandal, right: Toyota).

4.3 Lasso with Initial Weight Values Being 1

One difference between AdaFeaSE and the baselines is the initial weight of the features. In AdaFeaSE, all features are initially active and given weight 1. In the other methods, the initial weights of features are 0. Because we use features that are nominated by human workers as prospective, this difference may affect the performance of AdaFeaSE and the other methods at the very early rounds.

To examine if it is one of the reasons of the difference of the performance, we conducted another experiment simulating Lasso with the initial weights 1. In order to simulate it, we pre-train Lasso with 1, 10, or 50 positive samples that has all the features, and use this pre-trained classifier as the initial classifier.

Figure 6 shows the result. This result shows that this pre-training improves the performance of Lasso at the very early rounds, but it is still outperformed by AdaFeaSE up to around the 240th round.

4.4 Scalability of the Algorithm

As explained before, the time complexity of AdaFeaSE is $O(|\mathcal{F}|^2|\mathcal{X}|)$ in the worst case, where $O(|\mathcal{F}|)$ rows in the statistics table $\mathcal{S}$ are updated in every round. We measured how many rows were actually updated in our experiments. Figure 7 shows the numbers of updated rows in each round for Scandal (with $|\mathcal{F}| = 386$) and Toyota (with $|\mathcal{F}| = 656$) data sets. As shown in these graphs, except for the very beginning, less than 10 rows are updated. In addition, the comparison of Scandal case and Toyota case suggests that the numbers of updates do not increase proportionally to $|\mathcal{F}|$. Therefore, we conclude that AdaFeaSE scales well with the number of nominated features. In fact, the computation time of AdaFeaSE is far shorter than the other baseline methods in our experiment.

5 Conclusions

In this paper, we focus on the situations where we want to quickly find some number (not all) of items of the target class from a fixed data set, but we only have very noisy feature sets including some useful ones and many useless ones.

Noisy feature sets usually require more training data, and it conflicts with our goal of quickly finding some data. We proposed a learning method that can quickly discard useless features. We compare features in a pairwise way based on their discriminative power, and if a feature is inferior to any other feature with a statistically significant difference, we temporarily inactivate the inferior one.

Another issue in our problem setting is that the classifier is biased toward positive instances found earlier. To avoid that, when we run out of remaining items that have the superior features, we re-evaluate inactivated features. By this strategy, even when positive instances are distributed across several clusters, our method can focus on those clusters one by one.

We compared our method with several baselines using two data sets from a real application and found that our method is superior to them in the performance at the early rounds. When we want to quickly find 30 or 50 instances of the target class, our method can find them earlier than the baselines.

An interesting remaining issue is how to switch from our method, which is good at early rounds, to other methods, which are good at later rounds. Such a method would enable us to apply our method to a wider range of applications.

References

1. Beluch, W.H., Genewein, T., Nürnberger, A., Köhler, J.M.: The power of ensembles for active learning in image classification. In: Proceedings of CVPR, pp. 9368–9377 (2018)
2. Cai, J., Luo, J., Wang, S., Yang, S.: Feature selection in machine learning: a new perspective. Neurocomputing **300**, 70–79 (2018)
3. Chandrashekar, G., Sahin, F.: A survey on feature selection methods. Comput. Electr. Eng. **40**(1), 16–28 (2014)
4. Cheng, J., Bernstein, M.S.: Flock: hybrid crowd-machine learning classifiers. In: Proceedings of CSCW, pp. 600–611 (2015)
5. Durand, A., Gagné, C.: Thompson sampling for combinatorial bandits and its application to online feature selection. In: Proceedings of AAAI Conference Workshop on Sequential Decision-Making with Big Data, pp. 6–9 (2014)
6. Gal, Y., Islam, R., Ghahramani, Z.: Deep Bayesian active learning with image data. In: Proceedings of ICML, pp. 1183–1192. PMLR (2017)
7. Ganguly, D., Leveling, J., Magdy, W., Jones, G.J.: Patent query reduction using pseudo relevance feedback. In: Proceedings of CIKM, pp. 1953–1956 (2011)
8. Gupta, M., Bendersky, M.: Information retrieval with verbose queries. In: Proceedings of SIGIR, pp. 1121–1124 (2015)
9. Hall, M.A., Smith, L.A.: Feature selection for machine learning: comparing a correlation-based filter approach to the wrapper. In: Proceedings of International Florida Artificial Intelligence Research Society Conference, pp. 235–239. AAAI Press (1999)
10. Huang, P., Bu, J., Chen, C., Qiu, G.: An effective feature-weighting model for question classification. In: Proceedings of International Conference on Computational Intelligence and Security, pp. 32–36 (2007)
11. Jörger, P., Baba, Y., Kashima, H.: Learning to enumerate. In: Villa, A.E.P., Masulli, P., Pons Rivero, A.J. (eds.) ICANN 2016. LNCS, vol. 9886, pp. 453–460. Springer, Cham (2016). https://doi.org/10.1007/978-3-319-44778-0_53
12. Khalid, S., Khalil, T., Nasreen, S.: A survey of feature selection and feature extraction techniques in machine learning. In: Proceedings of Science and Information Conference, pp. 372–378 (2014)

13. Konyushkova, K., Sznitman, R., Fua, P.: Learning active learning from data. arXiv preprint arXiv:1703.03365 (2017)
14. Koopman, B., Cripwell, L., Zuccon, G.: Generating clinical queries from patient narratives: a comparison between machines and humans. In: Proceedings of SIGIR, pp. 853–856 (2017)
15. Kou, G., Yang, P., Peng, Y., Xiao, F., Chen, Y., Alsaadi, F.E.: Evaluation of feature selection methods for text classification with small datasets using multiple criteria decision-making methods. Appl. Soft Comput. **86**, 105836 (2020)
16. Lai, T.L., Robbins, H.: Asymptotically efficient adaptive allocation rules. Adv. Appl. Math. **6**(1), 4–22 (1985)
17. Lan, M., Tan, C.L., Low, H.: Proposing a new term weighting scheme for text categorization. In: Proceedings of National Conference on Artificial Intelligence, pp. 763–768 (2006)
18. Qi, Y., Zhang, J., Liu, Y., Xu, W., Guo, J.: CGTR: convolution graph topology representation for document ranking. In: Proceedings of CIKM, pp. 2173–2176 (2020)
19. Reddy, G.T., et al.: Analysis of dimensionality reduction techniques on big data. IEEE Access **8**, 54776–54788 (2020)
20. Remeseiro, B., Bolon-Canedo, V.: A review of feature selection methods in medical applications. Comput. Biol. Med. **112**, 103375 (2019)
21. Sánchez-Maroño, N., Alonso-Betanzos, A., Tombilla-Sanromán, M.: Filter methods for feature selection: a comparative study. In: Proceedings of IDEAL, pp. 178–187 (2007)
22. Settles, B.: Closing the loop: fast, interactive semi-supervised annotation with queries on features and instances. In: Proceedings of EMNLP, pp. 1467–1478. ACL (2011)
23. Settles, B.: Active learning. Synth. Lect. Artif. Intell. Mach. Learn. **6**(1), 1–114 (2012)
24. Sun, X., Tang, H., Zhang, F., Cui, Y., Jin, B., Wang, Z.: Table: a task-adaptive BERT-based listwise ranking model for document retrieval. In: Proceedings of CIKM, pp. 2233–2236 (2020)
25. Takahama, R., Baba, Y., Shimizu, N., Fujita, S., Kashima, H.: AdaFlock: adaptive feature discovery for human-in-the-loop predictive modeling. In: Proceedings of AAAI Conference, pp. 1619–1626 (2018)
26. Van Der Maaten, L., Postma, E., Van den Herik, J.: Dimensionality reduction: a comparative. J. Mach. Learn. Res. **10**(66–71), 13 (2009)
27. Weinberger, K.Q., Sha, F., Saul, L.K.: Learning a kernel matrix for nonlinear dimensionality reduction. In: Proceedings of ICML, p. 106 (2004)
28. Yahoo! crowdsourcing. http://crowdsourcing.yahoo.co.jp/
29. Yahoo! news. http://news.yahoo.co.jp/
30. Yu, H., Yang, X., Zheng, S., Sun, C.: Active learning from imbalanced data: a solution of online weighted extreme learning machine. IEEE Trans. Neural Netw. Learn. Syst. **30**(4), 1088–1103 (2018)
31. Zhang, J., Geng, Y.A.O., Li, Q., Shi, C.: More than one: a cluster-prototype matching framework for zero-shot learning. In: Proceedings of CIKM, pp. 1803–1812 (2020)
32. Zou, J.Y., Chaudhuri, K., Kalai, A.T.: Crowdsourcing feature discovery via adaptively chosen comparisons. In: Proceedings of AAAI HComp, pp. 198–205 (2015)

A Transformer-Based Object-Centric Approach for Date Estimation of Historical Photographs

Francesc Net(✉), Núria Hernández, Adriá Molina, and Lluis Gómez(✉)

Computer Vision Center, Universitat Autónoma de Barcelona, Catalunya, Spain
{fnet,nhernandez,amolina,lgomez}@cvc.uab.cat

Abstract. The accurate estimation of the creation date of cultural heritage photographic assets is a challenging and complex task, typically requiring the expertise of qualified archivists, with significant implications for archival and preservation purposes. This paper introduces a new dataset for image date estimation, which complements existing datasets, thus creating a more balanced and realistic training set for deep learning models. On this dataset, we present a set of modern strong baselines that outperform previous state-of-the-art methods for this task. Additionally, we propose a novel approach that leverages "dating indicators" or "dating clues" through object detection and a self-attention based Transformer encoder. Our experiments demonstrate that the proposed approach has promising applicability in real scenarios and that incorporating "dating indicators" through object detection can improve the performance of image date estimation models. The dataset and code of our models are publicly available at https://github.com/cesc47/DEXPERT.

Keywords: Historical photographs · Image retrieval · Date estimation

1 Introduction

Images are a fundamental source of information to understand both the past and present of humankind. They capture moments in time, preserve memories, and provide valuable perspectives on historical, social, and cultural events. The accurate estimation of the creation date of cultural heritage photographic assets is an important problem in the field of archivistics, with significant implications for archival and preservation purposes. It is a challenging and complex task that requires the expertise of qualified archivists.

To accurately determine the production date of a historical photograph, certain objects or details in the image, known as "dating clues" or "dating indicators," can prove particularly useful. These may include, for example, clothing styles and hairstyles worn by individuals in the photograph, the design and style of furniture and decor in the background, and any identifiable object that can help to pinpoint the photograph's time period. Other helpful dating clues may include photographic techniques, paper type, and inscriptions or markings on

N. Goharian et al. (Eds.): ECIR 2024, LNCS 14610, pp. 137–150, 2024.
https://doi.org/10.1007/978-3-031-56063-7_9

the photograph. These objects or details can provide critical context for the photograph, allowing experts to more accurately estimate its production date.

As illustrated in Fig. 1 Grad-CAM [23] class activation maps of a CNN trained for image date estimation show how the model also focuses on certain objects that play the role of "date indicators". Certain vehicles such as cars and trains, or peoples' clothing and hair cuts provide the visual features that the model relies on to make its predictions.

Fig. 1. Grad-CAM class activation maps of a model trained for image date estimation show how the model focuses on certain objects that play the role of "date indicators". Certain vehicles such as cars and trains, or peoples' clothing and haircuts provide important cues to estimate the creation date of an image correctly. The plot on the right illustrates the availability (number of images in log scale) of the most frequently detected object classes in the DEW [15] dataset.

In this paper, we present a new deep learning architecture called DEXPERT (Date Estimation using eXPERTs), which is designed to take full advantage of these "date indicators". The proposed DEXPERT architecture incorporates an ensemble of Convolutional Neural Networks (CNNs), an object detection model, and a Transformer encoder to achieve this goal. By using an object detector to identify specific object categories and leveraging their visual features, DEXPERT prevents loss of crucial information that traditional deep learning models can incur due to image resizing. Specifically, DEXPERT uses an ensemble of CNNs expert models that have been trained to date specific object classes and a Transformer encoder that aggregates information from the different experts to generate a final prediction of the photograph's creation date. As illustrated in Fig. 1 (plot on the right), the most frequent object classes found on a standard data set for image date estimation with a pre-trained object detection model include objects that can potentially act as "date indicators".

The automatic estimation of the creation date of historical images has been a topic of growing interest in recent years [1,7,14,15,20,24,25], with significant implications for cultural heritage preservation, archiving, and understanding. Prior research works have mostly employed the "Date Estimation in the Wild" (DEW) dataset [15], encompassing one million Flickr photographic records ranging from the 1930s until the 2000s. It should be noted that the restricted availability of photographs prior to the 1920s has prompted this dataset to focus predominantly on the 20th century. Despite its utility, the dataset is subject to a temporal distribution imbalance, manifested in an insufficient representation of images from the earlier decades of the 20th century. Consequently, this undermines the effectiveness of the dataset in the accurate estimation of image creation dates from that period. To overcome this obstacle, in this work, we present a new dataset, DEW-B, which has been curated to be well-balanced. Furthermore, our work includes an in-depth study of modern deep learning baseline models, which include ConvNeXt [11], ViT [5], and CLIP [18].

The rest of the paper is organized as follows: Sect. 2 provides an overview of recent related work in image date estimation. Section 3 describes the construction and collection of the contributed dataset, DEW-B. Section 4 presents the architecture of our DEXPERT object-centric method for image date estimation. Section 5 presents the experiments conducted and the results obtained. Finally, we close the paper in Sect. 6 with the conclusions.

2 Related Work

In recent years, there has been a growing interest in the problem of date estimation in collections of historical photographs from computer vision and digital humanities researchers, historical management institutions, and social platforms. Early contributions have focused on related but distinct challenges, such as analyzing the 3D geometry of historical photographs. For instance, Schindler *et al.* [21] used a Structure from Motion approach to temporally sort a collection of photos by estimating camera poses and reasoning about the visibility of 3D structures in each image.

Palermo *et al.* [17] explored for the first time the possibility of automatically estimating the date of photographs in relatively small collections. The authors used a classical computer vision approach based on SVMs, color histograms, and other hand-crafted features to tackle the problem. Their models were validated on a small-scale dataset consisting of 225 and 50 images per decade for training and testing, respectively. Fernando *et al.* [6] proposed an image date estimation technique based on device-dependent features, including color, color derivatives, and color angles. Martin *et al.* [13] extended the previous SVM-based methods by introducing an ordinal regression constraint. They formulated the problem as a one-vs-many binary classifier to predict whether an image was captured before or after each considered date.

More recent works have used deep learning models to estimate the date of a specific object/image from specific domains. Vittayakorn *et al.* [25] focused on clothing and museums' items, Salem *et al.* [20] analyzed faces and torsos in high school yearbooks, and Ginosar *et al.* [7] examined portraits.

Müller *et al.* [15] curated the "Date Estimation in the Wild" (DEW) dataset, consisting of 1 million images from Flickr, spanning from 1930 to 1999 and representing diverse visual concepts and image types. Thanks to its scale, DEW enabled rigorous training of deep learning models. Two baseline approaches for image date estimation were proposed using deep convolutional neural networks

(GoogLeNet), treating the task as a classification and regression problem. The experimental results showed that those baselines achieved super-human performance, surpassing the annotations made by untrained humans. Molina *et al.* [14] reformulated the date estimation problem on the DEW dataset as a retrieval task and proposed a new metric learning objective, where the closer the embedded image representations, the closer their creation dates. Their method was evaluated on two tasks - date estimation and date-sensitive image retrieval. Stacchio, L. *et al.* [24] presented a system that automatically catalogs vernacular pictures drawn from family photo albums, and introduced the IMAGO dataset, which is composed of 80,000 photographs conserved by the Department of the Arts of the University of Bologna. They investigated the effectiveness of selectively using objects (persons and faces) for image date estimation. The results showed that this approach outperformed analyzing the entire image with a single model. Ashida *et al.* [1], on the other hand, tackled the Date Estimation problem by performing a rank-consistent ordinal classification with an object-centered ensemble.

In this work, we adopt an approach similar to that of Ashida *et al.* [1] and Stacchio *et al.* [24] in developing object-centric models. However, unlike them, we create an ensemble of experts for multiple object classes (up to six), rather than just focusing on persons and faces [24] or persons and cars [1]. Furthermore, we introduce the use of a Transformer encoder to combine information from the various experts and generate a final prediction.

3 DEW-B Dataset

In this section, we present the DEW-B (Date Estimation in the Wild - Balanced) dataset, which aims to provide more training images for the years that are less represented in the DEW dataset [15], from 1930 to 1969. The contributed training data has been harvested from three sources: the LAION-5B [22] dataset, and the Flickr[1] and Europeana[2] web portals.

LAION-5B is an open and publicly available dataset of 5.8 billion image-text pairs. We have used the clip-retrieval [2] tool to search for historical photos based on a query text prompt using the CLIP text embedding and the publicly available LAION-5B pre-computed kNN indices.

We have manually inspected the retrieved LAION subsets for several text queries and selected the following prompt as our base text query template: `''A photograph from <YEAR> showing the daily life of the people''`, where the placeholder `<YEAR>` is replaced by a value in the range $\{1930, 1970\}$. The rationale for using this text prompt is that provides high variability of scene images and human activities, which correspond to real-world photo dating scenarios in contrast to other types of images such as landscapes, animals, plants, etc. in which photo dating is not reliable.

[1] https://www.flickr.com.
[2] https://www.europeana.eu.

We performed 40 queries using the base text query template (one for each year in the range of interest). From each query we kept the top 1M retrieved image-text pairs, which accounts for a total of 40 million pairs. However, since CLIP embeddings have only rough dating estimation capabilities (see Fig. 2 and the CLIP zero-shot dating performance in Sect. 5) we found many duplicate pairs in the retrieved sets of consecutive years. After de-duplication, we retain only ~ 3 million image-text pairs from the clip-retrieval system.

Fig. 2. Top-5 retrieved images from clip-retrieval for the query "A photograph from 1931 showing the daily life of the people". Since CLIP embeddings have only rough dating estimation capabilities, all five images are also among the Top-100 retrieved images when the query is for the year 1932.

In order to assign accurate and fine-grained date labels to our training images we search for the image-text pairs in which a specific year (in the range $\{1930, 1970\}$) is explicitly mentioned in one of its metadata fields (caption, or URL). We further filter the remaining image set by dropping those images with width or height < 200 pixels.

Still, the collected dataset is noisy and weakly annotated. By manual inspection, we found two main sources of noise: (1) there is a considerable number of document images (e.g. magazines' cover from a specific year); and (2) wrong labels are introduced by captions mentioning a specific year that does not correlate with the image content (e.g. the caption of a contemporary photo made in 2023 of the Empire State Building mentioning it was built on 1931). In order to further clean our dataset we apply two more filters addressing these issues. First, we use CLIP zero-shot image-text similarity prediction for each image and the following text prompts: {`''A piece of paper''`, `''Some handwritten text''`, `''A page of text''`, `''Cover of a magazine''`, `''A magazine''`, `''A newspaper''`}. If the CLIP similarity score of an image with any of those texts is larger than 0.2[3] we remove the image from the dataset. Second, we use a date estimation model trained on the DEW dataset [15] to predict the creation date of each image, if the model prediction differs from the assigned label by more than 15 years we remove the image from the dataset. The final LAION-DEW subset after the cleaning process consists of 257,896 images with their corresponding labels.

On the other hand, the Flickr API was used as in the original DEW dataset to download photos with a timestamp in the period from 1930 to 1970. We also cleaned the retrieved set with the same cleaning filters described above. In addition, we remove from this set any duplicate images already existing in the original DEW dataset. The Flickr_DEW subset consists of 57,570 images.

[3] The threshold value of 0.2 was found by manual inspection.

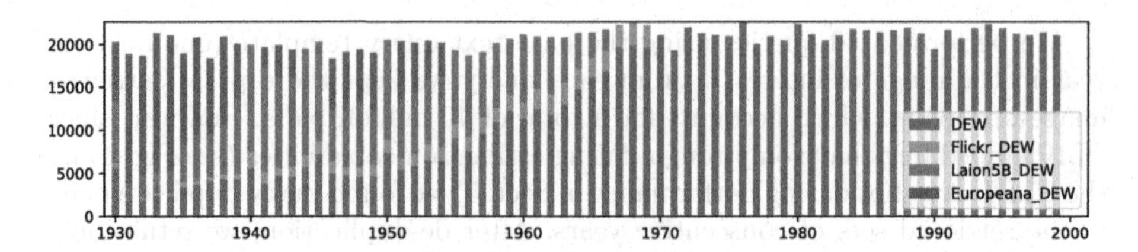

Fig. 3. Number of training images per year in the DEW-B dataset. The original DEW dataset [15] (blue bars in the plot) was highly unbalanced, with the number of images drastically decreasing with the age of the data. Our contributed data (Flickr_DEW, Laion5B_DEW, and Europeana_DEW) provide more training images for the less represented years. (Color figure online)

Finally, we used the Europeana Search API to download an additional set of photos from their Photography collection with a creation date that falls within the period of interest. Europeana is a web portal created by the European Union containing digitized cultural heritage collections of more than 3,000 institutions across Europe. Since the content of this portal is contributed by cultural heritage expert organizations we did not apply the DEW prediction filter in this subset, assuming the year provided in the metadata is accurate and reliable. The Europeana_DEW set has 171,390 photos. As shown in Fig. 3, when combined with the original DEW images, the DEW-B dataset becomes a well-balanced set of 1,444,587 annotated images. The dataset is publicly available at https://github.com/cesc47/DEXPERT.

4 Method

In this section, we present the DEXPERT architecture for image date estimation. As illustrated in Fig. 4 the proposed method consists of an object detection model, an ensemble of Convolutional Neural Network (CNN) experts, and a Transformer encoder that aggregates the information from the different experts to generate the final prediction.

4.1 Single Image Date Estimation Model

The first model in our CNN ensemble is a classification network trained for date estimation from a single image. In order to find the best possible model we have experimented with both a regression and a classification setting for this task, using different state-of-the-art architectures, such as the ConvNeXt [11] CNN, ViT [5], and CLIP [18]. Following the work of Müller *et al.* [15] in the classification setting for the DEW dataset the model output space consists of 14 classes, each representing a half-decade time interval. At inference time the estimated creation date y_E for a given image is computed as:

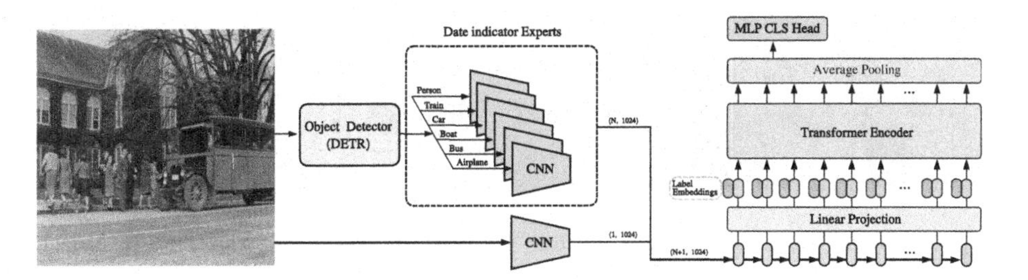

Fig. 4. The proposed DEXPERT model consists of an object detector, an ensemble of Convolutional Neural Network (CNNs) experts, and a Transformer encoder that aggregates the information from the different experts to generate a final prediction.

$$y_E = 1930 + \left\lfloor 0.5 + \frac{1999 - 1930}{|c| - 1} \cdot \sum_{i=0}^{|c|-1} i \cdot p(i) \right\rfloor, \quad \text{with} \sum_{i=0}^{|c|-1} p(i) = 1. \tag{1}$$

where p is the model's output probability distribution over classes, and $c = 14$ is the total number of classes.

In our baseline study (see Sect. 5 for more details) we have empirically found that the ConvNeXt classification model outperformed the rest of the implemented baseline methods. ConvNeXt is a convolutional neural network (CNN) that builds upon the ResNet [8] architecture while incorporating key components of vision Transformers, such as using larger kernel sizes, depthwise convolutions (which are similar to the weighted sum operation in self-attention), Layer Normalization, and the Gaussian Error Linear Unit (GELU). The resulting network competes favorably with Transformers in terms of accuracy and scalability, while maintaining the simplicity and efficiency of standard CNNs.

Once trained for single image date estimation, we incorporate the ConvNeXt$_{base}$ model to our ensemble as a "global feature extractor". This module processes the input image resized to 224×224 pixels and outputs the features after the global average pooling layer as a 1024-D representation of the image.

4.2 Leveraging Object Detection for Date Estimation

As mentioned before we aim at using an object detector to identify specific object categories and leveraging their visual features as "date indicators". In this way we expect our method to prevent the loss of crucial fine-grained information that the single image model can incur due to image resizing. To this end we first conducted a comparative study of two distinct state-of-the-art models for object detection: a Faster-RCNN [19] trained on Open Images V4 [9] with an ImageNet [4] pre-trained Inception ResNet V2 backbone; and a DETR detection Transformer [3] trained on MS-COCO [10] with an ImageNet pre-trained ResNet-101 backbone.

Fig. 5. Object detection results in four DEW-B images using the DETR object detector.

Since object-level annotations are not available for the DEW-B dataset, a qualitative comparative study was conducted via manual inspection of a randomly selected subset of images. Notably, both models exhibited remarkable proficiency in detecting common objects within the dataset, even for those that may exhibit significant variability across different time periods. Figure 5 shows some qualitative examples of the DETR model.

As a result of the qualitative study, we finally selected the DETR model [3] as the object detection block in our method, as it proved more consistent with the object detection confidence scores. The DETR model consists of a convolutional backbone followed by an encoder-decoder Transformer which can be trained end-to-end for object detection.

In our framework, the output of the object detection model is used to identify specific object categories that can potentially act as "date indicators". By analyzing the most frequent object classes found on the DEW-B dataset (see Fig. 1), we defined the set of "date indicators" object classes as: {`''person''`, `''train''`, `''car''`, `''boat''`, `''bus''`, `''airplane''`}.

4.3 DEXPERT Architecture

The object detection module in our framework generates a list of the identified objects' categories and their respective bounding box coordinates. This output is then utilized to crop the detected objects pertaining to any of the six date indicator objects, which are subsequently fed into an object-specific expert convolutional neural network (CNN) trained specifically for date estimation on each object class.

To train the date indicators' experts, we employed a fine-tuning approach using as the base model the pre-trained single image date estimation model presented in Sect. 4.1. The object crops of the DEW-B training set were used to fine-tune the model. Subsequently, the fine-tuned models were used to extract features for each date indicator object present in a given input image. The resulting set of N date indicator descriptors have a dimensionality of 1024, which together with the global descriptor of the image (Sect. 4.1) becomes the set of $N+1$ input tokens of the DEXPERT Transformer encoder.

The input tokens are projected linearly to the encoder input dimensionality of 256-D, followed by the addition of a set of learnable label embeddings specific to each of the six date indicator object classes. The label embeddings provide semantic information for each token type to the encoder. In case an image contains multiple objects of different classes, label embeddings corresponding to each of the object classes are added to their corresponding feature representations after the linear projection. For instance, in an image with two persons and a bus, label embeddings for the "person" and "bus" labels are added to their respective feature representations.

The final stage of our DEXPERT architecture involves the average pooling of the outputs of the Transformer encoder and a classification layer, which is responsible for predicting the date of the image. The encoder consists of 2 stacked Transformer-blocks with a hidden dimension size of 512, and the multi-head attention mechanism has 4 attention heads. The classification head has a single Linear layer with 14 output classes following the same configuration described in Sect. 4.1 for the single image classifier. The implementation details and training regimes for all DEXPERT modules are provided in Appendix A.

5 Experiments

This section presents the experimental results of the different evaluated models for the task of date estimation on the DEW and DEW-B datasets. The evaluation metric used is the Mean Absolute Error (MAE), which measures the average absolute difference between the predicted date and the ground truth annotation.

Table 1 summarizes the results obtained with several baselines and models in the original DEW test set when training with different datasets. We evaluate the performance of the ConvNeXt model described in Sect. 4.1 and the zero-shot performance of CLIP ViT-L/14. The CLIP zero-shot results were obtained with the publicly available official model[4] using image-text matching scores with query prompts such as: `''This image was taken in the year <YEAR>.''`. We provide more details about the prompt engineering in the Appendix A. It is worth noting that all the ConvNeXt models we evaluated demonstrated significant improvements over the previous state-of-the-art methods on the DEW dataset. The models trained using weighted cross entropy loss (a rescaling weight given to each class) achieve the best results when training data is unbalanced (DEW).

The results in Table 1 also show that our DEXPERT model outperforms both the best single image baseline (ConvNeXt$_{base}$) and the ensemble-based method proposed by Ashida *et al.* [1]. More remarkably, DEXPERT achieves a MAE that is less than half of the MAE reported for untrained humans in a study by Müller *et al.* [15]. Overall, these results demonstrate the efficiency of the proposed model and validate the idea of using an object-centric ensemble of date indicators and a Transformer-based aggregator.

We appreciate that when training with the additional data provided by the DEW-B dataset, the Mean Absolute Error (MAE) of the $ConvNeXt_{Base}$ model and the previous state of the art (ResNet-50, from Muller *et al.* [15] and Ashida *et al.* [1]) is clearly reduced. Notice that all results are directly comparable as the

[4] https://github.com/openai/CLIP.

Table 1. MAE performance comparison on the DEW test set for different baselines and methods trained on the original DEW dataset [15] and our DEW-B dataset. Methods with names in bold are contributions of this paper and/or implemented/trained by us.

Method	Training Data	MAE
Human performance [15]	n.a	10.90
Visual similarity baseline [14] ResNet50 (ImageNet KNN)	n.a	12.40
CLIP ViT-L/14 (zero-shot)	n.a	6.55
Müller *et al.* [15] GoogLeNet (regression)	DEW	7.50
Müller *et al.* [15] GoogLeNet (classification)	DEW	7.30
Müller *et al.* [15] ResNet50 (classification)	DEW	7.12
Smooth-nDCG [14] ResNet50 (metric learning)	DEW	7.48
Ashida *et al.* [1] ResNet50 (classification)	DEW	6.82
Ashida *et al.* [1] Object-centric Ensemble (weighted avg.)	DEW	6.14
ConvNeXt$_{Base}$	DEW	5.49
ConvNeXt$_{Large}$	DEW	5.48
ConvNeXt$_{Base}$ **(weighted cross entropy)**	DEW	4.75
ConvNeXt$_{Large}$ **(weighted cross entropy)**	DEW	4.72
DEXPERT Object-centric Transformer	DEW	**4.55**
Müller *et al.* [15] **ResNet50 (classification)**	DEW-B	5.72
ConvNeXt$_{Base}$	DEW-B	**4.10**
DEXPERT Object-centric Transformer	DEW-B	4.46

test partition of the DEW dataset remains unchanged, and the only modification is the additional training data. Both models show a significant improvement in MAE by utilizing the DEW-B dataset for training, which validates the value of the contributed data. The performance of the DEXPERT model is slightly improved with the additional training data, but unfortunately do not outperform the results of the $ConvNeXt_{Base}$. We hypothesize that this might be related to the domain gap of the original and contributed data at the object crops' level.

The results of the expert models on the set of object crops generated by the DETR object detector on the DEW test set are shown in Table 2, while Fig. 6 shows the confusion matrices for each of the date indicators' expert models in the classification setting with 14 classes.

Table 2. Object-specific experts' performance on the set of DEW object crops.

	Person	Train	Car	Boat	Bus	Airplane
MAE	6.76	4.26	5.41	8.62	4.37	5.58

In Fig. 6 we appreciate that for most object classes the predictions behave in a highly correlated manner with the ground truth while for classes such as bus or boat, they are less reliable. In DEXPERT the Transformer encoder learns to aggregate this information, taking into account the interactions between objects and their importance in the final prediction.

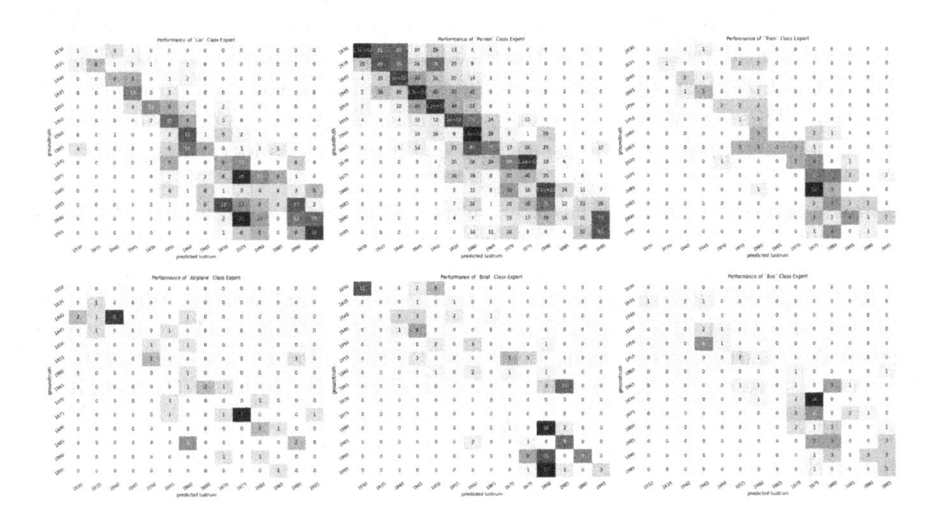

Fig. 6. Confusion matrices for each of the expert models in the classification setting.

Figures 1 and 7 show qualitative results of the DEXPERT model predictions for images of the DEW dataset.

Fig. 7. Qualitative examples with absolute error $\in \{0, 3\}$. We show the Ground-Truth date (GT), and predictions for three models: DEXPERT, ConvNeXt, and ResNet-50.

6 Conclusions

In this paper, we presented the DEW-B dataset, which complements existing datasets for the task of estimating the creation date of historical photographs. Our study explored several contemporary deep learning models and proposed a novel architecture that uses "dating indicators" by leveraging an object detector, an ensemble of expert CNNs, and a self-attention based Transformer encoder. Our experimental results demonstrate that our approach, along with the use of DEW-B, outperforms the previous state-of-the-art in the task of image date estimation. This research provides opportunities for more accurate analysis of cultural heritage photographic assets and has implications in various fields.

Acknowledgment. This work has been supported by the Spanish project PID2021-126808OB-I00, Ministerio de Ciencia e Innovación, the Departament de Cultura of the Generalitat de Catalunya, the CERCA Programme / Generalitat de Catalunya, and ACCIO INNOTEC 2021 project Coeli-IA (ACE034/21/000084), Adrià Molina is funded with the PRE2022-101575 grant provided by MCIN / AEI / 10.13039 / 501100011033 and by the European Social Fund (FSE+). Lluis Gomez is funded by the Ramon y Cajal research fellowship RYC2020-030777-I / AEI / 10.13039/501100011033.

Appendices

A Implementation details

Global CNN. We fine-tune the model for image date estimation using AdamW optimizer [12] with an base learning rate of 1e−4, a momentum of 0.9, and a weight decay of 0.05. We train the network for a total of 50 epochs with a batch size of 704. We dynamically set the learning rate of each parameter group using a cosine annealing schedule. During training we perform data-augmentation with random crop, horizontal flip, and wide TrivialAugment [16]. The model was trained on 8×NVIDIA Tesla T4 GPUs using distributed repeated augmentation sampling that restricts data loading to a subset of the dataset with repeated augmentation.

Date Indicator Experts. The date indicator expert models were trained using the same hyper-parameters described above except for a base learning rate of 1e−5. They were fine-tuned from the pre-trained global CNN model, on the respective set of object crops from the DEW/DEWB dataset. Each model was trained for 20 epochs with a batch size of 1024 on 2×NVIDIA A40 GPUs.

DEXPERT. The DEXPERT architecture aggregates the visual features of the global CNN and expert models by means of a Transformer Encoder. The following parameters were randomly initialized: input linear projection, class labels embeddings, Transformer encoder layers, and the classification head. We trained them for 10 epochs with a batch size of 512 on 8×NVIDIA Tesla T4 GPUs using the same hyper-parameters as before but with a base learning rate of 1e−3.

CLIP Zero-Shot Text Prompt Engineering. The CLIP [18] ViT-L/14 zero-shot results were obtained with the publicly available official model[5] using image-text matching scores with the following text prompts: `''An image of year <year>''`, `''This image was taken in the year <year>''`, `''This scene was shot in the year <year>''`, `''This moment was captured on camera in the year <year>''`, `''This image was taken between years <year> and <year+4>''`, `''This image was captured in the year <year>''`, `''This image was recorded in the year <year>''`. We conducted experiments with several CLIP backbones and prompts combinations, and the best-performing model was found to be the ViT-L-14 architecture when using the average prediction of all prompts.
The source code for training/evaluation of our models and pre-trained weights are publicly available at https://github.com/cesc47/DEXPERT.

References

1. Ashida, S., Jatowt, A., Doucet, A., Yoshikawa, M.: Determining image age with rank-consistent ordinal classification and object-centered ensemble. In: Proceedings of the 2nd ACM International Conference on Multimedia in Asia, pp. 1–8 (2021)
2. Beaumont, R.: Clip retrieval: easily compute clip embeddings and build a clip retrieval system with them (2022). https://github.com/rom1504/clip-retrieval
3. Carion, N., Massa, F., Synnaeve, G., Usunier, N., Kirillov, A., Zagoruyko, S.: End-to-end object detection with transformers. In: Vedaldi, A., Bischof, H., Brox, T., Frahm, J.-M. (eds.) ECCV 2020, Part I. LNCS, vol. 12346, pp. 213–229. Springer, Cham (2020). https://doi.org/10.1007/978-3-030-58452-8_13
4. Deng, J., Dong, W., Socher, R., Li, L.J., Li, K., Fei-Fei, L.: ImageNet: a large-scale hierarchical image database. In: CVPR 2009 (2009)
5. Dosovitskiy, A., et al.: An image is worth 16x16 words: transformers for image recognition at scale. arXiv preprint arXiv:2010.11929 (2020)
6. Fernando, B., Muselet, D., Khan, R., Tuytelaars, T.: Color features for dating historical color images. In: Proceedings of the International Conference on Image Processing, pp. 2589–2593 (2014)
7. Ginosar, S., Rakelly, K., Sachs, S., Yin, B., Efros, A.A.: A century of portraits: a visual historical record of American high school yearbooks. In: Proceedings of the IEEE Conference on Computer Vision and Pattern Recognition Workshops, pp. 1–7 (2015)
8. He, K., Zhang, X., Ren, S., Sun, J.: Deep residual learning for image recognition. In: Proceedings of the IEEE Conference on Computer Vision and Pattern Recognition, pp. 770–778 (2016)
9. Kuznetsova, A., et al.: The open images dataset v4: unified image classification, object detection, and visual relationship detection at scale. IJCV **128**, 1956–1981 (2020)
10. Lin, T.-Y., et al.: Microsoft COCO: common objects in context. In: Fleet, D., Pajdla, T., Schiele, B., Tuytelaars, T. (eds.) ECCV 2014, Part V. LNCS, vol. 8693, pp. 740–755. Springer, Cham (2014). https://doi.org/10.1007/978-3-319-10602-1_48

[5] https://github.com/openai/CLIP.

11. Liu, Z., Mao, H., Wu, C.Y., Feichtenhofer, C., Darrell, T., Xie, S.: A convNet for the 2020s. In: Proceedings of the IEEE/CVF Conference on Computer Vision and Pattern Recognition, pp. 11976–11986 (2022)
12. Loshchilov, I., Hutter, F.: Decoupled weight decay regularization. In: International Conference on Learning Representations (2017)
13. Martin, P.: Datation automatique de photographies à partir de caractéristiques textuelles et visuelles. Ph.D. thesis, Université Caen Normandie (2015)
14. Molina, A., Riba, P., Gomez, L., Ramos-Terrades, O., Lladós, J.: Date estimation in the wild of scanned historical photos: an image retrieval approach. In: Lladós, J., Lopresti, D., Uchida, S. (eds.) ICDAR 2021, Part II. LNCS, vol. 12822, pp. 306–320. Springer, Cham (2021). https://doi.org/10.1007/978-3-030-86331-9_20
15. Müller, E., Springstein, M., Ewerth, R.: "When was this picture taken?" – image date estimation in the wild. In: Jose, J.M., et al. (eds.) ECIR 2017. LNCS, vol. 10193, pp. 619–625. Springer, Cham (2017). https://doi.org/10.1007/978-3-319-56608-5_57
16. Müller, S.G., Hutter, F.: TrivialAugment: tuning-free yet state-of-the-art data augmentation. In: Proceedings of the IEEE/CVF International Conference on Computer Vision, pp. 774–782 (2021)
17. Palermo, F., Hays, J., Efros, A.A.: Dating historical color images. In: Fitzgibbon, A., Lazebnik, S., Perona, P., Sato, Y., Schmid, C. (eds.) ECCV 2012. Dating historical color images, vol. 7577, pp. 499–512. Springer, Heidelberg (2012). https://doi.org/10.1007/978-3-642-33783-3_36
18. Radford, A., et al.: Learning transferable visual models from natural language supervision. In: International Conference on Machine Learning, pp. 8748–8763. PMLR (2021)
19. Ren, S., He, K., Girshick, R., Sun, J.: Faster R-CNN: towards real-time object detection with region proposal networks. In: Advances in Neural Information Processing Systems, vol. 28 (2015)
20. Salem, T., Workman, S., Zhai, M., Jacobs, N.: Analyzing human appearance as a cue for dating images. In: Proceedings of the IEEE Winter Conference on Applications of Computer Vision, pp. 1–8 (2016)
21. Schindler, G., Dellaert, F., Kang, S.B.: Inferring temporal order of images from 3d structure. In: Proceedings of the IEEE Conference on Computer Vision and Pattern Recognition, pp. 1–7 (2007)
22. Schuhmann, C., et al.: LAION-5b: an open large-scale dataset for training next generation image-text models. In: Thirty-sixth Conference on Neural Information Processing Systems Datasets and Benchmarks Track (2022)
23. Selvaraju, R.R., Cogswell, M., Das, A., Vedantam, R., Parikh, D., Batra, D.: Grad-CAM: visual explanations from deep networks via gradient-based localization. In: Proceedings of the IEEE International Conference on Computer Vision, pp. 618–626 (2017)
24. Stacchio, L., Angeli, A., Lisanti, G., Calanca, D., Marfia, G.: Toward a holistic approach to the socio-historical analysis of vernacular photos. ACM Trans. Multimed. Comput. Commun. Appl. **18**(3s), 1–23 (2022)
25. Vittayakorn, S., Berg, A.C., Berg, T.L.: When was that made? In: Proceedings of the IEEE Winter Conference on Applications of Computer Vision, pp. 715–724 (2017)

Shallow Cross-Encoders for Low-Latency Retrieval

Aleksandr V. Petrov(✉), Sean MacAvaney, and Craig Macdonald

University of Glasgow, Glasgow, UK
a.petrov.1@research.gla.ac.uk,
{sean.macavaney,craig.macdonald}@glasgow.ac.uk

Abstract. Transformer-based Cross-Encoders achieve state-of-the-art effectivness in text retrieval. However, Cross-Encoders based on large transformer models (such as BERT or T5) are computationally expensive and allow for scoring only a small number of documents within a reasonably small latency window. However, keeping search latencies low is important for user satisfaction and energy usage. In this paper, we show that weaker shallow transformer models (i.e. transformers with a limited number of layers) actually perform *better* than full-scale models when constrained to these practical low-latency settings, since they can estimate the relevance of more documents in the same time budget. We further show that shallow transformers may benefit from the generalised Binary Cross-Entropy (gBCE) training scheme, which has recently demonstrated success for recommendation tasks. Our experiments with TREC Deep Learning passage ranking querysets demonstrate significant improvements in shallow and full-scale models in low-latency scenarios. For example, when the latency limit is 25 ms per query, MonoBERT-Large (a cross-encoder based on a full-scale BERT model) is only able to achieve NDCG@10 of 0.431 on TREC DL 2019, while TinyBERT-gBCE (a cross-encoder based on TinyBERT trained with gBCE) reaches NDCG@10 of 0.652, a +51% gain over MonoBERT-Large. We also show that shallow Cross-Encoders are effective even when used without a GPU (e.g., with CPU inference, NDCG@10 decreases only by 3% compared to GPU inference with 50 ms latency), which makes Cross-Encoders practical to run even without specialised hardware acceleration.

1 Introduction

The introduction of the Transformer [35] neural network architecture, and especially pre-trained language models that use Transformers (such as BERT [7]), has been transformative for the IR field; for example, Nogueira et al. [27] improved MRR@10 on the MS-MARCO dev set by 31% with the help of a BERT-based model. Although there are a variety of ranking architectures used within IR (e.g., dense Bi-Encoders [14,18,40], sparse Bi-Encoders [9,22], and late interaction models [13,15]), the best results for document re-ranking are typically achieved with the help of *Cross-Encoders* [14] – a family of models which encode both the query and the document simultaneously as a single textual input [41]. Aside from their high in-domain precision, Cross-Encoders tend to be more robust

N. Goharian et al. (Eds.): ECIR 2024, LNCS 14610, pp. 151–166, 2024.
https://doi.org/10.1007/978-3-031-56063-7_10

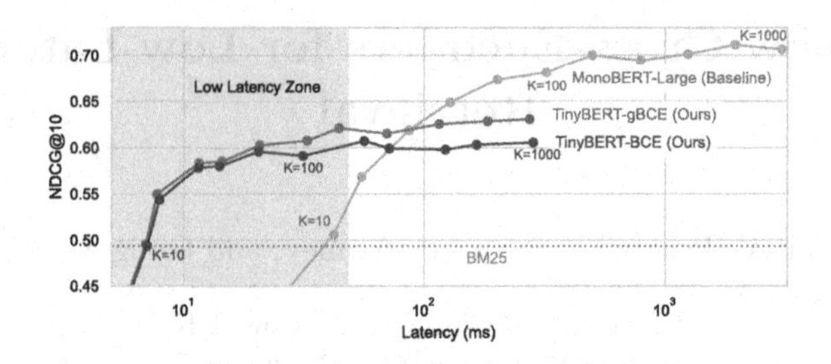

Fig. 1. Latency/NDCG tradeoffs on the TREC-DL2020 queryset when varying the number of retrieved candidates K.

when generalising across retrieval tasks/domains [33]. Although Cross-Encoders can only practically be used as re-ranking models, limitations in their first-stage recall can be efficiently mitigated using pseudo-relevance feedback [23]. Further, Cross-Encoders can typically be fine-tuned from scratch (i.e., starting from the checkpoint of a foundational model, such as BERT).

Despite these benefits, the application of Cross-Encoders in production retrieval systems is still limited. Cross-Encoders require a model inference for each query-document pair and, therefore, struggle with high computational complexity and high latency [24]. In real-world search systems, high latency negatively affects key performance metrics, such as the number of clicks, revenue, and user satisfaction [16, Ch. 5]. Further, high latencies tend to be correlated with higher energy usage, resulting in negative impacts on the climate [32].

The high computational complexity and resulting latency of Cross-Encoder models motivated researchers to investigate Bi-Encoder [14] models. These models separately encode the query and the document, and then estimate relevance score using an inexpensive operation over the encoded representations (e.g. cosine similarity [30] or the MaxSim operation [15]). By pre-computing the document representations offline and using a variety of approaches to accelerate retrieval [17], Bi-Encoders can achieve low retrieval latency. However, this comes at other costs. For instance, Bi-Encoders are markedly more complicated to train than Cross-Encoders, typically relying on knowledge distillation from other models (e.g., [19]), training data balancing (e.g., [12]), and/or hard negative mining (e.g., [40]). Further, Bi-Encoders must pre-encode all documents in the collection and keep the encoded versions of all documents in memory. This may be an inefficient strategy given the long-tail distribution of document popularity in a corpus. Indeed, if most documents are never retrieved, building their dense representations and keeping them in memory wastes resources. Moreover, document encoding costs for Bi-Encoders must also be incurred every time the retrieval model changes (e.g., when a model is re-trained to reflect new search trends.) Finally, Bi-Encoders often struggle to transfer across retrieval tasks and domains [33].

Therefore, in this paper, we investigate *shallow Cross-Encoders* (Cross-Encoders with a limited number of transformer layers) as a solution for low-latency search. Shallow Cross-Encoders are much smaller than full-scale models and require much fewer computations than full-scale models. Therefore, these models can score many times more documents within the low-latency window[1] than the full-scale models; this means that in low-latency scenarios, they can rerank more candidates and, ultimately, have better effectiveness than the full-scale models.

Note that there are a number of approaches for reducing the latency of ranking models, such as *dynamic pruning* or the use of *approximate nearest neighbour* indices (an overview of these methods can be found in [3]); however, most of these methods are not applicable to cross-encoder models. Indeed, in this paper, we focus on two main ways of reducing latency in the cross-encoder: (i) reducing the model's size and (ii) reducing the number of candidate documents, and we evaluate which of these ways achieves better effectiveness in the low-latency scenario.

Training effective Shallow Cross-Encoders, despite their promise of good efficiency, presents a significant challenge (this is in contrast to full-scale cross-encoders, which, as we previously mentioned, are relatively easy to train). MacAvaney et al. [24] first explored limiting the depths of a transformer network for ranking, but was only able to reduce the depth of the network to 5 layers without substantial effectiveness degradation. Other successful attempts to train shallower Cross-Encoder models have required applying complicated knowledge-distillation techniques. For example, the popular Sentence-Transformer package [30] provides several shallow models[2] and reports model performance competitive to larger models. Unfortunately, no academic paper is associated with these checkpoints, and the exact details of training these models are unclear. Our analysis of the training code shows that these models were trained using a knowledge distillation setup from an ensemble of full-sale models, loosely following the process described by Hofstätter et al. [11], which assumes the existence of such an ensemble in the first place. While resulting in effective checkpoints for the MSMARCO dataset, we argue that a training strategy that requires training an ensemble of full-scale language models before training the shallow model is hard to replicate for other settings (e.g. for different languages or other datasets). Indeed, in [37], the challenges of reproducing knowledge distillation models with "dependency chains" of models was found to be challenging - and hence, we argue that training shallow Cross-Encoders this way is rather art than science.

In contrast, this paper proposes a direct training method for shallow Cross-Encoders, which does not resort to complex techniques such as Knowledge Distillation. Our approach based on the gBCE training scheme [29] has been recently shown to improve the effectiveness of transformer-based models for recommender

[1] In [6], Google researchers argued that for a smooth user experience, total search latency should be kept under 100 ms. This includes time for network round-trips, page rendering, and other overheads. Therefore, this paper uses a 50 ms cutoff for defining *low-latency retrieval*, leaving the remainder of the time to these other overheads.

[2] https://www.sbert.net/docs/pretrained-models/ce-msmarco.html.

systems. The training scheme consists of two key components: (i) an increased number of sampled negative instances for each labelled positive instance and (ii) the gBCE loss function, which counters the effects of negative sampling (which is typically used to train Cross-Encoder models). Our experiments show that both of these components positively affect model training. Figure 1 summarises some of the main findings of this paper. The figure illustrates efficiency/effectiveness tradeoffs of two very small cross-encoders (2 layers, TinyBERT [34]) and a full-scale MonoBERT-Large model [27]. One of the small Cross-Encoders is trained using the gBCE training scheme, and the is trained using traditional BCE. As the figure shows, while MonoBERT-Large is more effective when allowed latency is high (i.e. it is allowed to re-rank many documents), within the low latency zone (<50 ms latency), shallow Cross-Encoders are more effective. Moreover, TinyBERT trained with gBCE is more effective compared to TinyBERT trained with traditional BCE.

Overall, the contributions of this paper can be summarised as follows: (i) We propose a simple and replicable method for training shallow Cross-Encoders based on the gBCE training scheme, which does not rely on knowledge distillation; (ii) We analyse the efficiency/effectiveness tradeoffs of Cross-Encoders of different sizes and demonstrate that shallow Cross-Encoders are preferable to full-size models under low-latency constraints; (iii) We demonstrate that shallow Cross-Encoders are efficient and effective even when used without a GPU.

We note that outside academia, there is interest in shallow Cross-Encoders, which is expressed in several industrial blog posts.[3,4] This interest makes us believe that our research has high potential to be adopted by industry and serve as a basis for further study.

The rest of the paper is organised as follows: Sect. 2 provides an overview of the Cross-Encoder architecture and Efficiency/Effectiveness tradeoffs arising from this architecture, Sect. 3 describes training scheme for shallow Cross-Encoders, Sect. 4 contains experimental evaluation of the efficiency/effectiveness tradeoffs for shallow Cross-Encoders, and Sect. 5 contains final remarks.

2 Efficiency/Effectiveness Tradeoffs in Cross-Encoders

A *Cross-Encoder* [14] is a model that jointly encodes a query-document pair using a single language model. Figure 2 illustrates a typical Transformer [35] encoder-based Cross-Encoder. The input to the Cross-Encoder model consists of a concatenation of the query with the document, joined with the help of some special tokens. In our example, there are three different special tokens: (i) *[CLS]* token is added to the beginning of the input; a contextualised representation of this token is then used for classification); (ii) *[SEP]* token is added at the end of both text and the document; these tokens separate help the model to separate different groups of text; (iii) a series of *[PAD]* tokens is added at the end of the

[3] https://blog.vespa.ai/pretrained-transformer-language-models-for-search-part-4/.

[4] https://towardsdatascience.com/tinybert-for-search-10x-faster-and-20x-smaller-than-bert-74cd1b6b5aec.

sequence to equalise the input length of each document in the batch. The input is then encoded using a standard Transformer encoder network, which consists of an embedding layer, positional embeddings, and a stack of Transformer blocks. For brevity, we omit details of the Transformer encoder network and refer to the original papers [7,35]. The output of the Transformer encoder consists of a sequence of embedding, where each embedding is a contextualised representation of each input token. In particular, for classification tasks, the representation of the *[CLS]* token is usually used to represent the whole input sequence. The task is usually cast as a binary classification for information retrieval with two possible outcomes (relevant/not relevant). Typically, these probabilities are obtained by passing the *[CLS]* representation through a simple Feed-Forward network with two outputs and then apply a Softmax operation over these two outputs.

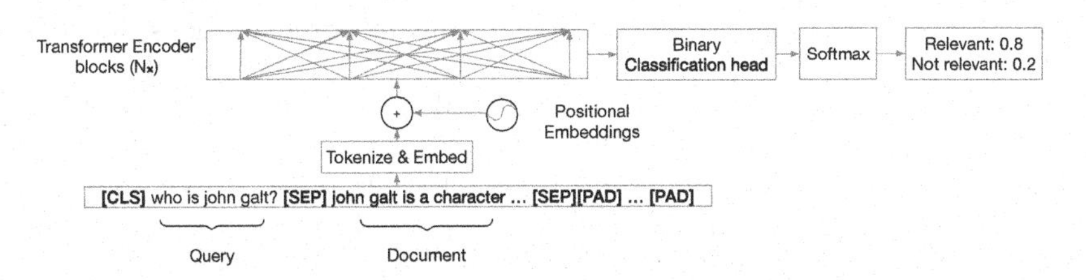

Fig. 2. A typical BERT-based [7] Cross-Encoder. Note that the structure can be adapted to other transformers with slight modification.

As Cross-Encoders jointly encode the query and the document, the scoring of K candidate documents requires applying the model for inference K times[5]. Assuming each inference requires λ milliseconds, we can infer an upper bound on the number of documents that can be scored within the latency window ω:

$$K \leq \frac{\omega}{\lambda} \tag{1}$$

Equation (1) defines the tradeoff between the latency window ω and the number of scored documents K. In practice, this tradeoff limits the use of Cross-Encoders to a *re-ranking* scenario, where the Cross-Encoder is applied to a relatively small number of candidates retrieved with a lightweight first-stage model, such as BM25 [31]. Moreover, reducing the latency window ω decreases the number of retrieved documents that can be scored, decreasing the result's recall and reducing overall model effectiveness. Overall, we can say that there exists a tradeoff between model performance $Q(K)$ and the latency window ω:

$$Q(K) \bowtie \omega \tag{2}$$

where the $\bowtie$ symbol denotes a dependency between Q and ω. The exact form of this dependency is unknown, and investigating the properties of this dependency is one of the main goals of this paper. Equation (1) also shows that the number

[5] For simplicity, we omit batching in this reasoning. With a slight tweaking, it remains valid for batching as well (e.g. inference time should be divided by the batch size).

of scored documents K can be increased if we decrease model inference time λ. In the case of Transformer-based Cross-Encoders, we can decrease the inference time by limiting the number of transformer blocks in the model and/or by limiting the computational complexity of each block by reducing embedding sizes and the number of attention heads (i.e. making the Cross-Encoder *shallow*). On the other hand, many recent publications [27,28,38] show that larger models are more effective for document re-ranking (without considering the latency).

In summary, decreasing the Cross-Encoder size has two effects on overall model effectiveness:

- **Positive effect:** Decreasing model size allows for scoring more documents, which leads to **increased model effectiveness** according to Eq. (1).
- **Negative effect:** Decreasing model size hinders accuracy on each individual (query/document) pair; as a result, overall **model effectiveness decreases**.

To the best of our knowledge, no published research has analysed which of these two effects dominates in a low-latency scenario. This work aims to close this gap. In particular, we aim to verify the following research hypothesis:

Hypothesis H1

When reducing the latency window ω, the ability of shallow Cross-Encoder to score and rank more candidate documents within the given latency window results in higher effectiveness than when ranking less canididate documents with more accurate full-scale models.

Note that hypothesis **H1** is somewhat counter-intuitive. Indeed, it has been consistently shown that larger language models are more effective for documents re-ranking [27,28,38] compared to the smaller ones. However, contrary to these findings, hypothesis **H1** states that smaller language models can be **more effective** compared to the larger ones when the latency window is limited.

To analyse whether or not this hypothesis holds, we need to establish an effective training scheme for shallow Cross-Encoders, which we do in the next section. We then analyse Hypothesis **H1** experimentally in Sect. 4.

3 Training Shallow Cross-Encoders with gBCE

Usually, Cross-Encoder models output estimated relevance *scores* (logits), which can be converted to *probabilities*. For example, the architecture of a BERT-based Cross-Encoder on Fig. 2 outputs positive and negative sccores $s+$ and s^-, and the probability of the document being relevant to the query is then computed using the Softmax$(\cdot)$ transformation:

$$p^+ = \frac{e^{s^+}}{e^{s^+} + e^{s^-}} \tag{3}$$

As the model outputs can be converted into probabilities, Binary Cross-Entropy Loss can be used to train the model:

$$\mathcal{L}_{BCE} = -\left[y \cdot \log(p^{+}) + (1 - y) \cdot \log(1 - p^{+})\right] \tag{4}$$

where y is the ground truth relevance judgment for a given query-document pair, BCE loss drives the model to minimise the KL divergence between ground truth relevancy and predicted probabilities $D_{KL}(y||p)$. Therefore, p^{+} will converge to the "real" probability (the probability that a user finds the document relevant before the judgement has been made; see [29] for proofs). BCE is a very popular choice and has been used to train many effective Cross-Encoder models [28,36]. However, an underlying assumption of BCE is that the distribution of positive/negative samples during training and inference match each other. In practice, there are many more negative documents than positives. Due to computational and memory constraints, it is not feasible to score all possible negatives with large-scale models and therefore, researchers employ *negative sampling* techniques. For example, the popular MonoT5 model [28] samples just one negative (non-relevant) document for each query during training.

A recent publication [29] has shown that negative sampling coupled with the BCE loss leads to the *overconfidence* problem: the estimated probability for positives becomes too high, and the model training becomes unstable, leading to poorer performance compared to the models trained without negative sampling.

Our initial experiments have shown that overconfidence does not cause effectiveness degradation of full-scale Cross-Encoder models, such as MonoT5; we speculate that the use of pre-trained checkpoints works as a strong model regulariser (see [10, Ch. 15] for an intuition). However, our experiments show (see Sect. 4) that overconfidence is indeed a problem in shallow cross-encoders, and it has to be mitigated to achieve high effectiveness.

To mitigate the overconfidence problem for recommender systems, the authors [29] propose the *gBCE* training scheme. The gBCE training scheme consists of two components:

1. Increased number of negative samples per positive.
2. Generalised Binary-Crossentropy (gBCE) loss function instead of classic BCE.

gBCE, parametrised by a parameter β, is defined as:

$$\begin{aligned}\mathcal{L}^{\beta}_{gBCE} &= -\left[y \cdot \log((p^{+})^{\beta}) + (1 - y) \cdot \log(1 - p^{+})\right] \\ &\qquad\qquad (\beta \text{ can be taken out of the log}) \\ &= -\left[y \cdot \beta \cdot \log(p^{+}) + (1 - y) \cdot \log(1 - p^{+})\right]\end{aligned} \tag{5}$$

Parameter β controls model confidence: when $\beta = 1$, gBCE becomes regular BCE; however, decreasing β decreases the model's tendency to predict high probabilities. The authors of [29] proposed to control β indirectly with the help of a calibration parameter t:

$$\beta(t) = \alpha\left(t\left(1 - \frac{1}{\alpha}\right) + \frac{1}{\alpha}\right) \tag{6}$$

where α is the negative sampling rate: $\alpha = \frac{|D_k^-|}{|D^-|}$ (number of sampled negatives as a fraction of the overall number of retrieved negative candidates). For example, if, at each training step, we use 10 sampled negatives for each positively labelled relevant document, and the overall number of retrieved negatives from the first stage retriever is 1000, then the sampling rate is $\alpha = 0.01$.

In summary, at each training step, we sample a batch of B positive (query/document) pairs, then use the first stage retriever model (we use BM25) to retrieve 1000 negatives. After that, we sample K negatives out of 1000 candidates and add these K negative (query/document) pairs for each query to the training batch. Overall, each training batch contains $B \cdot (K+1)$ (query/document) pairs with B positives and $B \cdot K$ negatives. We then perform a standard gradient descent update step using the gBCE loss function.

This concludes the description of the gBCE training scheme for Cross-Encoder models. We now turn to the experimental evaluation of shallow Cross-Encoders for low-latency retrieval.

Table 1. Salient characteristics of experimental models. * For the MonoT5 model, "Transformer Layers" is the combined number of encoder and decoder layers.

Model Type	Model	Transformer Layers	Embedding Size	Attention Heads	Vocab Size	Number of Parameters	Chekpoint File Size
Shallow Cross-Encoders	TinyBERT	2	128	2	30522	4,386,178	17 Mb
	MiniBERT	4	256	4	30522	11,171,074	43 Mb
	SmallBERT	4	512	8	30522	28,764,674	110 Mb
Full-Size baselines	MonoT5-Base	24*	768	12	32128	222,903,552	850 Mb
	MonoBERT-Large	24	1024	16	30522	335,143,938	1.2 Gb

4 Experiments

We design our experiments to answer the following research questions:

RQ1 How does the model size affect efficiency and effectiveness of Cross-Encders?

RQ2 What are the effects of an increased number of negatives and calibration parameter t in the gBCE training strategy applied to shallow Cross-Encoders?

RQ3 What is the efficiency/effectiveness tradeoff of shallow Cross-Encoders when using CPU-only inference?

4.1 Experimental Setup

Frameworks. We use PyTerrier [25] as our main experimental framework, the PyTerrier-Pisa [20, 26] plugin for a low-latency BM25 first-stage retriever (making use of a memory-mapped BlockMax-WAND index [8]), and the ir-measures [21] library for evaluation metrics. We use model implementations from the HuggingFace Transformers [39] library v4.30.2.[6]

Datasets. For training shallow Cross-Encoders, we use the large-scale MSMARCO dataset [1]. For training, we select queries from the *train* section of the dataset, for which the BM25 retriever retrieves at least one relevant document within the top 1000 results[7]. After pre-filtering, the dataset contains 436299 queries. Out of this pre-filtered dataset, we randomly select 200 queries into a *validation set*, which are held out from training and used for monitoring effectiveness metrics during training and for early stopping. For evaluation, we use TREC Deep Learning track (denoted TREC-DL) queries from the 2019 [5] and 2020 [4] tracks. Both querysets are based on queries from MSMARCO but with comprehensive assessments for reliable evaluation.

Hardware. All our experiments use a computer with a Ryzen 5950x CPU, 128Gb DDR-4 memory, NVIDIA RTX 4090 GPU, and a Samsung 980 SSD.

Models. As the backbone architecture for shallow Cross-Encoders, we use pre-trained versions of BERT [7]. Namely, we use TinyBERT, MiniBERT, and Small-BERT checkpoints[8] provided by Turc et al. [2,34]. The smallest model, Tiny-BERT, has just two Transformer Encoder layers and an embedding size of 128, and the largest model, SmallBERT, has four Transformer Layers and an embedding size of 512. We also use two full-size pre-trained Cross-Encoders as the baselines: MonoBERT-Large [27] and MonoT5-Base [28]. For both models, we use official pre-trained checkpoints provided by the authors[9][10]. Table 1 describes salient characteristics of all experimental models. For the inference of all BERT-based models, we use a batch size of 8 – the largest batch size with which we did not experience memory-related issues with the MonoBERT-Large model. For MonoT5, we use a batch size of 64.

During training, following best practices [10, Ch.7], we use an early stopping mechanism to ensure model convergence. In particular, after every 600 training batches, we measure NDCG@10 on the validation set and stop training when validation does not improve for 200 validation steps.

4.2 RQ1. Efficiency/Effectiveness Tradeoffs

Our first research question aims to verify Hypothesis H1 and test whether or not shallow Cross-Encoders provide better efficiency/effectiveness tradeoffs compared to the full-scale models.

[6] Source code for this paper can be found at https://github.com/asash/shallow-cross-encoders.

[7] We do not use the standard MSMARCO triplets file because it only contains one negative per query, and for gBCE training scheme we need up to 128 negatives.

[8] https://huggingface.co/prajjwal1/bert-tiny.

[9] https://huggingface.co/castorini/monot5-base-msmarco-10k.

[10] https://huggingface.co/castorini/monobert-large-msmarco-finetune-only.

To evaluate the tradeoffs, we train TinyBERT, MiniBERT and SmallBERT using the gBCE training scheme. Following the recommendations in [29], we use 128 negatives per positive, and we set calibration parameter $t = 0.75$ for training. We then evaluate shallow models and pre-trained full-scale models using a variable number of candidates from the first-stage BM25 retriever. In our experiments, we vary the number of candidate documents between 1 and 1000. For each number of candidates, we measure model effectiveness using the NDCG@10 metric and the efficiency by measuring latency in milliseconds.

Figure 3 illustrates the efficiency/effectiveness tradeoffs for all experimental models when varying the number of retrieved BM25 candidates, K. Note that the latency includes overheads, such as the BM25 retrieval time and tokenization. From the figure, we see that on both the TREC-DL2019 and TREC-DL2020 querysets, shallow Cross-Encoders (TinyBERT-gBCE, MiniBERT-gBCE, SmallBERT-gBCE) outperform the larger full-scale models in the low-latency zone (latency less than 50 ms). For example, on the TREC-DL2019 queryset, for the maximum latency of 25 ms, TinyBERT-gBCE achieves NDCG@10 of 0.652, a +13.7% improvement over MonoT5 (NDCG@10 0.573) and +51% improvement over MonoBERT-Large (NDCG@10 0.4316). However, if we allow large latencies, full-scale models outperform shallow Cross-Encoders: for example, the best NDCG@10 on TREC-DL2020 achieved by MonoBERT-Large is 0.711, whereas the best NDCG@10 achieved by TinyBERT-gBCE is 0.630 (−12%). Note that the difference between shallow models in the low-latency zone is relatively small; for example, at TREC-DL2019 at 10ms latency, all shallow models achieve an NDCG@10 of 0.60. However, the smallest model (TinyBERT-gBCE, which has just two transformer layers) consistently has better effectiveness than the larger models with latencies less than 10 ms; therefore, we argue that at very small latency requirements, the usage of the smallest model is preferable, as in addition to the best possible performance it also has lowest memory requirements.

Note that these effectiveness/efficiency tradeoffs are specific to our hardware. Also, the tradeoffs can be improved using better-optimised versions of the transformers or using engineering techniques, such as document pre-tokenisation. Nevertheless, these improvements are likely to take effect on all models, and overall, we still expect shallow models to outperform full-size models for low-latency retrieval (however, the point where they start to perform better may change).

Overall, for RQ1, we conclude that Hypothesis H1 holds, and shallow cross-encoders are more effective compared to full-size models for low-latency retrieval.

4.3 RQ2. Effect of the Training Scheme

To answer RQ2, we first analyse whether or not overconfidence is present in shallow Cross-Encoders and whether or not gBCE helps to mitigate it.

To do that, we analyse predicted probabilities at different ranks on queries from TREC-DL2019 and TREC-DL2020. Figure 4 shows the results of this analysis on a sample query from TREC-DL2019 (results for other queries looked similar). As we can see from the figure, TinyBERT-BCE (trained with BCE loss and 1 negative) predicts probabilities very close to 1 for ranks from 1 to 5.

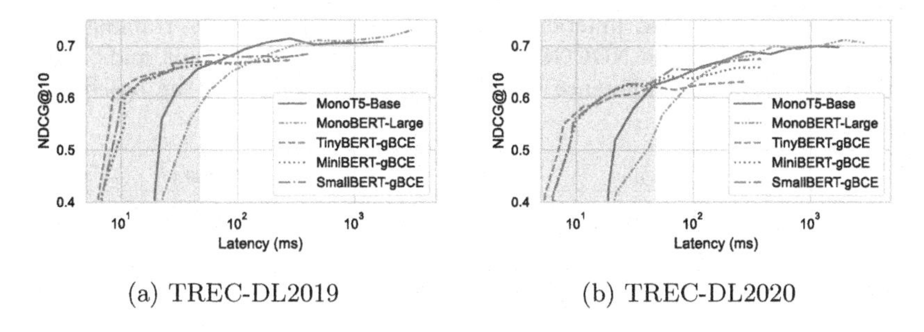

(a) TREC-DL2019

(b) TREC-DL2020

Fig. 3. Latency/NDCG tradeoffs of experimental models when varying the number of candidates from BM25 between 1 and 1000. The shaded area represents the low-latency zone (latency less than 50 ms).

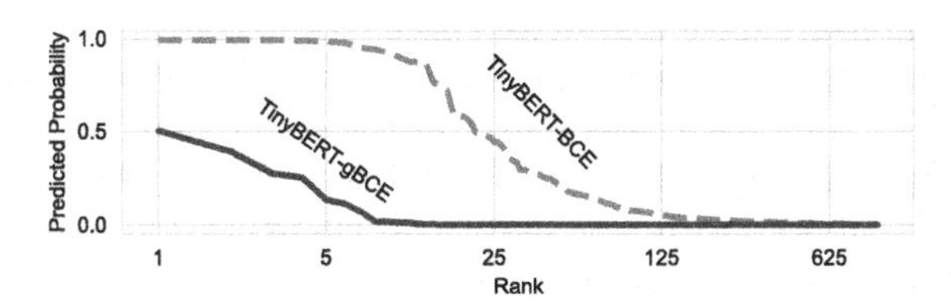

Fig. 4. Predicted probabilities at different ranks for TREC-DL2019 query 146187 "difference between a mcdouble and a double cheeseburger".

For this query, there was only one relevant (label = 3) document identified by the TREC assessors (among 138); therefore, most of these high probabilities are false positives – a clear sign of overconfidence. If such a false positive (query/document) pair appears in the training data, according to Eq. (4), the BCE loss function on the false positive sample will be computed as $\log(1 - p^{+})$, which tends to $-\infty$ when p^{+} tends to 1; this causes numerical instability (i.e. large gradients) during training. This is less problematic for full-size models: recall from Sect. 3 that the use of pre-trained checkpoints in large models works as a strong regulariser, which stabilises model training. Full-size models are also more robust because they have more other regularisations in their architecture, such as dropouts and layer normalisations in each transformer block.

In contrast, TinyBERT-gBCE (model with gBCE loss and 128 negatives) never predicts a probability higher than 0.5, showing more variation in scores in the top predicted results. This confirms that the gBCE training scheme is able to mitigate overconfidence in shallow Cross-Encoders.

We now analyse the effect of the gBCE training scheme on shallow Cross-Encoders. As we discussed in Sect. 3, the gBCE scheme has two key components: (i) a large number of negatives and (ii) gBCE loss. To better understand the effect of gBCE, we analyse these components independently: we train the TinyBERT model, selecting the number of negatives from (1, 128) and the loss function from (BCE, gBCE).

Table 2. Effect of the loss function and the number of negatives training on Tiny BERT-based Cross-Encoder NDCG@10. Bold indicates the best result, and * indicates a statistically significant difference ($pvalue < 0.05$) compared to the baseline (BCE loss, one negative).

(a) TREC-DL2019

Num Negatives→ Loss↓	1	2	8	32	128
BCE	0.6386	0.6640	**0.6735***	0.6622	0.6701
gBCE	0.6593	0.6607	0.6700*	0.6619	0.6721

(b) TREC-DL2020

Num Negatives→ Loss↓	1	2	8	32	128
BCE	0.6056	0.6203	0.6340	**0.6424***	0.6323
gBCE	0.6193	0.6150	0.6270	0.6381	0.6306

Table 2 summarises the results of our evaluation. As we can see from the table, compared to the standard TinyBERT-BCE model (BCE loss, one negative), both components have a positive effect. For example, on TREC-DL2019, an increased number of negatives improves the NDCG@10 metric from 0.638 to 0.670 (+4.9 %), and gBCE loss improves the result to 0.6593. The combination of these improvements leads to an improvement over the baseline (0.6721, +5.2%). As we can see, with 128 negatives, changing the loss function from BCE to gBCE does not have a big effect +0.3% on TREC-DL2019 and -0.3% on TREC-DL2020. This is in line with the original gBCE paper [29], which suggests that gBCE loss is less important with many negatives. We also observe from Table 2 that the effectiveness of the model only increases up to a certain number of negatives, after which the effectiveness fluctuates. Indeed, the best effectiveness is achieved with 8 negatives on TREC-DL2019 and with 32 negatives on TREC-DL2020. At this point, switching from BCE to gBCE is already unnecessary, and all the improvements come from the increased number of negatives. Note that these numbers depend on how many candidate documents are re-ranked. However, our analysis shows that TinyBERT trained with gBCE and 128 negatives consistently outperforms TinyBERT with BCE and one negative with a different number of candidates, and therefore with different latencies; the same observation can also be seen in Fig. 1 for varying sizes of K.

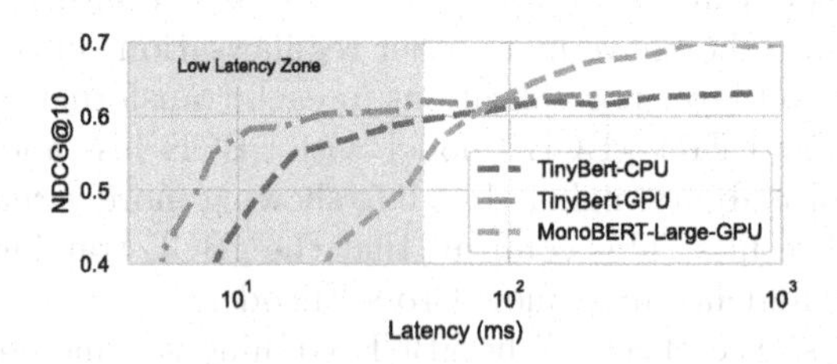

Fig. 5. Comparison of tradeoffs on CPU and GPU, TREC-DL2020 queryset.

We also performed the analysis of the gBCE training scheme on the MS-MARCO dev queryset and found that the evaluation results are in line with the results on the TREC querysets, but due to the larger size of the queryset, we achieved statistically significant improvements compared to the baseline in most of the experiments. For example, evaluation of TinyBERT on the dev queryset showed statistically significant improvements when switching to gBCE loss with one negative sample; see Appendix A for more details.

Overall, in answer to RQ2, we summarise that the gBCE training scheme leads to improvements on both experimental TREC-DL querysets; a large number of negatives is the most important component of the scheme for shallow Cross-Encoders.

4.4 RQ3. CPU Inference

Finally, we analyse the effectiveness/efficiency tradeoffs of a shallow Cross-Encoder without use of a GPU. For this experiment, we use the smallest model, TinyBERT-gBCE, because it has shown the best effectiveness in the low-latency zone in RQ1.

Figure 5 compares the tradeoffs of the model when it is used with CPU inference and with GPU inference. The plot also shows the tradeoff of a full-scale MonoBERT-Large model with GPU inference. The figure shows that GPU inference is better than CPU inference, especially within the low latency zone. For example, with a 10 ms latency window, the model with CPU inference only achieves NDCG@10 of 0.447, whereas the model with GPU inference achieves NDCG@10 of 0.573 (+28%). However, with a larger allowed latency, the difference decreases. At 50 ms (the upper bound of our "low latency zone"), the difference in effectiveness between GPU and CPU is just 3%. As we can see from the plot, in both cases, TinyBERT provides a better tradeoff than MonoBERT with GPU inference, and all 3 models intersect at approximately 100 ms latency.

The fact that TinyBERT allows us to achieve relatively high performance even using CPU-only inference allows us to use it in cases where GPU inference is not feasible, such as for on-device search in applications. The ability to efficiently perform inference on a CPU can also amount to cost savings. For instance, the AWS's GPU-equipped `g5.4xlarge` instance costs $1.624/hr at the time of writing, while a roughly equivalent instance without a GPU (`m6gd.4xlarge`) costs less than half as much, at $0.7232/hr.

Overall, in answer to RQ3, we conclude that while CPU inference is less efficient for shallow Cross-Encoders compared to GPU inference, it allows us to achieve relatively high effectiveness. Considering its low memory footprint (checkpoint is only 17 Mb), we argue that a TinyBERT-based encoder may be an effective solution for systems without a GPU (such as on-device search).

5 Conclusion

In this paper, we proposed shallow Cross-Encoders as a solution for low-latency information retrieval. We showed that shallow Cross-Encoders are more effective than full-size when latency is limited (e.g. TinyBERT model achieved +51%

NDCG@10 on TREC-DL2019 compared to MonoBERT-Large with latency limited by 25 ms; see Fig. 1). We adapted the gBCE training scheme to shallow Cross-Encoders and showed that it improves the effectiveness of shallow Cross-Encoder models (e.g. +5.2% NDCG@10 on TREC-DL2019; see Table 2). We also showed that shallow Cross-Encoders can be effective even for CPU-only inference (e.g., on TREC-DL2020, the difference in NDCG@10 is only 3% with 50 ms latency; see Fig. 5). We believe that shallow Cross-Encoders can be further optimised by applying engineering techniques, such as pre-tokenisation.

A Effect of gBCE Training Scheme on Tiny BERT-Based Cross-Encoder on the MS MARCO Dev Set

Table 3 reports the effectiveness of a Tiny BERT model on a 6,980 queries subset of the MS MARCO dev set (dataset `irds:msmarco-passage/dev/small` in PyTerrier). The evaluation follows the scheme described in Sect. 4.3, with the exception of using MRR@10, which is the official metric for this queryset, instead of NDCG@10. As we can see from the table, the overall trends follow the observations in Sect. 4.3. In particular, an increased number of negatives is more important than the loss function; gBCE loss improves results with a small number of negatives but has a moderate effect when the number of negatives increases. However, we observe that overall gBCE in this experiment is better than BCE loss in 5 out of 6 cases. With 1 negative, the improvement over BCE loss is statistically significant. Overall, the combination of gBCE loss and 128 number of negatives provides a significant improvement of MRR@10, from 0.2942 to 0.3200 (+8.76%), compared to the "standard" training scheme with 1 negative and BCE loss. Note that this result is lower compared to the larger models – e.g. Nogueira et al. [27] achieved MRR@10 of 0.36 on this queryset with a BERT-Large model. Lower effectiveness compared to the full-scale models is an expected result, as we do not control for latency in this experiment. When latency is limited, shallow Cross-Encoders are more effective (see Fig. 1).

Table 3. Effect of the loss function and the number of negatives training on Tiny BERT-based Cross-Encoder MRR@10 on the MS MARCO dev set. Bold indicates the best result, and * indicates a statistically significant difference ($pvalue < 0.05$) compared to the baseline (BCE loss, one negative).

Num Negatives→ Loss↓	1	2	8	32	128
BCE	0.2942	0.3032*	0.3057*	0.3128*	0.3172*
gBCE	0.3035*	0.2974	0.3109*	0.3183*	**0.3200***

References

1. Bajaj, P., et al.: MS MARCO: a human generated machine reading comprehension dataset. In: Proceedings of NeurIPS (2018)
2. Bhargava, P., Drozd, A., Rogers, A.: Generalization in NLI: Ways (Not) To Go Beyond Simple Heuristics (2021). http://arxiv.org/abs/2110.01518
3. Bruch, S., Lucchese, C., Nardini, F.M.: Efficient and effective tree-based and neural learning to rank. Found. Trends® Inf. Retrieval **17**(1), 1–123 (2023)
4. Craswell, N., Mitra, B., Yilmaz, E., Campos, D.: Overview of the TREC 2020 deep learning track. In: Proceedings of TREC (2020)
5. Craswell, N., Mitra, B., Yilmaz, E., Campos, D., Voorhees, E.M.: Overview of the TREC 2019 deep learning track. In: Proceedings of TREC (2019)
6. Dean, J., Barroso, L.A.: The tail at scale. Commun. ACM **56**(2), 74–80 (2013)
7. Devlin, J., Chang, M.W., Lee, K., Toutanova, K.: BERT: pre-training of deep bidirectional transformers for language understanding. In: Proceedings of of NAACL-HLT, pp. 4171–4186 (2019)
8. Ding, S., Suel, T.: Faster top-k document retrieval using block-max indexes. In: Proceedings of SIGIR, pp. 993–1002 (2011)
9. Formal, T., Piwowarski, B., Clinchant, S.: SPLADE: sparse lexical and expansion model for first stage ranking. In: Proceedings of SIGIR, pp. 2288–2292 (2021)
10. Goodfellow, I., Bengio, Y., Courville, A., Bach, F.: Deep Learning. MIT Press, Cambridge (2017)
11. Hofstätter, S., Althammer, S., Schröder, M., Sertkan, M., Hanbury, A.: Improving Efficient Neural Ranking Models with Cross-Architecture Knowledge Distillation (2021). http://arxiv.org/abs/2010.02666
12. Hofstätter, S., Lin, S.C., Yang, J.H., Lin, J., Hanbury, A.: Efficiently teaching an effective dense retriever with balanced topic aware sampling. In: Proceedings of SIGIR, pp. 113–122 (2021)
13. Hofstätter, S., Zlabinger, M., Hanbury, A.: Interpretable & time-budget-constrained contextualization for re-ranking. In: Proceedings of ECAI (2020)
14. Humeau, S., Shuster, K., Lachaux, M.A., Weston, J.: Poly-encoders: Transformer Architectures and Pre-training Strategies for Fast and Accurate Multi-sentence Scoring (2020). http://arxiv.org/abs/1905.01969
15. Khattab, O., Zaharia, M.: ColBERT: efficient and effective passage search via contextualized late interaction over BERT. In: Proceedings of SIGIR, pp. 39–48 (2020)
16. Kohavi, R., Tang, D., Xu, Y.: Trustworthy Online Controlled Experiments: A Practical Guide to A/B Testing. Cambridge University Press, Cambridge (2020)
17. Kulkarni, H., MacAvaney, S., Goharian, N., Frieder, O.: Lexically-accelerated dense retrieval. In: Proceedings of SIGIR, pp. 152–162 (2023)
18. Lin, S.C., Yang, J.H., Lin, J.: In-batch negatives for knowledge distillation with tightly-coupled teachers for dense retrieval. In: Proceedings of RepL4NLP, pp. 163–173 (2021)
19. Lu, W., Jiao, J., Zhang, R.: TwinBERT: distilling knowledge to twin-structured compressed BERT models for large-scale retrieval. In: Proceedings of CIKM, pp. 2645–2652 (2020)
20. MacAvaney, S., Macdonald, C.: A python interface to PISA! In: Proceedings of SIGIR, pp. 3339–3344 (2022)
21. MacAvaney, S., Macdonald, C., Ounis, I.: Streamlining evaluation with IR-measures. In: Proceedings of ECIR, pp. 305–310 (2022)

22. MacAvaney, S., Nardini, F.M., Perego, R., Tonellotto, N., Goharian, N., Frieder, O.: Expansion via prediction of importance with contextualization. In: Proceedings of SIGIR, pp. 1573–1576 (2020)
23. MacAvaney, S., Tonellotto, N., Macdonald, C.: Adaptive re-ranking with a corpus graph. In: Proceedings of SIGIR, pp. 1491–1500 (2022)
24. MacAvaney, S., Yates, A., Cohan, A., Goharian, N.: CEDR: contextualized embeddings for document ranking. In: Proceedings of SIGIR, pp. 1101–1104 (2019)
25. Macdonald, C., Tonellotto, N., MacAvaney, S., Ounis, I.: PyTerrier: declarative experimentation in python from BM25 to dense retrieval. In: Proceedings of CIKM, pp. 4526–4533 (2021)
26. Mallia, A., Siedlaczek, M., Mackenzie, J.M., Suel, T.: PISA: performant indexes and search for academia. In: Proceedings of OSIRRC@SIGIR 2019, vol. 2409, pp. 50–56 (2019)
27. Nogueira, R., Cho, K.: Passage Re-ranking with BERT (2020). http://arxiv.org/abs/1901.04085
28. Nogueira, R., Jiang, Z., Lin, J.: Document Ranking with a Pretrained Sequence-to-Sequence Model (2020). http://arxiv.org/abs/2003.06713
29. Petrov, A.V., Macdonald, C.: gSASRec: reducing overconfidence in sequential recommendation trained with negative sampling. In: Proceedings of RecSys, pp. 116–128 (2023)
30. Reimers, N., Gurevych, I.: Sentence-BERT: sentence embeddings using siamese BERT-networks. In: Proceedings of EMNLP (2019)
31. Robertson, S., Walker, S., Jones, S., Hancock-Beaulieu, M., Gatford, M.: Okapi at TREC 3. In: Proceedings of TREC (1994)
32. Scells, H., Zhuang, S., Zuccon, G.: Reduce, reuse, recycle: green information retrieval research. In: Proceedings of SIGIR, pp. 2825–2837 (2022)
33. Thakur, N., Reimers, N., Rücklé, A., Srivastava, A., Gurevych, I.: BEIR: A Heterogenous Benchmark for Zero-shot Evaluation of Information Retrieval Models (2021). http://arxiv.org/abs/2104.08663
34. Turc, I., Chang, M.W., Lee, K., Toutanova, K.: Well-Read Students Learn Better: On the Importance of Pre-training Compact Models (2019). http://arxiv.org/abs/1908.08962
35. Vaswani, A., et al.: Attention is all you need. In: Proceedings of NeurIPS (2017)
36. Wallat, J., Beringer, F., Anand, A., Anand, A.: Probing BERT for ranking abilities. In: Proceedings of ECIR, pp. 255–273 (2023)
37. Wang, X., MacAvaney, S., Macdonald, C., Ounis, I.: An inspection of the reproducibility and replicability of TCT-ColBERT. In: Proceedings of SIGIR, pp. 2790–2800 (2022)
38. Wang, X., Macdonald, C., Tonellotto, N., Ounis, I.: Reproducibility, replicability, and insights into dense multi-representation retrieval models: from ColBERT to Col*. In: Proceedings of SIGIR, pp. 2552–2561 (2023)
39. Wolf, T., et al.: HuggingFace's Transformers: State-of-the-art Natural Language Processing (2020). http://arxiv.org/abs/1910.03771
40. Xiong, L., et al.: Approximate Nearest Neighbor Negative Contrastive Learning for Dense Text Retrieval (2020)
41. Zhuang, H., et al.: RankT5: fine-tuning T5 for text ranking with ranking losses. In: Proceedings of SIGIR, pp. 2308–2313 (2023)

Asking Questions Framework for Oral History Archives

Jan Švec, Martin Bulín(✉), Adam Frémund, and Filip Polák

Department of Cybernetics, Faculty of Applied Sciences, University of West Bohemia, Pilsen, Czech Republic
{honzas,bulinm,afremund,polakf}@kky.zcu.cz

Abstract. The importance of oral history archives in preserving and understanding past experiences is counterbalanced by the challenges encountered in accessing and searching through them, primarily due to their extensive size and the diverse demographics of the speakers. This paper presents an approach combining ASR technology and Transformer-based neural networks into the Asking questions framework. Its primary function is to generate questions accompanied by concise answers that relate to the topics discussed in each interview segment. Additionally, we introduce a semantic continuity model that filters the generated questions, ensuring that only the most relevant ones are retained. This enables a real-time semantic search through thousands of hours of recordings, with the crucial benefit that the speakers' original words remain unaltered and still semantically align with the query. While the method is exemplified using a specific publicly available archive, its applicability extends universally to datasets of a similar nature.

Keywords: Oral archives · Spoken language understanding · Semantic search

1 Introduction

Oral history archives are an essential resource for understanding individuals' and communities' past and present experiences. Through preserving and documenting personal narratives and memories, oral history archives can provide valuable insights into the social, cultural, and political contexts of different historical periods. Speech recognition technology plays a crucial role in enhancing the accessibility and usability of these archives, as it enables the conversion of audio recordings into searchable and editable transcripts.

Over time, different institutions collect interviews and testimonies about major historical topics. Many of the well-known archives are related to Holocaust, for example, the University of Southern California Shoah Foundation Visual History Archive[1] (USC-SFI VHA) or the collection of US Holocaust Memorial Museum[2] (USHMM).

[1] https://sfi.usc.edu/what-we-do/collections.

[2] https://www.ushmm.org/.

This research was supported by the Czech Science Foundation (GA CR), project No. GA22-27800S, and by the grant of the University of West Bohemia, project No. SGS-2022-017.

N. Goharian et al. (Eds.): ECIR 2024, LNCS 14610, pp. 167–180, 2024.
https://doi.org/10.1007/978-3-031-56063-7_11

The archives are large-scale collections of audio or audiovisual interviews, often containing different accents and a broad range of speaker ages, but all the recordings follow a similar scenario. For example, the Holocaust witnesses giving testimonies for USC-SFI completed a 50-page-long questionnaire asking for names, dates, and experiences from before, after, and during the Holocaust and World War II. Interviewers are expected to familiarize themselves with the survivor's history so that relevant questions can be asked. The Foundation Interviewer Guidelines include "helpful hints" for interview questions and suggest attention to chronology over stream-of-consciousness narration. The instructions also propose ways the interviewer might deal with certain subjects [1].

Listening to huge quantities of audio materials is impractical for an everyday user. Also, the testimonies provided by the interviewees were intended to give evidence of historical events, and it is not ethical to change the meaning or cherry-pick single facts from a given interview. Many efforts to provide access to such archives were proposed in recent research works, including speech-to-text technologies [12] and spoken-term detection methods [20]. Other approaches used traditional information retrieval methods [11]. However, the most used method for accessing the archives is a keyword search in automatically or manually created transcripts.

This paper describes our initial efforts in addressing a significant challenge in the realm of oral history archives: *effectively navigating and engaging with the extensive monologues that often characterize these archives*. While automatically generated transcripts and subtitles assist researchers in locating relevant interviews, they fall short in facilitating a comprehensive understanding of the entire testimony, whether through speech or text. Our innovative approach, leveraging Transformer-based neural networks, seeks to bridge this gap. It not only aids in clearer navigation through lengthy testimonies but also transforms the listening experience from passive to interactive. By generating contextually relevant questions, our system enriches the interview monologues, allowing listeners to better orient themselves within the narrative and identify key segments of interest. These questions are designed to enhance understanding without altering the original meaning of the testimony, thereby maintaining the integrity of the historical record. Additionally, this method empowers users to engage more deeply with the material by posing their own inquiries, fostering a dynamic exploration of the rich narratives contained within the archives. This approach, employing advanced speech-to-text and neural network technologies, represents a significant stride forward in how we interact with and comprehend the wealth of knowledge stored in oral history archives, transforming traditional access methods into an actively engaging experience.

This paper proposes an innovative approach that generates a new question-answer structure on top of the existing interview transcripts. The questions can be indexed and time-aligned with the audio so the user can quickly get to interesting parts of the interview. The questions complement the interviewer and are useful in passages where only the interviewee speaks. In other words, the questions can be understood as "open-set topics" related to the testimony. It is important to stress that the questions do not change the meaning of the testimony because the related parts of the original interview are presented as an answer. On top of the pre-generated and indexed questions, we can perform a real-time semantic search in the interview. We provide a user-friendly web interface with a search field, and we use Sentence-BERT embeddings [16] to match the user's query with the most relevant pre-generated questions. As described in Sect. 5, we enhanced the interface with a real-time ASR and a language translator to demonstrate the capabilities of the developed models in a useful application. In contrast to the question-answering methods, we use the term *asking questions* for our approach (abbreviated as AQ in contrast to question answering – QA).

The primary motivation of our approach is similar to the recently published approach called Doc2Query-- [4]. It was proposed as a *query expansion* method for [5] information retrieval from textual sources. Since the spoken interviews contain mainly spontaneous speech without any inherent grammatical structure, we had to design methods for this use case. In line with this, the work of Yao et al. [19] presents an approach to generating conversational characters from text articles. Their methodology, which employs question generation tools like Question Transducer and OpenAryhpe, could be adapted to create interactive agents capable of navigating the complexities of text articles and reports. Such conversational agents could potentially interpret and respond to user queries in a contextually relevant manner, offering a more intuitive way to access and understand the unstructured data found in textual documents. Yoshikawa et al. [18] developed the ArchivalQA dataset, focusing on temporal news question answering. This dataset, with its emphasis on handling temporal ambiguities and diverse data sources, aligns well with the challenges of parsing and understanding oral history archives.

In this paper, our main contribution lies in creating a unique framework that generates synthetic question-answer pairs using a ChatGPT-generated dataset, coupled with a novel semantic continuity model designed to identify the best-matching question-answer pairs, thus facilitating enhanced search capabilities within large oral history archives while preserving the privacy of sensitive data. This work focused on English USC-SFI data, but the approach is not limited to a specific language. Our system aims to provide a more robust and contextually aware interface for retrieving information from extensive and grammatically unstructured oral history collections. To supplement our approach, we utilized "The American Life Podcasts dataset" [9] as a proxy for generating synthetic QA data. Leveraging OpenAI's ChatGPT, we crafted prompts to produce a dataset of questions and answers, reflecting the conversational and thematic diversity of the podcasts, thereby enriching our question-answering model's training and adaptability.

2 Speech Recognition for Oral Archives

We used the *Wav2Vec 2.0* architecture [2] as a speech recognizer. As we deal with English interviews, we started with English `wav2vec2-base` model. The model consists of 12 Transformer blocks with a hidden dimensionality of 768 and with 8 attention heads. It has 95 million parameters in total. We adopted a two-phase fine-tuning scheme from [8]. We used 12.5 thousand hours of transcribed data from the CommonVoice, English part of USC-SFI VHA, and GigaSpeech [3] datasets in the first phase. Then, in the second phase, we used only 245.7 h from USC-SFI VHA. The resulting Wav2Vec 2.0 end-to-end model was combined with a lower-case four-gram language model estimated from CommonCrawl data. To decrease the model's size, we pruned all unigrams with counts lower than ten and higher-order n-grams with counts lower than 100. For the English test dataset of Holocaust testimonies from USC-SFI VHA [15], we achieved a 12.9% word error rate (for comparison: OpenAI Whisper-large model achieved 17.3% WER). In addition, the raw output of the speech recognizer was post-processed using BERT-based automatic punctuation detection and casing reconstruction [17] to obtain segmentation into sentence-like units.

3 Asking Questions Framework

To generate the questions from interview transcripts, we first use a sentence-level sliding window to process the raw audio stream of recognized words (Fig. 2). The sliding window comprises a few sentences and defines a *context* for the asking-questions framework. Then, we use a *T5-based asking questions (AQ) model* for each context to generate one possible question related to this context. We also generate the answers to the questions because it helps to train the model to generate more specific questions. Because the T5 model always generates the question-answer pair for a given context – even if the context does not contain any meaningful information (e.g., it is only a discourse marker) – we used a second model to classify the *Semantical continuity* of the question-context pair. This way, only the questions which semantically precede the context can be presented to the user (the scoring is depicted in Fig. 1).

3.1 Asking Questions Model

Generating questions for a specific context is virtually an inverse task to question answering. Since the task is text-to-text, we employ the Text-to-text transfer transformer (T5) model [13]. The model was pre-trained on a self-supervised text-restoration task using the CommonCrawl web text data.

We used the `t5-base` pre-trained model, which is freely available for subsequent fine-tuning. The T5 model's architecture is fixed, and therefore, the overall performance of the model is given by the text-to-text dataset used to fine-tune the model. Since the T5 was successfully applied on the question-answering task in models like UnifiedQA [6], we decided to apply it also in the inverted asking-questions task. In this setup, the model takes the context as input and generates a question for a given context. During the experiments, we found that the model trained to generate a single question often generates non-specific and broad questions. Therefore, we also added the required answer as part of the training target to obtain questions of higher quality that are more topic-specific, and the answers are directly expressed in the input context.

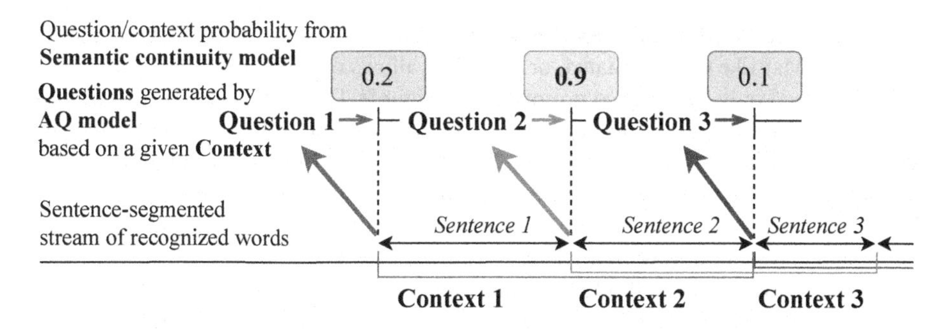

Fig. 1. Processing in the AQ pipeline. If we use indexation threshold 0.8, then only the *Question 2* (generated by Asking-questions model) semantically continued by *Context 2* (consisting of sentences 2&3) with semantic continuity score 0.9 (from Semantic continuity model) will be included in the resulting index.

First, we started with the *Stanford Question Answering Dataset* (SQuAD) [14] as our fine-tuning dataset. While the SQuAD 2.0 dataset effectively handles factual questions from descriptive texts extracted from Wikipedia, it is unsuited for spontaneous interviews in oral history archives. To overcome this limitation, we combine the abilities of the well-known ChatGPT model to perform tasks described in natural language with the proxy dataset for conversational speech to obtain the *American Life Podcasts synthetic QA dataset*. This new dataset provides more appropriate training data for handling conversational questions, and it can be used to improve the accuracy and performance of QA/AQ systems.

SQuAD Dataset. The SQuAD dataset is created as a sample of 536 from the top 10,000 Wikipedia articles. From such articles, 23,215 individual paragraphs were extracted. Each paragraph was used as a context, and the corresponding question-answer pairs were manually created. The dataset is split into training, development, and test containing 130,319/11,873/8,862 context-question-answer triplets. An example of one data point in the SQuAD dataset:

Context: *All police officers in the United Kingdom, whatever their actual rank, are 'constables' in terms of their legal position. This means that a newly appointed constable has the same arrest powers as a Chief Constable or Commissioner. However, certain higher ranks have additional powers to authorize certain aspects of police operations, such as a power to authorize a search of a suspect's house (section 18 PACE in England and Wales) by an officer of the rank of Inspector, or the power to authorize a suspect's detention beyond 24 h by a Superintendent.*

Question: *What is the legal status of UK police officers?*
Answer: *constables*

Question: *What can only Inspector-ranked UK officers do?*
Answer: *authorize a search of a suspect's house*

Question: *What can only Superintendent-ranked UK officers do?*
Answer: *authorize a suspect's detention beyond 24 h*

The American Life Podcasts Synthetic QA Dataset. The advent of large language models (LLMs) like GPT-3, Llama, or ChatGPT allows one to perform zero-shot NLP tasks described only by the used prompts. For example, OpenAI ChatGPT [10] trained on large amounts of data can generate far more coherent and contextually relevant responses and exhibit signs of reasoning. With the proper prompts, it can propose problem solutions, explain complex issues, and sometimes produce human-like behavior. Its advanced capabilities have made ChatGPT a game-changer in conversational AI and the field of zero-shot or few-shot NLP.

In this work, we were limited by the availability of labelled data for the task of asking questions. Its collection, if performed manually, is time-consuming and expensive to obtain. Therefore, we utilize the OpenAI ChatGPT to generate synthetic data, which is then used to fine-tune the T5 AQ model. Because the testimonies in oral history archives often contain sensitive information, we cannot use them to feed the closed ChatGPT API. Instead, we used a proxy dataset of interviews to generate a synthetic QA dataset.

As a proxy dataset, we used *The American Life Podcasts dataset* [9] consisting of transcripts for 663 podcasts from the This American Life radio program from 1995 to 2020, covering 637 h of audio (57.7 min per conversation) and an average of 18 unique speakers per conversation. To generate the T5 AQ training dataset, we instructed ChatGPT using a so-called *System prompt* in natural language and provided a *Context* to obtain *Output* containing the synthetic questions and related answers:

System Prompt: *You are a helpful assistant. Your task is to generate factual questions based on a provided interview context. You should aim to generate 1 to 3 general questions that can be truthfully and reasonably answered from the given context. In case the answer is not available in the context or is not mentioned in the interview, label it as '<not-known>'. For each generated question, please provide a straightforward answer based on the given context.*

Context: *Ed Ryder plays the trumpet. He was sentenced to Graterford Penitentiary in Pennsylvania for 20 years for a murder it was later shown he did not commit. He played jazz when he was in prison. He played jazz when he got out. And he says that it is a completely different experience playing jazz to inmates.*

Output: *What instrument does Ed Ryder play? - Ed Ryder plays the trumpet. – How long was Ed Ryder sentenced for? - Ed Ryder was sentenced for 20 years. – Was Ed Ryder convicted of the murder he was sentenced for? - No, it was later shown that Ed Ryder did not commit the murder he was sentenced for.*

This way, we generated over 15K samples. Each sample comprises a context equivalent to a single line in the podcast data and one to three relevant question-answer pairs. Any lines shorter than 50 characters were filtered out and not used in the generation process.

T5 AQ Model Training. Using the two above-mentioned QA datasets, we trained the T5 AQ model. For this purpose, we used the `t5s` library[3]. As the base model for

[3] https://github.com/honzas83/t5s.

fine-tuning, we used the English `t5-base` model provided by Google. The model *AQ1* was fine-tuned from 84,688 data points generated from the SQuAD dataset, while the model *AQ2* was trained from 35,077 samples from synthetic The American Life Podcasts synthetic QA dataset. We used the initial learning rate $5 \cdot 10^{-4}$ with a linear decrease to 0 on 40k update steps. We updated all T5 weights during training.

3.2 Semantic Continuity Model

The *semantic continuity model* is a novel BERT architecture producing two sentence-level embeddings: left- and right-embedding (Fig. 2). The idea of training objective is similar to Sentence-BERT [16]. In the shared embedding space, a trainable distance and probability metric are defined to compute the degree of semantic continuity.

The left- and right- embeddings are created using the stack of neural network layers. First, the sequence of sentence tokens $S = (s_1, s_2, \ldots s_T)$ is converted using the BERT model to the sequence of interim hidden vectors $H = (h_1, h_2, \ldots h_T)$:

$$H = \text{BERT}(S) \tag{1}$$

where $h_i \in \mathbb{R}^N$, N denotes the internal dimensionality of transformer blocks in BERT. For the `bert-base` architecture $N = 768$. Then, each BERT output h_i is processed using network layers with a GELU activation function. We use two fully-connected layers Dense_1 and Dense_2 to obtain two network outputs - sequences $L' = (l'_1, l'_2, \ldots l'_T)$ and $R' = (r'_1, r'_2, \ldots r'_T)$:

$$\begin{aligned} l'_i =& \text{GELU}(\text{Dense}_1(h_i)) \\ r'_i =& \text{GELU}(\text{Dense}_2(h_i)), \quad l'_i, r'_i \in \mathbb{R}^M \end{aligned} \tag{2}$$

where the dimensionality M is a meta parameter. In our architecture, we used $M = 3072$, the dimensionality used in feed-forward layers in `bert-base`. Then, the global average pooling together with a layer normalization (implemented as trainable layers $\text{LayerNorm}_1(\cdot)$ and $\text{LayerNorm}_2(\cdot)$) is used to obtain a single vector representing the input sentence S:

$$\begin{aligned} l'(S) =& \text{LayerNorm}_1\left(\frac{1}{T}\sum_{i=1}^{T} l'_i\right) \\ r'(S) =& \text{LayerNorm}_2\left(\frac{1}{T}\sum_{i=1}^{T} r'_i\right) \end{aligned} \tag{3}$$

Layer normalization ensures that l', r' have a mean close to 0 and a standard deviation close to 1. The vectors $l'(S)$ and $r'(S)$ are then used to produce the left- and right- embeddings using fully-connected layers with linear activations $\text{Dense}_3(.)$ and $\text{Dense}_4(.)$:

$$\begin{aligned} l(S) =& \text{Dense}_3\left(l'(S)\right) \\ r(S) =& \text{Dense}_4\left(r'(S)\right), \quad l_i, r_i \in \mathbb{R}^{N'} \end{aligned} \tag{4}$$

The dimensionality N' of $l(S)$ and $r(S)$ is also a meta parameter. We used the same value as for the hidden vectors of `bert-base`, i.e., $N = N' = 768$.

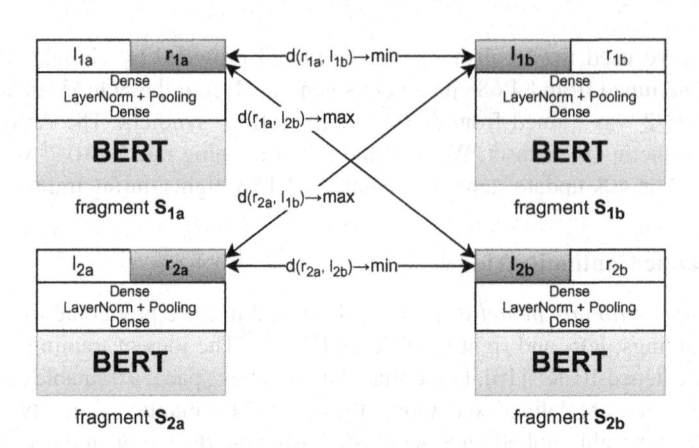

Fig. 2. Semantic continuity model based on BERT and contrastive loss function.

During training, we randomly sample two fragments of texts from the training textual data - S_1 and S_2. Each fragment is then split on the sentence boundary. This way, we obtain four fragments (S_{1a}, S_{1b}, S_{2a}, and S_{2b}). We say that the fragment S_{1b} is a semantic continuation of a corresponding fragment S_{1a}. In other words, the fragment S_{1b} semantically follows fragment S_{1a}. From the fragments, we compute the following four embeddings: right embeddings $r_{1a} = r(S_{1a})$, $r_{2a} = r(S_{2a})$ and left embeddings $l_{1b} = l(S_{1b})$, $l_{2b} = l(S_{2b})$.

We model the semantic continuity between segments a and b through the distance between the two. The closer the right embedding of the first fragment a to the left embedding of the second fragment b, the better the semantic continuity between a and b. The distance is computed from the dot-product of vectors l and r:

$$d(l, r) = \text{ELU}(\alpha_1 \cdot l^T r + \beta_1) + 1 \tag{5}$$

where α_1 and β_1 are trainable parameters and ELU(.) is an exponential linear activation function.

We also use binary (probabilistic) outputs from the network based on the dot-product $l^T r$:

$$p(l, r) = \sigma(\alpha_2 \cdot l^T r + \beta_2) \tag{6}$$

where again α_2 and β_2 are trainable parameters and $\sigma(.)$ is a sigmoid function.

The primary training loss used to optimize the network is based on the distances. We minimize distances between corresponding continuation pairs and maximize distances of two mismatching pairs (see Fig. 2). We use the contrastive loss function with a margin D defined as:

$$\begin{aligned} L_C = &\max\{0, D + d(r_{1a}, l_{1b}) - d(r_{1a}, l_{2b})\} \\ &+ \max\{0, D + d(r_{2a}, l_{2b}) - d(r_{2a}, l_{1b})\} \\ &+ \max\{0, D + d(r_{1a}, l_{1b}) - d(r_{2a}, l_{1b})\} \\ &+ \max\{0, D + d(r_{2a}, l_{2b}) - d(r_{1a}, l_{2b})\} \end{aligned} \tag{7}$$

Because the neural network is not constrained in the maximum values of $d(l,r)$, to prevent it from diverging, we use the following regularization loss to minimize the distance of two matching fragments:

$$L_E = d(r_{1a}, l_{1b}) + d(r_{2a}, l_{2b}) \tag{8}$$

To calibrate the probabilistic output of Eq. 6, we use two binary cross-entropy functions:

$$L_P = -\log[p(r_{1a}, l_{1b})] - \log[p(r_{2a}, l_{2b})] \tag{9}$$

$$L_N = -\log[1 - p(r_{1a}, l_{2b})] - \log[1 - p(r_{2a}, l_{1b})] \tag{10}$$

where L_P calibrates parameter α_2 and β_2 from Eq. 6 for the corresponding pairs and L_N for the non-corresponding pairs of text fragments.

Semantic Continuity Model Training. We trained three different semantic continuity models: *SC1* from SQuAD only dataset where fragment $S_{\cdot b}$ represents a context and fragment $S_{\cdot a}$ the corresponding question; *SC2* from USC-SFI data, where fragments $S_{\cdot a}$, $S_{\cdot b}$ represent two consecutive sequences of recognized sentence-like units; and *SC3* trained from the union of *SC1* and *SC2* training data.

The overall training loss in all three cases is the weighted sum of L_C, L_E, L_P, and L_N with the following weights:

$$L = 0.25 \cdot L_C + 0.01 \cdot L_E + 0.04 \cdot L_P + 0.16 \cdot L_N \tag{11}$$

The weights in the combined loss functions were chosen so that the optimization of the combined network converges. The value of the margin parameter was $D = 0.15$. We update all parameters of the network layers (BERT, Dense, LayerNorm) and also the calibration parameters α and β during the training. The learning rate was $1 \cdot 10^{-4}$ with a linear decrease to 0 in 500k update steps.

4 Experimental Evaluation

Evaluating the whole AQ framework is a challenging task. We started with the evaluation of the semantic continuity models *SC1*, *SC2*, and *SC3*. In Sect. 4.1, we first define a semantic continuity score and then use it to compare the performance of the models on two different datasets. Then, in Sect. 4.2, we evaluate the outputs of AQ models *AQ1* and *AQ2* using the semantic continuity models. Finally, we report quantitative metrics of applying the AQ framework to the oral history data.

4.1 Semantic Continuity Model Evaluation

The semantic continuity model is trained to assess the relevance of a question in relation to a given context. For a given context-question pair, the model provides a distance (which indicates higher relevance as it gets smaller, as defined in Eq. 5) and a probability (the higher, the more relevant, Eq. 6).

The model is evaluated on a reference set of context-question pairs. For each context, we score it against its corresponding relevant (reference, ground truth) question

Table 1. The SC scores for three variants of the semantic continuity model (mean value ± std).

Semantic continuity model	SQuAD dev	USC-SFI test
SC1 (SQuAD)	0.780 ± 0.004	0.278 ± 0.003
SC2 (USC-SFI)	0.486 ± 0.004	0.690 ± 0.002
SC3 (SQuAD ∪ USC-SFI)	**0.781 ± 0.004**	**0.710 ± 0.005**

and $N - 1$ randomly chosen questions related to other contexts and considered irrelevant to the context being scored. A sample is marked as positive if the relevant question has the lowest distance score out of the N candidates and negative otherwise. This allows us to determine the accuracy across the reference set. The semantic continuity score (SC) is the average accuracy over R realizations of the experiment. Let P be the number of context-question pairs in the testing set, c_i the i-th scored context from the test set, Q_i the set of N candidates scored against the c_i context, and for each context c_i, the reference (ground-truth) question is denoted as $q_i^* \in Q_i$. Then the SC score gives an estimate of the probability of selecting the ground-truth question from the set of N chosen questions, including the ground truth:

$$SC = \frac{1}{R} \sum_{r=1}^{R} \frac{1}{P} \sum_{i=1}^{P} \begin{cases} 1, & \text{if } \arg\min_{q \in Q_i} d(q, c_i) = q_i^* \\ 0, & \text{otherwise} \end{cases} \tag{12}$$

For evaluation, we used $R = 5$ different samplings and $N = 10$ as a set of candidates. Together with the average SC score over R realizations, we also evaluated the standard deviation of the scores. The values for the three semantic continuity models evaluated on two different reference sets (SQuAD and USC-SFI) are reported in Table 1.

The SQuAD dev set adheres to the desired structure of context-question pairs, whereas the USC-SFI set represents data from the target domain but does not contain any reference questions. The results indicate that including the question-based SQuAD data in training enhances its performance in scoring questions against contexts, as demonstrated by the higher accuracy of *SC3* (0.71) compared to *SC2* (0.69). Furthermore, evaluating the *SC1* and *SC3* models on the SQuAD set, achieving an accuracy score of 0.78, demonstrated that the USC-SFI data (without reference questions) did not impact the models' ability to score the SQuAD question-context pairs.

4.2 AQ Models Comparison

In this section, we focused on comparing two different AQ models: the model *AQ1* fine-tuned from the SQuAD dataset and the model *AQ2* trained from the American Life Podcasts synthetic QA dataset. Because there is an obvious obstacle in directly evaluating the AQ model on USC-SFI data, we used a method that compares the AQ models' performance on the well-known SQuAD dataset.

In our setup, the semantic continuity model acts as a filter, which selects the questions generated by the AQ model. Question-context pairs that received a probability score from the SC model below the set threshold of 95% were excluded. We first used

Table 2. Percentage of kept questions generated by the AQ model and filtered by the SC model.

	Model *SC1*	Model *SC2*	**Model** *SC3*
Model *AQ1*	34.21	42.03	54.75
Model *AQ2*	**34.95**	**49.58**	**56.39**
SQuAD reference	46.82	61.36	59.31

the SQuAD dataset (development subset, 6428 question-context pairs) and evaluated the percentage of question-context pairs the SC model classified as the semantic continuation for the ground-truth data. This way, we constituted a reference value on the percentage of kept questions (last row of Table 2). The percentage of kept questions varies across different SC models because the size of training data for *SC2* and *SC3* is an order of magnitude larger than for *SC1*, the vector space of left- and right-embeddings is better sampled, and the probabilities/distances over this space are better modelled.

Let's compare the percentage of kept questions from the SQuAD dataset for the *AQ1* and *AQ2* models. We observe that *AQ2* consistently achieves a higher percentage of kept questions (Table 2). In other words, it generates questions that better semantically match the contexts. This conclusion is independent of the SC model used for evaluation. But if we focus on *SC3* model (which came as the best model from Table 1) in combination with *AQ2*, the portion of kept questions is similar to the SQuAD reference.

QUESTION: **Did the interviewee's mother work?** (score: 0.99, timestamp: 0:07:44)
CONTEXT: No, she never worked. How would you describe your mother. It's very hard for me to describe it because she was a lovely, nice looking woman and she was the real Jewish mother.

QUESTION: **What is Pesach?** (score: 0.98, timestamp: 0:15:23)
CONTEXT: How about Pesach? Do you remember anything about Pesach? Oh, Pesach was a special holiday for Jewish people, not only myself, but for all the Jewish people. What was the for you?

QUESTION: **Did the speaker see the person killing somebody?** (score: 0.98, timestamp: 1:14:18)
CONTEXT: No he. I never saw him killing somebody but whipping. Yes, I did see him without reason with a reason. What this? I saw him. To my knowledge he was a very nice, intelligent man.

Fig. 3. Selected samples of asked questions regarding the testimony of Abraham Bomba [1] publicly available from `https://youtu.be/1eWo8j6uEow`. The questions are automatically generated for a given context (automatically recognized speech fragment), the semantic continuity model assigns the score, and the timestamp refers to the above-mentioned interview.

Finally, we applied the resulting combination of models *AQ2-SC3* to the USC-SFI VHA data. Because we do not have any ground-truth reference questions, we only report a quantitative evaluation. We used a test set of the USC-SFI VHA corpus [15],

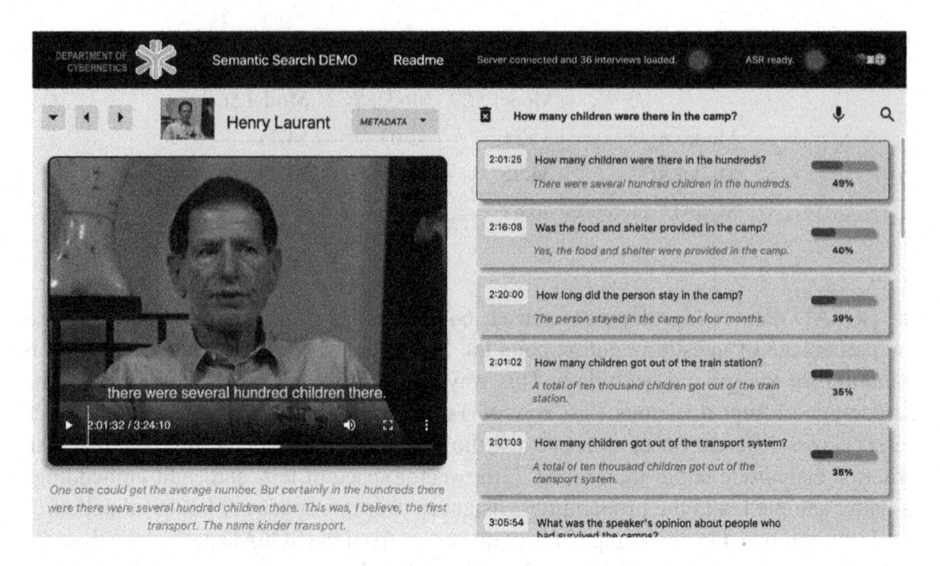

Fig. 4. Graphical user interface for the Asking Questions Framework. Interview segment of the testimony of Henry Laurant available from `https://youtu.be/VftHKcjIzmc`

which consists of interviews with 32 Holocaust survivors, 15.6 h in total. The sentence-level sliding window approach generated 8027 different contexts. For each context, the *AQ2* model generated exactly one question. Those questions were filtered using the *SC3* model, and the above-mentioned thresholding was used. We obtained 477 questions (5.94% out of 8027) that semantically precede the corresponding context. On average, we generated one question for every two minutes of an interview. A sample of generated questions is provided in Fig. 3, including the questioned context and the semantic continuity score.

5 Asking Questions Demonstrator

We have developed a user-friendly web interface[4] to showcase the capabilities of our novel "Asking Questions Framework". For the purpose of this demonstration, we employed the publicly available USC-SFI VHA videos from the corresponding YouTube channel[5], which contains Holocaust testimonies. In Fig. 4, the user posed the question: "*How many children were there in the camp?*". This query is instantly transformed into an embedding vector of dimension 768 using the SentenceTransformers library[6] (model `all-distilroberta-v1`) [16]. We use the cosine similarity to score this embedding against the embeddings of the set of questions already prepared in advance. This method allows us to identify the segments of the interview that are

[4] Demonstrator available at https://malach-aq.kky.zcu.cz.
[5] https://www.youtube.com/uscshoahfoundation.
[6] https://www.sbert.net.

semantically most relevant to the user's question. Moreover, as our questions are synchronized with the video timestamps, we are able to swiftly navigate to the section of the testimony where the survivor responds to the question in their own words. Should the first question not align well with the user's requirements, they have the option to select another one from the sorted list that better suits their needs.

Additionally, we have integrated our ASR engine into the interface, enabling users to input their queries through voice. Moreover, we have incorporated an online Czech-to-English LINDAT translator [7] to enhance accessibility. While the framework operates in English, aligning with the language spoken by the people in the video, users can also input their queries in Czech. In our forthcoming work, we aim to further enhance this integration by introducing a Czech Text-to-Speech (TTS) engine, thereby completing the cycle of accessibility and usability.

6 Conclusion and Future Work

This paper describes our initial effort to better understand interviews in oral history archives by generating synthetic questions answered in the corresponding context. We understand that interpreting oral history interviews is a very sensitive process. Therefore, we designed a pipeline that considers the privacy issues related to topics such as the Holocaust. We trained the asking-questions model for generating questions based on a proxy synthetic ChatGPT-generated QA dataset. Then, we trained the semantic continuity model using the in-domain data. This way, the sensitive transcripts do not leave the owners' possession. Finally, we have showcased a web interface that employs our models to enable real-time semantic searching within a large oral history archive.

We understand that our initial contribution only states the basic principles of the AQ framework, and it still requires a thorough evaluation using real, human-labeled data. Many scientific tasks remain for future research. For instance, when evaluating the SC model, it is important to consider that there may not be a single correct question for a given context but rather a range of acceptable questions. Another open question is how to train the AQ model and exploit different question-answering datasets and recent large language models. And finally, there is the challenge of evaluating the generation of open questions which should be answered in the context.

References

1. USC Shoah Foundation Oral History with Abraham Bomba — Experiencing History: Holocaust Sources in Context. https://perspectives.ushmm.org/. Accessed 12 Apr 2023
2. Baevski, A., Zhou, Y., Mohamed, A., Auli, M.: Wav2Vec 2.0: a framework for self-supervised learning of speech representations. In: Advances in Neural Information Processing Systems, vol. 33, pp. 12449–12460 (2020)
3. Chen, G., et al.: Gigaspeech: an evolving, multi-domain ASR corpus with 10,000 hours of transcribed audio. In: Proceedings of Interspeech 2021 (2021)
4. Gospodinov, M., MacAvaney, S., Macdonald, C.: Doc2query-: when less is more. In: Kamps, J., et al. (eds.) ECIR 2023. LNCS, vol. 13981, pp. 414–422. Springer, Cham (2023). https://doi.org/10.1007/978-3-031-28238-6_31

5. He, B., Ounis, I.: Studying query expansion effectiveness. In: Boughanem, M., Berrut, C., Mothe, J., Soule-Dupuy, C. (eds.) ECIR 2009. LNCS, vol. 5478, pp. 611–619. Springer, Heidelberg (2009). https://doi.org/10.1007/978-3-642-00958-7_57
6. Khashabi, D., et al.: UNIFIEDQA: crossing format boundaries with a single QA system. In: Findings of the Association for Computational Linguistics: EMNLP 2020, pp. 1896–1907. Association for Computational Linguistics, Online (2020)
7. Košarko, O., Variš, D., Popel, M.: LINDAT translation service (2019). http://hdl.handle.net/11234/1-2922. LINDAT/CLARIAH-CZ digital library at the Institute of Formal and Applied Linguistics (ÚFAL), Faculty of Mathematics and Physics, Charles University
8. Lehečka, J., Švec, J., Pražák, A., Psutka, J.V.: Exploring capabilities of monolingual audio transformers using large datasets in automatic speech recognition of Czech. In: Proceedings of Interspeech 2022, pp. 1831–1835 (2022)
9. Mao, H.H., Li, S., McAuley, J., Cottrell, G.W.: Speech recognition and multi-speaker diarization of long conversations. In: Proceedings of Interspeech 2020, pp. 691–695 (2020)
10. OpenAI: GPT-3 API (2021). https://beta.openai.com/docs/api-reference/introduction. Accessed 25 Mar 2023
11. Pecina, P., Hoffmannová, P., Jones, G.J.F., Zhang, Y., Oard, D.W.: Overview of the CLEF-2007 cross-language speech retrieval track. In: Peters, C., et al. (eds.) CLEF 2007. LNCS, vol. 5152, pp. 674–686. Springer, Heidelberg (2008). https://doi.org/10.1007/978-3-540-85760-0_86
12. Picheny, M., Tüske, Z., Kingsbury, B., Audhkhasi, K., Cui, X., Saon, G.: Challenging the boundaries of speech recognition: the MALACH corpus. In: Proceedings of Interspeech 2019, pp. 326–330 (2019)
13. Raffel, C., et al.: Exploring the limits of transfer learning with a unified text-to-text transformer. CoRR (2019). http://arxiv.org/abs/1910.10683
14. Rajpurkar, P., Jia, R., Liang, P.: Know what you don't know: unanswerable questions for SQuAD. In: Proceedings of ACL 2018, Melbourne, Australia, pp. 784–789. ACL (2018)
15. Ramabhadran, B., et al.: USC-SFI MALACH Interviews and Transcripts English LDC2012S05. Linguistic Data Consortium, Philadelphia (2012). https://catalog.ldc.upenn.edu/LDC2012s05
16. Reimers, N., Gurevych, I.: Sentence-BERT: sentence embeddings using siamese BERT-networks. In: Proceedings of the 2019 EMNLP-IJCNLP, Hong Kong, China, pp. 3982–3992. Association for Computational Linguistics (2019)
17. Švec, J., Lehečka, J., Šmídl, L., Ircing, P.: Transformer-based automatic punctuation prediction and word casing reconstruction of the ASR output. In: Ekštein, K., Pártl, F., Konopík, M. (eds.) TSD 2021. LNCS, vol. 12848, pp. 86–94. Springer, Cham (2021). https://doi.org/10.1007/978-3-030-83527-9_7
18. Wang, J., Jatowt, A., Yoshikawa, M.: Archivalqa: a large-scale benchmark dataset for open domain question answering over archival news collections. CoRR abs/2109.03438 (2021)
19. Yao, X., et al.: Creating conversational characters using question generation tools. Dialogue Discourse **3**(2), 125–146 (2012)
20. Švec, J., Šmídl, L., Psutka, J.V., Pražák, A.: Spoken term detection and relevance score estimation using dot-product of pronunciation embeddings. In: Proceedings of Interspeech 2021, pp. 4398–4402 (2021)

Improved Learned Sparse Retrieval with Corpus-Specific Vocabularies

Puxuan Yu[1], Antonio Mallia[2(✉)], and Matthias Petri[3]

[1] University of Massachusetts Amherst, Amherst, USA
pxyu@cs.umass.edu
[2] Pinecone, New York, Italy
antonio@pinecone.io
[3] Amazon AGI, Seattle, USA
mkp@amazon.com

Abstract. We explore leveraging corpus-specific vocabularies that improve both efficiency and effectiveness of learned sparse retrieval systems. We find that pre-training the underlying BERT model on the target corpus, specifically targeting different vocabulary sizes incorporated into the document expansion process, improves retrieval quality by up to 12% while in some scenarios decreasing latency by up to 50%. Our experiments show that adopting corpus-specific vocabulary and increasing vocabulary size decreases average postings list length which in turn reduces latency. Ablation studies show interesting interactions between custom vocabularies, document expansion techniques, and sparsification objectives of sparse models. Both effectiveness and efficiency improvements transfer to different retrieval approaches such as uniCOIL and SPLADE and offer a simple yet effective approach to providing new efficiency-effectiveness trade-offs for learned sparse retrieval systems.

Keywords: Learned sparse retrieval · Language model vocabulary

1 Introduction

Sparse term representations such as SPLADE [10], TILDE [41] or uniCOIL [12,17] establish competitive retrieval performance using existing sparse retrieval techniques underpinned by standard *inverted indexes* data structures [42]. The inverted index has been optimized to be highly scalable, cost-efficient, updateable in real-time, and continue to be one of the core first-stage retrieval components in most commercial search systems today.

One of the key distinctions of state-of-the-art learned sparse representations compared to traditional ranking functions such as BM25 [32] is the tight integration between the vocabulary of the inverted index and the one of the model producing term importance representations for each document. While BM25 based inverted indexes contain potentially millions of unique tokens, learned sparse indexes generally restrict the vocabulary to tokens occurring in the underlying

A. Mallia—Work partly done while working at Amazon Alexa.

N. Goharian et al. (Eds.): ECIR 2024, LNCS 14610, pp. 181–194, 2024.
https://doi.org/10.1007/978-3-031-56063-7_12

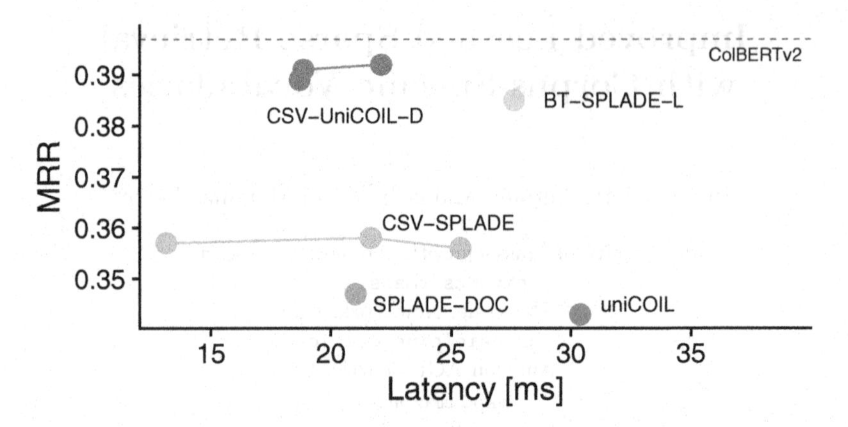

Fig. 1. Latency and effectiveness improvements achieved by leveraging Corpus Specific Vocabularies (CSV) (with different vocabulary sizes) compared to baseline learned sparse retrieval models.

BERT [5] vocabulary. This vocabulary is usually restricted to say 30,000 entries to improve model efficiency.

While some work has elucidated the link between term score distribution and learned sparse representations [18,24], in this work we explore the relationship between vocabulary selection, retrieval quality, and runtime efficiency of learned sparse representations.

Contribution. This work provides the following contributions:

- We show the benefit of creating corpus-specific vocabularies to pre-train underlying language models to retrieval quality.
- We explore trade-offs between vocabulary size, pre-training time, document expansion, and effectiveness improvements.
- We demonstrate that our approach is applicable to many state-of-the-art techniques such as SPLADE, and uniCOIL.
- We propose a corpus-specific modification to TILDE document expansion that leverages custom vocabularies as well as augmentation of hard negatives at training time.
- We analyze improvements in retrieval latency resulting from large corpus-specific vocabularies.

Overall our proposed approach is simple and offers new performance trade-offs for different learned sparse models (see Fig. 1).

2 Background and Related Work

Learned Sparse Models. Usage of pre-trained contextualized language models (LMs) has resulted in improvements to search effectiveness, albeit with higher retrieval costs than traditional lexical models [22]. While models such as BM25 leverage term frequency statistics to estimate term importance in a document, LMs can be leveraged to learn the importance of a term in a document by directly optimizing for the actual retrieval task. These term importance scores form the basis of many *learned sparse* retrieval techniques that still leverage the inverted index for query processing. Such models include SPLADE [10], TILDE [41], DeepImpact [23] or uniCOIL [12,17] which differ in their handling of document and query processing, vocabulary selection, and training objective but offer state-of-the-art retrieval performance while providing different efficiency and effectiveness trade-offs.

Pre-training. Pre-training refers to allowing a model to learn general language representations by performing tasks such as Masked Language Modeling (MLM) on large text corpora. In the search setting, techniques such as coCondenser [11] provide additional search-specific pre-training tasks to improve the performance of LMs on the actual retrieval task. Such pre-training objectives may operate on the target retrieval corpus, or larger potentially out-of-domain text corpora. Recent work has explored the relationship of vocabulary size in standard-pretraining arrangements [8] as well as the notion of rare-terms in pre-training requiring special consideration [39].

Document Expansion. To mitigate the vocabulary mismatch problem [40], learned sparse representations perform document expansion to augment the document with potentially relevant (future query) tokens. The DocT5Query [27] technique augments documents with tokens by appending generated queries from the source document, while TILDE [41] directly optimizes for both term importance estimation and document expansion.

Inverted Index and Dynamic Pruning. The *inverted index* stores one *postings list* for each unique term t produced by a ranking model. Each postings list comprises a sequence of the document ID and corresponding term importance score pairs [30,42]. During query processing, the posting lists of all query terms are processed to retrieve the top-k highest-scoring documents. Query processing algorithms such as the MaxScore [37] or BlockMaxWand [7] dynamic pruning mechanisms enable skipping of large sections of postings lists. However, a relationship still exists between the length of each postings list and overall query latency [36].

3 Corpus-Specific Vocabularies

This section introduces the notion of **Corpus-Specific Vocabularies** (CSV) and shows how it can be incorporated into different aspects of the overall training procedures of sparse retrieval models: vocabulary selection, pre-training, document expansion, and model training (see Fig. 2 for an overview). We find that CSV provides greater coverage of query terms, can be easily incorporated into training procedure of different models, and better correspond to the actual usage of the vocabulary entries in the downstream ranking task inside the inverted index.

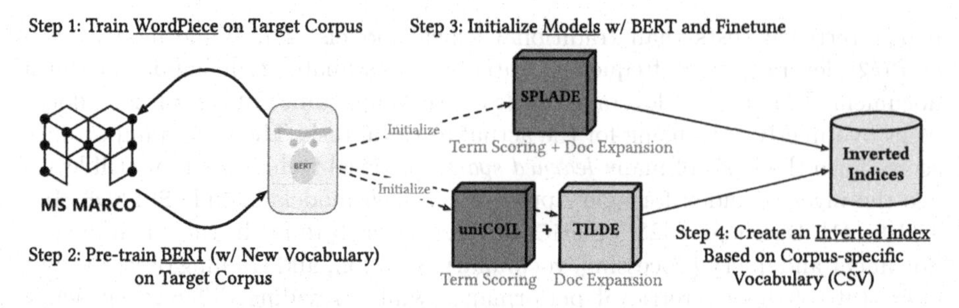

Fig. 2. A high-level overview of the workflow described in this work. As the vocabulary of the language model is learned on the target retrieval corpus, and that the sparse retrieval models (e.g., SPLADE and uniCOIL) and the document expansion models (e.g., TILDE) all use the language model as backbones, all the components in the learned sparse retrieval systems, including the acquired inverted index, are influenced by the corpus-specific vocabulary (CSV).

3.1 Vocabulary Selection

Before the advent of learned sparse models based on language models, it was common practice for an inverted index to contain lists for all unique tokens in the target corpus. Standard text collections such as Gov2 contain 25 million unique tokens [28] comprising parsing errors, named entities, numbers, etc. Indexing these unique tokens has the benefit of being able to precisely (and efficiently) retrieve documents containing rare tokens, a key benefit of sparse retrieval models over alternative dense retrieval systems.

On the contrary, due to computational restrictions (parameter memory usage, softmax inefficiencies among others) associated with Transformers, it is common practice to limit the number of unique tokens fed into such models to only tens of thousands (e.g., 30,000 in the case of standard BERT [5]). Algorithms such as byte-pair encoding (BPE) [35] or WordPiece [38] have been developed to tokenize text into sub-word units, to minimize the occurrence of out-of-vocabulary tokens during text processing with such limited vocabulary sizes. These vocabulary size restrictions are generally non-problematic, as these sub-word tokens are represented in the context of word sequences during standard NLP tasks such as machine translation.

However, in the context of sparse retrieval models such as uniCOIL, this contextualization of sub-word tokens only takes place at training time. At retrieval time when using a standard inverted index, each token is processed in isolation.

We propose to adjust the vocabulary used in sparse models such as uniCOIL (and the underlying language model) to better account for this mismatch. For simplicity, we train WordPiece tokenizers on our target corpus with varying, larger vocabulary sizes. While vocabulary selection could be enhanced by incorporating other signals such as query logs and term frequency counts into the learning process, we seek to isolate the effect of vocabulary size in this work and leave these extensions to future work. We refer to this process as leveraging Corpus-Specific Vocabularies (CSV). In this work, we specifically experiment with vocabulary sizes of 30,000, 100,000, and 300,000. As we will show in detail in Sect. 4, increasing vocabulary size has positive effects on both retrieval quality and runtime latency.

3.2 Pre-training Objectives

Since our model employs a different vocabulary from BERT, we cannot use pre-trained BERT checkpoints. BERT is pre-trained with two objectives in mind: Masked Language Modeling (MLM) and Next Sentence Prediction (NSP) [5]. Pre-training is usually performed on the BooksCorpus (800M words) and English Wikipedia (2,500M words) datasets. Models such as coCondenser [11] and SPLADE [9,10,15] begin with a pre-trained BERT checkpoint and undergo further pre-training on the retrieval corpus, which is sometimes referred to as "middle-training" [15].

In this study, we *bypass* the pre-training step on large, out-of-domain text corpora (e.g. BooksCorpus and Wikipedia) and *only* pre-train on the target retrieval corpus. We make this choice due to (1) the aim of mitigating the cost and environmental impact associated with pre-training multiple LMs using multiple different vocabularies on large corpora, and (2) empirical evidence suggesting that LMs pretrained from scratch on the retrieval corpus exhibit improved effectiveness in retrieval tasks [16].

Note that pre-training with large vocabulary sizes can be computationally expensive but methods such as hierarchical or sampled softmax are standard, drop-in replacements for softmax cross entropy which improves scalability with regard to the number of classes (vocabulary entries).

3.3 Sparse Retrieval Models

We purpose to use CSV to enhance the efficiency and effectiveness of sparse retrieval models that rely on the underlying language model vocabulary as their index vocabulary. To demonstrate this, we leverage uniCOIL [12,17] and SPLADE [9,10] as examples. uniCOIL assigns an impact score to each query and passage token, and discards tokens with a non-positive impact. It relies on an additional model for document expansion.

SPLADE, on the other hand, projects queries and passages into $|V|$-dimensional embeddings, where $|V|$ is the vocabulary size of the underlying LM, and calculates the matching score based on the dot product of these embeddings. To "sparsify" these dense representations for efficiency, SPLADE employs FLOPS regularizers [29] to restrict the number of tokens with non-zero weights. Unlike uniCOIL, SPLADE performs query and document expansion automatically, without the need for prior corpus expansion. Because query expansion significantly influences retrieval efficiency, which is not the focus of this study, we limit our experiments to SPLADE-DOC proposed in SPLADE-v2 [9]. SPLADE-DOC only performs token weighting and expansion on the passage side and assigns uniform weight to query tokens without expansion. Note that changing vocabulary size affects how SPLADE should be regularized, as the FLOPS loss is calculated by *summing* the square of a token's average absolute weight in a mini-batch *across the vocabulary.*

3.4 Document Expansion

We use TILDE [41] for document expansion with uniCOIL, as it requires fewer resources for both training and inference and provides comparable performance to DocT5Query [27]. TILDE is initially trained using labeled relevant query-passage pairs. However, the shallow labeling of MS MARCO [1] and the presence of false negatives [21] make this approach restricted. To enhance TILDE for document expansion, we propose to aggregate the rankings of 12 dense rankers [31] using the Borda Count [2] into a consolidated ranking, and then use the top 10 passages from this ranking as training signals to train TILDE. We choose the Borda Count for ranking aggregation due to its simplicity. Note that TILDE can also leverage CSV as it predicts additional document tokens over the underlying LM vocabulary space which we adjust and fine-tune to the target corpus. We refer to this approach as **Corpus-Specific Document Expansion** leveraging augmentation (TILDE-AUG-CSV) as both the vocabulary used during expansion and the underlying expansion model fit directly to the target corpus. The positive benefits of these enhancements will be explored in detail in Sect. 4.

3.5 Distillation-Based Training

Training a student model using the outputs of a trained teacher ranking model as training signals can considerably enhance learning outcomes [9,15,34]. We use KL Divergence [13,14] as the training loss and a standard cross-encoder [26] as the teacher to train the uniCOIL and SPLADE-DOC models as suggested by Lassance and Clinchant [15].

Overall we seek to apply CSV to a variety of state-of-the-art techniques, showing they are broadly applicable and generalize to different approaches, which we will show in Sect. 4.

4 Experiments

4.1 Setup

Datasets. We use the MS MARCO v1 (referred to as MSM in tables; $8.8M$ passages) and MS MARCO v2 ($138M$ passages) collections. For evaluation, we mainly use the 6,980 queries from the MS MARCO v1 Dev set. We also use the test queries of TREC 2019 [4] and 2020 [3] from the TREC Deep Learning track.

Latency Experiments. We use the PISA engine [25] which substantially outperforms Lucene in terms of space usage and runtime efficiency for retrieval over learned sparse indexes [22]. We use the state-of-the-art BlockMaxWand [7] dynamic-pruning based query algorithm. All our indexes leverage recursive-graph-bisection [6,19,20] to optimize efficiency. Our code and experimental setup is available at https://github.com/PxYu/CSV-for-LSR-ECIR24. We report latency as mean retrieval time (MRT) in ms averaged over 5 runs.

Hardwares. All models are trained on $8 \times A100$ GPUs, whereas our latency experiments are performed on an Intel Xeon 8375C CPU in single-threaded execution mode.

Models and Baselines. While adjusting the underlying LM and pretraining on the target corpus is general, we focus specifically on the impact on sparse retrieval models. Here we experiment with SPLADE (abbreviated as SPL) and uniCOIL (abbreviated as uCOIL). For each, we experiment with vocabulary sizes of 30,000, 100,000, and 300,000. We pre-train our CSV models using the MLM objective (with a 15% masking probability) on MS MARCO (497M words) (and MS MARCO v2 as indicated) for 10 epochs. uniCOIL models can also leverage TILDE-AUG-CSV (abbreviated as TILDE-A) document expansion by first training TILDE-AUG-CSV with query likelihood and document likelihood for 5 epochs and take the top-200 tokens predicted in the document likelihood distribution as expansion, ignoring stop-words, sub-words and tokens in the original passage [41]. We also train a distillation based version of uniCOIL (abbreviated uCOIL-D) as discussed in Sect. 3.

As baselines, we retrain a uniCOIL model with the standard BERT vocabulary (to compare to our 30,000 vocabulary models, to which it has similar vocabulary size). We additionally report numbers for existing BT-SPLADE-L model [15] as competitive efficient and effective baselines. We also compare to standard BM25, DocT5Query and ColBERTv2 [34] baselines.

4.2 Vocabulary Selection and Index Statistics

First, we explore the effect of vocabulary selection on different index and query statistics without document expansion. Table 1 shows query statistics for a uniCOIL index (trained only on MS MARCO; no document expansion) with different vocabularies. We observe that the mean number of query tokens decreases as the vocabulary size increases. The number of queries where any sub-word token (compared to only having full word tokens) is present decreases from 48% for the default BERT vocabulary to 35% for our custom vocabulary of the same size. Larger vocabularies further decrease the number of queries containing sub-word tokens to 11% and 2% respectively. We also observe that passage length, postings per query, and mean retrieval time (MRT) decrease as the vocabulary size increases. Also, note that a custom vocabulary with 30k tokens outperforms the regular BERT-30k vocabulary on all metrics. Overall, CSV-300k is 20% faster compared to a standard BERT-30k based uniCOIL model.

4.3 Retrieval Quality

Table 2 explores pretraining, document expansion, and model distillation. Note that we omit certain configurations and metrics that do not provide additional

Table 1. Mean query length ($|Q|$), percentage of split queries, passage length ($|D|$, in terms of tokens), postings per query, and MRT. Metrics are derived from uniCOIL models without document expansion.

Vocab	$\|Q\|$	%Split Qrys	$\|D\|$	Postings	MRT
BERT-30K	7.02	48.27%	47.6	6,482,729	22.88
CSV-30K	6.68	35.50%	46.2	6,207,331	19.70
CSV-100K	6.29	11.36%	42.5	5,502,811	18.66
CSV-300K	6.17	2.36%	41.3	5,118,462	18.62

Table 2. MRR and MRT for different pre-training, document-expansion and uniCOIL training objectives.

#	Vocab	Pretrain	D. Exp	Model	MRR	MRT
1	BERT	BERT	TILDE	uCOIL	0.354	33.88
2	BERT	MSM	TILDE	uCOIL	0.343	29.29
3	CSV-100K	MSM	-	uCOIL	0.332	18.66
4	CSV-100K	MSM	TILDE	uCOIL	0.353	22.65
5	CSV-100K	MSM+v2	TILDE	uCOIL	0.370	22.45
6	CSV-100K	MSM+v2	TILDE-A	uCOIL	0.376	19.88
7	CSV-100K	MSM+v2	TILDE-A	uCOIL-D	0.391	18.85

insights to simplify presentation. Rows #1 and #2 represent reproduced standard baselines for reference.

First, comparing CSV-100k with no document expansion (row #3) and CSV-100k with standard TILDE expansion (row #4), we observe that latency increases but retrieval quality improves. Both pre-training on MS MARCO v2 (row #5) and augmented document expansion (TILDE-A) improve retrieval quality while remaining latency neutral or improving latency. Finally, we replace regular uniCOIL with a version trained with distillation (uCOIL-D, row #7). uCOIL-D provides the best retrieval quality (0.391 MRR). Note that this is competitive to state-of-the-art late interaction models such as ColBERTv2 [34] (0.397 MRR on the same task). In subsequent experiments, we restrict our analysis and presentation to pre-training on MSM+v2, expansion using TILDE-A and uCOIL-D.

Table 3 shows the effect of increasing vocabulary sizes on uniCOIL based models. No substantial difference between the CSV-100k (row #10) and CSV-300k (rows #11-#13) can be observed, which indicates that 100k is a sufficient vocabulary size for MS MARCO v1. We also observe that latency *increases* for CSV-300k (row #13) compared to CSV-30k (row #9), which is contrary to the numbers reported in Table 1. We find TILDE-A expansion increases document size substantially with larger vocabulary size (90.71 extra tokens on average for CSV-300k, 41.63 for CSV-30k). This increase counteracts the decrease in postings list lengths we obtained through increasing vocabulary size. Adjusting the TILDE-A

Table 3. MRR and MRT for different custom vocabulary sizes and document expansion limits. # Kept Tokens refers to the number of expansion tokens provided by TILDE-A that are actually used for document expansion. It acts as a hyperparameter to control the balance between effectiveness and efficiency under the same vocabulary.

#	Vocab	# Tilde Tokens	# Kept Tokens	MRR	MRT
8	BERT-30K	37.5	37.5	0.379	20.00
9	CSV-30K	41.6	40.0	0.389	18.69
10	CSV-100K	46.6	40.0	0.389	17.24
11	CSV-300K	90.7	40.0	0.388	**17.06**
12	CSV-300K	90.7	50.0	0.391	18.72
13	CSV-300K	90.7	90.7	**0.392**	22.08

hyperparameter to only expanding with the top-40/50 tokens, in rows #10 - #12 we see latency in line with CSV-30k for larger vocabularies while showing a negligible improvement in retrieval performance. In summary, for MS MARCO v1, leveraging a custom vocabulary (rows #9 compared to #8) is more important to improving retrieval quality compared to increasing vocabulary size, which however has a positive impact on latency.

Table 4 shows the effect of corpus-specific vocabulary in different sizes on SPLADE. Similarly, as with uniCOIL (row #8 and #9), the CSV model (row #16) outperforms the model with BERT vocabulary in similar size (row #17) in terms of MRR and MRT. Again, retrieval quality does not increase with larger vocabulary sizes, however, the CSV-300k version (rows #20 and #21) is roughly 40% faster than the comparable SPL-DOC baseline (row #15). This effect is related to the FLOPS sparsity regularization leveraged by SPLADE interacting with vocabulary size. Experimenting with different regularization strengths (λ_d) while trying to keep MRR roughly constant (rows #17 - #21), we find larger vocabularies ($300k$) result in improved retrieval speed ($13ms$ vs $25ms$).

Similarly, Table 5 shows that our improvements also transfer to the TREC query sets. While standard BM25 and DocT5Query are still faster, CSV reduces mean latency relative to regular uniCOIL by 50% ($17.63ms$ vs $33.73ms$) and improves over state-of-the-art BT-SPLADE-L method. We conduct Bonferroni corrected pairwise t-tests, and report significance with $p < 0.05$.

4.4 Query Latency

Previous experiments show that CSV with both 100k and 300k tokens substantially reduces the latency of existing approaches. For example, standard uniCOIL with BERT vocabulary (row 1; Table 2) exhibit a mean response time of 33.88 ms, whereas our fastest method uniCOIL based method reduces mean response time to 17.06 ms (row 11; Table 3), a 50% reduction. Similarly, SPLADE enhanced by CSV exhibits similar latency improvements (see Table 4). For com-

Table 4. MRR and MRT for several SPLADE based methods.

#	Method	Vocab	λ_d	MRR	MRT
14	BT-SPLADE-L [15]			0.380	27.62
15†	SPL-DOC	BERT-30K	0.008	0.347	21.03
16	SPL-DOC	BERT-30K	0.008	0.339	27.28
17	SPL-DOC	CSV-30K	0.008	0.356	25.39
18	SPL-DOC	CSV-100K	0.009	0.358	21.65
19	SPL-DOC	CSV-300K	0.006	**0.359**	18.47
20	SPL-DOC	CSV-300K	0.007	0.357	**13.12**
21	SPL-DOC	CSV-300K	0.008	0.354	14.21

† This is initialized with a DistilBERT model that is further pretrained on MSMARCO using MLM+FLOPS [15]. In comparison, row #16 is initialized with a BERT model that is only pretrained on MSMARCO v2 using MLM.

parison, ColBERTv2 [34] accelerated by PLAID [33] provides similar effectiveness but is substantially slower (185 ms single CPU; not run by us).

We observe that CSV-300K results in more lists (due to having a larger vocabulary) with larger list max scores referring to maximum score a term is assigned in any document in the collection as shown in Fig. 3. This also creates more skewed list max score distribution (the score "band" in Fig. 3 is more narrow for the standard BERT vocabulary) which is essential as pruning algorithms use list max scores to skip over low-scoring documents [7].

This has a direct effect on runtime performance which can be observed in the run-time statistics of the MaxScore algorithm shown in Table 6. Note that methods uniCOIL and BT-SPLADE-L which leverage smaller vocabularies score substantially more documents. This would be especially impactful in the case where a non trivial scoring function (e.g. scoring discovered documents with a more expensive secondary model as in proposed by Mallia et al. [24]) is used to score documents.

Interestingly, operation Insert which counts the number of insertions into the final top-k result heap during processing are similar. While more documents are scored, the amount of documents inserted into the resulting heap stays similar. This is an artifact of list max scores (plotted in Fig. 3) being used to determine if a document should be scored and larger vocabularies provide more fine-grained "decision boundaries" as fewer high and low-scoring terms are conflated into a single vocabulary entry.

4.5 Pre-training Cost and Model Size

Not using existing LM checkpoints requires more time and resources for pre-training. We pretrain each LM on MS MARCO for 10 epochs with MLM. For

Table 5. nDCG@10 and MRT, for TREC 19&20 queries. The symbol $\triangledown$ denotes a sig. difference viz. uCOIL-D-CSV-300K (#13).

Strategy	TREC 2019		TREC 2020	
	nDCG	MRT	nDCG	MRT
BM25	$0.501^{\triangledown}$	4.93	$0.487^{\triangledown}$	7.94
DocT5Query	$0.643^{\triangledown}$	4.87	$0.607^{\triangledown}$	7.80
UniCOIL-TILDE	$0.660^{\triangledown}$	31.59	$0.647^{\triangledown}$	33.73
BT-SPLADE-L	0.703	26.91	0.698	27.60
uCOIL-D-CSV-100K (#10)	0.718	15.00	0.706	18.01
uCOIL-D-CSV-300K (#11)	0.722	**14.46**	0.708	**17.63**
uCOIL-D-CSV-300K (#13)	**0.729**	17.55	**0.728**	22.13

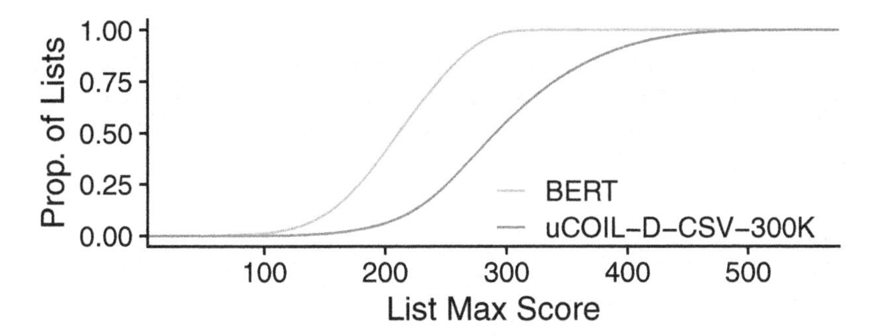

Fig. 3. Cumulative distribution of lists that have list max scores higher than a given value. BERT displaying less skew in list max scores which negatively affects performance.

experiments with uniCOIL, we train TILDE for 5 epochs for document expansion, and train uniCOIL for 5 epochs on the expanded corpus for retrieval. SPL-DOC is trained for 50k iterations using our pre-trained LM. We spend 4–13 hours pretraining LMs of different sizes due to the computational overhead of larger vocabulary sizes. Our pretraining does not currently leverage standard sampling/hierarchical softmax strategies used to deal with a large number of categories, which increases the cost. Increasing vocabulary size also increases model parameter size from 109M to 316M, similar to BERT-large with a 30k vocabulary. We experimented with pre-training BERT-large on our corpus to obtain a baseline with similar parameter size, but found that the MS MARCO corpora were too small to pre-train a model of this size. Note that search-specific pre-training tasks such as coCondenser [11] provide orthogonal benefits to vocabulary changes. We leave exploring potential interactions of these techniques to future work.

Table 6. Query processing statistics (avg per query) for MaxScore and three index varieties.

Strategy	SCORE	INSERT	NEXT	NEXT-GEQ
uniCOIL	7,004,022	301,650	6,575,301	2,048,790
uCOIL-D-CSV-300K	4,836,943	450,567	4,554,723	1,336,857
BT-SPLADE-L	6,400,831	556,829	6,110,824	1,175,790

5 Conclusion and Future Work

We demonstrate that corpus-specific vocabularies are effective at improving both retrieval quality and query latency of learned sparse retrieval systems. They are simple yet effective and can be applied to a variety of different modeling types.

We believe there is a large body of future work exploring the effect of the vocabulary on sparse retrieval models. Promising directions are developing more sophisticated vocabulary selection strategies and training and document expansion strategies that take underlying inverted index-based retrieval into account when assigning term weights.

References

1. Arabzadeh, N., Vtyurina, A., Yan, X., Clarke, C.: Shallow pooling for sparse labels. Inf. Retrieval **25**(4), 365–385 (2022)
2. Aslam, J., Montague, M.: Models for metasearch. In: Proceedings of ACM International Conference on Information and Knowledge Management (CIKM), pp. 276–284 (2001)
3. Craswell, N., Mitra, B., Yilmaz, E., Campos, D.: Overview of the TREC 2020 deep learning track. arXiv:2102.07662 (2021)
4. Craswell, N., Mitra, B., Yilmaz, E., Campos, D., Voorhees, E.M.: Overview of the TREC 2019 deep learning track. arXiv:2003.07820 (2020)
5. Devlin, J., Chang, M.W., Lee, K., Toutanova, K.: BERT: pre-training of deep bidirectional transformers for language understanding, pp. 4171–4186 (2019)
6. Dhulipala, L., Kabiljo, I., Karrer, B., Ottaviano, G., Pupyrev, S., Shalita, A.: Compressing graphs and indexes with recursive graph bisection. In: Proceedings of Conference on Knowledge Discovery and Data Mining (KDD), pp. 1535–1544 (2016)
7. Ding, S., Suel, T.: Faster top-k document retrieval using block-max indexes. In: Proceedings of ACM International Conference on Research and Development in Information Retrieval (SIGIR), pp. 993–1002 (2011)
8. Feng, Z., et al.: Pretraining without wordpieces: learning over a vocabulary of millions of words (2022)
9. Formal, T., Lassance, C., Piwowarski, B., Clinchant, S.: SPLADE v2: sparse lexical and expansion model for information retrieval. arXiv:2109.10086 (2021)
10. Formal, T., Piwowarski, B., Clinchant, S.: SPLADE: sparse lexical and expansion model for first stage ranking. In: Proceedings of ACM International Conference on Research and Development in Information Retrieval (SIGIR), pp. 2288–2292 (2021)

11. Gao, L., Callan, J.: Unsupervised corpus aware language model pre-training for dense passage retrieval, pp. 2843–2853 (2022)
12. Gao, L., Dai, Z., Callan, J.: COIL: revisit exact lexical match in information retrieval with contextualized inverted list, pp. 3030–3042 (2021)
13. Hinton, G., Vinyals, O., Dean, J.: Distilling the knowledge in a neural network. arXiv:1503.02531 (2015)
14. Kullback, S., Leibler, R.: On information and sufficiency. Ann. Math. Stat. **22**(1), 79–86 (1951)
15. Lassance, C., Clinchant, S.: An efficiency study for SPLADE models. In: Proceedings of ACM International Conference on Research and Development in Information Retrieval (SIGIR), pp. 2220–2226 (2022)
16. Lassance, C., Déjean, H., Clinchant, S.: An experimental study on pretraining transformers from scratch for IR. arXiv:2301.10444 (2023)
17. Lin, J., Ma, X.: A few brief notes on DeepImpact, COIL, and a conceptual framework for information retrieval techniques. arXiv:2106.14807 (2021)
18. Mackenzie, J., Mallia, A., Moffat, A., Petri, M.: Accelerating learned sparse indexes via term impact decomposition, pp. 18–27 (2022)
19. Mackenzie, J., Mallia, A., Petri, M., Culpepper, J.S., Suel, T.: Compressing inverted indexes with recursive graph bisection: a reproducibility study. In: Proceedings of European Conference on Information Retrieval (ECIR), pp. 339–352 (2019)
20. Mackenzie, J., Petri, M., Moffat, A.: Faster index reordering with bipartite graph partitioning. In: Proceedings of ACM International Conference on Research and Development in Information Retrieval (SIGIR), pp. 1910–1914 (2021)
21. Mackenzie, J., Petri, M., Moffat, A.: A sensitivity analysis of the MSMARCO passage collection. arXiv:2112.03396 (2021)
22. Mackenzie, J., Trotman, A., Lin, J.: Wacky weights in learned sparse representations and the revenge of score-at-a-time query evaluation. arXiv:2110.11540 (2021)
23. Mallia, A., Khattab, O., Tonellotto, N., Suel, T.: Learning passage impacts for inverted indexes. In: Proceedings of ACM International Conference on Research and Development in Information Retrieval (SIGIR), pp. 1723–1727 (2021)
24. Mallia, A., Mackenzie, J., Suel, T., Tonellotto, N.: Faster learned sparse retrieval with guided traversal. In: Proceedings of ACM International Conference on Research and Development in Information Retrieval (SIGIR), pp. 1901–1905 (2022)
25. Mallia, A., Siedlaczek, M., Mackenzie, J., Suel, T.: PISA: performant indexes and search for academia. In: Proceedings of OSIRRC at SIGIR 2019, pp. 50–56 (2019)
26. Nogueira, R., Cho, K.: Passage re-ranking with bert. arXiv preprint arXiv:1901.04085 (2019)
27. Nogueira, R., Lin, J.: From doc2query to docTTTTTquery (2019)
28. Ottaviano, G., Venturini, R.: Partitioned Elias-Fano indexes. In: Proceedings of ACM International Conference on Research and Development in Information Retrieval (SIGIR), pp. 273–282 (2014)
29. Paria, B., Yeh, C., Yen, I., Xu, N., Ravikumar, P., Póczos, B.: Minimizing flops to learn efficient sparse representations. arXiv:2004.05665 (2020)
30. Pibiri, G.E., Venturini, R.: Techniques for inverted index compression. ACM Comput. Surv. **53**(6), 125.1–125.36 (2021)
31. Reimers, N.: MS MARCO Passages Hard Negatives. In: HuggingFace, pp. 1747–1756 (2021). https://huggingface.co/datasets/sentence-transformers/msmarco-hard-negatives
32. Robertson, S.E., Zaragoza, H.: The probabilistic relevance framework: BM25 and beyond. Found. Trends Inf. Retrieval **3**, 333–389 (2009)

33. Santhanam, K., Khattab, O., Potts, C., Zaharia, M.: PLAID: an efficient engine for late interaction retrieval. In: Proceedings of ACM International Conference on Information and Knowledge Management (CIKM), pp. 1747–1756. ACM (2022)
34. Santhanam, K., Khattab, O., Saad-Falcon, J., Potts, C., Zaharia, M.: Colbertv2: effective and efficient retrieval via lightweight late interaction. In: NAACL, pp. 3715–3734 (2022)
35. Sennrich, R., Haddow, B., Birch, A.: Neural machine translation of rare words with subword units (2016)
36. Siedlaczek, M., Mallia, A., Suel, T.: Using conjunctions for faster disjunctive top-k queries. In: Proceedings of Conference on Web Search and Data Mining (WSDM), pp. 917–927 (2022)
37. Turtle, H.R., Flood, J.: Query evaluation: strategies and optimizations. Inf. Process. Manag. **31**(6), 831–850 (1995)
38. Wu, Y., et al.: Google's neural machine translation system: bridging the gap between human and machine translation. arXiv preprint arXiv:1609.08144 (2016)
39. Yu, W., et al.: Dict-BERT: enhancing language model pre-training with dictionary (2022)
40. Zhao, L.: Modeling and solving term mismatch for full-text retrieval. SIGIR Forum **46**(2), 117–118 (2012)
41. Zhuang, S., Zuccon, G.: TILDE: term independent likelihood moDEl for passage re-ranking. In: Proceedings of ACM International Conference on Research and Development in Information Retrieval (SIGIR), pp. 1483–1492 (2021)
42. Zobel, J., Moffat, A.: Inverted files for text search engines. ACM Comput. Surv. **38**(2), 6:1–6:56 (2006)

Interactive Topic Tagging in Community Question Answering Platforms

Radin Hamidi Rad[1(✉)], Silviu Cucerzan[2], Nirupama Chandrasekaran[2], and Michael Gamon[2]

[1] Toronto Metropolitan University, Toronto, ON, Canada
radin@torontomu.ca
[2] Microsoft Research, Redmond, WA, USA
silviu@microsoft.com, niruc@microsoft.com, mgamon@microsoft.com

Abstract. Community question-answering platforms offer new opportunities for users to share knowledge online. Such platforms allow building communities around areas of interest, and enable community members to post questions and have other members answer them. In this paper, we investigate a novel, interactive approach for tagging input questions with relevant topics, which are needed by community question-answering platforms for various tasks such as indexing and routing. Iteratively, we employ explicit feedback from the users who post questions to fine-tune further the tag suggestions for those questions. We show that our proposed method is able to suggest tags efficiently, and outperforms state-of-the-art methods applied to the tag suggestion task.

Keywords: Community Question Answering · Question Tagging · Topic Tagging

1 Introduction

Community question-answering (CQA) platforms provide users with the opportunity to share their knowledge and expertise online. Through these platforms, members of a community can ask and answer questions related to topics of interest to that community, which allows for the exchange of information and ideas in a collaborative learning environment. As communities grow, artificial intelligence techniques are needed to handle tasks such as routing new questions to community members who are likely to answer them and retrieving previously answered questions similar to a new question.

Because of the importance of CQA platforms in fostering such valuable online communities, in which members can learn from one another, multiple tasks for maintaining and improving CQA platforms have been investigated in AI/IR research, including: question routing [3,13,20], question duplication identification [22], topic evolution modelling [1], and question tagging [13,28].

The research was conducted during an internship at Microsoft Research in the summer of 2022.

N. Goharian et al. (Eds.): ECIR 2024, LNCS 14610, pp. 195–209, 2024.
https://doi.org/10.1007/978-3-031-56063-7_13

Question tagging, as one of the key tasks related to CQA platforms, concerns the assignment of relevant tags to questions. The tags employed by a community can be used to create a comprehensive overview of the topics of interest to that community. These tags can range from broad, general topics to more specific, detailed matters. They can be used by community members to identify discussions on topics they are interested in, as well as the topics they need help with.

Employing tags to categorize questions and answers not only makes it easier for community members to find or contribute information but also enables better automated *retrieval* and *routing*. Indeed, tagging a new question with relevant tags makes it easier to categorize it and then list it in relevant topic boards. This leads to a better organization of the community questions and improves the community experience by allowing members to focus more easily on their areas of interest. Moreover, having questions tagged, in particular when the tags are validated by the community members, can result in better retrieval of previous questions that are similar to a new question. In turn, this can help with the task of identifying duplicate questions and allows keeping a low content redundancy in the community. Similarly, question tagging is also very useful for routing questions in a CQA platform. By assigning distributions over favorite tags to members of the community based on their past interactions with tagged questions, the platform can direct questions to users who are likely to have the necessary knowledge to answer them. This helps to ensure that questions are answered by the most qualified individuals, leading not only to faster question answering but also to more accurate and helpful responses. Additionally, the CQA platform can provide daily feeds of new questions of interest to its members, which can substantially increase the effectiveness and usefulness of the platform.

While there have been several investigations of question tagging in the space of CQA platforms, interactive question tagging remains a novel task. In this setting, the model refines its tag suggestions based on the feedback received from the user.

In most scenarios, questions can be labeled with multiple tags. While each of the tags typically covers a separate aspect, positive correlations among sets of tags are likely to exist [15]. These can be utilized to improve tag suggestion performance through an iterative algorithm that exploits the observed correlations in conjunction with explicit feedback from users. At each step, a user is provided with a set of predicted tags for the new question and has the option to retain or reject any of the suggested tags. The user's feedback is employed for narrowing down the system's suggestion to better related tags in the next step. We refer to such an interaction as *positive feedback* when the user accepts a suggested tag as relevant. On the opposite, we consider an interaction as *negative feedback* when the user discards suggested tags. Note that both forms of feedback hold significant information that can be used to improve the tagging process, and we investigate each of the cases individually.

The key contributions of our work are: (1) we study different feedback scenarios and proposed a framework that can effectively leverage feedback to improve question tagging performance; (2) we propose a novel neural architecture that fuses user feedback into the network structure; (3) we use a transformer-based language model to generate contextualized embedding for questions and tags; (4) we conduct experiments on three real-world communities in order to assess the performance of our approach and compare it to a variety of strong baselines, including the current state-of-the-art.

The rest of the paper is structured as follows: after presenting a survey of related work in Sect. 2, we define the question tagging problem and explain our proposed method in Sect. 3. Experiments and findings are discussed in Sect. 4, followed by concluding remarks and proposed directions for future work in Sect. 5.

2 Related Work

One direct approach to tackling question tagging is to treat it as a text classification task over the set of topics treated as class labels. We categorize the available workarounds into two different groups: (1) advanced neural language models, and (2) interactive and adaptive models.

Advanced Neural Language Models: In recent years, a few deep-learning-based approaches have been shown as extremely effective for text classification. They use the deep neural network to learn the text representation, in conjunction with a fully connected layer, on which the softmax function is applied to predict a skewed distribution over class labels. Based on the strategies employed for text feature extraction, these models employ either Recurrent Neural Networks (RNN) or Convolutional Neural Networks (CNN).

The recursive neural network models (RNN) process the sequence of text as the input and generate representations based on that. Specifically, Liu et al. [9] proposed three RNN-based neural models that are trained on multiple tasks and use parameter sharing in order to train. Considering the structure of a corpus, for instance, a text document is composed of sentences, which are made up of words. More recently, Yang et al. [26] proposed a method namely Hierarchical Attention Network (HAN) that uses the Gate Recurrent Unit (GRU) and two levels of attention mechanisms to generate text representations. It is important to note that none of the mentioned methods consider the relationship between labels. This is especially important in multi-label text classification. In order to address the multi-label text classification challenge, Yang et al. [25] considered the task as a sequence generation problem and modelled the correlation between labels through the LSTM structure. Moreover, Zhao et al. [29] added dense links between recurrent units and modelled inherent hierarchical structures in the text to fortify the ability of the RNN models to capture the long-term dependency in text.

On the contrary, the Convolutional neural network (CNN)-based models focus on the relationship between the current word and its context. CNN models do this by adjusting their filter size. Particularly, Text-CNN [5] uses several convolutional filters to capture multiple feature maps and applies the max-pooling layer to them in order to achieve the sentence-level representation. Char-CNN [27] on the other hand, utilizes CNN to model the words to learn the text embedding vectors. However, there is a drawback with all the mentioned models: these models cannot learn variable N-gram features flexibly. Wang et al. [23] utilized densely connected CNN to obtain multi-scale N-gram features. Thereafter, an attention mechanism is used to select the most informative features for the text classification task. In traditional CNN models, filters remain constant regardless of different inputs. Because of that, they had less flexibility. To overcome this challenge, Choi et al. [2] designed filter-generating networks, that produce CNN filters dynamically with regard to the inputs. Although both CNN and RNN-based methods have achieved promising performance, they cannot tag text with new classes. Thereby, they cannot be applied to the CQA sites, where new topics appear constantly.

Interactive and Adaptive Models: Ma et al. [10] proposed an interactive attention model to consider the relation between the context and the target for classification. Moreover, Yang et al. [24] used an attention mechanism in the hierarchical discipline. The Ma et al. model is able to integrate factual knowledge to generate richer question-and-answer representations, while the Yang et al. model captures correlations between the question and answer.

Zhou et al. [30] in their KGQR platform propose an end-to-end deep reinforcement learning-based framework for interactive recommendation to address the sparsity issue. For the interactive recommender system, they use a knowledge graph which provides rich information for recommendation decision-making [30].

Lately, the multi-dimensional attention mechanism [19] is proposed in order to capture the fine-grained information. In this approach, a weight vector will be generated instead of a single representation to address the relationship between two samples. Self-attention as a variation of this approach is applied to many research studies including text mining and machine vision. In the self-attention approach, the model will encode the position of important information independently. They can also be used to generate sentence-level representations.

3 Proposed Method

In this section, we first define the question tagging problem in the context of community question-answering platforms and then explain the proposed approach in detail. Figure 1 shows the principal components of currently CQA platforms.

3.1 Problem Definition

Let $\mathcal{Q} = \left\{q^1, q^2, ..., q^i\right\}_{i=1}^{N}$ represent the set of N available questions in a CQA community and $\mathcal{T} = \left\{\tau^1, \tau^2, ..., \tau^j\right\}_{j=1}^{M}$ be the set of M unique topic tags that have been used to label those questions. $\mathcal{D} = \left\{(q^i, t^i)\right\}_{i=1}^{N}$ denotes the entire labeled dataset, where q^i represents i_{th} question in the dataset and t^i represents the set of tags assigned to q^i, where $t^i \subset \mathcal{T}$.

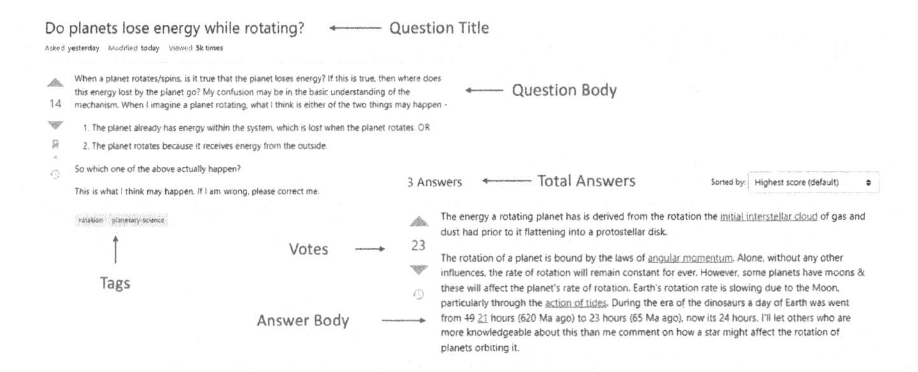

Fig. 1. An example of Stack Exchange user interface. Principle elements are: question title, question body, tags, total number of answers and for each answer, votes and

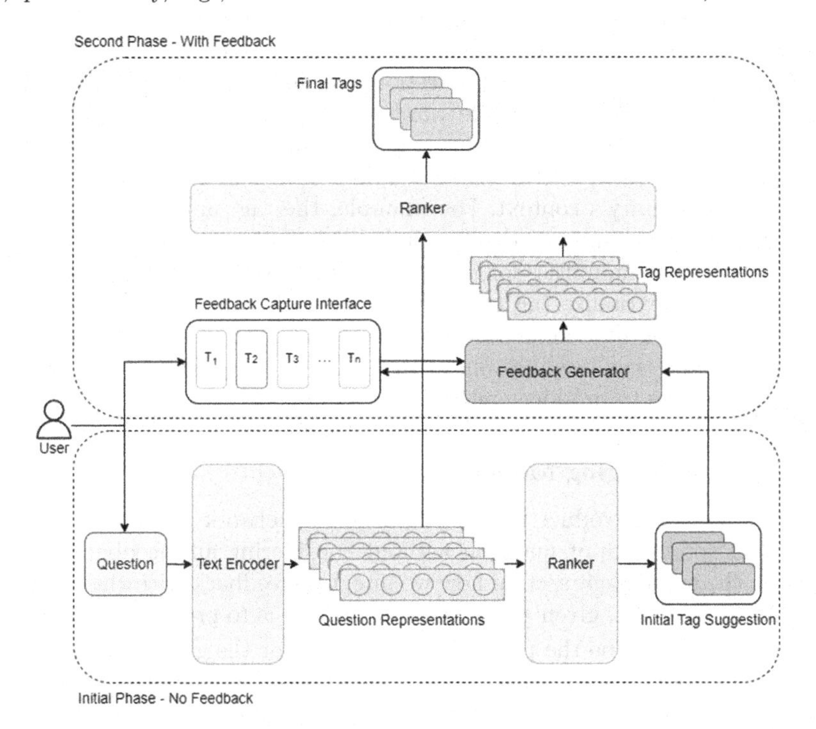

Fig. 2. Overall architecture of the proposed method.

In our interactive setup, users can share their feedback on suggested tags to be used as additional information for additional suggestions. We will explain further how our proposed method uses this additional piece of information to steer the prediction model toward correct tag predictions. Let the feedback with regard to suggested tags $\overline{t}^i$ be t^i_f; thus, the dataset in this interactive setup can be expressed as $\mathcal{D} = \left\{(q^i, t^i, t_f^i)\right\}_{i=1}^{N}$, where for each instance in the dataset, there is an additional tag as feedback.

3.2 Question and Tag Embedding

Figure 2 shows the overall architecture of our proposed method. In this study, we use embedding vectors to represent both the text of questions and the topic tags associated with them. For questions, the embedding vectors are generated in **Text Encoder** unit and for tags, embeddings are generated in **Feedback Generator** unit as shown in Fig. 2. While there exist numerous off-the-shelf embedding models (e.g. BERT [18], fastText [4], Word2Vec [11]), they do not necessarily provide optimum representations for questions and tags in the context of a community. To address this issue, we perform fine-tuning of a BERT model over the dataset corresponding to each community, and then compute the embedding vectors for each question. Fine-tuning the embedding model is even more important for tags than for questions because the vast majority of tags employed in CQA are either single words or very short phrases (two or three words). Thus, a large fraction of tags may be ambiguous outside of the community context and thus, pre-trained embeddings for them may not fit well their usage in a community's context. For example, the tag "apple" is used to refer to the fruit in a cooking community, while it would refer to the tech company in the context of communities for information technology. Therefore, it is highly beneficial to contextualize tag embeddings for each community. To do so, we employ for each tag in a community the questions labelled with that tag as the tag's context, and then, we fine-tune BERT using those questions along with the corresponding tags to produce embedding vectors.

3.3 Question Tagging Model

In this section, we introduce the proposed neural network architecture for predicting tags for the input question without considering any feedback. We will explain in the subsequent section how we integrate feedback with this model. As mentioned in Sect. 3.1, given a question q^i, the task is to predict tags from total known tags $\mathcal{T}$. Let $\overline{t}^i$ be the prediction set of tags for the given question q^i. We define the question tagging program as a low-dimensional mapping problem that transfers the representation vector of the input question into a probability for each individual tag. Let q^i_{emb} be the embedding vector corresponding to question q^i. We use d to annotate the embedding dimension size for queries. Therefore $q^i_{emb} \in \mathbb{R}^{N \times d}$. We can denote the mapping function as:

$$\phi(q^i) : q^i_{emb} \rightarrow \theta^i \tag{1}$$

where θ^i is the occurrence vector for all available tags in regards to the question q^i and $\theta^i \in \mathbb{R}^M$. Thus each element of this vector shows whether the tag is correct or not:

$$\theta^i = \left\{\vartheta_i^1, \vartheta_i^2, ..., \vartheta_i^j\right\}_{j=1}^{M} \tag{2}$$

where ϑ_i^j denote occurrence of jth tag i.e. τ^j for the question q^i. We construct θ^i as follows:

$$\vartheta_i^j = \begin{cases} 1 & \tau^j \in t^i \\ 0 & \tau^j \notin t^i \end{cases} \tag{3}$$

Now that we defined the input (q_{emb}^i) and output (θ^i) of the proposed model, we demonstrate the neural network architecture that serves as the mapping function $\phi(q^i)$.

$$X = q_{emb}^i \tag{4}$$

$$H_0 = \sigma(W^0 X + b^0) \tag{5}$$

$$H_k = \sigma(W^k H^{k-1} + b^k), k \in [1, l] \tag{6}$$

$$\overline{\theta}^i = \text{softmax}(W^l H^{l-1} + b^l) \tag{7}$$

where l is the number of hidden layers and X refers to the input layer which is the initial embedding q_{emb}^i synthesized by the BERT model. For each of the hidden layers $H_k, k \in [1, l]$, $W^k \in \mathbb{R}^{k-1 \times k}$ is the weight matrix and b^k is the bias of layer. σ denotes the sigmoid function as the activation function. We employ the commonly used softmax function for the activation function of the output layer $\overline{\theta}^i$. We fine-tune the embeddings during the training phase using Adam [6] for backpropagation.

The output of the softmax function will determine the confidence of whether each tag is relevant to the question or not. We rank all tags based on their softmax value and suggest top-k tags. The k value is determined based on the evaluation strategy or desired number of suggestions during the inference phase. These processes are marked as **Ranker** units in the proposed method architecture in Fig. 2. During the inference phase, we would first generate questions' embedding vectors using the same BERT model. Then we will pass the embedding to the model in Eqs. (4–7). The result will be predicted occurrence vector $\overline{\theta}^{new}$ for question q^{new}. The vector then will be sorted by their values and the result will be considered as a ranked list of tags.

3.4 Feedback-Aware Question Tagging

In Sect. 3.2, we explained how we computed representation vectors for both questions and tags and in Sect. 3.3, we explained how the proposed model is able to propose a set of tags for a new question using those representation vectors. In general, feedback information becomes available after the first interaction of a user with the model. The user feedbacks are captured in **Feedback Capture Interface** unit as shown in Fig. 2. Here, we introduce user feedback during interactive sessions and two different types of feedback information we can receive from them.

Positive Feedback. During an interaction, a user might find one or more tags from the list of suggestions related and be willing to select them as correct tags. S discussed in the introduction section, it is often seen that there is a correlation between some tags for each community. This means that by knowing a tag, we can make a better informed guess for other correct tags. In this case, selected tags will be considered as positive feedback to the initial suggested list of tags. This can help us improve our guessing ability to propose more related tags to users in the next iteration. Let $\overline{t}^i$ be the predicted tags for question q^i, we show correct predictions from this interaction as follows:

$$\overline{t}^i_{cor} = \overline{t}^i \cap t^i, \quad \overline{t}^i_{cor} \neq \varnothing \tag{8}$$

After the user selects relevant tags, we calculate the average embedding of correct tags as input. We show the average embedding of correct tags by $\overline{t}^i_{cor_{emb}}$. Following our prior discussion on considering user feedback for interactive question tagging, we propose a modification to neural architecture which makes it possible to fuse information as follows:

$$X = \text{concat}(q^i_{emb}, \overline{t}^i_{cor_{emb}}) \tag{9}$$

$$H_0 = \sigma(W^0 X + b^0) \tag{10}$$

$$H_k = \sigma(W^k H^{k-1} + b^k), k \in [1, l] \tag{11}$$

$$\overline{\theta}^i = \text{softmax}(W^l H^{l-1} + b^l) \tag{12}$$

Negative Feedback. Similar to positive feedback, we introduce a negative feedback scenario where users find one or more tags not relevant and ignore them. This scenario also provides the system with considerable insight as feedback can be used to steer the neural model in another direction. As mentioned before, in community question-answering platforms a portion of tags are significantly used more in comparison to the rest. The size of the portion may vary based on the community. We will discuss this phenomenon in detail in the experiments section. Moreover, it is also shown that some tags are used together very often. Therefore, by knowing which tags are not relevant, we can also infer that there is less chance for their frequent co-tags to be correct.

Table 1. Communities Statistics.

Community	# Unique Tags	Train		Validation		Test	
		# Q	T/Q	# Q	T/Q	# Q	T/Q
Chemistry	335	19232	2.36	2072	2.37	3865	2.51
History	737	7582	2.87	997	2.83	1728	2.89
Aviation	910	12599	2.55	1657	2.59	3090	2.68

4 Experiments

In this section, we present our experiments in terms of the dataset, baselines, evaluation strategy, performance and limitations.

4.1 Dataset

StackExchange is one of the most successful community question-answering platforms in recent years. This platform consists of more than 182 QA communities and over 100 million users every month. For better insight, in the past year, StackExchange communities cumulatively had 418.8 million views per month, 3.1 million questions asked and 3.5 million answers submitted with more than 13.5 million comments posted. We have used publicly available records of this platform from Internet Archive[1]. Considering the characteristics of communities (e,g, number of unique tags, number of posts), we have chosen three different communities for experiment purposes, namely chemistry[2], aviation[3] and history[4]. Community statistics for each of the datasets are shown in Table 1.

In selecting communities for our experiments, we considered several characteristics. The Chemistry community on StackExchange, while being one of the larger communities, has a notably smaller set of unique tags, with a low tag per question (T/Q) ratio. This is evident when compared to the History community, our second choice. Despite having fewer samples (questions), History has twice the number of unique tags, exemplifying a community with diverse tags but fewer instances and a higher T/Q ratio. The Aviation community, chosen as our final community, presents a domain with both a high number of instances and unique tags. This large community represents an active popular community to be studied for both the baselines and our proposed model in the task of question tagging.

We note important takeaways from Table 1: (1) the number of unique tags in each community does not necessarily correlate with the size of community questions, as it can be seen, that the history community has less than half of the questions in the chemistry community, however, it has almost 2 times more unique tags. (2) Despite the size of questions and unique tags for each community, the ratio of tags per question remains in a range with a weighted average of 2.54. This means that questions in any of the mentioned communities get assigned on average roughly 2.5 tags.

Moreover, the correlation between tag usage is shown in Table 2. We picked the top 5, 10, 20 and 50 most frequently used tags in each of the communities and calculated the percentage of questions that are tagged using those frequent tags. It can be concluded from Table 2 that not all the tags are used equally in communities and there are certain favourite tags that are dominant.

---

[1] https://archive.org/download/stackexchange.
[2] https://chemistry.stackexchange.com.
[3] https://aviation.stackexchange.com.
[4] https://history.stackexchange.com.

Table 2. Tag popularity in the investigated communities.

Community	Top 5	Top 10	Top 20	Top 50
Chemistry	49.24%	61.81%	75.08%	88.98%
History	37.14%	49.18%	63.4%	82.6%
Aviation	35.39%	48.29%	61.85%	81.3%

Table 3. Percentage of questions associated with at least 2 of the top frequently used tags.

Community	Top 5	Top 10	Top 20	Top 50
Chemistry	9.18%	16.9%	28.87%	50.57%
History	4.11%	12.45%	24.3%	45.79%
Aviation	5.16%	10.78%	22.36%	45.84%

We expanded our observation by testing the percentage of questions with at least two top tags. The results are shown in Table 3. As can be seen, there are a significant number of questions associated with at least 2 of the top tags. This means not only that the majority of community questions get assigned top tags but also that some pairs of top tags are often used together. Thus, knowing one of the top tags can be helpful as it helps in suggesting other tags.

In our study, we simulated user behaviour for interactive question tagging by employing actual relevancy values of tags as feedback. Upon the model's initial prediction, tags that were correctly predicted were labelled as positive feedback, while those incorrectly predicted were labelled as negative feedback.

4.2 Baselines

We compared our method with baselines that use state-of-the-art language models for generating embedding vectors for questions and tags. The list of baselines includes:

Majority. We used the most frequently used tags as predictions in this method. This baseline helps us understand how well question tagging can be done only if we focus on statistics of tag usage and suggest tags regardless of questions.

FastText. uses cosine similarity of embedding vectors generated by the FastText [4] to find the most similar question to the target question and their tag are used as the prediction.

Matching. Similar to FastText embedding mode, we found similar tags to the question based on cosine similarity between question and tag embedding vectors.

PROFIT. is based on the question tagging method proposed by Nie et al. [13]. This uses a deep interactive embedding model to jointly learn the embeddings of questions and topics.

TextCNN. Uses convolutional neural network on top of word vectors that are generated using unsupervised neural language model [5].

TextRNN. This method uses a recurrent neural network to generate word representation that later will be used for tagging [8].

KGQR. This method specifically proposes an end-to-end deep reinforcement learning-based framework for interactive recommendation to address the sparsity issue. For the interactive recommender system, it uses a knowledge graph which provides rich information for recommendation decision-making [30].

4.3 Evaluation Strategy

We framed question tagging as a learning-to-rank task, typically evaluated using the Recall@k metric. However, because of the employment of feedback by our model and the interactive nature of our system, we introduce the Hit@k metric as well. This metric specifically focuses on true positives to measure the system's efficacy. In the Hit@k metric, the model gets a score of 1 if at least one of the top k suggested tags is among the ground truth tags, and 0 otherwise.

Table 4. Hit @k

Dataset	Chemistry					History					Aviation				
Models	Hit@1	Hit@2	Hit@3	Hit@4	Hit@5	Hit@1	Hit@2	Hit@3	Hit@4	Hit@5	Hit@1	Hit@2	Hit@3	Hit@4	Hit@5
Majority	19.74	32.85	44.23	48.66	49.85	19.74	32.85	44.23	48.66	49.85	10.70	16.67	27.02	29.75	33.16
FastText	47.83	61.85	68.85	73.28	76.44	37.10	49.76	57.02	60.80	65.09	36.76	51.10	60.12	64.89	68.52
Matching	10.98	18.13	21.79	24.04	26.48	9.40	15.91	20.32	23.34	25.72	5.32	7.65	10.12	11.64	13.36
TextCNN	41.91	54.86	60.58	62.79	65.98	32.70	49.86	57.09	60.98	62.98	33.12	47.71	55.07	59.17	62.22
TextRNN	37.67	51.23	57.46	61.17	64.67	30.34	46.28	52.97	55.08	57.16	30.08	42.38	48.07	52.84	55.47
PROFIT	49.38	62.49	69.12	73.92	77.81	38.47	52.61	59.75	62.89	63.29	42.48	57.15	63.85	67.26	70.96
Proposed Method	**55.66**	**69.18**	**76.23**	**80.57**	**83.28**	**46.43**	**60.08**	**66.04**	**70.02**	**73.18**	**47.99**	**62.16**	**69.96**	**74.11**	**77.04**

Table 5. Hit @k for models with feedback.

Dataset	Chemistry					History					Aviation				
Models	Hit@1	Hit@2	Hit@3	Hit@4	Hit@5	Hit@1	Hit@2	Hit@3	Hit@4	Hit@5	Hit@1	Hit@2	Hit@3	Hit@4	Hit@5
Best without Feedback	55.66	69.18	76.23	80.57	83.28	46.43	60.08	66.04	70.02	73.18	47.99	62.16	69.96	74.11	77.04
FastText w/ Feedback	53.12	67.37	74.03	77.99	81.03	41.92	55.16	61.84	66.20	69.62	43.97	57.85	63.39	67.04	69.13
PROFIT w/ Feedback	54.10	69.11	75.01	78.38	81.16	47.44	60.51	65.67	68.87	71.71	48.21	72.80	74.53	77.18	78.05
KGQR	54.91	69.73	76.18	79.21	82.69	47.68	61.13	67.04	70.98	73.74	52.87	68.49	73.98	76.08	80.93
Proposed Method w/ Feedback	**56.82**	**70.64**	**77.99**	**82.53**	**85.51**	**47.79**	**62.13**	**69.22**	**73.80**	**77.23**	**56.10**	**68.69**	**74.47**	**77.79**	**85.51**

4.4 Results and Findings

Tables 4, 5 and 6 report the results for our proposed method as well as baselines based on the `Hit` and `Recall` metrics. We have evaluated the performance of our proposed method without considering the feedback information and reported the scores in Table 4. Based on the results, we have several observations:

Table 6. Recall @k for models with feedback.

Dataset	Chemistry					History					Aviation				
Models	Recall@1	Recall@2	Recall@3	Recall@4	Recall@5	Recall@1	Recall@2	Recall@3	Recall@4	Recall@5	Recall@1	Recall@2	Recall@3	Recall@4	Recall@5
Best without Feedback	20.00	31.69	38.07	42.45	45.55	15.66	24.13	28.57	31.70	34.42	20.00	31.69	36.07	40.45	42.55
FastText w/ Feedback	24.24	28.20	33.77	43.06	51.79	16.86	25.75	30.93	34.61	37.26	17.60	27.00	33.53	37.95	41.38
PROFIT w/ Feedback	24.39	29.73	32.08	40.63	47.27	17.42	26.65	32.05	35.79	38.13	19.73	28.03	34.58	38.15	42.79
KGQR	24.97	35.85	42.62	47.25	50.70	18.82	27.13	31.44	35.98	38.02	23.14	32.60	36.94	39.70	41.29
Proposed Method w/ Feedback	**25.36**	**38.57**	**45.53**	**50.18**	**53.62**	**19.36**	**28.81**	**33.89**	**37.37**	**40.48**	**27.52**	**36.08**	**39.94**	**42.42**	**45.67**

1. Among the baselines, the PROFIT method shows superior results in comparison to the other methods, which is likely due to richer information being stored in embedding vectors compared to other baselines. We hypothesize that this is a result of the fact that tag embedding vectors are often decontextualized and further contextualization of tag representations improves their quality in general.
2. While FastText and Matching are based on the same embeddings, it can be seen that the similarity between question embeddings performs better than question-tag embedding.
3. Our proposed method consistently performs better across all tested communities, i.e. it is the only method that has a high performance at a considerable and practical level over all the Hit@k cut-offs and StackExchange communities. Moreover, we have tested the interactive performance by designing an experiment where users share feedback with the system while choosing tags. We have chosen the best baseline from previous experiments i.e. FastText embedding for this part. Results are reported in Tables 5 and 6.
4. KGQR [30] shows a relatively better performance compared to the other baselines. We only studied its performance for feedback-enabled experiments since the model is proposed for interactive recommendation specifically. As expected, additional information in the form of feedback helped our proposed method and other baselines achieve higher performance. It is important to note that the mentioned increase in performance indicates that feedback information is utilized properly, however, we highlight that in higher levels of Hit rates, it is more difficult to improve the performance as the prediction set is close to ground truth and amount of insight from feedback drops in compare to methods that have inferior performance.

5 Concluding Remarks

We proposed an interactive paradigm for topic tagging of questions in community question-answering platforms, which employs user feedback as positive and negative signals in an iterative tag suggestion approach. In the three investigated communities, we found that tag usage is skewed and that there is a strong correlation with regard to the co-occurrence of some tags as question labels. Our proposed tag prediction method takes advantage of these findings and is able to improve the tagging performance by fusing the user feedback information into its prediction model. Large Language Models (LLMs) e.g. GPT 4 [14] have been performing very well for various tasks since this research was conducted. For future work, it would be interesting to study the performance of LLMs for the topic question tagging in CQAs. Additionally, a promising direction for future research is to integrate more sophisticated user feedback mechanisms into the method. This enhancement aims to maximize the benefits of user interactions as prompts for generating more accurate and relevant tags. It is noteworthy to mention that graph-based representation learning approaches [7,12,16,17,21] can also be tested as alternative group of approaches in interactive question tagging task.

References

1. Chen, H., Coogle, J., Damevski, K.: Modeling stack overflow tags and topics as a hierarchy of concepts. J. Syst. Softw. **156**, 283–299 (2019). https://doi.org/10.1016/j.jss.2019.07.033
2. Choi, B., Park, J., Lee, S.: Adaptive convolution for text classification. In: Burstein, J., Doran, C., Solorio, T. (eds.) Proceedings of the 2019 Conference of the North American Chapter of the Association for Computational Linguistics: Human Language Technologies, NAACL-HLT 2019, Minneapolis, MN, USA, 2–7 June 2019, Volume 1 (Long and Short Papers), pp. 2475–2485. Association for Computational Linguistics (2019). https://doi.org/10.18653/v1/n19-1256
3. Ji, Z., Wang, B.: Learning to rank for question routing in community question answering. In: He, Q., Iyengar, A., Nejdl, W., Pei, J., Rastogi, R. (eds.) 22nd ACM International Conference on Information and Knowledge Management, CIKM 2013, San Francisco, CA, USA, 27 October–1 November 2013, pp. 2363–2368. ACM (2013). https://doi.org/10.1145/2505515.2505670
4. Joulin, A., Grave, E., Bojanowski, P., Douze, M., Jégou, H., Mikolov, T.: Fasttext.zip: compressing text classification models. CoRR abs/1612.03651 (2016). http://arxiv.org/abs/1612.03651
5. Kim, Y.: Convolutional neural networks for sentence classification. In: Moschitti, A., Pang, B., Daelemans, W. (eds.) Proceedings of the 2014 Conference on Empirical Methods in Natural Language Processing, EMNLP 2014, 25–29 October 2014, Doha, Qatar, A meeting of SIGDAT, a Special Interest Group of the ACL, pp. 1746–1751. ACL (2014). https://doi.org/10.3115/v1/d14-1181
6. Kingma, D.P., Ba, J.: Adam: a method for stochastic optimization. arXiv preprint arXiv:1412.6980 (2014)
7. Kipf, T.N., Welling, M.: Semi-supervised classification with graph convolutional networks. In: 5th International Conference on Learning Representations, ICLR 2017, Toulon, France, 24–26 April 2017, Conference Track Proceedings. OpenReview.net (2017). https://openreview.net/forum?id=SJU4ayYgl
8. Lai, S., Xu, L., Liu, K., Zhao, J.: Recurrent convolutional neural networks for text classification. In: Bonet, B., Koenig, S. (eds.) Proceedings of the Twenty-Ninth AAAI Conference on Artificial Intelligence, 25–30 January 2015, Austin, Texas, USA, pp. 2267–2273. AAAI Press (2015). http://www.aaai.org/ocs/index.php/AAAI/AAAI15/paper/view/9745
9. Liu, P., Qiu, X., Huang, X.: Recurrent neural network for text classification with multi-task learning. In: Kambhampati, S. (ed.) Proceedings of the Twenty-Fifth International Joint Conference on Artificial Intelligence, IJCAI 2016, New York, NY, USA, 9–15 July 2016, pp. 2873–2879. IJCAI/AAAI Press (2016). http://www.ijcai.org/Abstract/16/408

10. Ma, D., Li, S., Zhang, X., Wang, H.: Interactive attention networks for aspect-level sentiment classification. In: Sierra, C. (ed.) Proceedings of the Twenty-Sixth International Joint Conference on Artificial Intelligence, IJCAI 2017, Melbourne, Australia, 19–25 August 2017, pp. 4068–4074. ijcai.org (2017). https://doi.org/10.24963/ijcai.2017/568
11. Mikolov, T., Chen, K., Corrado, G., Dean, J.: Efficient estimation of word representations in vector space. In: Bengio, Y., LeCun, Y. (eds.) 1st International Conference on Learning Representations, ICLR 2013, Scottsdale, Arizona, USA, 2–4 May 2013, Workshop Track Proceedings (2013). http://arxiv.org/abs/1301.3781
12. Nguyen, H., Rad, R.H., Zarrinkalam, F., Bagheri, E.: Dyhnet: learning dynamic heterogeneous network representations. Inf. Sci. **646**, 119371 (2023). https://doi.org/10.1016/J.INS.2023.119371
13. Nie, L., Li, Y., Feng, F., Song, X., Wang, M., Wang, Y.: Large-scale question tagging via joint question-topic embedding learning. ACM Trans. Inf. Syst. **38**(2), 20:1–20:23 (2020). https://doi.org/10.1145/3380954
14. OpenAI: GPT-4 (2023). https://www.openai.com/gpt-4. [Software]
15. Pal, K.K., Gamon, M., Chandrasekaran, N., Cucerzan, S.: Modeling tag prediction based on question tagging behavior analysis of communityqa platform users (2023)
16. Rad, R.H., Fani, H., Bagheri, E., Kargar, M., Srivastava, D., Szlichta, J.: A variational neural architecture for skill-based team formation. ACM Trans. Inf. Syst. **42**(1), 7:1–7:28 (2024). https://doi.org/10.1145/3589762
17. Rad, R.H., et al.: Learning heterogeneous subgraph representations for team discovery. Inf. Retr. J. **26**(1), 8 (2023). https://doi.org/10.1007/S10791-023-09421-6
18. Reimers, N., Gurevych, I.: Sentence-bert: sentence embeddings using siamese bert-networks. In: Inui, K., Jiang, J., Ng, V., Wan, X. (eds.) Proceedings of the 2019 Conference on Empirical Methods in Natural Language Processing and the 9th International Joint Conference on Natural Language Processing, EMNLP-IJCNLP 2019, Hong Kong, China, 3–7 November 2019, pp. 3980–3990. Association for Computational Linguistics (2019). https://doi.org/10.18653/v1/D19-1410
19. Shen, T., Zhou, T., Long, G., Jiang, J., Pan, S., Zhang, C.: Disan: directional self-attention network for RNN/CNN-free language understanding. In: McIlraith, S.A., Weinberger, K.Q. (eds.) Proceedings of the Thirty-Second AAAI Conference on Artificial Intelligence, (AAAI-18), the 30th Innovative Applications of Artificial Intelligence (IAAI-18), and the 8th AAAI Symposium on Educational Advances in Artificial Intelligence (EAAI-18), New Orleans, Louisiana, USA, 2–7 February 2018, pp. 5446–5455. AAAI Press (2018). https://www.aaai.org/ocs/index.php/AAAI/AAAI18/paper/view/16126
20. Trienes, J., Balog, K.: Identifying unclear questions in community question answering websites. In: Azzopardi, L., Stein, B., Fuhr, N., Mayr, P., Hauff, C., Hiemstra, D. (eds.) ECIR 2019, Part I. LNCS, vol. 11437, pp. 276–289. Springer, Cham (2019). https://doi.org/10.1007/978-3-030-15712-8_18
21. Velickovic, P., Cucurull, G., Casanova, A., Romero, A., Liò, P., Bengio, Y.: Graph attention networks. In: 6th International Conference on Learning Representations, ICLR 2018, Vancouver, BC, Canada, 30 April–3 May 2018, Conference Track Proceedings. OpenReview.net (2018). https://openreview.net/forum?id=rJXMpikCZ
22. Wang, L., Zhang, L., Jiang, J.: Duplicate question detection with deep learning in stack overflow. IEEE Access **8**, 25964–25975 (2020). https://doi.org/10.1109/ACCESS.2020.2968391

23. Wang, S., Huang, M., Deng, Z.: Densely connected CNN with multi-scale feature attention for text classification. In: Lang, J. (ed.) Proceedings of the Twenty-Seventh International Joint Conference on Artificial Intelligence, IJCAI 2018, Stockholm, Sweden, 13–19 July 2018, pp. 4468–4474. ijcai.org (2018). https://doi.org/10.24963/ijcai.2018/621
24. Yang, M., Chen, L., Chen, X., Wu, Q., Zhou, W., Shen, Y.: Knowledge-enhanced hierarchical attention for community question answering with multi-task and adaptive learning. In: Kraus, S. (ed.) Proceedings of the Twenty-Eighth International Joint Conference on Artificial Intelligence, IJCAI 2019, Macao, China, 10–16 August 2019, pp. 5349–5355. ijcai.org (2019). https://doi.org/10.24963/ijcai.2019/743
25. Yang, P., Sun, X., Li, W., Ma, S., Wu, W., Wang, H.: SGM: sequence generation model for multi-label classification. In: Bender, E.M., Derczynski, L., Isabelle, P. (eds.) Proceedings of the 27th International Conference on Computational Linguistics, COLING 2018, Santa Fe, New Mexico, USA, 20–26 August 2018, pp. 3915–3926. Association for Computational Linguistics (2018). https://aclanthology.org/C18-1330/
26. Yang, Z., Yang, D., Dyer, C., He, X., Smola, A.J., Hovy, E.H.: Hierarchical attention networks for document classification. In: Knight, K., Nenkova, A., Rambow, O. (eds.) NAACL HLT 2016, The 2016 Conference of the North American Chapter of the Association for Computational Linguistics: Human Language Technologies, San Diego California, USA, 12–17 June 2016, pp. 1480–1489. The Association for Computational Linguistics (2016). https://doi.org/10.18653/v1/n16-1174
27. Zhang, X., Zhao, J.J., LeCun, Y.: Character-level convolutional networks for text classification. In: Cortes, C., Lawrence, N.D., Lee, D.D., Sugiyama, M., Garnett, R. (eds.) Advances in Neural Information Processing Systems 28: Annual Conference on Neural Information Processing Systems 2015, Montreal, Quebec, Canada, 7–12 December 2015, pp. 649–657 (2015). https://proceedings.neurips.cc/paper/2015/hash/250cf8b51c773f3f8dc8b4be867a9a02-Abstract.html
28. Zhang, X., Liu, M., Yin, J., Ren, Z., Nie, L.: Question tagging via graph-guided ranking. ACM Trans. Inf. Syst. **40**(1), 12:1–12:23 (2022). https://doi.org/10.1145/3468270
29. Zhao, Y., Shen, Y., Yao, J.: Recurrent neural network for text classification with hierarchical multiscale dense connections. In: Kraus, S. (ed.) Proceedings of the Twenty-Eighth International Joint Conference on Artificial Intelligence, IJCAI 2019, Macao, China, 10–16 August 2019, pp. 5450–5456. ijcai.org (2019). https://doi.org/10.24963/ijcai.2019/757
30. Zhou, S., et al.: Interactive recommender system via knowledge graph-enhanced reinforcement learning. In: Huang, J.X., et al. (eds.) Proceedings of the 43rd International ACM SIGIR Conference on Research and Development in Information Retrieval, SIGIR 2020, Virtual Event, China, 25–30 July 2020, pp. 179–188. ACM (2020). https://doi.org/10.1145/3397271.3401174

Yes, This Is What I Was Looking For! Towards Multi-modal Medical Consultation Concern Summary Generation

Abhisek Tiwari[1(✉)], Shreyangshu Bera[1], Sriparna Saha[1], Pushpak Bhattacharyya[2], and Samrat Ghosh[1]

[1] Department of Computer Science and Engineering, Indian Institute of Technology Patna, Dayalpur Daulatpur, India
{abhisek_1921cs16,sriparna}@iitp.ac.in

[2] Department of Computer Science and Engineering, Indian Institute of Technology Bombay, Mumbai, India
pb@cse.iitb.ac.in

Abstract. Over the past few years, the use of the Internet for healthcare-related tasks has grown by leaps and bounds, posing a challenge in effectively managing and processing information to ensure its efficient utilization. During moments of emotional turmoil and psychological challenges, we frequently turn to the internet as our initial source of support, choosing this over discussing our feelings with others due to the associated social stigma. In this paper, we propose a new task of multi-modal medical concern summary (*MMCS*) generation, which provides a short and precise summary of patients' major concerns brought up during the consultation. Nonverbal cues, such as patients' gestures and facial expressions, aid in accurately identifying patients' concerns. Doctors also consider patients' personal information, such as age and gender, in order to describe the medical condition appropriately. Motivated by the potential efficacy of patients' personal context and visual gestures, we propose a transformer-based multi-task, multi-modal intent-recognition, and medical concern summary generation (*IR-MMCSG*) system. Furthermore, we propose a multitasking framework for intent recognition and medical concern summary generation for doctor-patient consultations. We construct the first multi-modal medical concern summary generation (*MM-MediConSummation*) corpus, which includes patient-doctor consultations annotated with medical concern summaries, intents, patient personal information, doctor's recommendations, and keywords. Our experiments and analysis demonstrate (a) the significant role of patients' expressions/gestures and their personal information in intent identification and medical concern summary generation, and (b) the strong correlation between intent recognition and patients' medical concern summary generation (The dataset and source code are available at https://github.com/NLP-RL/MMCSG).

Keywords: Clinical Conversation · Concern Summary · Multi-modality · Modality Fusion · Multi-tasking · Summary Generation

N. Goharian et al. (Eds.): ECIR 2024, LNCS 14610, pp. 210–225, 2024.
https://doi.org/10.1007/978-3-031-56063-7_14

1 Introduction

In the past few years, tele-health has grown immensely with the advancement of information & communication technologies (ICTs) and artificial intelligence-based applications for healthcare activities [18]. With the COVID-19 pandemic, internet utilization for healthcare activities has reached its peak and has become a new normal [31]. The outbreak has caused a striking 25% increase[1] in anxiety and depression, which are severely straining the mental healthcare systems. Tele-health usage is being actively encouraged by healthcare providers, and patients are adopting it at the same pace. Consequently, a massive amount of medical data became available for the first time over the internet [3]. Thus, arranging this data efficiently is essential for proper referencing and facilitating their potential for reuse.

Fig. 1. Utility of multi-modal medical concern summary generation; generated medical concern summary for a video helps in selecting a proper video relevant to user (right side)

In moments of mental distress, individuals often turn to the internet to seek similar cases and recovery insights. However, they have to go through several lengthy videos before getting a relevant case. While many videos do provide summaries, these summaries combine information from both the patient and the doctor, rendering them ineffective as valuable references for finding pertinent cases that align closely with patients' main concerns (Fig. 1). Motivated by this, we propose a new task of generating Multi-modal Medical Concern Summary (*MMCS*). *MMCS* is a synopsis of a patient's key concerns discussed in a patient-doctor interaction. Its potential benefits extend to both patients and clinicians, serving various purposes, including (a) aiding in the organization of consultations and enhancing search ranking for reference by other patients, (b) facilitating follow-up recommendations, and (c) contributing to resource allocation and planning.

When we consult with doctors, they also consider our facial expressions and gestures to analyze our concerns and plan treatment accordingly. Visual expression and

[1] https://www.who.int/news/item/02-03-2022-covid-19-pandemic-triggers-25-increase-in-prevalence-of-anxiety-and-depression-worldwide.

audio tone are also affected by user personality, so one behavior may be triggering for one personality while being normal for another. If the patient's key concern is known, it is easy to identify the patient's intent and vice-versa. Thus, we hypothesize there is a significant correlation between intent and medical concern summary. Moreover, we anticipate that visual and audio features are strongly associated with demographic information such as age and gender (context), and they can significantly influence the understanding of users' behavior and concerns with such context-attended features. Hence, we propose contextualized M-modality fusion, a new modality fusion technique that incorporates an adapter-based module into traditional transformer architecture to effectively infuse different modalities and end-user demographic information.
Research Questions. The paper aims to investigate the following research questions: (a) Does the visual appearance and expression of a patient aid in determining his/her key medical concerns? (b) Is there a correlation between medical concern summary generation and user intent identification? (c) Can patients' personal characteristics, such as age and gender information, contribute to adequately understanding their medical issues and providing appropriate medical advice?
Key Contributions The key contributions of the work are four-fold, which are as follows:

- We propose a new task of multi-modal medical concern summary (*MMCS*) generation, which generates a precise summary of key medical concerns discussed during doctor-patient consultations, resulting in better content searchability and organization.
- We first curated a Medical Concern Summary annotated multi-modal medical dataset named (*MM-MediConSummation*), which consists of patient-doctor counseling sessions annotated with a precise medical concern summary (MCS), intent, patient's personal attributes, doctor's key points, and keywords.
- We present a multitask, multi-modal intent recognition, and multi-modal medical concern summary generation (*IR-MMCSG*) model incorporated with an adapter-based contextualized M-modality fusion mechanism that evaluates audio tone and visual expression in conjunction with user demographic context.
- The proposed contextualized M-modality fusion incorporated *IR-MMCSG* outperforms existing state-of-the-art multi-modal text generation model across all evaluation metrics, including human evaluation.

2 Related Works

The work is mainly relevant to the following two research areas: Medical dialogue understanding and Medical dialogue summarization. We have summarized the relevant works in the following paragraphs.

Medical Dialogue Understanding: Diagnosis and treatment of diseases begin with patient-doctor interaction. Therefore, understanding patients' concerns from their utterances is critical to diagnosis and treatment outcomes [7]. The existing works on medical dialogue understanding can broadly be grouped into two categories: (a) pre-trained transformer-based joint intent and entity detection model [30] and (b) multi-label entity classification [22]. The existing works on medical entity extraction [4,6,26] are limited to medical attributes like symptoms and medicine, which are functions of an utterance rather than a conversation. The work [13] proposed a pipelined machine learning system that identifies intent, symptoms, and disease from a patient-doctor conversation. In [25], the authors demonstrated that considering patients' education level, health literacy, and emotional state greatly improves the likelihood of recommending a relevant medical document.

Medical Dialogue Summarization: Joshi et al., [14] proposed a summarization model based on a pointer network generator. The model takes dialogues as input and generates a summary for each turn (doctor-patient) of the interaction. The work [24] proposed a hierarchical encoder-tagger for summarizing medical patient-doctor conversations by identifying important utterances. Multi-modal summarization aims to generate coherent and important information from data having multiple modalities [32]. In the last few years, the main focus of multi-modal summarization has been to find co-relation among different modalities: text, audio, and image for video data [1]. In [27], the authors investigated the impact of external medical knowledge infusion in text-based dialogue summarization and showed the efficacy of modeling the knowledge in generating medical terms preserving dialogue summaries. An important segment of a video is a subjective concern and may also vary among consumers. In [12], the authors have proposed a new task of user constraint-based summarization and proposed an attention mechanism to summarize the query-relevant content. Shang et al., [20] proposed a time-aware multi-modal transformer (TAMT) that leverages time stamps across image, text, and audio to generate an adequate and coherent video summary.

3 Dataset

We first extensively scrutinized the existing benchmark video summarization datasets, and the summary is presented in Table 1. The most relevant dataset for our proposed task is HOPE [17], which contains patient-doctor counseling sessions. The therapy sessions have been collected from open-source platforms such as YouTube, which are credible and authentic, recorded by psychiatrists and clinicians. The consultations cover therapy sessions for various mental distress like anxiety, depression, and post-traumatic stress disorder (PTSD). It contains only doctor-patient conversation transcripts and dialogue acts corresponding to each utterance. With the guidance of two psychiatrists and a doctor, we incorporate the following attributes into the existing HOPE dataset: medical concern summary, primary intent, secondary intent, doctor suggestion, focal point, and patient's personal context (gender and age group).

Table 1. Characteristics of some of the most relevant existing medical dialogue datasets and comparison with the curated dataset

Dataset	Description	Size	Video	Transcript	Intent	MMCS	Focal point	Keywords	Patient Personal Context
DAIC-WOZ [10]	Patient- Psychiatrist conversation	189	✓	✓	✓	×	×	×	×
Dr. Summarize [14]	Patient-Doctor conversations	1690	×	✓	×	×	✓	✓	×
GPT3-ENS SS [5]	Patient-Doctor conversations	210	×	✓	×	✓	×	✓	×
CoDEC [23]	Patient- Psychiatrist conversation	30	✓	✓	×	×	×	×	×
HOPE [17]	Patient- Psychiatrist conversation	212	✓	✓	×	×	×	×	×
MM-MediConSummation	Patient- Psychiatrist conversation	467	✓	✓	✓	✓	✓	✓	✓

3.1 Data Collection and Annotation

We, along with the three medical experts, first analyzed a few therapy sessions of the HOPE dataset. The clinicians viewed a small subset of the dataset, curated a sample dataset of 25 therapy sessions, and annotated it with a multi-modal medical concern summary, patients' intent, and other crucial information, namely the summary of the doctor's suggestion, focal point, and keywords. Due to the dataset's limited number of samples, we expanded our search to include additional similar samples from open sources and platforms such as YouTube, focusing on credible channels. We employed two biology graduates and one medical student to scale up the dataset and annotation. We provided the sample dataset with a set of guidelines to annotate these tags. They first watched full videos and annotated the details based on a comprehensive understanding of all three modalities involved in a video, i.e., text, audio, and visual, for identification. The process of annotating a video involves crafting a medical concern summary (MCS), identifying the intent, and delineating a specific segment within the video that effectively conveys essential details regarding the patient's concerns and the doctor's recommendations. They manually generated transcripts for specific therapy sessions in cases where video captions were either unavailable or subject to restrictions. In order to ensure annotation agreement among the annotators, we calculated Fleiss' kappa [9]. It was found to be 0.71, indicating a significant uniform annotation. The *MM-MediConSummation* statistics and its word cloud are provided in Table 2 and Fig. 2, respectively.

Table 2. *MM-MediConSummation* dataset statistics

Entries	Value
# of doctor-patient sessions	467
avg duration of video	10 min 58 s
# of utterances	74473
avg. conversation length (utterance)	159
# of unique words	13639
# of intents	7
avg length of MMCS (# of words)	21
# of age-groups	6
tags	MMCS, keywords, doctor suggestion, focus points, patient gender, patient age, primary intent, secondary intent, transcript

Fig. 2. Word cloud of the *MM-MediConSummation* corpus

3.2 Qualitative Aspects

To generate an adequate summary of patient concerns, we analyze different qualitative characteristics of therapy sessions and incorporate them accordingly. The different characteristics are analyzed and illustrated below.

Role of Medical Concern Summary. Medical concern summary aims to aid online healthcare users in recognizing whether the content (particularly therapy session) contains the information they are looking for or not. For example, Fig. 1 illustrates that the patient easily finds out a relevant video when the therapy sessions are tagged with *MMCS*, intent, and keywords. Additionally, we have labeled each session with a focal point, indicating the specific segment where the user articulates the primary purpose of the discussion.

Role of Intent. The user's intent in a video pertains to the purpose behind the patient's interaction with the clinician. It is desirable to comprehend users' intent to serve them effectively. For example, a user's goal might be to get a suggestion for a medical condition. We can use it to effectively locate the relevant span in the concerned video. Sometimes, we observed patients having multiple intentions for consultation; thus, we tagged primary and secondary intentions (Fig. 3 and Fig. 4) to each session. We have used only primary intent for multi-tasking intent identification and MMCS generation.

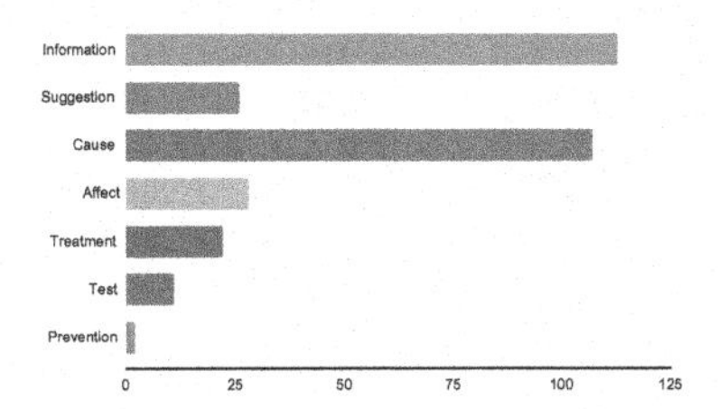

Fig. 3. Primary intent distribution

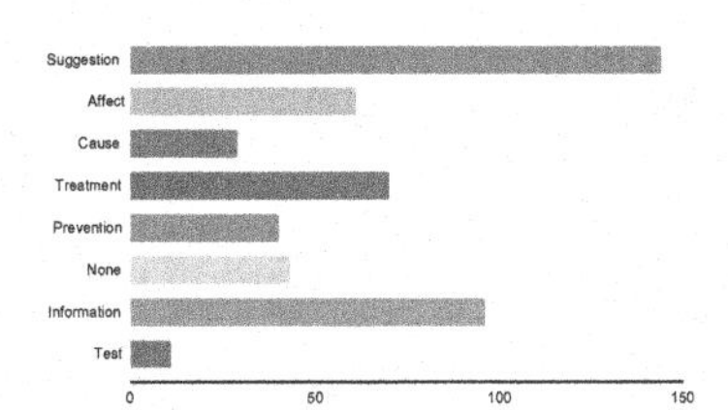

Fig. 4. Secondary intent distribution

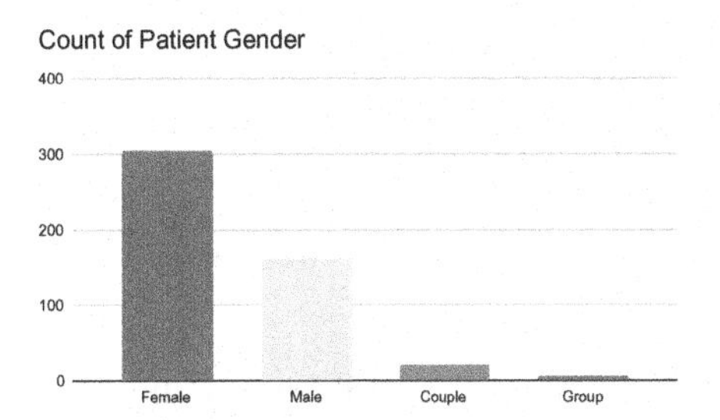

Fig. 5. Gender distribution

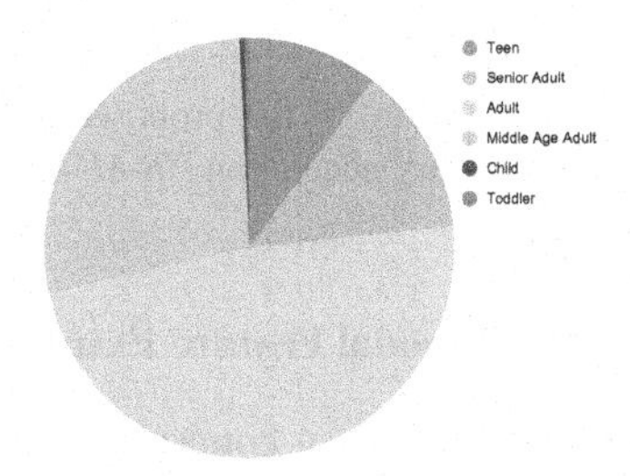

Fig. 6. Age group distribution

Role of Patient's Personal Information. In real life, doctors often consider patients' personal context for determining a medical condition and plan treatment accordingly. Thus, we also annotated patients' personal information for each counseling session. Figures 5 and 6 demonstrate the distribution of gender and age among the patients in the dataset.

4 Methodology

We anticipate that *MMCSG* is affected by (a) patient's visual expression, (b) patient's intent of communication, and (c) patient's personal information. Thus, we propose a multi-tasking, multi-modal intent recognition, and multi-modal medical concern summary generation (*IR-MMCSG*) framework. The proposed architecture is illustrated in Fig. 7. There are three key stages: Multi-modal feature extraction, Contextualized M-modality fusion, and Medical concern summary generation. The explanation and illustration of each stage and the flow are described below.

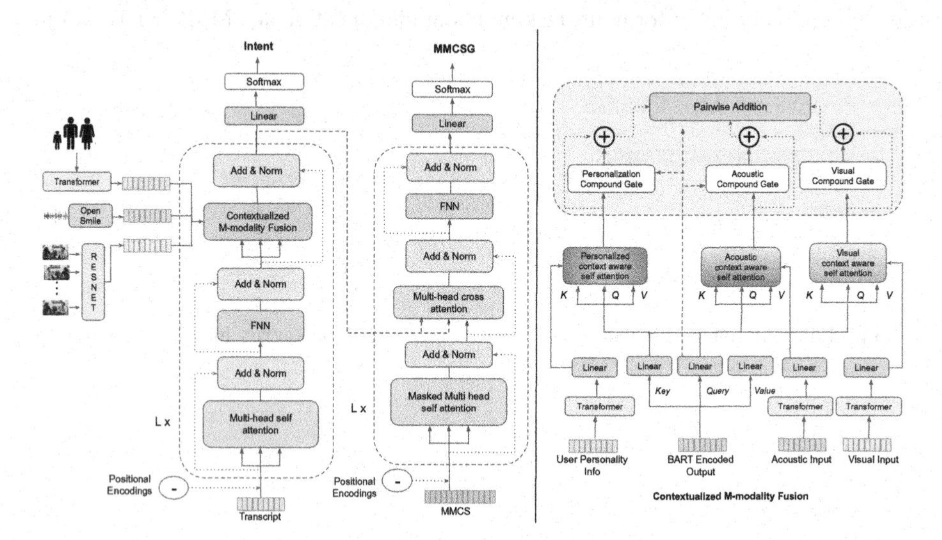

Fig. 7. Architecture of the proposed Multitasking multi-modal intent recognition and medical concern summary generation (*IR-MMCSG*) framework.

4.1 Multi-modal Feature Extraction

In order to encode a counseling session, we have considered all three modalities of video, i.e., audio, text (transcript), and image (video frame). We extract the features of different modalities as follows:

Textual Features: The textual features of a therapy session represent the text embedding of its transcript. We use existing T5 [19] and BART [16] tokenizers for transcript embedding.

Audio Features: We extracted the audio feature from one of the most popular audio processing platforms named openSMILE [8]. The feature representation considers maxima dispersion quotients, glottal source parameters, low-level descriptors (LLD), voice quality, MFCC, pitch, and their statistics.

Video Features: We extracted frames from each counseling session at ten frames per second (fps). The frames extracted from the video are analyzed using Katna's approach [28], aiming to identify frames with distinctive features. This process yielded a set of ten highly pertinent frames. Subsequently, these frames were passed to the pre-trained ResNet 152 model [11] to acquire embeddings of these frames. Finally, we computed the average of these embeddings to form a representation vector of the video.

4.2 Contextualized M-Modality Fusion

We propose a novel multi-modal adapter-based infusion mechanism called contextualized M-modality fusion. It generates context- and modality-conditioned key and value vectors and produces a scaled dot product attention vector. The contextualized modality attention vector is utilized to calculate the global information attended over audio, visual, and personal context, which is utilized for intent identification and medical concern summary generation. It takes the hidden state (H) and calculates the contextualized modality attention as follows:

$$[QKV] = H[W_Q, W_K, W_V] \tag{1}$$

where $Q, K, V \in \mathbb{R}^{lxd}$ are query, key, and value, respectively. Here, l and d denote the sequence length and the dimension of the hidden state (H), respectively. The term W_Q, W_K, and W_v are the learnable parameters corresponding to the query, key, and value vectors, with the dimension of $\mathbb{R}^{dxd}$.

To understand the medical consultation effectively, we generate different modalities and patients' personal context-conditioned key ($\hat{K}$) and value ($\hat{V}$) vectors. The attention vectors transpose the query vector (video transcript) to generate a contextualized, multimodal, coherent information vector. The key and value pairs are calculated as follows:

$$\begin{bmatrix}\hat{K}\\ \hat{V}\end{bmatrix} = (1 - \begin{bmatrix}\lambda_k\\ \lambda_v\end{bmatrix})\begin{bmatrix}K\\ V\end{bmatrix} + \begin{bmatrix}\lambda_k\\ \lambda_v\end{bmatrix}(M\begin{bmatrix}U_k\\ U_v\end{bmatrix}) \tag{2}$$

where $\lambda \in \mathbb{R}^{lx1}$ is the learnable parameter that determines how much information from the textual modality should be retained and how much other modality information should be integrated. Here, M denotes modality, which could be audio, video, and personal information. U_k and U_v are the learnable parameters. The modality controlling parameters (λ) are calculated using the gating mechanism as follows:

$$\begin{bmatrix}\lambda_k\\ \lambda_v\end{bmatrix} = \sigma(\begin{bmatrix}K\\ V\end{bmatrix}\begin{bmatrix}W_{k_1}\\ W_{v_1}\end{bmatrix} + M\begin{bmatrix}U_k\\ U_v\end{bmatrix}\begin{bmatrix}W_{k_2}\\ W_{v_2}\end{bmatrix}) \tag{3}$$

where $W_{k_1}, W_{k_2}, W_{v_1}$ and W_{v_2} ($\in \mathbb{R}^{dx1}$) are trainable weight matrices. Finally, the modality aware attentions (H_a, H_v and H_p) and the final attended vector ($\hat{H}$) are calculated as follows:

$$\begin{aligned} H_a &= Softmax(\frac{Q\hat{K}_a^T}{\sqrt{d_k}})\hat{V}_a \\ H_v &= Softmax(\frac{Q\hat{K}_v^T}{\sqrt{d_k}})\hat{V}_v \\ H_p &= Softmax(\frac{Q\hat{K}_p^T}{\sqrt{d_k}})\hat{V}_p \end{aligned} \tag{4}$$

Fusion. In order to infuse and control the amount of information transmitted from the different modalities (user personality information, audio, and visual), we build three compound gates: personalization (g_p), acoustic (g_a), and visual (g_v). The context information is transmitted via the gates as follows:

$$\begin{aligned} g_a &= [H \oplus H_a]W_a + b_a \\ g_v &= [H \oplus H_v]W_v + b_v \\ g_p &= [H \oplus H_p]W_p + b_p \end{aligned} \tag{5}$$

where $\oplus$ denotes a concatenation operation. W_a, W_v, W_p ($\in \mathbb{R}^{2dXd}$) and b_a, b_v, b_p ($\in R^{dX1}$) are parameters. The final contextualized attended vector ($\hat{H}$) is computed as follows:

$$\hat{H} = H + g_p \odot H_p + g_a \odot H_a + g_v \odot H_v \tag{6}$$

4.3 MMCS Generation

MMCS is our primary task, which is being comprehended with the other task, intent recognition. We take the attended multi-modal encoder representation vector ($\hat{H}$) and pass it to a linear layer for intent identification. The vector ($\hat{H}$) is fed to the decoder's multi-head attention layer as key and value, with the key as the hidden representation of the medical concern summary. The infused information is processed with the traditional transformer's layers and computes the vocabulary's probability distribution. We have utilized a joint categorical cross-entropy loss function, which is the sum of loss functions of classification (CL) and generation (GL) tasks, i.e., $L = \alpha_1 * CL + \alpha_2 * GL$ and $\alpha_1 (= 0.2) + \alpha_2 (= 0.8) = 1$.

4.4 Experimental Details

We have utilized the PyTorch framework for implementing the proposed model. The generation models have been trained, validated, and evaluated with 80%, 6%, and 14% samples of the *MM MediConSummation* dataset, respectively. The hyperparameter values, which are selected empirically, are as follows: sequence length (480), output max len (50), learning rate (3e-05), batch size (16), and activation function (ReLU). The different baselines and state-of-the-art models are listed as follows:

- **Seq2Seq-Transformer** It is a transformer-based sequence-to-sequence model [21], which takes a combined representation of transcript, audio, and video features as input and generates a medical concern summary.
- **BART** It is a denoising autoencoder model that is trained to reconstruct corrupted sentences [16].
- **T5** It [19] is a versatile text-to-text model that combines encoder-decoder architecture with pre-training on a mixture of unsupervised and supervised tasks.
- **MAF** MAF [15] is a fusion model that incorporates an additional adapter-based layer in the encoder of BART to infuse information from different modalities.
- **MMCSG** is the proposed model with the proposed contextualized M-modality fusion mechanism only (without the multi-task intent recognition and MMCG generation setting).
- ***IR-MMCSG*** is the proposed multi-tasking, multi-modal intent identification, and medical concern summary generation model, incorporated with contextualized M-modality fusion mechanism.

5 Result and Discussion

The purpose of the proposed multi-task framework is to enhance the performance of the primary task, MCSG, by utilizing the additional task of intent recognition. Thus, the results and analysis of MCSG are emphasized as the main focus in all task combinations.

5.1 Experimental Results

The obtained performance by different multi-modal medical concern summary generation models are reported in Table 3. Furthermore, we also investigated these models' efficacies for doctor summary generation, and results are presented in Table 4. We ran experiments for ten iterations with different random seeds and reported the average values. The reported values in the following tables are statistically significant as the p-values obtained from Welch's t-test [29] at 5% significance level are less than 0.05.

Table 3. Performances of different models for multi-modal medical concern summary generation. Here, † indicates statistical significant findings ($p < 0.05$ at 5% significance level).

Model	BLEU-1	BLEU-2	BELU-3	BELU-4	BLEU	ROUGE - 1	ROUGE - 2	ROUGE- L	METEOR
Seq2Seq-Transformer [21]	18.65	10.93	3.85	1.23	8.67	24.03	8.62	22.16	21.36
T5 [19]	19.67	11.73	5.92	1.93	9.81	28.23	11.01	26.48	23.44
BART [16]	19.46	12.47	6.90	2.62	10.36	26.85	12.92	26.01	32.46
MAF [15]	21.89	13.05	6.05	3.26	11.06	30.04	12.43	26.93	30.10
MMCSG w/o visual	21.63	13.57	7.33	2.97	11.37	30.60	14.27	27.56	34.44
MMCSG	23.26	14.13	7.80	**3.81**†	12.25	31.47	14.16	28.51	33.60
IR-MMCSG	**23.72**†	**14.68**†	**7.88**†	2.98	**12.31**†	**32.16**†	**14.41**†	**29.61**†	**35.87**†

Table 4. Performances of different models for doctors' impression summary generation. Here, † indicates statistical significant findings ($p < 0.05$ at 5% significance level).

Model	BLEU-1	BLEU-2	BELU-3	BELU-4	BLEU	ROUGE - 1	ROUGE - 2	ROUGE- L	METEOR
T5 [19]	17.37	9.18	2.51	0.088	7.28	26.70	7.83	23.60	18.44
BART [16]	20.87	9.96	1.22	0.417	8.12	26.86	8.04	22.84	23.35
MAF [15]	20.88	11.53	3.69	2.69	9.70	26.99	9.83	24.03	23.89
MMCSG w/o visual	21.10	11.66	2.52	0.934	9.05	28.59	10.17	25.17	26.88
MMCSG	23.28	12.38	3.71	**3.13**†	10.63	30.46	10.41	26.35	27.52
IR-MMCSG	**23.78**†	**12.67**†	**4.01**†	2.57	**10.76**†	**31.23**†	**10.86**†	**26.43**†	**29.17**†

Ablation Study. In order to understand the effectiveness of various components within the proposed model, we carried out an ablation study involving different combinations of these components. The obtained result has been reported in Table 5. The results show that the model's performance is improving as the various different modalities are infused together to represent the context.

Human Evaluation. We have also conducted a human evaluation of all test samples. In this assessment, two medical domain experts and one researcher (other than the authors) were

Table 5. Performance of the proposed model with different modalities. Here, T, P, A, and V indicate transcript, personality information, audio, and video features, respectively. Here, † indicates statistical significant findings ($p < 0.05$ at 5% significance level).

Model	BLEU	ROUGE-1	ROUGE-2	ROUGE- L	METEOR
T	11.00	28.88	14.05	26.88	33.71
T + A	11.37	30.60	14.27	27.56	35.71
T + P	11.95	30.78	14.64	27.80	34.91
T + V	11.73	30.89	14.44	28.18	35.80
T + A + V	11.56	30.46	**14.84**†	28.48	35.82
T + A + V + P	**12.31**†	**32.16**†	14.41	**29.61**†	**35.87**†

employed to evaluate the generated medical concern summaries (twenty samples for each model) without revealing the models' names. The samples are assessed based on the following five metrics: *domain relevance (DR), adequacy, fluency, informativeness (Info), and patient's personal context coherence (PC)* on a scale of 0 to 5. The obtained scores are presented in Table 6.

Table 6. Human evaluation for medical concern summary generation models

Model	DR	Adequecy	Fluency	PC	Info	Avg.
Seq2Seq [21]	2.86	2.72	3.88	2.40	3.61	3.09
BART [16]	3.10	3.16	4.22	2.65	3.80	3.37
MAF [15]	3.28	3.56	4.38	2.80	3.94	3.59
MMCSG	3.44	3.81	4.55	**3.14**	3.84	3.76
IR-MMCSG	**3.82**	**3.94**	**4.74**	3.08	**4.06**	**3.93**

Key Observations. The main observations and insights are as follows: **(i)** The proposed medical concern summary generation model outperforms traditional sequence-to-sequence and transfer-learning-based generation models by a large margin, highlighting the importance of (a) task-specific conditioning and (b) the incorporated contextualized M-modality fusion (BART with early fusion- a simple concatenation of modalities vectors). **(ii)** For MCSG, visual modality (movements and expressions) infusion with text was more important than audio and demographic information (Table 5). **(iii)** In human evaluation, we observed that the multi-task model incorporating intent representation generates a significantly more contextualized and informative summary of medical concerns (Table 6 and Fig. 8).

5.2 Findings to Research Questions

RQ 1: *Do visuals and patient expression help in identifying patient key concern and generating adequate medical advice for the same?* The proposed MMCSG model outperforms the medical concern summary generation without visual information across all evaluation metrics (Table 3 and Table 5 - T vs. $T + V$ & $T + A$ vs. $T + A + V$). Furthermore, a similar trend has been observed for doctor suggestion summary generation (Table 4). The enhanced findings strongly establish the effectiveness of utilizing visual cues, such as patients' facial expressions and body movements, in accurately identifying their medical conditions and responding accordingly.

RQ 2: *Can the patient's demographic context aid in generating an appropriate and relevant medical concern summary /suggestion?* Some behaviors, namely facial expressions/body movement and medical conditions, are heavily influenced by demographic information such as age and gender. Therefore, the proposed contextualized M modality fusion aided *IR-MMCSG* (Table 3 - MAF vs. MMCSG, Table 4 - MAF vs. MMCSG and Table 5 - T vs. $T + P$ & $T + A + V$ vs. $T + A + V + P$) that exploits the information to analyze and constrain different modalities performed significantly better than the non-context aware models.

RQ 3: *Does any correlation exist between medical concern summary generation and user intent identification?* If a psychiatrist is aware of the reason for a patient's visit, it becomes more straightforward to identify the patient's primary mental issue and suggest him/her accordingly. The observed results firmly support the hypothesis, revealing that the multi-tasking *IR-MMCSG* clearly outperforms the single-task model, MMCSG (Table 3 and Table 4 - MMCSG vs. *IR-MMCSG*). Furthermore, we also observed a significant improvement by *IR-MMCSG* model in human evaluation (Table 6) across different metrics.

6 Analysis

We have analyzed the proposed model's generated medical concern summaries and different models' behavior on some test cases, which are presented in Fig. 8. The comprehensive analyses lead to the following key observations: (i) The proposed *IR-MMCSG* generates medical concern summaries (Fig. 8) that include (a) a comprehensive and contextualized understanding of the patient's concerns and (b) a sense of the discussion that will be undertaken during the session. **(ii)** During the human evaluation, we found a significant number of cases where our model has also generated a cause of abnormality in the MCS (Table 7). **(iii)** We observed that the models added additional words in the medical concern summary, leading to low evaluation scores despite being relevant and informative (Table 7).

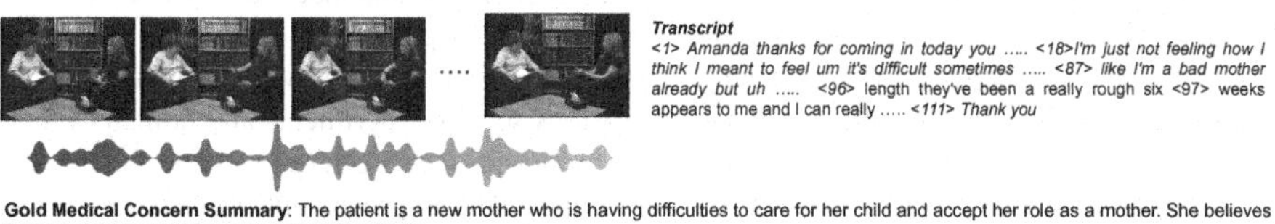

Gold Medical Concern Summary: The patient is a new mother who is having difficulties to care for her child and accept her role as a mother. She believes that she is not receiving enough support from family.

Generated Multimodal Medical Concern Summaries (MMCS)

BART	MAF	MMCSG	IR-MMCSG
The patient finds difficulty in accepting her as a mother.	The patient is a new mother and finds herself in a complex situation. She needs suggestions to be a good mother.	The patient is a new mother who is having difficulty caring for her child. She believes that no one is supporting her.	The patient is a new mother and she finds it hard to take care of the child. She did not consider herself as a good mother. She is not getting support from her family.

Fig. 8. Performances of different baselines and the proposed *IR-MMCSG* model for a test case

Limitations. We have observed the following key limitations of the proposed medical concern summary generation model: **(i)** In the dataset, there were few instances where sets of symptoms

Table 7. Some generated medical concern summaries by the proposed IR-MMCSG model and their corresponding gold summaries.

No.	Gold MMCS	Generated MMCS
1	The patient is feeling uncomfortable in his job	The patient is feeling increasingly uncomfortable in his new job
2	The patient has self-doubt about herself and she is worried about being judged	The patient feels depressed, she thinks that everyone is judging her, and it makes her upset
3	The patient realizes that he is overgeneralizing the fact that he is not going to fit in with their colleague. Meanwhile, he does not know much about his colleague.	The patient has the issue of feeling non-existing. He feels that he is different from other people at work.

and conditions were displayed on the screen, along with their status. Using OCR technology in these instances could lead to more accurate and precise summaries of the patient's medical condition. **(ii)** The transcripts of counselling sessions lack speaker type (psychiatrist and patient) information, which could have played an important role in effectively identifying and summarising patient concerns. **(iii)** Due to the limited demographic information (age and gender) and the small size of the dataset, there is a risk that the model may become biased towards certain age groups and gender that are more frequently represented.

7 Conclusion

In this paper, we introduce a new task of generating a succinct summary of the primary concerns and expectations expressed by the initiator of a conversation. We curated the first multi-modal medical concern summary generation (*MM-MediConSummation*) corpus annotated with medical concern summary, user demographic information, user intent, and summary of doctors' suggestions. We proposed a multi-tasking multi-modal intent recognition and medical concern summary generation (*IR-MMCSG*) model incorporated with a novel adapter-based contextualized multi-modality fusion mechanism for analyzing acoustics and visual features with demographic and personality context. With the obtained results of various sets of experiments and human evaluation, we found firm evidence of the efficacy of the proposed *IR-MMCSG* model and the infused fusion mechanism over existing state-of-the-art methods. The improvements obtained establish the crucial role of facial expression/movement behavior and demographic context in identifying patients' medical concerns and generating an adequate summary for them. In the future, we would like to build an explainable multi-modal medical concern summary generation that generates medical concern summary generation along with evidence highlighting video spans.

8 Ethical Consideration

We strictly followed the medical research's ethical and regulatory guidelines particular to psychiatrist research [2] during the dataset curation process. We have not added or removed any utterances/medical entity from the conversation. The curated dataset does not reveal users' identities, such as their names. The names have been replaced with some synthetic names. The annotation guidelines were provided by two psychiatrists, and the curated dataset is thoroughly checked and corrected by them. Furthermore, we have also obtained approval from our institute's health-care committee and institutional ethical review board (ERB) to use the dataset and conduct the research. Thus, we confidently assert that the dataset, along with its comprehensive creation protocol, fully complies with the ethical and clinical imperatives of our discipline.

Acknowledgment. Abhisek Tiwari expresses sincere gratitude for being honored with the Prime Minister Research Fellowship (PMRF) Award by the Government of India. This grant has played a crucial role in supporting this research endeavor. Dr. Sriparna Saha extends sincere appreciation for receiving the Young Faculty Research Fellowship (YFRF) Award. This recognition is supported by the Visvesvaraya Ph.D. Scheme for Electronics and IT under the Ministry of Electronics and Information Technology (MeitY), Government of India, and executed by Digital India Corporation (formerly Media Lab Asia), has been invaluable in advancing the progress of the research. We want to express our gratitude to clinicians Dr. Pankaj Kumar (All India Institute of Medical Sciences Patna) and Dr. Minkashi Dhar (All India Institute of Medical Sciences Rishikesh) for their valuable contributions to this project.

References

1. Apostolidis, E., Adamantidou, E., Metsai, A.I., Mezaris, V., Patras, I.: Video summarization using deep neural networks: a survey. Proc. IEEE **109**(11), 1838–1863 (2021)
2. Avasthi, A., Ghosh, A., Sarkar, S., Grover, S.: Ethics in medical research: general principles with special reference to psychiatry research. Indian J. Psychiatry **55**(1), 86 (2013)
3. Barnes, R.K.: Conversation analysis of communication in medical care: description and beyond. Res. Lang. Soc. Interact. **52**(3), 300–315 (2019)
4. Bay, M., Bruneß, D., Herold, M., Schulze, C., Guckert, M., Minor, M.: Term extraction from medical documents using word embeddings. In: 2020 6th IEEE Congress on Information Science and Technology (CiSt), pp. 328–333. IEEE (2021)
5. Chintagunta, B., Katariya, N., Amatriain, X., Kannan, A.: Medically aware GPT-3 as a data generator for medical dialogue summarization. In: Machine Learning for Healthcare Conference, pp. 354–372. PMLR (2021)
6. Dreisbach, C., Koleck, T.A., Bourne, P.E., Bakken, S.: A systematic review of natural language processing and text mining of symptoms from electronic patient-authored text data. Int. J. Med. Inf. **125**, 37–46 (2019)
7. Enarvi, S., et al.: Generating medical reports from patient-doctor conversations using sequence-to-sequence models. In: Proceedings of the First Workshop on Natural Language Processing for Medical Conversations, pp. 22–30 (2020)
8. Eyben, F., Wöllmer, M., Schuller, B.: Opensmile: the munich versatile and fast open-source audio feature extractor. In: Proceedings of the 18th ACM International Conference on Multimedia, pp. 1459–1462 (2010)
9. Fleiss, J.L.: Measuring nominal scale agreement among many raters. Psychol. Bull. **76**(5), 378 (1971)
10. Gratch, J., et al.: The distress analysis interview corpus of human and computer interviews. University of Southern California Los Angeles, Technical Report (2014)
11. He, K., Zhang, X., Ren, S., Sun, J.: Deep residual learning for image recognition. In: Proceedings of the IEEE Conference on Computer Vision and Pattern Recognition, pp. 770–778 (2016)
12. Huang, J.H., Murn, L., Mrak, M., Worring, M.: Gpt2mvs: generative pre-trained transformer-2 for multi-modal video summarization. In: Proceedings of the 2021 International Conference on Multimedia Retrieval, pp. 580–589 (2021)

13. Jeblee, S., Khattak, F.K., Crampton, N., Mamdani, M., Rudzicz, F.: Extracting relevant information from physician-patient dialogues for automated clinical note taking. In: Proceedings of the Tenth International Workshop on Health Text Mining and Information Analysis (LOUHI 2019), pp. 65–74 (2019)
14. Joshi, A., Katariya, N., Amatriain, X., Kannan, A.: Dr. summarize: global summarization of medical dialogue by exploiting local structures. In: Findings of the Association for Computational Linguistics, EMNLP 2020, pp. 3755–3763 (2020)
15. Kumar, S., Kulkarni, A., Akhtar, M.S., Chakraborty, T.: When did you become so smart, oh wise one?! sarcasm explanation in multi-modal multi-party dialogues. In: Proceedings of the 60th Annual Meeting of the Association for Computational Linguistics (Volume 1: Long Papers), pp. 5956–5968 (2022)
16. Lewis, M., et al.: Bart: denoising sequence-to-sequence pre-training for natural language generation, translation, and comprehension. arXiv preprint arXiv:1910.13461 (2019)
17. Malhotra, G., Waheed, A., Srivastava, A., Akhtar, M.S., Chakraborty, T.: Speaker and time-aware joint contextual learning for dialogue-act classification in counselling conversations. In: Proceedings of the Fifteenth ACM International Conference on Web Search and Data Mining, pp. 735–745 (2022)
18. Nittari, G., et al.: Telemedicine practice: review of the current ethical and legal challenges. Telemedicine e-Health **26**(12), 1427–1437 (2020)
19. Raffel, C., et al.: Exploring the limits of transfer learning with a unified text-to-text transformer. J. Mach. Learn. Res. **21**(140), 1–67 (2020)
20. Shang, X., Yuan, Z., Wang, A., Wang, C.: Multimodal video summarization via time-aware transformers. In: Proceedings of the 29th ACM International Conference on Multimedia, pp. 1756–1765 (2021)
21. Shi, T., Keneshloo, Y., Ramakrishnan, N., Reddy, C.K.: Neural abstractive text summarization with sequence-to-sequence models. ACM Trans. Data Sci. **2**(1), 1–37 (2021)
22. Shi, X., Hu, H., Che, W., Sun, Z., Liu, T., Huang, J.: Understanding medical conversations with scattered keyword attention and weak supervision from responses. In: Proceedings of the AAAI Conference on Artificial Intelligence, vol. 34, pp. 8838–8845 (2020)
23. Singh, G.V., Ghosh, S., Ekbal, A., Bhattacharyya, P.: DeCoDE: detection of cognitive distortion and emotion cause extraction in clinical conversations. In: Kamps, J., et al. Advances in Information Retrieval, ECIR 2023, LNCS, vol. 13981, pp. 156–171 Springer, Cham (2023). https://doi.org/10.1007/978-3-031-28238-6_11
24. Song, Y., Tian, Y., Wang, N., Xia, F.: Summarizing medical conversations via identifying important utterances. In: Proceedings of the 28th International Conference on Computational Linguistics, pp. 717–729 (2020)
25. Stratigi, M., Kondylakis, H., Stefanidis, K.: Multidimensional group recommendations in the health domain. Algorithms **13**(3), 54 (2020)
26. Tiwari, A., Manthena, M., Saha, S., Bhattacharyya, P., Dhar, M., Tiwari, S.: Dr. can see: towards a multi-modal disease diagnosis virtual assistant. In: Proceedings of the 31st ACM International Conference on Information & Knowledge Management, CIKM 2022, pp. 1935–1944. Association for Computing Machinery, New York, NY, USA (2022). https://doi.org/10.1145/3511808.3557296
27. Tiwari, A., Saha, A., Saha, S., Bhattacharyya, P., Dhar, M.: Experience and evidence are the eyes of an excellent summarizer! towards knowledge infused multi-modal clinical conversation summarization. In: Proceedings of the 32nd ACM International Conference on Information and Knowledge Management, CIKM 2023, pp. 2452–2461. Association for Computing Machinery, New York, NY, USA (2023). https://doi.org/10.1145/3583780.3614870
28. Wang, Y., Liang, W., Huang, H., Zhang, Y., Li, D., Yu, L.F.: Toward automatic audio description generation for accessible videos. In: Proceedings of the 2021 CHI Conference on Human Factors in Computing Systems, pp. 1–12 (2021)

29. Welch, B.L.: The generalization of student's' problem when several different population variances are involved. Biometrika **34**(1/2), 28–35 (1947)
30. Weld, H., Huang, X., Long, S., Poon, J., Han, S.C.: A survey of joint intent detection and slot filling models in natural language understanding. ACM Comput. Surv. (CSUR) **55**(8), 1–38 (2021)
31. Wosik, J., et al.: Telehealth transformation: Covid-19 and the rise of virtual care. J. Am. Med. Inform. Assoc. **27**(6), 957–962 (2020)
32. Zhu, J., Li, H., Liu, T., Zhou, Y., Zhang, J., Zong, C.: Msmo: multimodal summarization with multimodal output. In: Proceedings of the 2018 Conference on Empirical Methods in Natural Language Processing, pp. 4154–4164 (2018)

Mitigating Data Sparsity via Neuro-Symbolic Knowledge Transfer

Tommaso Carraro[1,2(✉)], Alessandro Daniele[2], Fabio Aiolli[1], and Luciano Serafini[2]

[1] Department of Mathematics, University of Padova, Via Trieste, 63, 35121 Padova, Italy

[2] Data and Knowledge Management Unit, Fondazione Bruno Kessler, Via Sommarive, 18, 38123 Povo, Italy

tcarraro@fbk.eu

Abstract. Data sparsity is a well-known historical limitation of recommender systems that still impacts the performance of state-of-the-art approaches. The literature proposed various ways to mitigate the problem by providing additional information to the model (e.g., hybrid recommendation, knowledge graph-based systems). In particular, one promising technique involves transferring information from other domains or tasks to compensate for sparsity in the target domain, where the recommendations must be performed. Following this idea, we propose a novel approach based on Neuro-Symbolic computing designed for the knowledge transfer task in recommender systems. In particular, we use a Logic Tensor Network (LTN) to train *vanilla* Latent Factor Models (LFMs) for rating prediction. We show how the LTN can be used to regularize the LFMs using axiomatic knowledge that permits injecting pre-trained information learned by Collaborative Filtering on a different task or domain. Extensive experiments comparing our models with different baselines on two versions of a novel real-world dataset prove our proposal's potential in the knowledge transfer task. In particular, our models outperform the baselines, including those that can encode additional information, suggesting that the knowledge is effectively transferred to the target domain via logical reasoning. Moreover, an experiment that drastically decreases the density of user-item ratings shows that the benefits of the acquired knowledge increase with the sparsity of the dataset, showing the importance of exploiting knowledge from a denser source of information when training data is scarce in the target domain.

Keywords: knowledge transfer · matrix factorization · neuro-symbolic integration · logic tensor networks · rating prediction · explicit feedback · data sparsity

1 Introduction

Recommender systems (RSs) have recently become essential for e-services (e.g., Amazon, Netflix, Spotify). Given the user's historical data, these tools mitigate

N. Goharian et al. (Eds.): ECIR 2024, LNCS 14610, pp. 226–242, 2024.
https://doi.org/10.1007/978-3-031-56063-7_15

information overload by suggesting novel items (e.g., products, movies, songs) that match the user's preferences [29]. Since the beginning of the RSs literature, Collaborative Filtering (CF) [1,19,26,31] has been one of the most successful recommendation approaches. Latent Factor Models (LFMs), in particular Matrix Factorization (MF) and Factorization Machine (FM), have dominated the CF scene [20,25,27] for years, and this has been further emphasized with the deep learning rise [8,22,23,36].

Despite their success in improving recommendation performance, state-of-the-art models still suffer from a historical issue, i.e., data sparsity, that limits their applicability in real-world scenarios. Data sparsity is due to the *long-tail* distribution[1] of user-item interactions and, hence, intrinsic in recommendation datasets. The literature proposed various mitigations for data sparsity. On the one hand, hybrid recommendation [8,27] leverages content (e.g., movie genres) or contextual (i.e., weather, time) side information to inject additional knowledge into the model and compensate for scarcity. On the other hand, knowledge graph-based approaches [34] exploit paths and connections between users and items in heterogeneous ontologies to model higher-order interactions. This allows the models to learn better representations in sparse scenarios. One alternative to address the issue consists of leveraging models pre-trained on other sources of information (i.e., source domains) to make the final model more accurate in the target domain, where the recommendations must be performed. In the context of recommender systems, knowledge transfer (and cross-domain recommendation[2]) techniques [14,17,24,37] can be subdivided into two categories: *feature-based* and *fine-tuning* [38]. In *feature-based*, pre-trained models are used to learn features (e.g., embeddings) from side information for users and/or items. Then, these features are integrated into the recommendation framework. Instead, *fine-tuning* firstly trains a deep transferable neural model on user-item interactions taken from a source domain. Then, this pre-trained model is fine-tuned on the downstream recommendation task, namely the target domain. The first category inspires the approach we propose in this paper. In particular, instead of directly learning features for users and items using side information in the source domain, we learn to predict ratings via Collaborative Filtering using an LFM. Then, this knowledge (encoded in the embeddings of the LFM) is transferred to the target domain via logical reasoning.

Despite Neuro-Symbolic (NeSy) [15] approaches having been successfully applied in many AI fields [4,12,30], including RSs [7,9,10,16,21,32,35], they have not yet been investigated in the task of knowledge transfer for recommender systems [6], where we believe their application is particularly suited. In

[1] Users naturally tend to concentrate ratings on a few popular items. In fact, the more an item becomes popular, the more interactions (and ratings) it will generate. However, most items are unknown (i.e., unpopular) to users and lack ratings. In addition, the more popular an item is, the more users it will attract, giving fewer chances for less popular items to be reached. This makes the distribution even more skewed towards popular items. These phenomena clearly cause data sparsity.

[2] Note that cross-domain recommendation is a specific case of knowledge transfer aimed at mitigating data sparsity.

particular, NeSy aims at integrating knowledge, usually expressed as a set of logical axioms, with neural networks. The integration has shown to be particularly beneficial in tasks with poor training data [11], giving insights that this paradigm can help deal with data sparsity in recommendation datasets.

Following this intuition, we propose using a Logic Tensor Network (LTN) [2] to encode axiomatic knowledge to enable effective knowledge transfer and injection for two LFMs[3] (i.e., MF and FM) trained on movie ratings. LTN is a NeSy framework that effectively integrates logical reasoning and neural networks. Our approach uses it as the interface between the model pre-trained on the source domain and the final model trained on the target domain. We called it an interface as it allows the explicit transfer of information between domains via logical reasoning. To perform our experiments, we use MindReader [3], a novel recommendation dataset containing explicit ratings from real users on movies and *non-recommendable* entities, such as movie genres, actors, and producers. In particular, we use ratings on movie genres to learn our pre-trained model via CF and ratings on movies to train the final model to provide accurate recommendations in the target domain. The pipeline of the proposed approach consists of two steps. In the first step, we train a genre classifier using an MF model to learn which genres the users like and dislike in the source domain. In the second step, we use LTN to transfer the pre-trained knowledge to an LFM model trained on the target domain for the movie rating prediction task.

We compare our approach with different baselines to understand if knowledge transfer is successfully reached thanks to Neuro-Symbolic integration. The results show that our models outperform the baselines, proving our proposal's ability to transfer knowledge across domains. In addition, an experiment that drastically reduces the density of user-item ratings shows that the benefits of the knowledge increase with the sparsity of the dataset.

2 Related Works

The seminal approach that applied NeSy to RSs has been Neural Collaborative Reasoning (NCR) [10]. In NCR, the sequential recommendation task is formalized as a logical reasoning problem. In particular, the user's ratings are represented using propositional variables, and logical operators are used to construct propositional formulas that express sequential patterns between them. Then, NCR maps the variables to *logical* embeddings and the operators to neural networks that act on those embeddings. By doing so, the formulas can be organized as a neural network to conduct logical reasoning and prediction in a continuous space.

Another approach that successfully integrated logical reasoning and learning has been HYbrid Probabilistic Extensible Recommender (HyPER) [21]. In

[3] Note we selected LFMs for our experiments because, despite their simplicity, they are still some of the most powerful state-of-the-art approaches [13]. This is not to be intended as a limit of our approach, as any other state-of-the-art model could be used in principle.

particular, HyPER exploits the expressiveness of First-Order Logic (FOL) to encode knowledge from a wide range of information sources, such as multiple user and item similarity measures, content, and social information. Then, Hinge-Loss Markov Random Fields [18] are used to learn how to balance the different information types. HyPER is highly related to our work with LTN since the logical formulas we use resemble the ones used in HyPER.

One of the last published approaches [7] is highly related to ours. Specifically, they try to mitigate data sparsity by using LTN to inject *content* information into an MF model. In particular, they encode FOL formulas to use side information (i.e., movie genres) as a regularizer for the latent factors of the MF model. They show that the proposed NeSy approach can outperform the MF. However, the improvement is poor, and the model has some scalability issues due to the number of times the formulas have to be evaluated during training. Our approach differs from [7] on how axiomatic knowledge is used. In [7], the knowledge extends the MF model using content information, while our model uses it to enable the transfer of pre-trained knowledge learned via Collaborative Filtering on another task (i.e., movie genre rating prediction).

3 Background

This section provides useful notation and terminology used in the remainder of the paper.

3.1 Notation

Bold notation differentiates vectors, e.g., $\mathbf{x} = [3.2, 2.1]$, and scalars, e.g., $x = 5$. Matrices and tensors are denoted with upper case bold notation, e.g., $\mathbf{X}$. $\mathbf{X}_i$ is used to denote the i-th row of $\mathbf{X}$, while $\mathbf{X}_{i,j}$ to denote the item at row i and column j. We refer to the set of users of an RS with $\mathcal{U}$, where $|\mathcal{U}| = n$. Similarly, the set of items is denoted as $\mathcal{I}$, such that $|\mathcal{I}| = m$. We use $\mathcal{D}$ to denote a dataset. $\mathcal{D}$ is defined as a set of N triples $\mathcal{D} = \{(u, i, r)^{(j)}\}_{j=1}^{N}$, where $u \in \mathcal{U}$, $i \in \mathcal{I}$, and $r \in \{0, 1\}$ is a binary explicit rating. $\mathcal{D}$ can be reorganized in the so-called user-item matrix $\mathbf{R} \in \mathbb{N}^{n \times m}$, such that $\mathbf{R}_{u,i} = r$ if $(u, i, r) \in \mathcal{D}$, 0 otherwise. Then, since we work with binary feedback (i.e., $\mathbf{R} \in \{0, 1\}^{n \times m}$), we refer to $\mathcal{D}_+$ (resp. $\mathcal{D}_-$) as the dataset of positive (resp. negative) user-item pairs. $\mathcal{D}_+$ is defined as a set of N_+ couples $\mathcal{D}_+ = \{(u, i)^{(j)} | (u, i, r)^{(j)} \in \mathcal{D}, r^{(j)} = 1\}_{j=1}^{N}$. Similarly, $\mathcal{D}_-$ is defined as a set of N_- couples $\mathcal{D}_- = \{(u, i)^{(j)} | (u, i, r)^{(j)} \in \mathcal{D}, r^{(j)} = 0\}_{j=1}^{N}$. Finally, $\mathcal{D}_?$ denotes the dataset of user-item pairs for which the rating is unknown, i.e., $\mathcal{D}_? = \{(u, i) \in \mathcal{U} \times \mathcal{I} | (u, i, r) \notin \mathcal{D}\}$. Clearly, $N_? = n \cdot m - N$.

3.2 Matrix Factorization

Matrix Factorization (MF) [20] is an LFM that aims at factorizing the user-item matrix $\mathbf{R}$ into the product of two lower-dimensional rectangular matrices, $\mathbf{U} \in \mathbb{R}^{n \times k}$ and $\mathbf{I} \in \mathbb{R}^{m \times k}$, containing the users' and items' latent factors, respectively.

k represents the number of latent factors. More formally, the objective of MF is to find $\mathbf{U}$ and $\mathbf{I}$ such that $\mathbf{R} \approx \mathbf{U} \cdot \mathbf{I}^\top$. An effective way to learn the latent factors is by using gradient-descent optimization. Given the dataset $\mathcal{D}$, an MF model seeks to minimize the Mean Squared Error (MSE) between predicted and target ratings, defined as $\frac{1}{N}\sum_{(u,i,r)\in\mathcal{D}} ||\tilde{r} - r||^2 + \lambda||\boldsymbol{\theta}||^2$. In the formulation, $\tilde{r} = \mathbf{U}_u \cdot \mathbf{I}_i^\top + \mathbf{u}_u + \mathbf{i}_i$, where $\mathbf{u}_u$ and $\mathbf{i}_i$ are bias terms for user u and item i, respectively, and $\boldsymbol{\theta} = \{\mathbf{U}, \mathbf{I}, \mathbf{u}, \mathbf{i}\}$. λ is a hyper-parameter to set the strength of the $L2$ regularization.

In our setting, we use a different implementation of MF since we treat the recommendation problem as a binary classification task. Specifically, we need to recommend if a user likes (1) or dislikes (0) an item. Hence, the focal loss is used in place of MSE for the training, and the logistic function is applied to the prediction of MF to restrict the output between 0 and 1. Focal loss is defined as

$$\frac{1}{N}\sum_{(u,i,r)\in\mathcal{D}} -\alpha_r(1-p_r)^\gamma \log p_r + \lambda||\boldsymbol{\theta}||^2$$
$$\alpha_r = \begin{cases} \alpha & \text{if } r = 1 \\ 1-\alpha & \text{if } r = 0 \end{cases} \quad p_r = \begin{cases} p & \text{if } r = 1 \\ 1-p & \text{if } r = 0 \end{cases} \tag{1}$$

where α is a hyper-parameter to give different weights to the two classes, γ is a hyper-parameter that represents the penalty assigned to the examples that are hard to classify, and $p = \sigma(\mathbf{U}_u \cdot \mathbf{I}_i^\top + \mathbf{u}_u + \mathbf{i}_i)$, where σ is the logistic function.

3.3 Factorization Machine

Factorization Machine (FM) [27] is an LFM that models all nested variable interactions with the target using factorized parametrization. In the context of RSs, these variables are users, items, and content (e.g., movie genres) or contextual [28] side information (e.g., time or location). The input of an FM is a vector $\mathbf{x} \in \mathbb{R}^q$, representing categorical variables for users, items, and side information. q is the total number of variables. The model equation for the usual FM of second degree is defined as $\tilde{r}(\mathbf{x}) = w_0 + \sum_{i=1}^q w_i x_i + \sum_{i=1}^q \sum_{j=i+1}^q \langle \mathbf{v}_i \mathbf{v}_j \rangle x_i x_j$. The parameters to be estimated are the global bias w_0, the biases for the variables $\mathbf{w} \in \mathbb{R}^q$, and the latent factors for the variables $\mathbf{V} \in \mathbb{R}^{q \times k}$, where k the number of latent factors, and $\langle \cdot, \cdot \rangle$ represents the dot product. In this paper, users and items are represented using one-hot encoding, while side information is represented using multi-hot encoding, as we use movie genres, and each movie belongs to more than one genre. Specifically, $\mathbf{x} = \mathbf{x}^{(u)}||\mathbf{x}^{(i)}||\mathbf{x}^{(g_i)}$, where $\mathbf{x}^{(u)} \in \{0,1\}^n$ is the one-hot vector of user u, $\mathbf{x}^{(i)} \in \{0,1\}^m$ is the one-hot vector of item i, $\mathbf{x}^{(g_i)} \in \{0,1\}^{N_g}$ is the multi-hot vector of movie genres associated with item i, and $||$ represents the concatenation between vectors. N_g is the number of movie genres in the dataset. Clearly, $q = n + m + N_g$. FMs are general predictors that can be used for regression, classification, or ranking. In this paper, FM is used for classification. For this reason, the logistic function is applied to the prediction, and the training is performed using focal loss. Specifically, for FM, Equation (1) slightly changes. $(\mathbf{x}, r) \in \mathcal{D}$ is used in place of $(u, i, r) \in \mathcal{D}$, since for FMs, every user-item interaction in the dataset is represented as a vector $\mathbf{x}$ of categorical variables. Then, $p = \sigma(\tilde{r}(\mathbf{x}))$ and $\boldsymbol{\theta} = \{w_0, \mathbf{w}, \mathbf{V}\}$.

3.4 Logic Tensor Networks

Logic Tensor Networks (LTN) [2] is a NeSy framework that allows using a knowledge base composed of a set of FOL axioms as the objective of a neural model. LTN uses a specific first-order language, called Real Logic, that is fully differentiable and has concrete semantics that allows mapping every symbolic expression into the domain of real numbers. We refer to the term *grounding*, formally denoted by $\mathcal{G}$, as the function that defines this mapping. Thanks to $\mathcal{G}$, LTN can ground logical formulas into computational graphs, enabling gradient-based optimization based on logical reasoning.

In particular, $\mathcal{G}$ maps individuals (e.g., users) to tensors of real features (e.g., user demographics), functions (e.g., $\mathrm{Score}(user, item)$) as real functions (e.g., inner product), and predicates (e.g., $\mathrm{Likes}(user, item)$) as real functions with output in $[0, 1]$. Then, a variable y is mapped to a *sequence* of n_y individuals (e.g., some items of the dataset), with $n_y \in \mathbb{N}^+, n_y > 0$. As a consequence, a term $t(y)$ or a formula $\mathrm{P}(y)$, will be mapped to a sequence of n_y values too. Afterward, connectives are grounded using fuzzy semantics (i.e., operators dealing with fuzzy values), while quantifiers are grounded as special aggregation functions (e.g., generalized means). This paper uses the *product configuration*, best suited for gradient-based optimization [33]. It is defined as follows.
In the notation, $v, z, v_1, \dots, v_n \in [0, 1]$ and $p \geq 1$.

$$\mathcal{G}(\wedge) = \mathrm{T}_{prod}(v, z) = v * z \qquad \mathcal{G}(\neg) = \mathrm{N}_S(v) = 1 - v$$

$$\mathcal{G}(\implies) = \mathrm{I}_R(v, z) = 1 - v + v * z$$

$$\mathcal{G}(\forall) = \mathrm{ME}_p(v_1, \dots, v_n) = 1 - (\frac{1}{n}\sum_{i=1}^{n}(1 - v_i)^p)^{\frac{1}{p}}$$

$$\mathcal{G}(\exists) = \mathrm{M}_p(v_1, \dots, v_n) = (\frac{1}{n}\sum_{i=1}^{n} v_i^p)^{\frac{1}{p}})$$

The intuition behind the choice of hyper-parameter p is that the higher that p is, the more weight that M_p (resp. ME_p) will give to *true* (resp. *false*) truth-values, converging to the max (resp. min) operator. Real Logic also provides a special type of quantification, called *diagonal quantification*, denoted as $\mathrm{Diag}(y_1, \dots, y_n)$. It allows quantifying over specific tuples of individuals of variables $y_1, \dots, y_n$[4], such that the i-th tuple contains the i-th individual of each variable.

Given a Real Logic knowledge base $\mathcal{K} = \{\phi_1, \dots, \phi_n\}$, where $\phi_1, \dots, \phi_n$ are closed formulas, LTN allows learning the grounding of constants, functions, and

[4] Note that variables $y_1, \dots, y_n$ must have the same number of individuals (i.e., $n_{y_1} = \dots = n_{y_n}$) in order to apply Diag.

predicates appearing in them. In particular, if constants are grounded as embeddings and functions/predicates onto neural networks, their grounding $\mathcal{G}$ depends on some learnable parameters $\boldsymbol{\theta}$. We denote a parametric grounding as $\mathcal{G}(\cdot|\boldsymbol{\theta})$. In LTN, the learning of parametric groundings is obtained by finding parameters $\boldsymbol{\theta}^*$ that maximize the satisfaction of $\mathcal{K}$, namely $\boldsymbol{\theta}^* = \text{argmax}_{\boldsymbol{\theta}} \text{SatAgg}_{\phi \in \mathcal{K}} \mathcal{G}(\phi|\boldsymbol{\theta})$, where SatAgg : $[0,1]^* \mapsto [0,1]$ is a formula aggregating operator, generally defined using ME_p.

4 Method

Our approach uses an LTN to enable domain adaptation for effective knowledge transfer. Specifically, the LTN is trained using a Real Logic knowledge base containing facts designed to intuitively transfer information about movie genre preferences (i.e., the source domain) to an LFM trained on movie ratings (i.e., the target domain). In the next subsections, we will present our knowledge base, how $\mathcal{G}$ is used to convert it into a computational graph suitable for gradient-based optimization, and how the learning of the LTN takes place.

4.1 Real Logic Knowledge Base

The objective of our LTN model is the satisfaction of the following Real Logic knowledge base.

$$\forall \text{Diag}(user_+, movie_+) \text{Likes}(user_+, movie_+) \tag{2}$$

$$\forall \text{Diag}(user_-, movie_-) \neg \text{Likes}(user_-, movie_-) \tag{3}$$

$$\begin{aligned} &\forall \text{Diag}(user_?, movie_?)(\exists genre \neg \text{LikesGenre}(user_?, genre) \\ &\wedge \text{HasGenre}(movie_?, genre)) \implies \neg \text{Likes}(user_?, movie_?) \end{aligned} \tag{4}$$

Specifically, $user_+$ and $movie_+$ are variable symbols denoting positive user-item pairs, $user_-$ and $movie_-$ are variable symbols denoting negative user-item pairs, $user_?$ and $movie_?$ are variable symbols denoting user-item pairs for which the rating is unknown, and $genre$ is a variable symbol denoting the genres of the movies. Then, $\text{Likes}(u, i)$ is a predicate symbol denoting whether a user u likes a movie i, $\text{LikesGenre}(u, g)$ is a predicate symbol denoting whether a user u likes a movie genre g, and $\text{HasGenre}(i, g)$ is a predicate symbol denoting whether a movie i belongs to genre g.

Intuitively, Axiom (2), Axiom (3), and Axiom (4) are applied to user-item pairs in $\mathcal{D}_+$, $\mathcal{D}_-$, and $\mathcal{D}_?$, respectively. Diag is used to quantify over the desired user-item pairs rather than quantifying over all possible combinations of user and item indexes in the dataset.

4.2 Grounding of Symbols

The grounding $\mathcal{G}$ defines how logical symbols are mapped onto the real field and hence how the axioms in the knowledge base define the computational graph of the LTN model. In this work, $\mathcal{G}(user_*) = \langle u^{(j)} | (u,i)^{(j)} \in \mathcal{D}_* \rangle_{j=1}^{N_*}$ and $\mathcal{G}(movie_*) = \langle i^{(j)} | (u,i)^{(j)} \in \mathcal{D}_* \rangle_{j=1}^{N_*}$, namely $user_*$ and $movie_*$ are grounded as a sequence of the N_* user and movie indexes in $\mathcal{D}_*$, with $* \in \{+, -, ?\}$. Instead, $\mathcal{G}(genre) = \langle 1, \ldots, N_g \rangle$, namely $genre$ is grounded as a sequence of N_g genre indexes, where N_g is the number of movie genres in the dataset. Afterward, $\mathcal{G}(\text{Likes}\,|\boldsymbol{\theta}) : u, i \mapsto \sigma(f(u,i))$, namely Likes is grounded onto a function that takes as input a user index u and a movie index i and returns the prediction in $[0,1]$ of the LFM model f for the given user-item pair. In the case MF is used as predictor, $f(u,i) = \mathbf{U}_u \cdot \mathbf{I}_i^\top + \mathbf{u}_u + \mathbf{i}_i$. Intead, if FM is used, $f(u,i) = \tilde{r}(\mathbf{x}^{(u)} || \mathbf{x}^{(i)} || \mathbf{x}^{(g_i)})$. Then, $\mathcal{G}(\text{LikesGenre}) : u, g \mapsto \mathbf{G}_{u,g}$, where $\mathbf{G} \in \{0,1\}^{n \times N_g}$, namely LikesGenre is grounded onto a function that takes as input a user index u and a genre index g and returns the prediction contained in matrix $\mathbf{G}$ for user u and genre g. In particular, $\mathbf{G}$ can be seen as a *lookup* table containing the *binarized*[5] predictions of a pre-trained genre classifier. LTN has shown to work better with *binarized* outputs as the classifier was returning predictions too near the decision boundary for LTN to understand[6] the difference between *like* and *dislike*. Finally, $\mathcal{G}(HasGenre) : i, g \mapsto \{0,1\}$, namely HasGenre is grounded onto a function that takes as input a movie index i and a genre index g and returns one if the movie i belongs to genre g, zero otherwise. Intuitively, $\mathcal{G}(\text{LikesGenre})$ contains the knowledge pre-trained on the source domain. In contrast, $\mathcal{G}(\text{Likes}\,|\boldsymbol{\theta})$ represents the LFM model we need to train on the target domain.

Intuitively, Axiom (2) forces Likes to be true for each positive user-item pair in $\mathcal{D}_+$, while Axiom (3) forces Likes to be false for each negative user-item pair in $\mathcal{D}_-$. In other words, by maximizing the satisfaction of Axiom (2) and Axiom (3), the model learns to factorize the user-item matrix using the ground truth. In contrast, Axiom (4) is designed to transfer knowledge from the source domain to the target domain through logical reasoning. Specifically, it forces Likes to be false whenever a user u does not[7] like at least one genre g of a movie i. Note this axiom is applied only to *unknown* user-item pairs in $\mathcal{D}_?$. In fact, when no movie ratings are available on the target domain, knowing something about movie genre preferences is better than knowing nothing. In other words,

[5] A *binarized* prediction is obtained by using the decision boundary of the classifier on the output of the model to get values in $\{0,1\}$.

[6] For a binary classifier, 0.45 and 0.55 are predictions belonging to different classes. For the LTN framework, those values represent similar truth values.

[7] Note the negated formula is used on purpose, as it is likely that a user dislikes the majority of movies belonging to a genre she dislikes. This has also been confirmed by an empirical study that showed a performance boost when adding the negated formula and a performance drop when adding the standard (i.e., non-negated) one. This study has not been included in the paper due to space constraints.

we believe transferring knowledge from the source domain is crucial when data is missing in the target domain.

4.3 Learning of the LTN

The objective of our model is to learn $\mathcal{G}(\text{Likes}\,|\boldsymbol{\theta})$ by maximizing the satisfaction of the knowledge base. In other words, LTN seeks to minimize the following loss function:

$$\boldsymbol{L}(\boldsymbol{\theta}) = (1 - \text{SatAgg}_{\phi \in \mathcal{K}}\, \mathcal{G}_{\substack{(user_+, movie_+) \leftarrow \mathcal{B}_+ \\ (user_-, movie_-) \leftarrow \mathcal{B}_- \\ (user_?, movie_?) \leftarrow \mathcal{B}_?}} (\phi|\boldsymbol{\theta})) + \lambda ||\boldsymbol{\theta}||^2 \tag{5}$$

where $\mathcal{B}_*$, with $* \in \{+, -, ?\}$, denotes a batch of training examples randomly sampled from the corresponding dataset $\mathcal{D}_*$. The notation $(user_*, movie_*) \leftarrow \mathcal{B}_*$ denotes that variables $user_*$ and $movie_*$ are grounded with actual user-movie pairs coming from the corresponding batch $\mathcal{B}_*$. Notice the loss does not specify how the variable *genre* is grounded. At each training step, we ground it with the sequence of all the movie genre indexes in the dataset. Note $\mathcal{B}_?$ is created by uniformly sampling user-item pairs from $\mathcal{D}_?$ at each training step. While all the user-item pairs in $\mathcal{D}_+$ and $\mathcal{D}_-$ are iterated at each epoch, going through all the possible *unknown* pairs is unnecessary and would be unfeasible. In this sense, $\mathcal{B}_?$ has not to be considered a *mini-batch* in the usual sense.

5 Experiments

This section presents the experiments we performed with our method. They have been executed on an Apple MacBook Pro (2019) with a 2,6 GHz 6-Core Intel Core i7. The models have been implemented in Python using PyTorch. In particular, we used the `LTNtorch`[8] library [5]. Moreover, we used Weights and Biases (WandB) for hyper-parameter optimization. Our source code is freely available[9].

5.1 Datasets

To perform our experiments, we selected MindReader[10] (`MR`) [3], a novel dataset containing ratings from real users for movies and *non-recommendable* entities, such as movie genres, actors, and producers. We performed experiments on both `MR-100k` and `MR-200k`, the two available versions of the dataset. We used the ratings on movie genres as the source domain and the movie ratings as the target domain. To guarantee the users in the source and target domains totally overlapped, we removed the users that only rated movie genres or movies. After

[8] https://github.com/logictensornetworks/LTNtorch.
[9] https://github.com/tommasocarraro/NESYKnowledgeTransfer.
[10] https://mindreader.tech/dataset/.

this pre-processing, MR-100k (resp. MR-200k) comprised 962 (resp. 2,182) users, 3,034 (resp. 3,806) movies, and 140 (resp. 159) movie genres. The density of user-movie ratings was 0.62% (resp. 0.58%), while for user-genre ratings was 8.09% (resp. 6.37%). Selecting ratings on movie genres as the source domain allowed us to use particularly dense information for knowledge transfer. When $density(source) \gg density(target)$, knowledge transfer is more likely to be effective [39]. This, plus the availability of plentiful information to encode logical axioms, has been the reason behind the choice of MindReader for our experiments. The analysis of other potential datasets is left for future work.

MR provides three types of ratings: *likes* (1), *unknown* (0), and *dislikes* (–1). As in [3], we removed the unknown ratings. After that, we changed the label for negative ratings from –1 to 0. As the dataset provides binary explicit feedback, we treated the recommendation problem as a binary classification task, where one has to predict whether a user likes or dislikes an item[11]. This choice allowed us to use the focal loss (Equation (1)) to train the LFM baselines and the *F-measure* as an evaluation metric. This helped in dealing with class imbalance. The class imbalance ratio in MR-100k (resp. MR-200k) is 21%(-)/79%(+) (resp. 20%(-)/80%(+)) for movie genres, and 38%(-)/62%(+) (resp. 36%(-)/64%(+)) for movies. In both cases, the negative class is the minority one. Hence, we used it as the positive one to compute evaluation metrics.

As the *splitting* strategy for the target domain, we randomly sampled 20% of the movie ratings from each user to construct the test set. Then, we randomly sampled 10% of the remaining movie ratings from each user to construct the validation set. Instead, for the source domain, we only created the validation set by randomly sampling 20% of the movie genre ratings from each user. The test set is not needed in the source domain, as we only need a validation set to find the optimal hyper-parameters for the pre-trained model.

5.2 Experimental Setting

Our experiment compares the proposed Neuro-Symbolic approach with different LFM and LTN-based baselines to check if our proposal can effectively transfer knowledge from source to target domain and improve the performance when training data becomes scarce. The models included in the comparison are: (1) MF, a *vanilla* MF model as introduced in Sect. 3.2. (2) $\text{MF}_{\text{genres}}$, an MF model which also learns movie genre latent factors by treating movie genres as additional items, namely $\mathbf{I} = \mathbf{I}||_v\mathbf{I}_g$, where $\mathbf{I}_g \in \mathbb{R}^{N_g \times k}$ is the matrix of movie genre embeddings, and $||_v$ indicates the vertical stack of two matrices. Note the training set for this model also includes user-genre interactions to learn movie genre latent factors. (3) FM, an FM model that uses movie genres as content information, as specified in Sect. 3.3. (4) LTN_{MF}, the model presented in [7], which uses LTN to regularize an MF model by logical reasoning on relationships between

[11] Note that the choice of working on a rating prediction task should not be intended as a limitation of our approach, as modifications can be made to adapt our model to work in top-n recommendation scenarios too.

users, items, and movie genres. This model is learned using the same knowledge base[12] used by our approach (Sect. 4.1), without pre-training of LikesGenre(u, g). (5) NESY_{MF}, our proposed model that uses MF to implement $\text{Likes}(\text{u}, \text{i}|\boldsymbol{\theta})$. Note the only difference between NESY_{MF} and LTN_{MF} is that matrix $\mathbf{G}$ is sparse in the latter, and dense (i.e., complete) in the former, as filled with the predictions of the pre-trained model. (6) LTN_{FM}, same as LTN_{MF} but with FM to implement $\text{Likes}(\text{u}, \text{i}|\boldsymbol{\theta})$. This is an extension of [7] to FM and, hence, forms another contribution of this paper. (7) NESY_{FM}, same as NESY_{MF} but with FM to implement $\text{Likes}(\text{u}, \text{i}|\boldsymbol{\theta})$. We decided to propose both NESY_{MF} and NESY_{FM} to show the framework is flexible in the choice of the recommendation model.

Specifically, the experiment consists of the following pipeline: (1) additional training sets are generated by randomly sampling the 50%, 20%, 10%, and 5% of the movie ratings from the entire training set, referred to as 100%. Notice the ratings on movie genres are not reduced. Moreover, ratings are sampled independently from the user, differently from the splitting strategy explained previously. Then, (2) for each training set $Tr \in \{100\%, 50\%, 20\%, 10\%, 5\%\}$ and for each model ml: (2*a*) ml is trained on Tr using hyper-parameters found through a bayesian search. Finally, (2*b*) ml is evaluated on the test set. Note that for NESY_{MF} and NESY_{FM}, step (2*a*) consists of two steps: (*i*) a standard MF model is pre-trained on the source domain to populate matrix $\mathbf{G}$, then (*ii*) NESY_{MF} (or NESY_{FM}) is trained on the target domain, namely Tr. We repeated the entire procedure 30 times using seeds from 0 to 29. The test metrics have been averaged across these runs and reported in Table 1.

5.3 Training Details

All the models have been trained for 1000 epochs using the `Adam` optimizer. *Early stopping* has been used to stop the training if no improvements were found on the validation set for ten epochs. For all the models, the parameters $\boldsymbol{\theta}$ have been randomly initialized using the *Glorot* initialization. The MF and FM models have been trained using Equation (1), while LTN_{MF}, LTN_{FM}, NESY_{MF}, and NESY_{FM} using Equation (5). For NESY_{MF} and NESY_{FM}, Axiom (4) has been added to the loss from epoch five[13], allowing LTN to learn something about the latent factors before starting reasoning on the acquired knowledge.

We used Bayesian optimization to find the optimal hyper-parameters for our models. We executed every hyper-parameter search for 150 runs and selected the configuration that led to the best validation score. Due to computational time, the searches have been conducted only for the first seed of the experiment and just for the complete dataset (i.e., 100% ratings). The best hyper-parameters found for the models have been then used in the rest of the experiment. For all the models, we tried a number of latent factors

[12] Note the Real Logic formalization presented in [7] is designed for ranking optimization. Here, we propose a reformulation of it for rating prediction.

[13] Notice this is an arbitrary choice. In our experiments, this choice provided better improvement compared to adding the axiom from the first epoch.

$k \in \{5, 10, 25, 50\}$, regularization coefficient $\lambda \in \{0.01, 0.001, 0.0001, 0.00005\}$, learning rate $\eta \in \{0.01, 0.001, 0.0001\}$, training batch size $\beta \in \{64, 128, 256\}$. For the MF and FM models, we additionally tried focal loss hyper-parameters $\alpha \in \{0.05, 0.1, 0.2, 0.3, 0.4, 0.5\}$ and $\gamma \in \{0, 1, 2, 3\}$. For the MF model trained on the source domain, we also tried different thresholds for the decision boundary $t \in \{0.3, 0.4, 0.5, 0.6, 0.7\}$. Due to the huge class imbalance in the source domain, we preferred finding a threshold to maximize *precision* rather than *recall*[14]. Finally, for LTN_{MF}, LTN_{FM}, NESY_{MF}, and NESY_{FM}, we additionally tried different values for hyper-parameter $p \in \{2, 4, 6, 8, 10\}$ of ME_p and M_p. The best hyper-parameter configurations have been included in the provided code.

For the MF model trained on the source domain, we used *F0.5-measure* as the validation metric, while in all the other cases, we used *F1-measure*.

6 Results

The results obtained with our experiments are summarized in Table 1[15]. By looking at the *F1-measure*, it is possible to observe that knowledge transfer models are the best-performing or second-best ones in all the experiments. In particular, the performance gap between the best model and the second one on `MR-100k` (resp. `MR-200k`) increases with the sparsity of the user-item ratings, starting from a 0.47% (resp. 0.11%) improvement on the full dataset and ending with a 4.68% (resp. 2.11%) improvement on the most sparse dataset. This shows the benefits of transferring knowledge from a denser domain when training data is poor. Moreover, it suggests our proposal can effectively transfer knowledge.

Then, by looking at the LFM baselines, $\text{MF}_{\text{genres}}$ always outperforms MF. As expected, adding movie genre latent factors helps the model learn more accurate embeddings for the users, which is particularly helpful when sparsity increases. This is also evidenced by *Precision*, where $\text{MF}_{\text{genres}}$ is often the best-performing. However, especially for the most challenging scenarios (i.e., 10%, 5% folds), the FM model outperforms the other LFM baselines, always in terms of *F1-measure*. This is probably due to the modeling of second-order interactions.

Finally, by looking at the LTN-based baselines, we observe that logical reasoning can slightly improve the performance of LFMs. In particular, LTN_{MF} is often better than MF, suggesting logical regularization is effective. Moreover, for `MR-100k`, LTN_{MF} always outperforms $\text{MF}_{\text{genres}}$, although the latter can explicitly learn movie-genre preferences. However, this is not enough to beat FM, which remains the best for the most sparse scenarios. But when pre-training (i.e., knowledge transfer) is combined with logical reasoning (i.e., NeSy), the models always obtain their best. NESY_{MF} (resp. NESY_{FM}) is always better than LTN_{MF} (resp. LTN_{FM}). As expected, NESY_{FM} is the best-performing model, combining the second-order interactions modeling of FM with NeSy knowledge transfer.

---

[14] By maximizing *precision*, we obtained a more accurate pre-trained model for predicting user-genre preferences. This helped in transferring knowledge more effectively.

[15] For computational time results, all the models have been trained for 50 epochs, without early stopping, $k = 50$, $\beta = 256$, and $\eta = 0.001$ for fair comparison.

Table 1. Models comparison on `MR-100k` (top) and `MR-200k` (bottom). The test metrics are averaged across 30 runs. The standard deviation is between brackets. **Bold notation** indicates the best model overall. Underlined notation indicates the second-best overall. Computational time is included just for the full dataset and first run (seed 0).

MR-100k								
		LFM baselines			LTN-based baselines		Knowledge transfer	
Fold	Metric	MF	$\text{MF}_{\text{genres}}$	FM	LTN_{MF}	LTN_{FM}	NESY_{MF}	NESY_{FM}
100%	Precision	0.5595 (0.0592)	0.5666 (0.0621)	0.5450 (0.0572)	**0.5688** (0.0552)	0.5373 (0.0485)	0.5432 (0.0559)	0.5478 (0.0527)
	Recall	0.7377 (0.0452)	0.7342 (0.0451)	0.7797 (0.0375)	0.7349 (0.0275)	**0.7929** (0.0234)	0.7848 (0.0395)	0.7792 (0.0234)
	F1-measure	0.6328 (0.0322)	0.6357 (0.0311)	0.6386 (0.0376)	0.6390 (0.0318)	0.6392 (0.0358)	0.6392 (0.0349)	**0.6415** (0.0357)
	Time (sec)	15.53	23.99	43.26	38.62	82.09	41.45	85.09
50%	Precision	0.5410 (0.0580)	**0.5496** (0.0579)	0.4761 (0.0571)	0.5449 (0.0533)	0.4807 (0.0485)	0.5297 (0.0512)	0.4866 (0.0551)
	Recall	0.6723 (0.0459)	0.6717 (0.0419)	0.7857 (0.0630)	0.6816 (0.0281)	**0.8096** (0.0448)	0.7054 (0.0260)	0.7979 (0.0413)
	F1-measure	0.5956 (0.0276)	0.6008 (0.0256)	0.5881 (0.0369)	**0.6033** (0.0294)	0.6013 (0.0404)	0.6030 (0.0307)	0.6014 (0.0379)
20%	Precision	0.5081 (0.0558)	**0.5277** (0.0556)	0.4350 (0.0543)	0.5062 (0.0525)	0.4596 (0.0490)	0.4946 (0.0532)	0.4568 (0.0547)
	Recall	0.5948 (0.0411)	0.5961 (0.0351)	0.7877 (0.0885)	0.6283 (0.0255)	0.7527 (0.0284)	0.6529 (0.0249)	**0.8036** (0.0421)
	F1-measure	0.5444 (0.0261)	0.5565 (0.0235)	0.5544 (0.0362)	0.5584 (0.0296)	0.5689 (0.0380)	0.5605 (0.0311)	**0.5789** (0.0376)
10%	Precision	0.4856 (0.0545)	**0.5083** (0.0555)	0.4158 (0.0496)	0.4789 (0.0506)	0.4461 (0.0505)	0.4689 (0.0545)	0.4355 (0.0500)
	Recall	0.5468 (0.0357)	0.5402 (0.0355)	0.8108 (0.0768)	0.6154 (0.0235)	0.7052 (0.0282)	0.6456 (0.0322)	**0.8291** (0.0519)
	F1-measure	0.5112 (0.0261)	0.5204 (0.0229)	0.5449 (0.0347)	0.5365 (0.0301)	0.5441 (0.0350)	0.5407 (0.0349)	**0.5687** (0.0442)
5%	Precision	0.4606 (0.0476)	**0.4972** (0.0519)	0.4078 (0.0501)	0.4528 (0.0500)	0.4349 (0.0544)	0.4474 (0.0522)	0.4246 (0.0472)
	Recall	0.5284 (0.0250)	0.5022 (0.0288)	0.7962 (0.0757)	0.6294 (0.0233)	0.6710 (0.0603)	0.6584 (0.0286)	**0.8278** (0.0688)
	F1-measure	0.4897 (0.0213)	0.4970 (0.0216)	0.5343 (0.0347)	0.5244 (0.0314)	0.5246 (0.0427)	0.5304 (0.0358)	**0.5594** (0.0474)
MR-200k								
		LFM baselines			LTN-based baselines		Knowledge transfer	
Fold	Metric	MF	$\text{MF}_{\text{genres}}$	FM	LTN_{MF}	LTN_{FM}	NESY_{MF}	NESY_{FM}
100%	Precision	0.5640 (0.0560)	0.5550 (0.0561)	0.5760 (0.0483)	**0.5887** (0.0480)	0.5839 (0.0444)	0.5531 (0.0503)	0.5705 (0.0459)
	Recall	0.7555 (0.0333)	0.7500 (0.0336)	0.7515 (0.0209)	0.7380 (0.0187)	0.7542 (0.0259)	**0.7931** (0.0207)	0.7800 (0.0223)
	F1-measure	0.6431 (0.0341)	0.6350 (0.0333)	0.6505 (0.0296)	0.6536 (0.0301)	0.6569 (0.0298)	0.6498 (0.0334)	**0.6576** (0.0308)
	Time (sec)	51.86	73.29	135.48	114.80	252.67	125.88	260.35
50%	Precision	0.5510 (0.0554)	0.5428 (0.0550)	0.4829 (0.0456)	**0.5667** (0.0496)	0.5070 (0.0454)	0.5415 (0.0491)	0.4844 (0.0463)
	Recall	0.6893 (0.0332)	0.7057 (0.0337)	0.7792 (0.0337)	0.6597 (0.0204)	0.7087 (0.0193)	0.7240 (0.0195)	**0.7934** (0.0292)
	F1-measure	0.6095 (0.0300)	0.6106 (0.0304)	0.5943 (0.0350)	0.6080 (0.0283)	0.5898 (0.0329)	**0.6178** (0.0306)	0.5998 (0.0368)
20%	Precision	0.5093 (0.0499)	**0.5169** (0.0523)	0.4422 (0.0463)	0.5131 (0.0478)	0.4714 (0.0446)	0.5050 (0.0474)	0.4599 (0.0421)
	Recall	0.5838 (0.0302)	0.6318 (0.0298)	**0.7663** (0.0678)	0.5707 (0.0166)	0.7011 (0.0196)	0.6459 (0.0171)	0.7650 (0.0254)
	F1-measure	0.5413 (0.0231)	0.5657 (0.0261)	0.5570 (0.0333)	0.5385 (0.0240)	0.5623 (0.0324)	0.5650 (0.0286)	**0.5731** (0.0349)
10%	Precision	0.4806 (0.0469)	**0.4938** (0.0478)	0.3953 (0.0421)	0.4795 (0.0416)	0.4502 (0.0427)	0.4725 (0.0441)	0.4427 (0.0409)
	Recall	0.5359 (0.0283)	0.5726 (0.0292)	**0.8276** (0.0764)	0.5484 (0.0250)	0.6846 (0.0150)	0.6067 (0.0193)	0.7342 (0.0238)
	F1-measure	0.5041 (0.0202)	0.5276 (0.0215)	0.5313 (0.0354)	0.5098 (0.0215)	0.5418 (0.0314)	0.5295 (0.0258)	**0.5511** (0.0338)
5%	Precision	0.4508 (0.0438)	**0.4740** (0.0481)	0.3867 (0.0457)	0.4506 (0.0418)	0.4394 (0.0450)	0.4422 (0.0428)	0.4251 (0.0478)
	Recall	0.5049 (0.0257)	0.5281 (0.0253)	**0.8266** (0.0926)	0.5784 (0.0410)	0.6456 (0.0394)	0.5953 (0.0203)	0.7222 (0.0393)
	F1-measure	0.4740 (0.0193)	0.4971 (0.0217)	0.5216 (0.0346)	0.5045 (0.0285)	0.5206 (0.0309)	0.5057 (0.0273)	**0.5325** (0.0359)

7 Conclusions and Future Works

In this paper, we proposed an effective application of NeSy to knowledge transfer in RSs. In particular, an experiment that drastically reduced the density of the user-item ratings showed our proposal's ability to mitigate data sparsity. In the future, we would like to extend our model to the cross-domain recommendation task. In particular, instead of using movie-genre preferences, one could use more usual source domains (e.g., books, songs). Then, we plan to include some knowledge transfer and cross-domain approaches in our set of state-of-the-art baselines for comparison. Finally, the experiments must be extended to other datasets where knowledge transfer can be applied (e.g., Amazon datasets), and ranking performance must be evaluated to complete the analysis. We believe these additions will help better investigate the effectiveness of our contributions.

References

1. Aiolli, F.: Efficient top-n recommendation for very large scale binary rated datasets. In: Proceedings of the 7th ACM Conference on Recommender Systems, RecSys 2013, pp. 273–280. Association for Computing Machinery, New York, NY, USA (2013). https://doi.org/10.1145/2507157.2507189
2. Badreddine, S., d'Avila Garcez, A., Serafini, L., Spranger, M.: Logic tensor networks. Artif. Intell. **303**, 103649 (2022) https://doi.org/10.1016/j.artint.2021.103649, https://www.sciencedirect.com/science/article/pii/S0004370221002009
3. Brams, A.H., Jakobsen, A.L., Jendal, T.E., Lissandrini, M., Dolog, P., Hose, K.: Mindreader: recommendation over knowledge graph entities with explicit user ratings. In: Proceedings of the 29th ACM International Conference on Information & Knowledge Management, CIKM 2020, pp. 2975–2982. Association for Computing Machinery, New York, NY, USA (2020). https://doi.org/10.1145/3340531.3412759
4. Campari, T., Lamanna, L., Traverso, P., Serafini, L., Ballan, L.: Online learning of reusable abstract models for object goal navigation. In: 2022 IEEE/CVF Conference on Computer Vision and Pattern Recognition (CVPR), pp. 14850–14859. IEEE Computer Society, Los Alamitos, CA, USA, June 2022. https://doi.org/10.1109/CVPR52688.2022.01445
5. Carraro, T.: LTNtorch: PyTorch implementation of logic tensor networks, March 2023. https://doi.org/10.5281/zenodo.7778157
6. Carraro, T.: Overcoming recommendation limitations with neuro-symbolic integration. In: Proceedings of the 17th ACM Conference on Recommender Systems, RecSys 2023, pp. 1325–1331. Association for Computing Machinery, New York, NY, USA (2023). https://doi.org/10.1145/3604915.3608876
7. Carraro, T., Daniele, A., Aiolli, F., Serafini, L.: Logic tensor networks for top-n recommendation. In: AIxIA 2022 - Advances in Artificial Intelligence: XXIst International Conference of the Italian Association for Artificial Intelligence, AIxIA 2022, Udine, Italy, November 28 - December 2, 2022, Proceedings, pp. 110–123. Springer, Berlin (2023). https://doi.org/10.1007/978-3-031-27181-6_8
8. Carraro, T., Polato, M., Bergamin, L., Aiolli, F.: Conditioned variational autoencoder for top-n item recommendation. In: Pimenidis, E., Angelov, P., Jayne, C., Papaleonidas, A., Aydin, M. (eds.) Artificial Neural Networks and Machine Learning - ICANN 2022, pp. 785–796. Springer, Cham (2022). https://doi.org/10.1007/978-3-031-15931-2_64
9. Chen, H., Li, Y., Shi, S., Liu, S., Zhu, H., Zhang, Y.: Graph collaborative reasoning. In: Proceedings of the Fifteenth ACM International Conference on Web Search and Data Mining, WSDM 2022, pp. 75–84. Association for Computing Machinery, New York, NY, USA (2022). https://doi.org/10.1145/3488560.3498410
10. Chen, H., Shi, S., Li, Y., Zhang, Y.: Neural collaborative reasoning. In: Proceedings of the Web Conference 2021, pp. 1516–1527. WWW 2021, Association for Computing Machinery, New York, NY, USA (2021). https://doi.org/10.1145/3442381.3449973

11. Daniele, A., Serafini, L.: Knowledge enhanced neural networks. In: Nayak, A.C., Sharma, A. (eds.) PRICAI 2019. LNCS (LNAI), vol. 11670, pp. 542–554. Springer, Cham (2019). https://doi.org/10.1007/978-3-030-29908-8_43
12. Donadello, I., Serafini, L., d'Avila Garcez, A.: Logic tensor networks for semantic image interpretation. In: Proceedings of the Twenty-Sixth International Joint Conference on Artificial Intelligence, IJCAI-17, pp. 1596–1602 (2017). https://doi.org/10.24963/ijcai.2017/221
13. Ferrari Dacrema, M., Cremonesi, P., Jannach, D.: Are we really making much progress? a worrying analysis of recent neural recommendation approaches. In: Proceedings of the 13th ACM Conference on Recommender Systems, RecSys 2019, pp. 101–109. Association for Computing Machinery, New York, NY, USA (2019). https://doi.org/10.1145/3298689.3347058
14. Gao, C., Chen, X., Feng, F., Zhao, K., He, X., Li, Y., Jin, D.: Cross-domain recommendation without sharing user-relevant data. In: The World Wide Web Conference, WWW 2019, pp. 491–502. Association for Computing Machinery, New York, NY, USA (2019). https://doi.org/10.1145/3308558.3313538
15. d'Avila Garcez, A.S., Broda, K., Gabbay, D.M.: Neural-symbolic learning systems - foundations and applications. In: Perspectives in Neural Computing (2012). https://doi.org/10.1007/978-1-4471-0211-3
16. Ji, J., et al.: Counterfactual collaborative reasoning. In: Proceedings of the Sixteenth ACM International Conference on Web Search and Data Mining, WSDM 2023, pp. 249–257. Association for Computing Machinery, New York, NY, USA (2023). https://doi.org/10.1145/3539597.3570464
17. Kanagawa, H., Kobayashi, H., Shimizu, N., Tagami, Y., Suzuki, T.: Cross-domain recommendation via deep domain adaptation. In: Azzopardi, L., Stein, B., Fuhr, N., Mayr, P., Hauff, C., Hiemstra, D. (eds.) ECIR 2019. LNCS, vol. 11438, pp. 20–29. Springer, Cham (2019). https://doi.org/10.1007/978-3-030-15719-7_3
18. Kimmig, A., Bach, S., Broecheler, M., Huang, B., Getoor, L.: A short introduction to probabilistic soft logic, pp. 1–4. Mansinghka, Vikash (2012). https://lirias.kuleuven.be/retrieve/204697
19. Koren, Y., Bell, R.: Advances in collaborative filtering. In: Ricci, F., Rokach, L., Shapira, B., Kantor, P.B. (eds.) Recommender Systems Handbook, pp. 145–186. Springer, Boston, MA (2011). https://doi.org/10.1007/978-0-387-85820-3_5
20. Koren, Y., Bell, R., Volinsky, C.: Matrix factorization techniques for recommender systems. Computer **42**(8), 30–37 (2009). https://doi.org/10.1109/MC.2009.263
21. Kouki, P., Fakhraei, S., Foulds, J., Eirinaki, M., Getoor, L.: Hyper: a flexible and extensible probabilistic framework for hybrid recommender systems. In: Proceedings of the 9th ACM Conference on Recommender Systems, RecSys 2015, pp. 99–106. Association for Computing Machinery, New York, NY, USA (2015). https://doi.org/10.1145/2792838.2800175
22. LeCun, Y., Bengio, Y., Hinton, G.: Deep learning. Nature **521**, 436–444 (2015). https://doi.org/10.1038/nature14539
23. Liang, D., Krishnan, R.G., Hoffman, M.D., Jebara, T.: Variational autoencoders for collaborative filtering. In: Proceedings of the 2018 World Wide Web Conference, WWW 2018, pp. 689–698. International World Wide Web Conferences Steering Committee, Republic and Canton of Geneva, CHE (2018). https://doi.org/10.1145/3178876.3186150
24. Man, T., Shen, H., Jin, X., Cheng, X.: Cross-domain recommendation: an embedding and mapping approach. In: Proceedings of the Twenty-Sixth International Joint Conference on Artificial Intelligence, IJCAI-17, pp. 2464–2470 (2017). https://doi.org/10.24963/ijcai.2017/343

25. Ning, X., Karypis, G.: Slim: Sparse linear methods for top-n recommender systems. In: 2011 IEEE 11th International Conference on Data Mining, pp. 497–506 (2011). https://doi.org/10.1109/ICDM.2011.134
26. Polato, M., Aiolli, F.: Boolean kernels for collaborative filtering in top-n item recommendation. Neurocomput. **286**(C), 214–225 (2018). https://doi.org/10.1016/j.neucom.2018.01.057
27. Rendle, S.: Factorization machines. In: 2010 IEEE International Conference on Data Mining, pp. 995–1000 (2010). https://doi.org/10.1109/ICDM.2010.127
28. Rendle, S., Gantner, Z., Freudenthaler, C., Schmidt-Thieme, L.: Fast context-aware recommendations with factorization machines. In: Proceedings of the 34th International ACM SIGIR Conference on Research and Development in Information Retrieval, SIGIR 2011, pp. 635–644. Association for Computing Machinery, New York, NY, USA (2011). https://doi.org/10.1145/2009916.2010002
29. Ricci, F., Rokach, L., Shapira, B.: Recommender systems: introduction and challenges. In: Ricci, F., Rokach, L., Shapira, B. (eds.) Recommender Systems Handbook, pp. 1–34. Springer, Boston, MA (2015). https://doi.org/10.1007/978-1-4899-7637-6_1
30. Sarker, M.K., Zhou, L., Eberhart, A., Hitzler, P.: Neuro-symbolic artificial intelligence: current trends. ArXiv abs/2105.05330 (2021)
31. Sarwar, B., Karypis, G., Konstan, J., Riedl, J.: Item-based collaborative filtering recommendation algorithms. In: Proceedings of the 10th International Conference on World Wide Web, WWW 2001, pp. 285–295. Association for Computing Machinery, New York, NY, USA (2001). https://doi.org/10.1145/371920.372071
32. Spillo, G., Musto, C., De Gemmis, M., Lops, P., Semeraro, G.: Knowledge-aware recommendations based on neuro-symbolic graph embeddings and first-order logical rules. In: Proceedings of the 16th ACM Conference on Recommender Systems, RecSys 2022, pp. 616–621. Association for Computing Machinery, New York, NY, USA (2022). https://doi.org/10.1145/3523227.3551484
33. van Krieken, E., Acar, E., van Harmelen, F.: Analyzing differentiable fuzzy logic operators. Artif. Intell. **302**, 103602 (2022). https://doi.org/10.1016/j.artint.2021.103602, https://www.sciencedirect.com/science/article/pii/S0004370221001533
34. Wang, X., He, X., Cao, Y., Liu, M., Chua, T.S.: Kgat: knowledge graph attention network for recommendation. In: Proceedings of the 25th ACM SIGKDD International Conference on Knowledge Discovery & Data Mining, KDD 2019, pp. 950–958. Association for Computing Machinery, New York, NY, USA (2019). https://doi.org/10.1145/3292500.3330989
35. Xian, Y., et al.: Cafe: Coarse-to-fine neural symbolic reasoning for explainable recommendation. In: Proceedings of the 29th ACM International Conference on Information & Knowledge Management, CIKM 2020, pp. 1645–1654. Association for Computing Machinery, New York, NY, USA (2020). https://doi.org/10.1145/3340531.3412038
36. Xue, H.J., Dai, X.Y., Zhang, J., Huang, S., Chen, J.: Deep matrix factorization models for recommender systems. In: Proceedings of the 26th International Joint Conference on Artificial Intelligence, IJCAI 2017, pp. 3203–3209. AAAI Press (2017). https://doi.org/10.24963/ijcai.2017/447
37. Yuan, F., Yao, L., Benatallah, B.: Darec: deep domain adaptation for cross-domain recommendation via transferring rating patterns. In: Proceedings of the Twenty-Eighth International Joint Conference on Artificial Intelligence, IJCAI-19, pp. 4227–4233. International Joint Conferences on Artificial Intelligence Organization, July 2019. https://doi.org/10.24963/ijcai.2019/587

38. Zeng, Z., et al.: Knowledge transfer via pre-training for recommendation: a review and prospect. Frontiers Big Data **4** (2021). https://doi.org/10.3389/fdata.2021.602071
39. Zhu, F., Wang, Y., Chen, C., Zhou, J., Li, L., Liu, G.: Cross-domain recommendation: Challenges, progress, and prospects. In: Zhou, Z.H. (ed.) Proceedings of the Thirtieth International Joint Conference on Artificial Intelligence, IJCAI-21, pp. 4721–4728. International Joint Conferences on Artificial Intelligence Organization, August 2021. https://doi.org/10.24963/ijcai.2021/639, survey Track

InDi: Informative and Diverse Sampling for Dense Retrieval

Nachshon Cohen[1(✉)], Hedda Cohen-Indelman[2], Yaron Fairstein[1], and Guy Kushilevitz[1]

[1] Amazon, Haifa, Israel
nachshonc@gmail.com, yyfairstein@gmail.com, guyk@amazon.com
[2] Technion, Haifa, Israel
heddacohenind@gmail.com

Abstract. Negative sample selection has been shown to have a crucial effect on the training procedure of dense retrieval systems. Nevertheless, most existing negative selection methods end by randomly choosing from some pool of samples. This calls for a better sampling solution. We define desired requirements for negative sample selection; the samples chosen should be *informative*, to advance the learning process, and *diverse*, to help the model generalize. We compose a sampling method designed to meet these requirements, and show that using our sampling method to enhance the training procedure of a recent significant dense retrieval solution (coCondenser) improves the obtained model's performance. Specifically, we see a $\sim 2\%$ improvement in MRR@10 on the MS MARCO dataset (from 38.2 to 38.8) and a $\sim 1.5\%$ improvement in $Recall$@5 on the Natural Questions dataset (from 71% to 72.1%), both statistically significant. Our solution, as opposed to other methods, does not require training or inferencing a large model, and adds only a small overhead ($\sim$ 1% added time) to the training procedure. Finally, we report ablation studies showing that the objectives defined are indeed important when selecting negative samples for dense retrieval.

Keywords: Dense Retrieval · Contrastive Learning · Hard Negative Selection

1 Introduction

Duel Encoder (DE) solutions have recently become a common practice for information retrieval systems [5,8,18,38,40]. These solutions use text encoders to encode queries and passages[1] separately as dense vectors. The query-passage relevance is modeled as a simple dot product, resulting in low latency during inference even for huge corpora. This makes DE-based solutions a great fit for production systems.

The standard practice for training these systems is using the contrastive learning paradigm. In this paradigm, during training the model is shown a query along with a

[1] In this work we focus on retrieving passages, but dense retrieval methods are used for retrieving other items as well such as products, documents, images etc.

N. Cohen, H. Cohen-Indelman, Y. Fairstein and G. Kushilevitz–Contributed equally to this work
H. Cohen-Indelman–Work done as an intern at Amazon.

N. Goharian et al. (Eds.): ECIR 2024, LNCS 14610, pp. 243–258, 2024.
https://doi.org/10.1007/978-3-031-56063-7_16

positive passage and some negative passages. The model is encouraged by the loss to push the query embedding and the positive passage embeddings closer together than the query embedding and the negative passage embeddings. During inference, the passages closest to the query in the embedding space are retrieved.

While collecting a positive passage for a query can be done manually, labeling all passages as positive or negative is practically impossible due to the huge size of modern corpora. It is a common practice for datasets to provide a few positive examples per query, but no negative examples [21,29].

Since training with contrastive learning methods requires more negatives than are usually labeled in the dataset, the challenge of selecting which passages to use as negative samples during training arises (referred to as *negative sampling*). Previous work has shown that training with hard negative samples (i.e. irrelevant passages that are similar to a relevant passage) helps achieve better models [18,38]. Therefore, standard practice is to randomly sample passages from the top-retrieved passages for the query using some base retrieval method (e.g. DPR [18] uses BM25 retrieval, while ANCE [38] uses a slightly older version of the training model for retrieval). While this assures the passages chosen are fairly hard, it does not assure they are negative. Some filtering methods can be used to make sure the passages chosen are not actually positive, but this is not trivial since methods that filter out false negatives are prone to also filter out the *hardest* true negatives. RocketQA [34] deals with this by using a large Cross Encoder model for the filtering task, but this solution is costly in resources.

While there are many different techniques and heuristics to select the pool from which the hard negatives are sampled, ultimately most existing works sample the actual negatives used during training randomly from this pool. This common practice raises the following question: after forming a pool of passages considered as hard negatives for a train query, can we choose the passages to be used as the actual negatives during training in a way better than random sampling?

In this work, we try to answer this question. Specifically, we propose two important qualities we want the selected negatives to hold: *Informativeness* and *Diversity* (InDi). By requiring *diversity* we strive for a sampling method that can select multiple negatives for a single query that are not very similar, so that the model does not overfit. By requiring *informativeness*, we aim at choosing examples that the model is unsure whether they are negative or positive, as opposed to easier negatives that the model already knows are irrelevant for the given query.

We devise a sampling method aiming to optimize these qualities by utilizing the gradient embeddings of the model given the candidate samples. Our method selects the negative samples that will have the best effect on the training process and uses them during training. This enhancement to the training process allows us to achieve better performance, as tested on MS-MARCO [30] and Natural Questions [20] datasets, without adding any labeling costs and just a minor added compute. We publish our code and models for reproducibility and to encourage future research in this area[2].

[2] https://github.com/amzn/informative-diverse-hard-negative-sampling.

2 Preliminaries and Notation

In the passage retrieval task, given a query q and a corpus of M passages, the task is to retrieve a set of passages that are most relevant to q. In dense retrieval systems, this is done by encoding each passage and query into a dense, low-dimensional embedding. For every query, relevant passages are expected to be mapped close in the embedding space using some similarity measure. In most settings, an index of passages is built offline using a kNN tool such as FAISS [16] or ScaNN [10]. Then, when a query arrives, it is encoded online into a dense vector and the kNN index is queried for similar passages. While multiple similarity metrics are possible, most systems use a simple dot product similarity, which we denote by $s(a, b)$.

Encoding of passages and queries is done using a BERT-like [2] architecture (encoder-only transformer). The model is trained to encode query-passage relevance in the embedding space via a contrastive loss. The training dataset consists of tuples of a query (q), a positive passage (p^+), and a set of negative passages (N). The loss encourages the model to push queries and their relevant passages closer together in the embedding space, while pushing the query and irrelevant passages further apart. Given a query encoder E_q and a passage encoder E_p, the constrastive loss can be written as $\ell(q, p^+, N) =$

$$-\log \frac{e^{s(E_q(q), E_p(p^+)}}{e^{s(E_q(q), E_p(p^+))} + \sum_{p^- \in N} e^{s(E_q(q), E_p(p^-))}},$$

The selection of the negative passage set N has a crucial impact on the quality of the model achieved [40]. Following recent work, we assume an early version of the DE model, denoted $\mathcal{S}_1$, which is used to retrieve a pool of hard negative passages $\mathcal{T} = \{p_1, p_2, \ldots, p_t\}$. Out of these t passages, a subset of passages N of size k ($k << t$) are selected to serve as the negative passages during training (also referred to as *negative sampling*). Our work focuses on how these k samples are chosen out of the t available samples, and does not alter with the curation process of the pool $\mathcal{T}$. In this work, following our main baseline coCondenser [7], we use $t = 200$ and $k = 21$.

3 Methods

In this section, we present two methods used for selecting hard negative samples. The first is our main contribution; a sampling method focused on the Informativeness and Diversity (InDi) of the negative samples selected, replacing the currently widely-used random approach. The second is a complementary method addressing the redundancy of negative samples when using in-batch training.

3.1 Informative and Diverse Sampling

For each train query, existing systems sample k hard negatives randomly out of t top results retrieved for this query with some existing retrieval method. We argue that this approach does not realize the full potential of the hard negative examples. Therefore, we devise a new selection method aiming to select the samples that will bring the most benefit to the model. **Requirements** of the hard negative set chosen:

1. Be *diverse*; given we can choose k negative examples per query, we don't want them to be similar so that the model can generalize well.
2. Be *informative*[3]; contain passages that the model is *uncertain* about and cannot easily distinguish from a positive passage.

Inspiration. The Active Learning (AL) setting (see Sect. 6.3) shares our objectives as it also aims at selecting diverse and informative examples. A recent work in this field, BADGE [1], shows that clustering the gradient vectors of textual embeddings is beneficial for selecting diverse and informative examples. While BADGE is tailored for the classification use case and can not be directly applied for dense retrieval, it inspired our sampling method.

Method. We compute for each passage the gradient of its embedding with respect to the contrastive learning loss, assuming it is selected as a negative passage. We then take all gradient vectors and cluster them, using the K-means algorithm with Euclidean distance metric. The number of clusters k is selected as the number of negatives we use throughout training. From each cluster, we select the medoid (i.e., the example whose gradient embedding is closest to the cluster's center) as a representative negative example. An illustration of this method appears in Fig. 1.

In the contrastive loss setting, it is not possible to compute the actual gradient vectors. This is due to the fact that the loss depends on multiple negative passages, which are unknown at the selection time. Therefore, we need to find a way to approximate the loss. We do this by computing a point-wise approximation of the loss, depending on just the query q, the positive example p^+, and a single negative p_j^- from the pool $\mathcal{T}$:

$$\tilde{g}(p_j^-) = \frac{\partial \ell(q, p^+, p_j^-)}{\partial E_p(p_j^-)}.$$

We note that most existing dense retrieval systems work in multiple phases. In the first step, an early version of the model, denoted $\mathcal{S}_1$, is trained. Then, the pool of negatives (from which the actual training negatives are later selected) is created by computing the embedding of all passages in the corpus, and running a kNN retrieval algorithm. Later, the final model is trained using the acquired negative samples. Since the passages embeddings are required regardless of our technique (to run the kNN retrieval algorithm), the gradient vector of each passage can be computed while incurring only a small overhead on top of the existing method.

Intuition. We provide some intuition into why the described method should meet the aforementioned requirements. First, it is intuitive that taking a representative from each cluster will assure *diversity*. As for *informativeness*, since short vectors (i.e., with a low norm) fall in a dense area, when clustering using Euclidean distance many of them can be efficiently clustered in only a few clusters[4]. Therefore, only a few short representatives are likely to be chosen by our method, resulting in a set of negative samples

[3] Informativeness measures the ability of a sample to reduce the uncertainty of a model. Informativeness is commonly approximated by measuring the loss a sample causes the model [15,23]. High loss means high uncertainty, suggesting high informativeness of the sample.

[4] For example, many close-to-zero gradient vectors, pointing in different directions, will all be in the same cluster.

containing mostly long vectors. Long gradient vectors represent negative samples the model is uncertain about, which are considered more informative to the model [15,23].

The informativeness and diversity requirements are revisited after the main experimental results are presented (Sects. 5.7 and 5.5).

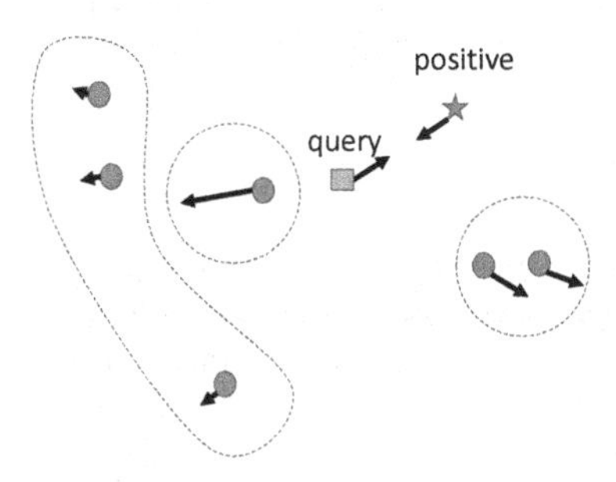

Fig. 1. Illustrating our negative selection process. The blue dots are the negative example. In arrows we mark the approximation of the gradient. We group negatives that, if chosen, will have a similar impact on the model. We then select one representative from each group. (Color figure online)

3.2 In-Batch Deduplication

Repeating training samples in general and especially in the same training batch was shown to negatively affect the performance of the trained model [22]. Luckily, in the standard training scheme used for dense retrieval, this is unlikely. This scheme, utilizes in-batch negatives, i.e., passages selected (as either positive or negative) for one query are also used as negative passages for other queries in the same batch. While this is expected to to increase the probability of duplication, in existing training methods, the negative passages of each query are selected at random from a large pool (e.g., of size 200) of top relevant passages. Thus, even if similar queries with intersecting pools belong to the same batch, it is still unlikely for the randomly selected passage to fall within this intersection.

When using our method this is no longer true. By focusing on highly informative negatives the method removes some of the randomness in the selection, making some passages more likely to be selected. As a result, the event of two similar queries belonging to the same batch may suffice to have the same passage selected more than once. Thus, addressing this potential hazard becomes essential.

We deduplicate each batch by iterating over the batch queries and removing passages already selected for the batch from the query's pool before selecting the hard negatives. The extent of this phenomenon and the effect of this method is demonstrated and discussed in Sect. 5.6.

Table 1. Corpus size is measured in passages, Train/Test size is measured in queries. Following DPR [18], We use the NQ dataset version published in the Tevatron toolkit.

Dataset	Corpus size	Train/Test size
MS-MARCO	8.8M	500K/6,980
NQ	21M	58K/3,610

4 Experimental Setup

We implemented our method in the Tevatron toolkit [8] for dense retrieval. For pre-training, we used the corpus-specific pre-trained checkpoint of coCondenser [7] for MS-MARCO and Natural Questions, which is trained in an unsupervised manner. Following coCondenser, training proceeds in two steps. First, we train the dense retrieval model using BM25 negatives. Based on this model, we gather top 200 results per query and select negatives from this pool either randomly (for the baseline), or via our sampling technique. The model is then trained from scratch using the hard negative examples.
Datasets. We follow many previous works and train and test our methods on the MS-MARCO dataset [30]. To demonstrate robustness and further validate our method, we also evaluate our sampling method on the Natural Questions dataset (NQ) [20]. Datasets information can be found in Table 1. Note, while there are passages published as negative in some versions of these datasets, there are no passages actually verified (i.e. human annotated) as being negative. In both datasets, the given annotation task is to mark/generate an answer for questions and not to annotate passages for relevance. Sometimes negatives are later generated with some heuristic (e.g. passages that do not contain the answer) but are not verified as actually being negative or to being hard.
Measure. Following previous work and official metrics, for MS-MARCO we report $MRR@10$ as the main metric and $Recall@50/@1000$ as complementary metrics and for NQ we report $Recall@5/@20/@1000$. The top results in the same column is marked **bold**, and results that are also better than all other results in the same column in a statistically significant manner, using a relative t-test with $p < 0.01$, are marked with an asterisk (*).
Baselines. Our proposed method is a train-procedure improvement method. Therefore, we mainly want to show using it during training of some base-model leads to a better model compared to training the same base-model without our method. We choose coCondenser as our base model for the following reasons: 1. It is a highly popular dense retrieval solution achieving competitive results. 2. It is not a distillation model which is very expensive to train. 3. The authors publish their code and framework, making the training procedure for this model relatively easy to reproduce. For reference, we also report results on some additional retrieval solutions. These include the classic BM25 lexical matching results, doc2query [32], DeepCT [28], docT5query [31] and SPLADEv2 [3], which are solutions that combine lexical matching and a transformer-based model. We also include dense retrieval methods. These include ANCE [38] that collects hard negatives in an online manner by training a model and building a kNN index in parallel, ME-BERT [27], which uses multiple vectors for representing passages, and ColBERT [19], a multistage model that represents both queries and passages using multiple

queries. Last, we include RocketQA [34], which filters out hard negatives suspected as positives based on a Cross Encoder. The latter four models do not start from a corpus-aware pretraining, and therefore show lower-quality results compared to coCondenser.

Table 2. Main results on MS-MARCO.

Method	MRR@10	R@50	R@1000
BM25	18.7	59.2	85.7
doc2query	21.5	64.4	89.1
DeepCT	24.3	69.0	91.0
docT5query	27.7	75.6	94.7
SPLADEv2	36.8	–	97.9
ANCE	33.0	–	95.9
ME-BERT	33.8	–	–
ColBERT	36.0	82.9	96.8
RocketQA	37.0	85.5	97.9
coCondenser	38.2	86.5	98.4
InDi	**38.8***	**86.6**	**98.5**

5 Results

5.1 Main Results

Main results are presented in Table 2. Our Informative and Diverse (InDi) negative sample selection method improves the training procedure of coCondenser reaching statistically significant better results compared to all baselines on the main official metric ($MRR@10$). Results on other measures ($Recall@50/@1000$) are comparable, this is further discussed and analyzed in Sect. 5.2.

5.2 Fine-Grained Topical Understanding

In the Dual Encoder (DE) training procedure, it is a common practice to use in-batch negatives. When computing the loss for a query q, apart from the negatives chosen specifically for q (i.e., hard negatives) passages chosen for other batch queries are also considered as negatives for q. Since batch queries are usually selected randomly, the passages selected for other queries are likely easy negatives (i.e. very different from positive passages) for q. It is a common belief that easy negatives teach the model a coarse-grained topical understanding while the hard negatives chosen specifically for q teach the model a fine-grained understanding of relevance. For example, easy negatives might teach the model that passages discussing cats or trees are irrelevant for a query about dogs, while hard negatives might teach it that passages discussing giant poodles are not relevant to queries asking about toy poodles.

InDi only affects the hard negative selection. Thus, it is expected for InDi to mainly affect the top-of-the-list results. Indeed, this is demonstrated in the main results in Table 2. InDi provides a statistically significant improvement to $MRR@10$ results, while a numerically higher, but not statistically different result in $Recall@50/1000$. To further corroborate this, we show additional $Recall@k$ values in Table 3, focusing on the top of the list. We find that the improvement of InDi is focused on the top-of-the-list retrieved passages, suggesting it mainly improves the fine-grained understanding of the model, and not the general, more coarse-grained topical understanding.

Table 3. MS-MARCO Recall@k with (InDi) and without (coCondenser) our selection method. Statistical significance in this table is calculated with respect to the rows.

R@k	coCondenser	InDi	Improvement
1	25.5	**26.2***	0.7
2	37.5	**38.7***	1.2
3	45.8	**46.5***	0.7
4	51.4	**52.3***	0.9
5	58.9	**59.2**	0.3
10	67.3	67.3	0.0
20	77.0	77.0	0.0
50	86.5	**86.6**	0.1

5.3 Additional Dataset

To further validate our sampling method, we apply it on the Natural Question (NQ) dataset, and report the results in Table 4. The main theme observed on MS-MARCO (and discussed in Sects. 5.1 and 5.2) is maintained in NQ as well; our sampling method improves results of top-of-the-list retrieved passages, and preserves results when looking at a larger portion of top retrieved passages.

Table 4. Results with (InDi) and without (coCondenser) our sampling method on the Natural Questions dataset.

Method	R@5	R@20	R@100
coCondenser	71.0	82.6	88.7
InDi	**72.1***	**83.2***	88.7

5.4 Diversity and Informativeness Importance

In this section, we aim to verify that both informativeness and diversity are indeed important when selecting hard negatives for Dense Retrieval. We do this by devising two more selection methods, one focusing solely on diversity and the other solely on informativeness. In a diversity-focused paradigm, the different samples are chosen to be as diverse as possible. We achieve this with **D-sampling** by first clustering passages using their embeddings (as generated by $\mathcal{S}_1$) and then choosing a representative from each cluster. This approach is similar to InDi, but it uses the contextual embeddings directly as opposed to the gradient embeddings. While the gradient vectors incorporate the informativeness of a sample (i.e. the model's uncertainty) in them, this is not the case when using the model embedding directly. In an informativeness-focused paradigm the samples that the model is least certain about are chosen. To achieve this we devise **I-sampling**, an informativeness-based method that selects the negative samples that produce the highest loss, i.e., those that are most similar to the query. Therefore, for this method, we take the top k passages retrieved for the query as the hard negative samples.

Results are shown in Table 5. It can be seen that InDi, combining both paradigms, performs best. I-sampling achieved subpar results. We hypothesize this is actually due to a data issue: the MS-MARCO dataset, like many other retrieval datasets, contains many unlabeled positives. I-sampling is prone to choosing these samples as false negatives since they are likely to be ranked high. Other methods may also select unlabeled positives, but are unlikely to select many of them since in their selection they take diversity into consideration, and positives are likely to be somewhat similar to each other.

To verify this hypothesis, we combine I-sampling with a precursory filtering step, where we attempt to eliminate false-negative samples. We follow the line of RocketQA [34], which uses a strong, capable Cross Encoder (CE) model to filter out false negatives[5]. Unlike the DE model, where the query-passage interaction is limited to a single dot product computation, CEs concatenate the query and passage and pass them through transformer layers. This allows the query and passage to interact via the self-attention mechanism, resulting in a more capable model that can be used for negative selection, but naturally comes with a vast computational cost (see Sect. 5.8). CE filtering vastly improves I-sampling (Table 5), suggesting that false negatives are indeed a problem. Still, results are inferior to InDi, showing both diversity and informativeness are important.

5.5 Diversity and Informativeness Measures

We measure informativeness and diversity and compare methods discussed in this paper.
Informativeness. Informativeness is considered as the ability of a sample to reduce the uncertainty of a model [15]. It is commonly approximated by measuring the loss a sample causes the model [23]. High loss means that currently the model is highly uncertain about this sample, suggesting it contains information that the model was not exposed to. Thus, teaching the model using this sample is highly informative.

[5] We find that the best result is achieved with a threshold of 0.8 on the CE score.

Table 5. MRR@10 on MS-MARCO of different methods for negative selection.

Selection Method	MRR@10
I-sampling	32.22
D-sampling	34.67
InDi	**38.84***
I-sampling + CE	37.81

Table 6. Diversity of different selection methods.

Selection Method	Diversity
I-sampling	0.187
Random	0.203
InDi	0.203
D-sampling	0.223

Building on this intuition, we measure informativeness by calculating for each query the contrastive loss of its selected negatives using the embeddings generated by S_1. Then, for each query we assess which selection method produced the highest loss. Finally, we used the fraction of queries for which the loss of one method is higher than the other as a proxy-metric for informativeness. Unsurprisingly, the I-sampling selection method always presents the highest loss. It is expected as this method selects the samples which S_1 is most uncertain about. Setting I-sampling aside, our method showed higher informativeness compared to the random (67% of the queries) and D-sampling (57% of the queries) selection methods. This suggests that it tends to select negatives that are highly informative to the model.

Diversity. We calculate the diversity w.r.t. each query separately and compare the average value across queries. Following Zhdanov [42], given a query q we define the diversity of a subset N selected out of T[6], as follows.

$$D(N) = \left(\frac{1}{|T|} \sum_{x_i \in T} \min_{x_j \in N} d(x_i, x_j) \right)^{-1},$$

where $d(x_i, x_j)$ is the Euclidean distance[7] between the embeddings of x_i and x_j (as calculated by S_1). Results appear in Table 6. D-sampling, as expected, produces the most diverse negatives. I-sampling produces the most homogeneous results. Our method is able to achieve a diversity score equal to that of the random selection. We note that random selection, especially when samples are relatively uniformly distributed, leads to diverse selections. For dense retrieval models using contrastive loss, this is usually the case as shown by Wang et al. [37]. InDi preserves the diversity compared to the already diverse baseline while improving informativeness.

5.6 Ablation

In Sect. 3 we presented the methods on which InDi is based. In this section, we evaluate the contribution of each of the methods separately. The results appear in Table 7. As can be seen, the lion's share of the improvement produced by InDi originates from our selection method.

[6] T represents the top t retrieved samples for q.

[7] We opt to use Euclidean distance as it is desired that the distance to a sample is minimized by the sample itself.

Nevertheless, we see that our in-batch deduplication method is helpful. When the negative samples are selected randomly $\sim 0.05\%$ of batches include at least one duplicated sample. Even though this is only a small fraction of the batches, deduplicating these samples led to a non-negligible performance improvement. In the case of negative samples selected by InDi this phenomenon is more common as $\sim 1.6\%$ of batches include at least one duplicated sample. Indeed, deduplication with InDi leads to an increased improvement in performance.

Table 7. Ablation study comparing MRR@10 on MS-MARCO of each of our methods.

	w/o deduplication	deduplication
Random	38.2	38.33
InDi	**38.66***	**38.84***

5.7 Cross Encoder Filtering with InDi

In Sect. 5.4 we saw that the I-sampling method, which prioritizes informativeness, vastly benefits from CE-filtering where false-negative samples are filtered. As InDi, like I-sampling, also selects highly informative samples, it raises the question whether this filtering step can further improve our results.

We enhance the filtering step by defining two thresholds, ce_{min} and ce_{max}, bounding the allowed relevance score of the hard negative passages. In some cases insufficient number of passages pass the filter[8], which makes our clustering step degenerate, leading to subpar performance. To alleviate this issue, we expand the acceptable relevance range by first increasing ce_{max} up to 0.95 until there are enough passages to choose from. If this does not suffice, we also decrease ce_{min} downward (down to 0) until there is a sufficient number of candidates.

In Table 8 we show results using CE-filtering. Results show that performance can be improved by using CE filtering, but only slightly. Further, results suggest that while using $ce_{min} > 0$ consistently improves results, showing again the importance of selecting hard negatives, for ce_{max} improvement is less clear. This is likely due to the following trade-off: as increasing ce_{max} should reduce the number of false-negative passages, it also decreases the hardness and informativeness of the passages surviving the filter.

[8] To make sure the number of passages is sufficient, we define a minimal ratio between the number of passages that pass the CE-filtering and the number of negatives selected.

Table 8. Parameter study for InDi with and without CE filtering. Min and Max are ce_{min} and ce_{max}.

CE-filtering		MS-MARCO
Min	Max	MRR@10
–	–	38.84
0.0	0.5	38.25
0.0	0.6	38.42
0.0	0.7	38.66
0.1	0.5	38.99
0.1	0.6	39.03
0.1	0.7	39.03
0.2	0.5	39.07
0.2	0.6	39.11
0.2	0.7	39.02

5.8 Computational Surplus

Our sampling solution, while demonstrating strong results, is also very efficient. It does not require training a new model or running inference (as opposed to using a CE[9]) and adds only a small overhead to the training procedure of the DE. While training a DE on the MS-MARCO train set takes roughly 48 hours[10], sampling negatives using our method adds only $\sim$ 0.5 hours ($\sim$ 1% degradation)[11]. Negative sampling using a CE (assuming a trained model is already available) adds more than 50 hours ($\sim$ 100% degradation).

6 Related Work

6.1 Retrieval

Retrieving relevant passages from a textual corpus given a textual query is a highly studied problem with many practical use-cases. Early solutions, such as IDF [17], TF-IDF [36] and BM-25 [35] are based on lexical matching. These solutions suffer from the vocabulary mismatch problem [14], where a query does not share any token with a relevant passage. Modern retrieval systems use SOTA embedding models to embed both queries and passages and search for passages similar to the query in the embedding space [6,7,11,12,18,34,38]. Training such models is a highly active area of research [41] due to the many benefits in terms of quality and efficiency. Today, SoTA results are achieved using distillation from a CE model [24,26]. However, these solutions suffer from the need to train (possibly multiple) CE models and continue training them during

[9] RocketQA trains and inferences an ERNIE-large model.
[10] Times are measured using an NVIDIA T4 GPU.
[11] Our sampling method requires only a CPU. Time was measured on a 4-core machine.

distillation, which is very costly. Hence, the lightweight training scenario (where we cannot run CE training or inference) remains an important challenge.

6.2 Negative Selection for Dense Retrieval

Since dense retrieval solutions have been introduced, much effort has been put into negative sample selection. It was shown that the number of negatives, as well as their difficulty, has a large impact on the model's quality [18,34,38]. DPR [18] uses in-batch negatives to increase the number of negatives per query, as well as BM25-retrieved hard negatives. In TAS [13], authors try to make in-batch negatives harder by clustering similar queries together and sampling queries from the same cluster to build batches. RocketQA [34] introduced cross-batch negatives; When using multiple GPUs for training it's possible to share information and use passages from other GPUs as negatives also. This increases the number of negative examples available compared to the in-batch technique, but does not assure the selected negatives are hard. While multiple works suggest different hard-negative sampling heuristics [18,25,38] as far as we know, they all end with some random sampling.

6.3 Active Learning

As stated in Sect. 3.1, the Active Learning (AL) [4,33] setting shares the uncertainty and diversity objectives discussed in our work. In this line-of-work, samples are selected for labeling out of a pool of unlabeled samples. The selected samples are then labeled and used for training a model. The goal is to maximize the performance of the model while minimizing the number of needed annotations. Recent AL works in the NLP domain mainly focus on optimizing uncertainty and diversity [1,9,39]. BADGE [1], the algorithm inspiring our selection method, combines the two by using the gradient vectors each sample generates for the model. These embeddings are clustered and a representative from each cluster is sent to annotators.

7 Conclusions

In this work we define two desired qualities for negatives selected for training dense retrieval models: *Informativeness* and *Diversity*. We present InDi, a negative sampling technique aiming at optimizing these qualities. We show the aforementioned qualities are indeed important (Sect. 5.4) and that InDi is able to balance between them (Sect. 5.5). This results in improved fine-grained understanding of the model, directly leading to a statistically significant increase in the MRR@10 result on MS MARCO as well as *Recall*@5 and *Recall*@20 on NQ. InDi is also superior to filtering negatives using CE scores, while being more than $100x$ faster. Both of these techniques can be combined for further benefit. Overall, we believe that InDi should be considered as an alternative to the existing random method for negative selection.

References

1. Ash, J.T., Zhang, C., Krishnamurthy, A., Langford, J., Agarwal, A.: Deep batch active learning by diverse, uncertain gradient lower bounds. In: 8th International Conference on Learning Representations, ICLR 2020, Addis Ababa, Ethiopia, April 26–30, 2020. OpenReview.net (2020)
2. Devlin, J., Chang, M., Lee, K., Toutanova, K.: BERT: pre-training of deep bidirectional transformers for language understanding. In: Burstein, J., Doran, C., Solorio, T. (eds.) Proceedings of the 2019 Conference of the North American Chapter of the Association for Computational Linguistics: Human Language Technologies, NAACL-HLT 2019, Minneapolis, MN, USA, June 2–7, 2019, Volume 1 (Long and Short Papers). pp. 4171–4186. Association for Computational Linguistics (2019)
3. Formal, T., Lassance, C., Piwowarski, B., Clinchant, S.: Splade v2: Sparse lexical and expansion model for information retrieval. arXiv preprint arXiv:2109.10086 (2021)
4. Fu, Y., Zhu, X., Li, B.: A survey on instance selection for active learning. Knowl. Inf. Syst. **35**(2), 249–283 (2013)
5. Gao, L., Callan, J.: Condenser: a pre-training architecture for dense retrieval. In: Moens, M., Huang, X., Specia, L., Yih, S.W. (eds.) Proceedings of the 2021 Conference on Empirical Methods in Natural Language Processing, EMNLP 2021, Virtual Event / Punta Cana, Dominican Republic, 7–11 November, 2021. pp. 981–993. Association for Computational Linguistics (2021)
6. Gao, L., Callan, J.: Condenser: a pre-training architecture for dense retrieval. arXiv preprint arXiv:2104.08253 (2021)
7. Gao, L., Callan, J.: Unsupervised corpus aware language model pre-training for dense passage retrieval. In: Proceedings of the 60th Annual Meeting of the Association for Computational Linguistics (Volume 1: Long Papers). pp. 2843–2853. Association for Computational Linguistics, Dublin, Ireland (May 2022)
8. Gao, L., Ma, X., Lin, J., Callan, J.: Tevatron: An efficient and flexible toolkit for dense retrieval. CoRR abs/2203.05765 (2022)
9. Gissin, D., Shalev-Shwartz, S.: Discriminative active learning. CoRR abs/1907.06347 (2019)
10. Guo, R., Sun, P., Lindgren, E., Geng, Q., Simcha, D., Chern, F., Kumar, S.: Accelerating large-scale inference with anisotropic vector quantization. In: International Conference on Machine Learning (2020)
11. Guu, K., Lee, K., Tung, Z., Pasupat, P., Chang, M.: Retrieval augmented language model pre-training. In: International Conference on Machine Learning. pp. 3929–3938. PMLR (2020)
12. Hofstätter, S., Lin, S.C., Yang, J.H., Lin, J., Hanbury, A.: Efficiently teaching an effective dense retriever with balanced topic aware sampling. In: Proceedings of the 44th International ACM SIGIR Conference on Research and Development in Information Retrieval. pp. 113–122 (2021)
13. Hofstätter, S., Lin, S., Yang, J., Lin, J., Hanbury, A.: Efficiently teaching an effective dense retriever with balanced topic aware sampling. In: Diaz, F., Shah, C., Suel, T., Castells, P., Jones, R., Sakai, T. (eds.) SIGIR '21: The 44th International ACM SIGIR Conference on Research and Development in Information Retrieval, Virtual Event, Canada, July 11–15, 2021. pp. 113–122. ACM (2021)
14. Huang, P.S., He, X., Gao, J., Deng, L., Acero, A., Heck, L.: Learning deep structured semantic models for web search using clickthrough data. In: Proceedings of the 22nd ACM international conference on Information & Knowledge Management. pp. 2333–2338 (2013)
15. Huang, S.J., Jin, R., Zhou, Z.H.: Active learning by querying informative and representative examples. Advances in neural information processing systems 23 (2010)

16. Johnson, J., Douze, M., Jégou, H.: Billion-scale similarity search with GPUs. IEEE Transactions on Big Data **7**(3), 535–547 (2019)
17. Jones, K.S.: A statistical interpretation of term specificity and its application in retrieval. Journal of documentation (1972)
18. Karpukhin, V., Oguz, B., Min, S., Lewis, P.S.H., Wu, L., Edunov, S., Chen, D., Yih, W.: Dense passage retrieval for open-domain question answering. In: Webber, B., Cohn, T., He, Y., Liu, Y. (eds.) Proceedings of the 2020 Conference on Empirical Methods in Natural Language Processing, EMNLP 2020, Online, November 16–20, 2020. pp. 6769–6781. Association for Computational Linguistics (2020)
19. Khattab, O., Zaharia, M.: Colbert: Efficient and effective passage search via contextualized late interaction over bert. In: Proceedings of the 43rd International ACM SIGIR conference on research and development in Information Retrieval. pp. 39–48 (2020)
20. Kwiatkowski, T., Palomaki, J., Redfield, O., Collins, M., Parikh, A., Alberti, C., Epstein, D., Polosukhin, I., Devlin, J., Lee, K., Toutanova, K., Jones, L., Kelcey, M., Chang, M.W., Dai, A.M., Uszkoreit, J., Le, Q., Petrov, S.: Natural questions: A benchmark for question answering research. Transactions of the Association for Computational Linguistics **7**, 452–466 (2019)
21. Kwiatkowski, T., Palomaki, J., Redfield, O., Collins, M., Parikh, A.P., Alberti, C., Epstein, D., Polosukhin, I., Devlin, J., Lee, K., Toutanova, K., Jones, L., Kelcey, M., Chang, M., Dai, A.M., Uszkoreit, J., Le, Q., Petrov, S.: Natural questions a benchmark for question answering research. Trans. Assoc. Comput. Linguistics **7**, 452–466 (2019)
22. Lee, K., Ippolito, D., Nystrom, A., Zhang, C., Eck, D., Callison-Burch, C., Carlini, N.: Deduplicating training data makes language models better. arXiv preprint arXiv:2107.06499 (2021)
23. Lewis, D.D.: A sequential algorithm for training text classifiers: Corrigendum and additional data. In: Acm Sigir Forum. vol. 29, pp. 13–19. ACM New York, NY, USA (1995)
24. Lin, Z., Gong, Y., Liu, X., Zhang, H., Lin, C., Dong, A., Jiao, J., Lu, J., Jiang, D., Majumder, R., et al.: Prod: Progressive distillation for dense retrieval. arXiv preprint arXiv:2209.13335 (2022)
25. Lu, J., Ábrego, G.H., Ma, J., Ni, J., Yang, Y.: Multi-stage training with improved negative contrast for neural passage retrieval. In: Moens, M., Huang, X., Specia, L., Yih, S.W. (eds.) Proceedings of the 2021 Conference on Empirical Methods in Natural Language Processing, EMNLP 2021, Virtual Event / Punta Cana, Dominican Republic, 7–11 November, 2021. pp. 6091–6103. Association for Computational Linguistics (2021)
26. Lu, Y., Liu, Y., Liu, J., Shi, Y., Huang, Z., Sun, S.F.Y., Tian, H., Wu, H., Wang, S., Yin, D., et al.: Ernie-search: Bridging cross-encoder with dual-encoder via self on-the-fly distillation for dense passage retrieval. arXiv preprint arXiv:2205.09153 (2022)
27. Luan, Y., Eisenstein, J., Toutanova, K., Collins, M.: Sparse, dense, and attentional representations for text retrieval. Transactions of the Association for Computational Linguistics **9**, 329–345 (2021)
28. Mackenzie, J., Dai, Z., Gallagher, L., Callan, J.: Efficiency implications of term weighting for passage retrieval. In: Proceedings of the 43rd International ACM SIGIR Conference on Research and Development in Information Retrieval. pp. 1821–1824 (2020)
29. Nguyen, T., Rosenberg, M., Song, X., Gao, J., Tiwary, S., Majumder, R., Deng, L.: MS MARCO: A human generated machine reading comprehension dataset. In: Besold, T.R., Bordes, A., d'Avila Garcez, A.S., Wayne, G. (eds.) Proceedings of the Workshop on Cognitive Computation: Integrating neural and symbolic approaches 2016 co-located with the 30th Annual Conference on Neural Information Processing Systems (NIPS 2016), Barcelona, Spain, December 9, 2016. CEUR Workshop Proceedings, vol. 1773. CEUR-WS.org (2016)

30. Nguyen, T., Rosenberg, M., Song, X., Gao, J., Tiwary, S., Majumder, R., Deng, L.: MS MARCO: A human generated machine reading comprehension dataset. In: Besold, T.R., Bordes, A., d'Avila Garcez, A.S., Wayne, G. (eds.) Proceedings of the Workshop on Cognitive Computation: Integrating neural and symbolic approaches 2016 co-located with the 30th Annual Conference on Neural Information Processing Systems (NIPS 2016), Barcelona, Spain, December 9, 2016. CEUR Workshop Proceedings, vol. 1773. CEUR-WS.org (2016)
31. Nogueira, R., Lin, J., Epistemic, A.: From doc2query to docttttquery. Online preprint 6 (2019)
32. Nogueira, R., Yang, W., Lin, J., Cho, K.: Document expansion by query prediction. arXiv preprint arXiv:1904.08375 (2019)
33. Prince, M.: Does active learning work? a review of the research. J. Eng. Educ. **93**(3), 223–231 (2004)
34. Qu, Y., Ding, Y., Liu, J., Liu, K., Ren, R., Zhao, W.X., Dong, D., Wu, H., Wang, H.: Rocketqa: An optimized training approach to dense passage retrieval for open-domain question answering. In: Toutanova, K., Rumshisky, A., Zettlemoyer, L., Hakkani-Tür, D., Beltagy, I., Bethard, S., Cotterell, R., Chakraborty, T., Zhou, Y. (eds.) Proceedings of the 2021 Conference of the North American Chapter of the Association for Computational Linguistics: Human Language Technologies, NAACL-HLT 2021, Online, June 6–11, 2021. pp. 5835–5847. Association for Computational Linguistics (2021)
35. Robertson, S.E., Walker, S., Jones, S., Hancock-Beaulieu, M.M., Gatford, M., et al.: Okapi at trec-3. Nist Special Publication Sp **109**, 109 (1995)
36. Salton, G., Buckley, C.: Term-weighting approaches in automatic text retrieval. Information processing & management **24**(5), 513–523 (1988)
37. Wang, T., Isola, P.: Understanding contrastive representation learning through alignment and uniformity on the hypersphere. In: Proceedings of the 37th International Conference on Machine Learning, ICML 2020, 13–18 July 2020, Virtual Event. Proceedings of Machine Learning Research, vol. 119, pp. 9929–9939. PMLR (2020)
38. Xiong, L., Xiong, C., Li, Y., Tang, K., Liu, J., Bennett, P.N., Ahmed, J., Overwijk, A.: Approximate nearest neighbor negative contrastive learning for dense text retrieval. In: 9th International Conference on Learning Representations, ICLR 2021, Virtual Event, Austria, May 3–7, 2021. OpenReview.net (2021)
39. Yuan, M., Lin, H., Boyd-Graber, J.L.: Cold-start active learning through self-supervised language modeling. In: Webber, B., Cohn, T., He, Y., Liu, Y. (eds.) Proceedings of the 2020 Conference on Empirical Methods in Natural Language Processing, EMNLP 2020, Online, November 16–20, 2020. pp. 7935–7948. Association for Computational Linguistics (2020)
40. Zhao, W.X., Liu, J., Ren, R., Wen, J.R.: Dense text retrieval based on pretrained language models: A survey. arXiv preprint arXiv:2211.14876 (2022)
41. Zhao, W.X., Liu, J., Ren, R., Wen, J.R.: Dense text retrieval based on pretrained language models: A survey. arXiv preprint arXiv:2211.14876 (2022)
42. Zhdanov, F.: Diverse mini-batch active learning. arXiv preprint arXiv:1901.05954 (2019)

Short Papers

Enhancing Legal Named Entity Recognition Using RoBERTa-GCN with CRF: A Nuanced Approach for Fine-Grained Entity Recognition

Arihant Jain and Raksha Sharma(✉)

Indian Institute of Technology Roorkee, Roorkee, India
{arihant_j,raksha.sharma}@cs.iitr.ac.in
https://www.iitr.ac.in/ CSE/Raksha_Sharma

Abstract. Accurate identification of named entities is pivotal for the advancement of sophisticated legal Artificial Intelligence (AI) applications. However, the legal domain presents distinct challenges due to the presence of fine-grained, domain-specific entities, including *lawyers, judges, courts*, and *precedents*. This necessitates a nuanced approach to Named Entity Recognition (NER).

In this paper, we introduce a novel NER approach tailored to the legal domain. Our system combines Robustly Optimized BERT (RoBERTa) with a Graph Convolutional Network (GCN) to harness two distinct types of complementary information related to words in the data. Furthermore, the application of a Conditional Random Field (CRF) at the output layer ensures global consistency in data labeling by considering the entire sequence when predicting a named entity. RoBERTa captures contextual information about individual words, while GCN allows us to exploit the mutual relationships between words, resulting in more precise named entity identification. Our results indicate that RoBERTa-GCN (CRF) outperforms other standard settings, such as, RoBERTa, textGCN, and BiLSTM, including state-of-the-art for NER in the legal domain.

Keywords: Legal Domain · Pretrained Language Models · Named Entity Recognition · Conditional Random Fields

1 Introduction

Populous countries like India are grappling with the challenge of high case pendency, with a substantial backlog of pending cases. As of 2023, the Indian judicial system continues to face a significant burden, with millions of cases awaiting resolution. According to the latest available data from the National Judicial Data Grid (NJDG)[1], as of May 2023, there are over 43 million pending cases in Indian

[1] https://njdg.ecourts.gov.in/njdgnew/index.php.

N. Goharian et al. (Eds.): ECIR 2024, LNCS 14610, pp. 261–267, 2024.
https://doi.org/10.1007/978-3-031-56063-7_17

courts, highlighting the magnitude of the issue. This pressing issue necessitates the adoption of Artificial Intelligence (AI) to alleviate the strain on the judicial system and enhance access to justice. By harnessing the power of AI, there is potential to improve efficiency, reduce pendency, and facilitate more streamlined legal processes. Standard NLP-based NER tools like NLTK and spaCy are designed to identify and label generic named entities such as *person*, *organization*, *location*, *date etc.*. However, when it comes to legal texts, these generic entities do not provide the level of detail and specificity required. Legal texts contain a wide range of fine-grained entities that are unique to the legal domain, such as *court names, judge names, legal statutes, case citations, legal principles*, and specific roles like *lawyers* and *defendants*. Following is a preamble text from a legal document having three domain-specific named entities. "State of Kerela - versus - Raneef - Judgement - Markandey Katju". Specifically, these entitites are *Appellant* (State of Kerala), *Respondent* (Raneef) and *Judge* (Markandey Katju). However, we observed that spaCy's NER system failed to identify even generic named entities, such as, *person* and *place* for this text belonging to legal domain. This emphasises the challenge of using standard NLP tools for fine-grained categorization in legal texts.

This paper presents a novel approach for conducting named entity recognition within the legal domain in the Indian context. Our approach is a hybrid system that combines the power of Robustly Optimized BERT (RoBERTa) with a Graph Convolutional Network (GCN). RoBERTa effectively captures contextual information related to individual words, while GCN enables us to leverage the inherent relationships between words. The ultimate output is generated through the use of Conditional Random Field (CRF). The results demonstrate that our RoBERTa-GCN (CRF) approach surpasses the current state-of-the-art named entity recognition system for the legal domain as reported in Kalamkar et al., 2022 [4]. Additionally, our system exhibits a noteworthy performance improvement compared to other standard methods such as RoBERTa, textGCN [12], and BiLSTM when applied to named entity recognition tasks within the legal domain.

2 Related Work

Named Entity Recognition (NER) is widly studied among Natural Language Processing (NLP) researchers [1,2,5,6,8,11]. However, it is worth noting that the nature and nomenclature of entities vary according to specific domains, hence NER is a domain-specific task. Considering the significance of applying NLP to advance the judicial system, NER in legal domain has recently grabbed the attention of researchers. Huang et al., 2015 [3] proposed Bidirectional LSTM-CRF Models for Sequence Tagging, which have been widely used in various NER tasks. They demonstrated the effectiveness of these models in capturing contextual information and modeling the sequential nature of legal texts. Leitner et al. 2019 [7] presented a valuable resource, a dataset of German legal documents which focuses on NER in German legal texts. The dataset provided by their work serves as a valuable benchmark for evaluating NER models in the Gernam legal domain. Vasile Pais et al., 2019 [9] explored the challenges of NER in the Romanian legal domain. Their study, *Named entity recognition in the Romanian legal domain*, highlighted the need for domain-specific approaches to handle the unique characteristics and fine-grained entities present in legal texts. Nils Reimers et al., 2014 [10] introduced the GermEval-2014 dataset, which focused on nested named entity recognition using neural networks. Their work addressed the complex nature of nested entities and showcased the effectiveness of neural network-based approaches in capturing hierarchical structures.

Our research takes inspiration from Kalamkar et al.'s *NER in Indian Court Judgements* [4]. They introduced a transformer-based model (RoBERTa, an acronym for Robustly Optimized BERT) for NER in Indian court judgments, setting the stage for further developments within this domain. In this paper, we propose a system that builds upon the insights and methodologies of prior research to contribute to the field of Named Entity Recognition (NER) within the legal domain, with a specific focus on Indian legal texts. Our approach yields an F-1 score of 87.84%, a significant improvement over the current state-of-the-art system [4].

3 Dataset

The dataset utilized in our study is the result of Kalamkar et al.'s 2022 endeavor in Named Entity Recognition (NER) within the Indian judicial system [4]. This dataset comprises a collection of 14,444 sentences extracted from Indian court judgments, along with 2,126 judgment preambles, all meticulously annotated with 14 distinct legal named entities. This dataset was curated with the specific purpose of facilitating tasks related to legal named entity recognition (NER). The dataset facilitates fine-grained annotations for the legal named entities, *viz.*, COURT, PETITIONER, RESPONDENT, JUDGE, LAWYER, DATE, ORG, GPE, STATUTE, PROVISION, PRECEDENT,CASE-NUMBER, WITNESS, and OTHER-PERSON. In addition, a representative sample of Indian high court and supreme court judgments was selected for training and testing. The sample consists of 11,970 judgments, covering eight case types across 29 Indian courts. The training and development data, along with a trained baseline model for extracting legal named entities from judgment text, is available on the GitHub repository.[2]

4 Methodology

In this paper, our objective is to extend the boundaries of named entity recognition within the context of the Indian legal domain. We experimented with two variations of the RoBERTa-GCN model for named entity recognition: one with a softmax classifier and another with a CRF classifier(as shown in Fig. 1).

[2] https://github.com/Legal-NLP-EkStep/legal_NER.

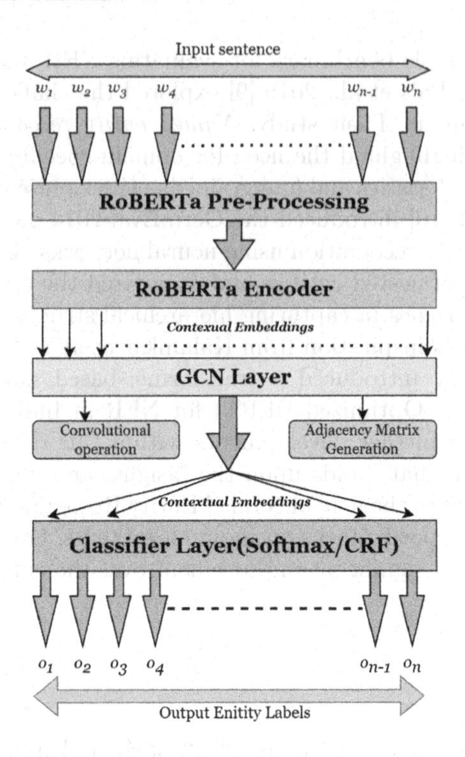

Fig. 1. Ensemble architecture of RoBERTa-GCN model with softmax/CRF as classifier

For the RoBERTa-GCN model with a softmax classifier, we started with the RoBERTa model, which is a robust pre-trained language model known for its ability to capture contextual information. RoBERTa excels in the legal domain due to its fine-grained contextual understanding, making it adept at recognizing complex entities and handling the intricate language often found in legal documents. Its large-scale pre-training and adaptability to legal terminology contribute to superior performance in tasks such as Named Entity Recognition within the legal domain. We enhanced its performance in the Indian legal domain by incorporating the Graph Convolutional Network (GCN) architecture. The incorporation of the Graph Convolutional Network (GCN) after obtaining RoBERTa embeddings enhances accuracy by capturing contextual relationships and dependencies between words. This allows for a more nuanced understanding of semantic connections within legal text, leading to improved named entity identification. We then employed a softmax classifier to assign entity labels to the identified entities based on their contextual features. In addition to this version, we also explored the integration of a Conditional Random Field (CRF) classifier into the RoBERTa-GCN architecture. The CRF classifier considers the dependencies between neighboring labels, ensuring the sequential coherence of the named entity predictions. By incorporating the CRF layer, we aimed to capture the sequential patterns and constraints within the named entity labels, further improving the model's performance in recognizing named entities. Both adaptations of the RoBERTa-GCN model were meticulously trained using a dataset specifically tailored for Named Entity Recognition within Indian court judgments. These models underwent fine-tuning using deep learning frameworks, allowing for the optimization of model parameters and hyper-parameters.

5 Experimental Setup

This section presents the experimental details of our RoBERTa-GCN model for named entity recognition (NER) on the legal dataset. We used a dataset consisting of 46,545 labeled examples. The RoBERTa-GCN model architecture was built by integrating the RoBERTa language model with a Graph Convolutional Network (GCN) layer. We utilized the Hugging Face Transformers library for RoBERTa implementation and Pytorch for Graph Convolutional networks. During training, we employed the Adam optimization algorithm with a learning rate of 0.001, with momentum (beta1) and variance (beta2) set to 0.9 and 0.999, respectively. We used a batch size of 32 and trained the model for 50 epochs. We applied a dropout rate of 0.2 to mitigate over-fitting and employed early stopping based on the validation performance. Hyperparameter settings for our RoBERTa-GCN model included a single GCN layer with a hidden dimension of 256. In the evaluation phase, we measured performance using precision, recall, and the F1 score as our primary metrics. To assess the effectiveness of our approach, we conducted a comparative analysis against established baselines, including fine-tuning the RoBERTa transformer, Text-GCN, and BiLSTM-CRF models from previous research works. The results, as presented in Table 1, underscore our model's ability to leverage the advantages of both GCN and BERT, leading to a robust NER system tailored for the legal domain.

Table 1. Comparative Analysis of F1 Scores Among Various Models

S.No	Model	F1-score
1	RoBERTa ([4])	78.80%
2	Text-GCN (softmax)	83.50%
3	RoBERTa-GCN (softmax)	86.40%
4	RoBERTa-GCN-CRF (Our Approach)	87.84%
5	BiLSTM-CRF	85.34%

6 Results

Table 1 offers a comprehensive comparison of the results obtained using various configurations and our approach for Named Entity Recognition (NER)

within the legal domain. State-of-the-art, the fine-tuned RoBERTa transformer by Kalamkar et al., 2022, [4], is based on a pre-trained language model and has achieved an F1-score of 78.80%. Additionally, the Text-GCN model with a softmax classifier showcased promising outcomes, achieving a notable F1-score of 83.50%. In a similar vein, the BiLSTM-CRF model, utilizing Corollo-Marcell embeddings introduced by Vasile Pais et al., 2022, [9], attained an F1-score of 85.34%.

However, our primary objective was to enhance performance further by integrating the Graph Convolutional Network (GCN) architecture with RoBERTa. Our proposed model, RoBERTa-GCN with a softmax classifier, demonstrated significant improvements, resulting in a noteworthy F1-score of 86.40%. This underscores the value of incorporating GCN, which captures contextual relationships among words, thereby enhancing the model's capability to recognize named entities within the Indian judicial domain. Furthermore, we extended the RoBERTa-GCN architecture by integrating a Conditional Random Field (CRF) classifier. This variant, RoBERTa-GCN with a CRF classifier, achieved the highest F1-score of 87.84%. The inclusion of the CRF layer takes into account the sequential dependencies between adjacent labels, thus enhancing the overall coherence of predictions related to named entities. Table 2 provides the F1 score by RoBERTa-GCN-CRF for each entity type individually.

Table 2. Entity wise results for RoBERTa-GCN-CRF (Our Approach)

Entity	Precision	Recall	F1 Score
COURT	88.15%	86.25%	90.15%
PETITIONER	87.79%	92.12%	83.84%
RESPONDANT	70.65%	71.32%	69.99%
JUDGE	89.12%	88.14%	90.13%
LAWYER	89.42%	90.60%	88.27%
DATE	84.59%	83.24%	85.98%
ORG	81.57%	83.21%	80.00%
GPE	78.83%	79.77%	77.91%
STATUTE	92.05%	92.24%	91.87%
PROVISION	89.97%	89.16%	90.80%
PRECEDENT	73.88%	75.76%	72.10%
CASE_NUMBER	81.24%	82.45%	80.07%
WITNESS	85.22%	86.19%	84.28%
OTHER_PERSON	87.08%	88.15%	86.03%
Average F1 Score			87.84%

7 Conclusion

Recognizing Named Entities (NER) within the legal domain is a domain-specific task that demands fine-grained classification. This paper introduces an NER system tailored for the legal domain in the context of India. Our approach entails the fusion of RoBERTa with Graph Convolutional Networks (GCN) to facilitate entity recognition. Our system exhibits the capability to identify a total of 14 distinct entity types commonly found in Indian legal judgments. The results demonstrate the efficacy of our model, RoBERTa-GCN (CRF), surpassing all other baseline models by a substantial margin.

References

1. Chiu, J.P., Nichols, E.: Named entity recognition with bidirectional LSTM-CNNs. Trans. Assoc. Comput. Linguist. **4**, 357–370 (2016)
2. Finkel, J.R., Manning, C.D.: Nested named entity recognition. In: Proceedings of the 2009 Conference on Empirical Methods in Natural Language Processing, pp. 141–150 (2009)
3. Huang, Z., Xu, W., Yu, K.: Bidirectional LSTM-CRF models for sequence tagging. arXiv preprint arXiv:1508.01991 (2015)
4. Kalamkar, P., Agarwal, A., Tiwari, A., Gupta, S., Karn, S., Raghavan, V.: Named entity recognition in Indian court judgments. In: Proceedings of the Natural Legal Language Processing Workshop 2022, pp. 184–193. Association for Computational Linguistics, Abu Dhabi, United Arab Emirates (Hybrid), December 2022. https://aclanthology.org/2022.nllp-1.15
5. Kuru, O., Can, O.A., Yuret, D.: Charner: Character-level named entity recognition. In: Proceedings of COLING 2016, the 26th International Conference on Computational Linguistics: Technical Papers, pp. 911–921 (2016)
6. Lample, G., Ballesteros, M., Subramanian, S., Kawakami, K., Dyer, C.: Neural architectures for named entity recognition. arXiv preprint arXiv:1603.01360 (2016)
7. Leitner, E., Rehm, G., Moreno-Schneider, J.: A dataset of German legal documents for named entity recognition. In: Proceedings of the 22nd Nordic Conference on Computational Linguistics, pp. 381–386 (2019)
8. Marrero, M., Urbano, J., Sánchez-Cuadrado, S., Morato, J., Gómez-Berbís, J.M.: Named entity recognition: fallacies, challenges and opportunities. Comput. Stand. Interfaces **35**(5), 482–489 (2013)
9. Pais, V., Mitrofan, M., Gasan, C.L., Coneschi, V., Ianov, A.: Named entity recognition in the romanian legal domain. In: Proceedings of the 2019 Conference on Empirical Methods in Natural Language Processing and the 9th International Joint Conference on Natural Language Processing (EMNLP-IJCNLP), pp. 157–168 (2019)
10. Reimers, N., Eckle-Kohler, J., Schnober, C., Kim, J., Gurevych, I.: Germeval-2014: Nested named entity recognition with neural networks. In: Proceedings of the 4th Workshop on Semantic Evaluation (SemEval-2014), pp. 244–250 (2014)
11. Ritter, A., Clark, S., Etzioni, O., et al.: Named entity recognition in tweets: an experimental study. In: Proceedings of the 2011 Conference on Empirical Methods in Natural Language Processing, pp. 1524–1534 (2011)
12. Yao, L., Mao, C., Luo, Y.: Graph convolutional networks for text classification. arXiv preprint arXiv:2008.05959 (2020)

A Novel Multi-Stage Prompting Approach for Language Agnostic MCQ Generation Using GPT

Subhankar Maity(✉), Aniket Deroy, and Sudeshna Sarkar

IIT Kharagpur, Kharagpur, India
{subhankar.ai,roydanik18}@kgpian.iitkgp.ac.in, sudeshna@cse.iitkgp.ac.in

Abstract. We introduce a multi-stage prompting approach (MSP) for the generation of multiple choice questions (MCQs), harnessing the capabilities of GPT models such as `text-davinci-003` and GPT-4, renowned for their excellence across various NLP tasks. Our approach incorporates the innovative concept of *chain-of-thought* prompting, a progressive technique in which the GPT model is provided with a series of interconnected cues to guide the MCQ generation process. Automated evaluations consistently demonstrate the superiority of our proposed MSP method over the traditional single-stage prompting (SSP) baseline, resulting in the production of high-quality distractors. Furthermore, the one-shot MSP technique enhances automatic evaluation results, contributing to improved distractor generation in multiple languages, including English, German, Bengali, and Hindi. In human evaluations, questions generated using our approach exhibit superior levels of *grammaticality*, *answerability*, and *difficulty*, highlighting its efficacy in various languages.

Keywords: Question Generation (QG) · Multiple Choice Questions (MCQs) · Chain-of-Thought(CoT) · Prompt · GPT

1 Introduction and Background

Multiple choice questions (MCQs) are a common way to assess learner knowledge, but manually creating them is time-consuming and demanding for educators [27]. This underscores the importance of automating MCQ generation. The prompting approach involves enhancing the performance of a large language model (LLM) in downstream tasks by providing additional information, known as a "prompt" to condition its generation [12]. Recently, the use of prompts has gained prominence in several natural language generation (NLG) tasks, such as interview question generation (QG) [18], summarization [1], machine translation (MT) [24], etc. These approaches take advantage of state-of-the-art GPT models. However, there is limited exploration of prompt-based GPT models for generating MCQs, which presents a unique challenge in harnessing the full potential of these GPT models. Our work seeks to address this gap by investigating the

N. Goharian et al. (Eds.): ECIR 2024, LNCS 14610, pp. 268–277, 2024.
https://doi.org/10.1007/978-3-031-56063-7_18

application of prompt-based GPT models in the creation of MCQs. Very few works explore prompt-based LLMs for the generation of MCQs in multilingual settings (especially for low-resource languages) [10]. This approach is crucial to addressing language barriers, promoting accessibility, and advancing education in underserved communities. Inspired by recent advances in *chain-of-thought* (CoT) prompting, we seek to bridge the gap between human-like reasoning and the ability of GPT models to generate MCQs. In the work by Wei et al. [28], few-shot CoT prompting demonstrates the potential to explicitly generate intermediate reasoning steps, facilitating a more accurate prediction of the final answer with step-by-step manual reasoning demonstrations. Additionally, zero-shot CoT, as introduced by Kojima et al. [7], remarkably improves the performance of LLMs without requiring manually crafted examples by simply appending "*Let's think step by step*" to the prompt. Moreover, a work by Tan et al. [24] utilizes the multi-stage prompting (MSP) approach to improve the task of MT using GPT. In this paper, we extend the idea of CoT to the domain of MCQ generation using MSP [20]. Our approach involves four essential stages: paraphrase generation, keyword extraction, QG, and distractor generation. By guiding the GPT models through these sequential interactions, we encourage iterative reasoning, which creates well-crafted MCQs. By extending MSP to multiple languages, our approach enables GPT models to create linguistically diverse MCQs.

Our contributions are as follows: (i) We propose a lightweight language agnostic multi-stage prompting (MSP) approach to generate zero-shot MCQs using GPT models such as `text-davinci-003` [2] and GPT-4 [17] without fine-tuning the models. We also try the one-shot technique in our proposed method to improve the accuracy of our method; (ii) Since automated metrics have their own limitations, in terms of evaluating questions [16], we perform a human evaluation of the generated questions.

State-of-the-Art. Kumar et al. [8] proposed a hybrid method using ontology-based and machine learning-based techniques to automatically generate MCQ stems for educational purposes. Hadifar et al. [5] fine-tuned T5 [21] for the generation of educational MCQs in their proposed EduQG dataset. Rodriguez-Torrealba et al. [23] used fine-tuned T5 for the generation of questions, answers, and associated distractors. Wang et al. [27] explored the use of a Text2Text formulation to generate cloze-style MCQs using LLMs such as BART [13] and T5. Vachev et al. [26] presented a system called Leaf, which used T5 to generate questions, answers, and distractors from the given text. Kalpakchi et al. [6] explored the ability of GPT-3 to create Swedish prompt-based MCQs in a zero-shot manner. All these prior works (except [6]) focused on the generation of English MCQs. Moreover, no existing MCQ generation research explores the capability of GPT models, such as `text-davinci-003` and GPT-4, for prompt-based MCQ generation in multiple languages.

2 Method

Given a set of contexts C = $\{c_1, c_2, ..., c_N\}$, the goal is to generate diverse and high-quality MCQs for each context $c_i \in C$ along with their corresponding answer options. We aim to demonstrate the efficacy of our proposed prompt-based MCQ generation approach using `text-davinci-003` (abbreviated as `Davinci`) and GPT-4 to provide high-quality, diverse, and language-agnostic MCQs.

(A) Multi-Stage Prompting (MSP) Approach
The proposed method called MSP (see Fig. 1) involves four steps as follows:

Paraphrase Generation(PG). We aim to generate multiple paraphrases for each context $c_i \in C$ to ensure diversity in the QG process. Let $P_i = \{p_{i1}, p_{i2}, ..., p_{iM}\}$ be the set of paraphrases for context c_i, where M is the number of paraphrases generated. The prompt we use M (M = 3) times is "*Paraphrase the given context* $<$ *context* $>$ *in language x*", where $x \in \{English, German, Hindi, Bengali\}$.

Keyword Extraction(KE). Next, we extract keywords from each paraphrase $p_{ij} \in P_i$ to represent important information in the generated questions. Let $K_{ij} = \{k_{ijk} | 1 \leq k_{ijk} \leq V\}$ be the set of extracted keywords for paraphrase p_{ij}, where V represents the size of the vocabulary. The prompt we utilize is, "*Extract keywords from the paraphrased context* <*paraphrased context*> *in language x*".

Question Generation(QG). Using the extracted keywords K_{ij} for each p_{ij}, our objective is to create questions that capture the essential information from the paraphrased context p_{ij}. Let $Q_{ij} = \{q_{ijk} | 1 \leq q_{ijk} \leq Q\}$ be the set of questions generated for paraphrase p_{ij}, where Q is the number of questions generated for each paraphrase. We feed the prompt as "*Generate a question based on the paraphrased context* <*paraphrased context*> *and the correct answer* $<$ *keyword* $>$ *in language x*". In addition, we have tried the one-shot technique to improve the quality of the question. The one-shot technique for QG involves the following prompt: "*For the paraphrased context* $<$*paraphrased context*$_i$ $>$ *and the correct answer* $<$ *keyword*$_i$ $>$, *the question is* $<$ *Question*$_i$ $>$ *in language x. Generate a question based on the paraphrased context* $<$*paraphrased context*$_j$ $>$ *and the correct answer* $<$ *keyword*$_j$ $>$ *in language x*".

Distractor Generation (DG). To form plausible answer options for the generated questions, we introduce the generation of distractors using relevant information from the corresponding question and keyword. Let $D_{ijk} = \{d_{ijkl} | 1 \leq d_{ijkl} \leq D\}$ be the set of distractors for question q_{ijk}, where D (D = 3) represents the number of distractors per question. The prompt we use is, "*Create three plausible distractors for the question* $<$ *question* $>$ *and the correct answer* $<$ *keyword* $>$ *in language x*". In addition, we have tried the one-shot technique to improve the quality of the distractors. The one-shot technique for DG involves the following prompt: "*The distractors for the question* $< question_i >$ *and the correct answer* $< correct\,answer_i >$ *are* $< distractor_{i1}, distractor_{i2}, distractor_{i3} >$ *in language* x*. Create three plausible distractors for the question* $< question_j >$ *and the correct answer* $< keyword_j >$ *in language* x".

(B) Baseline: Single-Stage Prompting (SSP) Approach
There is existing work [23] for the English language in which a T5 model is fine-tuned on the DG-RACE dataset [11] to generate distractors for MCQs. However, there is no existing work in multilingual settings, such as German, Hindi, and Bengali, where an encoder-decoder-based model is explored for DG.

Therefore, we compare our proposed MSP approach, inspired by CoT prompting [28] and involving multiple reasoning steps, with the basic (single-stage) prompting approach to investigate the effectiveness of our method. Here, the prompt we use is, "*Generate MCQs based on the given context* $< context >$ *along with the correct answer and three distractors in language* x".

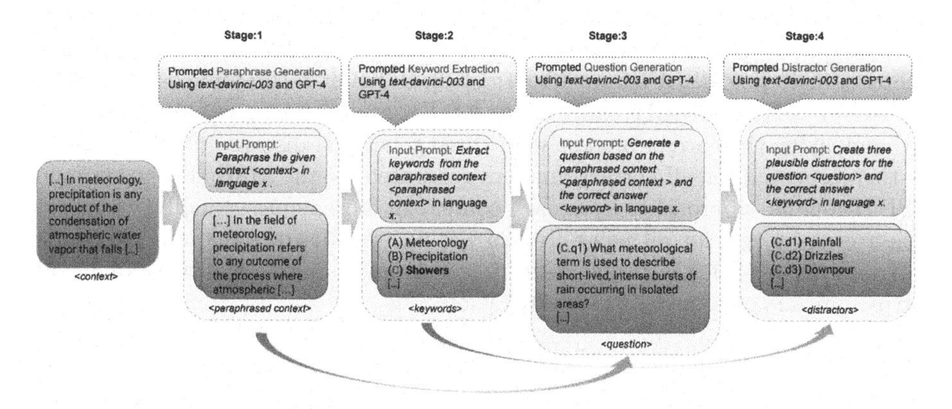

Fig. 1. Overview of proposed MSP approach for MCQ generation in various languages. $x \in \{\textbf{English}, German, Hindi, Bengali\}$. In this example, $x = \textbf{English}$.

3 Experiments

We evaluate our proposed MSP-based approach in zero-shot and one-shot settings. We compare the results of proposed MSP approach with those of SSP baseline in a zero-shot setting. Additionally, we rigorously examine the effectiveness of our approach in multilingual scenarios, incorporating both high-resource languages such as English and German, as well as low-resource languages such as Hindi and Bengali. This section covers details about the dataset used in our experiments followed by details of the implementation with evaluation metrics.

Dataset. Generally, the datasets used for the QG task are taken from the datasets used for question answering (QA). These QA datasets include triples of <*Context*, *Question*, *Answer*>. We use *SQuAD* [22] as the English (En) MCQ generation dataset. It consists of more than 100K carefully crafted questions in English. We use *GermanQuAD* [15] as the German (De) MCQ generation data. It consists of more than 13K human-generated QA pairs in German Wikipedia articles. *HiQuAD* [9] is a special dataset to generate questions in Hindi (Hi). It consists of 6,555 pairs of questions and answers. The questions were created based on a Hindi storybook. We use *BanglaRQA* [4] as Bengali (Bn) MCQ generation data. It consists of more than 14K human-generated QA pairs in Bengali Wikipedia articles. We randomly sampled 850 contexts from each of the above four QA datasets for our experiments.

Experimental Setup. We use `Davinci` with *max tokens* = 50, *presence penalty* = 1.0, *frequency penalty* = 0.0, and *temperature* = 0.7. We use GPT-4 with *maximum tokens* = 50 and *temperature* = 0.7. We use the paid APIs of the GPT models[1]. We use the 850 samples present in a particular language for automatic evaluation. Following the approach of Rodriguez-Torrealba et al. [23], we evaluate the quality of the generated distractors using several automated evaluation metrics, including BLEU (from 1 to 4 g) [19], ROUGE-L [14] and cosine similarity (CS) based on mBERT embeddings [3]. For each sample, we calculate scores by comparing the correct answer with three generated distractors. This process is then repeated for all correct answers generated by the MSP (Proposed) and SSP (Baseline) approaches from each context in our samples. Finally, we report the average scores for all pairs of <correct answer, distractor> in a particular language in Table 1. Our implementation code and datasets will be available at https://github.com/my625/CoT-MCQGen.

Results. Table 1 demonstrates the impressive performance of `Davinci` (w/MSP) in almost all automated evaluation metrics for one-shot DG tasks in English, German, and Bengali. However, for the Hindi language, GPT-4 (w/MSP) surpasses all others, exhibiting the best performance across all automated metrics in the one-shot setting. Among all languages, English (w/`Davinci`) emerges as the best performer in all automated metrics. Furthermore, German (w/ `Davinci`) shows a commendable performance. However, Hindi (w/GPT-4) and Bengali (w/ `Davinci`) have exhibited lower results in automated metrics compared to other languages. This can be attributed to the fact that most of the `Davinci` and GPT-4's training data come from high-resource languages such as English, German, etc. Interestingly, for Hindi and Bengali, the improvement in the one-shot setting (w/MSP) is higher than that observed in English and German in most cases. It shows that our proposed MSP approach, whether using `Davinci` or GPT-4, generates higher quality distractors than the SSP approach for the four languages.

[1] https://platform.openai.com/docs/models.

Table 1. Automatic evaluation results for various GPT models for DG. We selected the GPT model that performs best among `Davinci` and `GPT-4` in 0-shot (w/MSP) and then explored the same model in 0-shot (w/SSP) and 1-shot (w/MSP). Each cell in this table represents the average value of BLEU-1, BLEU-2, BLEU-3, BLEU-4, ROUGE-L, and cosine similarity (CS) based on mBERT embeddings between every <correct answer, distractor> pair generated by the corresponding GPT model and language setting. A higher average value denotes the better quality of the generated distractors. Best results are marked in **bold** and values marked with * are statistically significant on performing student *t*-test at the 95% confidence interval w.r.t the best performing GPT model (w/SSP).

Model (Approach)	**Setting**	**BLEU-1**	**BLEU-2**	**BLEU-3**	**BLEU-4**	**ROUGE-L**	**CS**
			English				
`Davinci` (w/ MSP)	0-shot	14.75	7.68	**4.47**	2.47	13.47	0.73*
`GPT-4` (w/ MSP)	0-shot	14.47	7.45	4.28	2.27	13.18	0.69
`Davinci` (w/ SSP) (Baseline)	0-shot	**14.92**	7.17	4.45	2.29	12.94	0.68
`Davinci` (w/ MSP)	1-shot	14.64	**7.69**	4.15	**2.49**	**13.54**	**0.74***
			German				
`Davinci` (w/ MSP)	0-shot	12.14*	**6.79**	3.77	2.58	12.88	0.70
`GPT-4` (w/ MSP)	0-shot	11.45	6.45	3.48	2.11	12.12	0.66
`Davinci` (w/ SSP) (Baseline)	0-shot	10.89	6.11	**3.79**	2.43	12.49	0.69
`Davinci` (w/ MSP)	1-shot	**12.15***	6.42	3.78	**2.79**	**12.98**	**0.71**
			Hindi				
`Davinci` (w/ MSP)	0-shot	9.58*	4.98	2.13	1.36	11.12	0.64*
`GPT-4` (w/ MSP)	0-shot	9.74	5.01	2.44	1.59	11.82	0.66*
`GPT-4` (w/ SSP) (Baseline)	0-shot	8.14	4.27	2.48	1.08	11.28	0.60
`GPT-4` (w/ MSP)	1-shot	**10.47***	**5.22***	**3.47***	**1.81**	**12.47**	**0.67***
			Bengali				
`Davinci` (w/ MSP)	0-shot	9.44	5.41	**3.97***	2.41	9.45	0.63*
`GPT-4` (w/ MSP)	0-shot	9.21	4.98	2.88	2.31	9.19	0.62*
`Davinci` (w/ SSP) (Baseline)	0-shot	8.45	4.81	2.94	2.19	9.34	0.59
`Davinci` (w/ MSP)	1-shot	**9.70***	**5.48**	3.87*	**2.97**	**11.78***	**0.67***

Table 2. Human evaluation results of various best-performing GPT models (w/MSP) for QG across four languages.

Language (GPT Model)	Setting	Grammaticality	Answerability	Difficulty
English (`Davinci`)	0-shot	4.38	4.29	3.68
	1-shot	**4.47**	**4.42**	3.75
German (`Davinci`)	0-shot	4.21	4.30	3.78
	1-shot	4.32	4.37	**3.89**
Hindi (`GPT-4`)	0-shot	3.21	3.04	2.62
	1-shot	3.34	3.17	2.78
Bengali (`Davinci`)	0-shot	3.10	2.78	2.54
	1-shot	3.27	2.98	2.62

Human Evaluation. Taking into account the limitations associated with automated metrics in terms of the evaluation of generated questions [16], we conduct a human evaluation by appointing an educated native speaker for each respective language. Each human evaluator[2] was asked to rate a total of 400 questions, considering the best-performing GPT model (both for the zero-shot and one-shot settings). The rating scale used ranged from 1 (worst) to 5 (best) based on three criteria: *grammaticality* [25], *answerability* [25], and *difficulty* [6]. As shown in Table 2, English (w/ `Davinci`) showed the best performance for *grammaticality* and *answerability*. And for *difficulty*, German (w/ `Davinci`) outperforms other languages. On the other hand, Hindi (w/GPT-4) and Bengali (w/ `Davinci`) showed poorer performance than high-resource languages such as English and German. This is because GPT models[3] such as `text-davinci-003` and GPT-4 have been pre-trained on smaller amounts of data taken from low-resource languages, such as Hindi and Bengali. We also observed improvements in *grammaticality*, *answerability*, and *difficulty* across all languages in the one-shot setting compared to the zero-shot setting. However, the improvement in *grammaticality*, *answerability*, and *difficulty* is greater for low-resource languages, such as Hindi and Bengali, than for high-resource languages, such as English and German. This occurs because showing a one-shot example to the GPT model helps to improve human evaluation criteria more for low-resource languages than for high-resource languages. This is due to the fact that GPT models have been pre-trained on a smaller amount of data from low-resource languages, as discussed earlier. Figure 2 shows an example of a *grammatically* incorrect question with lower *answerability*, along with its options generated by `Davinci` (w/MSP) in one-shot setting in Bengali.

Context: স্টারশিপ হল মার্কিন মহাকাশ সংস্থা স্পেসএক্স কর্তৃক বিকশিত সম্পূর্ণ পুনঃব্যবহারযোগ্য উৎক্ষেপণ যানের একটি শ্রেণি। উৎক্ষেপণ যানে মরিচাবিহীন ইস্পাত থেকে তৈরি একটি পুনঃব্যবহারযোগ্য অত্যন্ত ভারী বুস্টার ও স্টারশিপ মহাকাশযান রয়েছে,[...]
(Starship is a class of fully reusable launch vehicle developed by the US space agency SpaceX. The launch vehicle consists of a reusable very heavy booster and Starship spacecraft made of stainless steel,[...])
Generated MCQ (Davinci w/ MSP + 1-shot): কোন মার্কিন মহাকাশ সংস্থা দ্বারা বিকশিত সম্পূর্ণ পুনঃব্যবহারযোগ্য উৎক্ষেপণ যান স্টারশিপ সম্পর্কে কী বলা হয়েছে?
(What is said about the Starship, a fully reusable launch vehicle developed by which US space agency?)
(a) স্পেসএক্স (SpaceX) (b) নাসা (NASA) (c) ব্লু অরিজিন (Blue Origin) (d) র‍্যাপ্টর (Raptor)

Fig. 2. An example of a generated *grammatically* incorrect, low-*answerability* question in Bengali, along with the generated highlighted correct answer option and associated distractors.

4 Conclusion and Future Work

GPT-4 and `text-davinci-003`, when used with the MSP method, show promise in generating MCQs in multiple languages. They demonstrate potential benefits

[2] We use Surge AI as our evaluation platform.
[3] Hyperlink to a CSV file containing training data statistics for GPT-3.

in terms of performance and quality, both for high-resource and low-resource languages. Our proposed MSP approach outperforms the baseline SSP method, indicating that the MSP approach effectively produces higher-quality distractors. Additionally, the one-shot technique enhances the automatic evaluation results for low-resource languages, addressing the challenges in MCQ generation across various languages. In human evaluation, we observed that the questions generated for high-resource languages exhibit better *grammaticality*, *answerability*, and *difficulty*. Furthermore, the one-shot setting improves the *grammaticality*, *answerability*, and *difficulty* of the generated MCQs compared to the zero-shot setting in all languages. However, further research and fine-tuning may be necessary to improve results for low-resource languages and narrow the gap with high-resource languages in terms of automated and human evaluation metrics.

References

1. Bhaskar, A., Fabbri, A., Durrett, G.: Prompted opinion summarization with GPT-3.5. In: Findings of the Association for Computational Linguistics: ACL 2023, pp. 9282–9300. Association for Computational Linguistics, Toronto, Canada, July 2023. https://doi.org/10.18653/v1/2023.findings-acl.591, https://aclanthology.org/2023.findings-acl.591
2. Brown, T., et al.: Language models are few-shot learners. In: Larochelle, H., Ranzato, M., Hadsell, R., Balcan, M., Lin, H. (eds.) Advances in Neural Information Processing Systems, vol. 33, pp. 1877–1901. Curran Associates, Inc. (2020). https://proceedings.neurips.cc/paper_files/paper/2020/file/1457c0d6bfcb4967418bfb8ac142f64a-Paper.pdf
3. Devlin, J., Chang, M.W., Lee, K., Toutanova, K.: BERT: pre-training of deep bidirectional transformers for language understanding. In: Proceedings of the 2019 Conference of the North American Chapter of the Association for Computational Linguistics: Human Language Technologies, Volume 1 (Long and Short Papers), pp. 4171–4186. Association for Computational Linguistics, Minneapolis, Minnesota, June 2019. https://doi.org/10.18653/v1/N19-1423, https://aclanthology.org/N19-1423
4. Ekram, S.M.S., et al.: BanglaRQA: a benchmark dataset for under-resourced Bangla language reading comprehension-based question answering with diverse question-answer types. In: Findings of the Association for Computational Linguistics: EMNLP 2022, pp. 2518–2532. Association for Computational Linguistics, Abu Dhabi, United Arab Emirates, December 2022. https://aclanthology.org/2022.findings-emnlp.186
5. Hadifar, A., Bitew, S.K., Deleu, J., Develder, C., Demeester, T.: Eduqg: a multi-format multiple-choice dataset for the educational domain. IEEE Access **11**, 20885–20896 (2023)
6. Kalpakchi, D., Boye, J.: Quasi: a synthetic question-answering dataset in Swedish using GPT-3 and zero-shot learning. In: Proceedings of the 24th Nordic Conference on Computational Linguistics (NoDaLiDa), pp. 477–491. University of Tartu Library, Tórshavn, Faroe Islands, May 2023. https://aclanthology.org/2023.nodalida-1.48
7. Kojima, T., Gu, S.S., Reid, M., Matsuo, Y., Iwasawa, Y.: Large language models are zero-shot reasoners. In: Koyejo, S., Mohamed, S., Agarwal, A., Belgrave, D., Cho, K., Oh, A. (eds.) Advances in Neural

Information Processing Systems, vol. 35, pp. 22199–22213. Curran Associates, Inc. (2022). https://proceedings.neurips.cc/paper_files/paper/2022/file/8bb0d291acd4acf06ef112099c16f326-Paper-Conference.pdf
8. Kumar, A.P., Nayak, A., K, M.S., Chaitanya, Ghosh, K.: A novel framework for the generation of multiple choice question stems using semantic and machine-learning techniques. Int. J. Artif. Intell. Educ. 1–44 (2023). https://doi.org/10.1007/s40593-023-00333-6
9. Kumar, V., Joshi, N., Mukherjee, A., Ramakrishnan, G., Jyothi, P.: Cross-lingual training for automatic question generation. In: Proceedings of the 57th Annual Meeting of the Association for Computational Linguistics, pp. 4863–4872. Association for Computational Linguistics, Florence, Italy, July 2019. https://doi.org/10.18653/v1/P19-1481, https://aclanthology.org/P19-1481
10. Kurdi, G., Leo, J., Parsia, B., Sattler, U., Al-Emari, S.: A systematic review of automatic question generation for educational purposes. Int. J. Artif. Intell. Educ. **30**, 121–204 (2020)
11. Lai, G., Xie, Q., Liu, H., Yang, Y., Hovy, E.: RACE: Large-scale ReAding comprehension dataset from examinations. In: Palmer, M., Hwa, R., Riedel, S. (eds.) Proceedings of the 2017 Conference on Empirical Methods in Natural Language Processing, pp. 785–794. Association for Computational Linguistics, Copenhagen, Denmark, September 2017. https://doi.org/10.18653/v1/D17-1082, https://aclanthology.org/D17-1082
12. Lester, B., Al-Rfou, R., Constant, N.: The power of scale for parameter-efficient prompt tuning. In: Proceedings of the 2021 Conference on Empirical Methods in Natural Language Processing, pp. 3045–3059. Association for Computational Linguistics, Online and Punta Cana, Dominican Republic, November 2021. https://doi.org/10.18653/v1/2021.emnlp-main.243, https://aclanthology.org/2021.emnlp-main.243
13. Lewis, M., et al.: BART: denoising sequence-to-sequence pre-training for natural language generation, translation, and comprehension. In: Proceedings of the 58th Annual Meeting of the Association for Computational Linguistics, pp. 7871–7880. Association for Computational Linguistics, July 2020. https://doi.org/10.18653/v1/2020.acl-main.703, https://aclanthology.org/2020.acl-main.703
14. Lin, C.Y.: ROUGE: a package for automatic evaluation of summaries. In: Text Summarization Branches Out, pp. 74–81. Association for Computational Linguistics, Barcelona, Spain, July 2004. https://aclanthology.org/W04-1013
15. Möller, T., Risch, J., Pietsch, M.: Germanquad and germandpr: improving non-English question answering and passage retrieval. arXiv preprint arXiv:2104.12741 (2021)
16. Nema, P., Khapra, M.M.: Towards a better metric for evaluating question generation systems. In: Proceedings of the 2018 Conference on Empirical Methods in Natural Language Processing, pp. 3950–3959. Association for Computational Linguistics, Brussels, Belgium, Oct-Nov 2018. https://doi.org/10.18653/v1/D18-1429, https://aclanthology.org/D18-1429
17. OpenAI: Gpt-4 technical report (2023)
18. Pal, S., Khan, K., Singh, A.K., Ghosh, S., Nayak, T., Palshikar, G., Bhattacharya, I.: Weakly supervised context-based interview question generation. In: Proceedings of the 2nd Workshop on Natural Language Generation, Evaluation, and Metrics (GEM), pp. 43–53. Association for Computational Linguistics, Abu Dhabi, United Arab Emirates (Hybrid), December 2022. https://aclanthology.org/2022.gem-1.4

19. Papineni, K., Roukos, S., Ward, T., Zhu, W.J.: Bleu: a method for automatic evaluation of machine translation. In: Proceedings of the 40th Annual Meeting of the Association for Computational Linguistics, pp. 311–318. Association for Computational Linguistics, Philadelphia, Pennsylvania, USA, July 2002. https://doi.org/10.3115/1073083.1073135, https://www.aclweb.org/anthology/P02-1040
20. Qiao, S., et al.: Reasoning with language model prompting: a survey. In: Proceedings of the 61st Annual Meeting of the Association for Computational Linguistics (Volume 1: Long Papers), pp. 5368–5393. Association for Computational Linguistics, Toronto, Canada, July 2023. https://doi.org/10.18653/v1/2023.acl-long.294, https://aclanthology.org/2023.acl-long.294
21. Raffel, C., et al.: Exploring the limits of transfer learning with a unified text-to-text transformer. J. Mach. Learn. Res. **21**(1), 5485–5551 (2020)
22. Rajpurkar, P., Zhang, J., Lopyrev, K., Liang, P.: SQuAD: 100,000+ questions for machine comprehension of text. In: Proceedings of the 2016 Conference on Empirical Methods in Natural Language Processing, pp. 2383–2392. Association for Computational Linguistics, Austin, Texas, November 2016. https://doi.org/10.18653/v1/D16-1264, https://aclanthology.org/D16-1264
23. Rodriguez-Torrealba, R., Garcia-Lopez, E., Garcia-Cabot, A.: End-to-end generation of multiple-choice questions using text-to-text transfer transformer models. Expert Syst. Appl. **208**, 118258 (2022)
24. Tan, Z., Zhang, X., Wang, S., Liu, Y.: MSP: Multi-stage prompting for making pre-trained language models better translators. In: Proceedings of the 60th Annual Meeting of the Association for Computational Linguistics (Volume 1: Long Papers), pp. 6131–6142. Association for Computational Linguistics, Dublin, Ireland, May 2022. https://doi.org/10.18653/v1/2022.acl-long.424, https://aclanthology.org/2022.acl-long.424
25. Ushio, A., Alva-Manchego, F., Camacho-Collados, J.: Generative language models for paragraph-level question generation. In: Proceedings of the 2022 Conference on Empirical Methods in Natural Language Processing, pp. 670–688. Association for Computational Linguistics, Abu Dhabi, United Arab Emirates, December 2022. https://aclanthology.org/2022.emnlp-main.42
26. Vachev, K., Hardalov, M., Karadzhov, G., Georgiev, G., Koychev, I., Nakov, P.: Leaf: multiple-choice question generation. In: Hagen, M., et al. (eds.) ECIR 2022. LNCS, vol. 13186, pp. 321–328. Springer, Cham (2022). https://doi.org/10.1007/978-3-030-99739-7_41
27. Wang, H.J., et al.: Distractor generation based on Text2Text language models with pseudo Kullback-Leibler divergence regulation. In: Findings of the Association for Computational Linguistics: ACL 2023, pp. 12477–12491. Association for Computational Linguistics, Toronto, Canada, July 2023. https://doi.org/10.18653/v1/2023.findings-acl.790, https://aclanthology.org/2023.findings-acl.790
28. Wei, J., et al.: Chain-of-thought prompting elicits reasoning in large language models. Adv. Neural. Inf. Process. Syst. **35**, 24824–24837 (2022)

An Adaptive Framework of Geographical Group-Specific Network on O2O Recommendation

Luo Ji[1], Jiayu Mao[2], Hailong Shi[3](✉), Qian Li[2], Yunfei Chu[1], and Hongxia Yang[1]

[1] DAMO Academy, Alibaba Group, Hangzhou, China
{jiluo.lj,fay.cyf,yang.yhx}@alibaba-inc.com
[2] Alibaba Group, Hangzhou, China
{jiayumao.mjy,lq167324}@alibaba-inc.com
[3] Institute of Microelectronics, Chinese Academy of Sciences, Beijing, China
shihailong2010@gmail.com

Abstract. Online to offline recommendation strongly correlates with the user and service's spatiotemporal information, therefore calling for a higher degree of model personalization. The traditional methodology is based on a uniform model structure trained by collected centralized data, which is unlikely to capture all user patterns over different geographical areas or time periods. To tackle this challenge, we propose a geographical group-specific modeling method called GeoGrouse, which simultaneously studies the common knowledge as well as group-specific knowledge of user preferences. An automatic grouping paradigm is employed and verified based on users' geographical grouping indicators. Offline and online experiments are conducted to verify the effectiveness of our approach, and substantial business improvement is achieved.

Keywords: O2O Recommendation · Personalized Network · Reinforcement Learning · Expectation Maximization

1 Introduction

Online to offline (O2O) platforms such as Uber and Meituan map online users with offline service providers on users' smartphones. This mapping is naturally geographically and temporal influenced, which is significantly different from traditional e-commerce platforms like Amazon/Taobao. Examples of this spatiotemporal influence include 1) for a specific user, only services within his/her adjacent area are applicable candidates according to the order fulfillment possibility, resulting in an extremely sparse user-item interaction matrix (and this sparsity inevitably happens when user and item from different areas); 2) users' interests may vary dramatically in different time periods (*e.g.* food orders in the

L. Ji and J. Mao—The first two authors contributed equally to this research.

N. Goharian et al. (Eds.): ECIR 2024, LNCS 14610, pp. 278–286, 2024.
https://doi.org/10.1007/978-3-031-56063-7_19

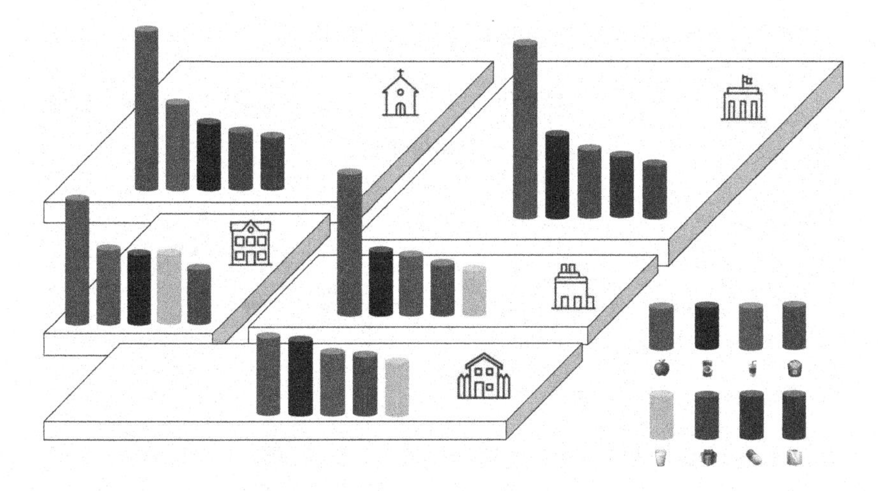

Fig. 1. Geographic influence on order distributions of an O2O retail delivery platform. Top-Five item categories and their fractions are exhibited for five different functional regions (residence, working, business, education, and hospital).

morning or evening; traveling options in workdays or weekends); 3) users from different geographical areas could have varied food tastes and therefore distinct behavior patterns (see Fig. 1 as an illustrative example). These characteristics pose additional challenges for reasonable servicing personalization with respect to user spatiotemporal information. For the conventional unified model architecture [11], user data across all time periods and geographical areas are leveraged together to study a uniform model representation, which may suffer performance degradation given non-uniform data distribution, as shown in Fig. 1. On the contrary, one can choose to train a distinct model on each different geographical area and time period, to better capture local data distributions. Nevertheless, one needs to arbitrarily determine the model granularity, while fail to capture the user behavior commonality [9]. The training dataset of each local model is also much more sparse than the uniform framework.

In this work, we propose a novel **Geo**graphic **Grou**p-**s**pecific (GeoGrouse) model framework to tackle the aforementioned challenges, on Ele.me[1], a world-leading O2O food delivery application. Similar to the STAR architecture [9], our model includes a shared-central network, as well as group-specific networks each of which tailored to a specific user group. During training, the central network is trained on the entire data scope to capture the user commonality; while the group-specific network provides the group-level specializations by finetuning with data of the corresponding group. The user grouping indicator is determined by a trainable latent embedding function with the user geographical features as input. This methodology can be generalized to different types of user grouping specifications. The main contributions of this paper include:

[1] https://www.ele.me/.

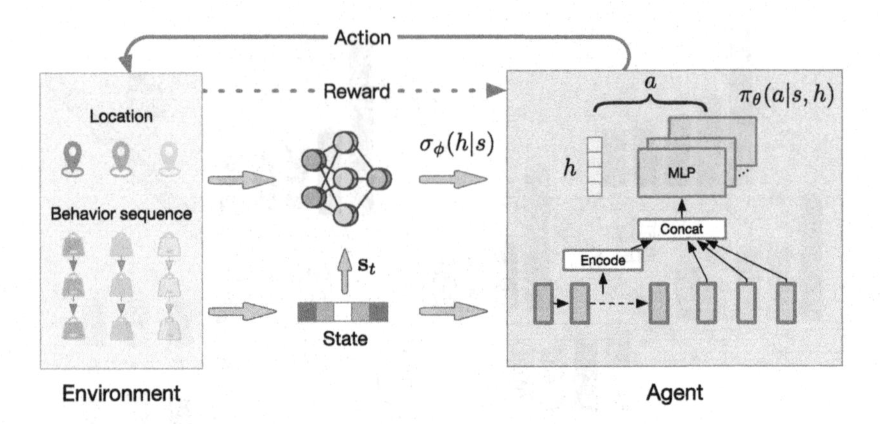

Fig. 2. The framework of GeoGrouse. User states are processed with a centered module and group-specific modules, while σ_ϕ generates the user grouping latent variable which determines the active group-specific module.

- To the best of our knowledge, this is the first time to incorporate the idea of group-specific modeling with O2O recommendation, for better personalization of spatiotemporal influences.
- We design an adaptive user grouping mechanism instead of arbitrary user grouping.
- Performances of GeoGrouse on different business indicators are verified by realistic live experiments.

2 Method

2.1 Framework

Reinforcement Learning (RL) is an interactive learning algorithm between the agent and the environment. The agent observes the state s, acts with the action a, and receives the reward r from the environment. An episode with length t can be denoted as $\tau_t := \{s_0, a_0, r_0, s_1, a_1, r_1, \cdots, s_t, a_t, r_t\}$. The state transits by $\mathcal{T}(s_{t+1}|s_t, a_t)$. The objective is the discounted accumulated rewards $G_t = \sum_t^\infty \gamma^t r_t$ with $\gamma \in (0, 1]$ as the discounted factor, and the agent aims to find an optimal policy $\pi(a|s)$ which maximizes the expected G_t.

Here we employ this RL framework to solve the Top-K recommendation problem, with a system configuration similar to [2]. Nonetheless, motivated by the spatial-temporal dependency of O2O, we model our policy by explicit user grouping. In this work, we further assume the distribution of states is implicitly determined by a latent grouping variable h, with the likelihood recognition function $\sigma(h|s)$. Accordingly, the original policy $\pi(a|s)$ becomes a latent space policy $\pi(a|s, h)$. Below are the detailed definitions of system variables:

- s: the user profiles, historical behavior sequences, and context features including the season, weather, and geographic info (denoted by g).
- a: embedding of recommended items.
- r: the immediate reward obtained after a recommendation, assigned as 1 with a click or add-to-cart, and 0 otherwise.
- h: the grouping indicator as a learnable embedding of g.

Similar to the STAR topology [9], our policy network is a combination of one group-shared module and multiple group-specification modules. Grouping is achieved by parametric recognition model $\sigma_\phi(h|s)$ which is jointly learned with the parametric policy $\pi_\theta(a|s,h)$. We name this recommendation method as Geographic Group-Specific (GeoGrouse) network, as indicated by Fig. 2.

2.2 Implementation of Group-Specification

As stated in Sect. 2.1, the policy network π includes the group-shared module at the bottom and the group-specification module at the top. The group-shared module and the group-specification module are then denoted by

$$a_s = \mathrm{DIN}_\mu(s), \quad a = \mathrm{GS}_\eta(a_s, h)$$

in which a_s is the shared part of action and DIN is the Deep Interest Network [12] tower. The policy can then be re-expressed as $\pi_\theta(a|s,h) = \mathrm{GS}_\eta(\mathrm{DIN}_\mu(s), h)$ with θ as union of $\{\mu, \eta\}$. The embedding of the grouping indicator can be further expressed as the parametric form of $h = \sigma_\phi(g)$. In the following subsections we propose three possible group-specification implementations of $\sigma_\phi(g)$ and $\mathrm{GS}_\eta(a_s, h)$, with their architectural comparison shown in Fig. 3.

K-Means. K-Means is a classic clustering method and is tightly correlated with MLE and EM [6]. With the number of clusters K as key hyper-parameter, K-Means acts as σ_ϕ which first learns K cluster centroid $\{g_k\}_{k=1}^K$, then determine the most nearby cluster from the current g

$$h = \hat{k} = \arg\min_{k\in[1,K]} \|g - g_k\|_2$$

Then K identical MLP towers are implemented to form GS_η

$$a^k = \mathrm{MLP}_{\eta_k}(a_s), \quad k = 1, \cdots, K$$

then a is simply the output selection of the $\hat{k}$th tower, $a = a^{\hat{k}}$ with $\eta = \{\eta_1, \cdots, \eta_k\}$. During training, MLP_{η_k} is only trained with samples of the kth cluster to achieve the group-specialization.

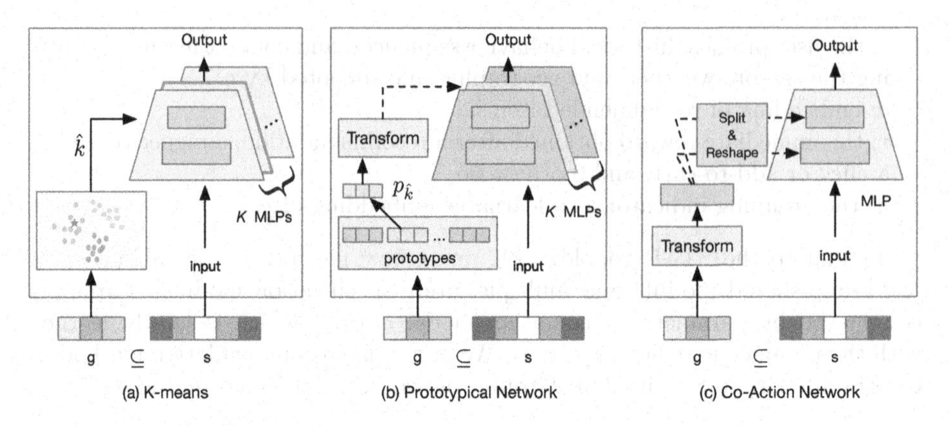

Fig. 3. Comparison of Grouping Implementations. g is the geography-related part among the s attributes.

Prototypical Networks. Similar to K-means, the prototypical method [7] also intrigues K towers MLP_{η_k} but in a more automatic manner. First h is represented by K learned prototype vectors, i.e., $\{p_k\}_{k=1}^{K}$, using method in [7], the current optimal prototype is determined by

$$\hat{k} = \mathop{\arg\max}_{k \in [1,K]} \cos(g, p_k)$$

where $\cos(\cdot, \cdot)$ is the cosine similarity. Then for each k, p_k can be further transformed to η_k with a uniform expression

$$\eta_k = \tanh(W p_k + b),$$

where W and b are trainable and $\eta = [W, b]$.

Co-Action Network. Co-Action Network (CAN) [1] is a feature-cross processing technique that provides an automatic manner of group specification, without the inclusion of explicit K separated towers. By linearly transforming g to h and directly utilizing it as the weight & bias parameter of micro-MLP tower,

$$h = L_{\phi} g, \quad a = \mathrm{MLP}_{\eta=h}(a_s)$$

a uniform-structured group-specification module is then obtained which can be automatically adapted to different g.

2.3 Algorithm

We approximate our solution by the famous Expectation-Maximization method (EM) [3]. During the Expectation stage, the latent variable is recognized by maximizing the likelihood of ϕ with the fixed θ:

$$L(\phi) := \log P(h|\tau) \sim \frac{1}{N} \sum_{s \sim \mu(\pi_\theta)}^{|s|=N} \log q_\phi(h|s) \tag{1}$$

On the Maximization stage, the policy parameter θ is updated given the current best estimate $\hat{h}$. Analogous to the original REINFORCE (Sect. 13.3 in [10]) derivation, we have

$$\nabla J(\theta) \sim E_\pi \sum_a q_\pi(s,a) \nabla \pi(a|s,\hat{h}) = E_\pi[G_t \nabla \ln \pi(a|s,\hat{h})] \tag{2}$$

3 Experiment

We launch GeoGrouse on the Ele.me platform for the retail product-instore recommendation. CAN in Sect. 2.2 is adopted as the default group-specification logic since it has the best experimental result. Codes have been made public[2].

3.1 Experimental Configurations

We obtain the geographic features g by concatenating embeddings of spatiotemporal features, such as city, GPS, area-of-interests (AOI), hour, and season.

The model is trained with data extracted from 60 days' logs. The average session length is 35 while the maximum is 586. We compare GeoGrouse with several baselines including (1) **StEN** [8] has state-of-the-art performance on O2O recommendation which encodes spatiotemporal information by specially designed activation and attention. (2) **DIN** (Deep Interest Network) [13] has a local activation that captures the user interest with the target item, but with no specific spatiotemporal logic. (3) **DeepFM** [5] is a classical cross-feature technique for deep neural networks.

3.2 Offline Experiment and Sensitivity Analysis

Data from the very last day is used as the test set. Experiments are repeated 10 times. Widely-used metrics such as Area Under Curve (AUC), Normalized Discounted Cumulative Gain (NDCG), and Hit Rate are used for evaluation. Table 1 shows the offline results. GeoGrouse outperforms baselines on metrics. Among the baselines, StEN is obviously better than DIN and DeepFM, indicating the importance of spatiotemporal considerations. We also perform a sensitivity analysis of AUCs according to the choice of AOI level (and its vocabulary size), which is one of the key geographic indicators of g. Figure 4 indicates the optimal AOI level is 3 therefore we adopt this grouping granularity in formal experiments.

[2] https://github.com/AaronJi/GeoGrouse.

Table 1. Result of Offline Experiment

Model	StEN	DIN	DeepFM	GeoGrouse
AUC	0.820 ± 0.004	0.658 ± 0.005	0.778 ± 0.006	$\mathbf{0.832} \pm 0.007$
NDCG@3	0.672 ± 0.012	0.504 ± 0.010	0.575 ± 0.011	$\mathbf{0.674} \pm 0.012$
NDCG@5	0.695 ± 0.014	0.536 ± 0.011	0.606 ± 0.015	$\mathbf{0.696} \pm 0.015$
NDCG@10	0.728 ± 0.015	0.583 ± 0.017	0.651 ± 0.015	$\mathbf{0.730} \pm 0.017$
NDCG@20	0.759 ± 0.016	0.627 ± 0.018	0.691 ± 0.018	$\mathbf{0.760} \pm 0.017$
NDCG@50	0.783 ± 0.015	0.665 ± 0.015	0.721 ± 0.017	$\mathbf{0.784} \pm 0.018$
Hit Rate@10	0.959 ± 0.006	0.893 ± 0.005	0.932 ± 0.008	$\mathbf{0.960} \pm 0.009$

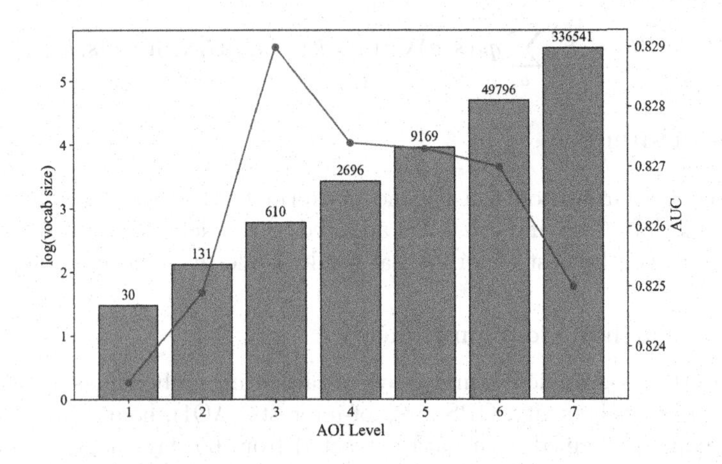

Fig. 4. Sensitivity Analysis of AOI Levels.

3.3 Online A/B Test and Ablation Test

The online A/B test lasts for 7 days. The key performance index (KPI) includes the click-through rate (CTR), the add-to-cart rate (ACR), the number of users with impressions (impress-UV), the number of users with clicks (click-UV), and the number of users with add-to-cart behavior (cart-UV). Due to online industrial constraints, only StEN is deployed as the live baseline. Compared with StEN and Geogrouse with group-specification of K-means and Prototypical (denoted as 'GeoGrouse-K' and 'GeoGrouse-P'), GeoGrouse improves all KPIs substantially as shown in Table 2.

Table 2. Result of Live Experiment. Results of GeoGrouse-K and GeoGrouse-P are relative numbers to GeoGrouse.

Model	StEN	GeoGrouse	GeoGrouse-K	GeoGrouse-P
CTR	13.08%	**13.20%**	−0.50%	−0.05%
ACR	9.99%	**10.06%**	−0.03%	−0.02%
impress-UV	313,206	**313,920**	+0.97%	−0.04%
click-UV	40,980	**41,439**	−0.81%	−0.43%
cart-UV	31,277	**31,579**	−0.67%	+0.03%

4 Conclusion

In this paper, we propose a novel GeoGrouse method that applies self-adaptive user group-specification to O2O recommendation, for better personalization. Our approach is not limited to geographical factors but can be generalized to any grouping considerations. One limitation is the increased mode size due to multiple group-specific modules, which can be alleviated by split-deployment on edge devices [4]. In the future, it would be interesting to examine the broader scope of user grouping possibilities and attempt different levels of grouping granularity.

References

1. Bian, W., et al.: Can: Feature co-action for click-through rate prediction. In: Proceedings of the 15th ACM International Conference on Web Search and Data Mining. WSDM 2022 (2022)
2. Chen, M., Beutel, A., Covington, P., Jain, S., Belletti, F., Chi, E.H.: Top-k off-policy correction for a reinforce recommender system. In: Proceedings of the Twelfth ACM International Conference on Web Search and Data Mining, pp. 456–464. WSDM 2019 (2019)
3. Dempster, A., Laird, N., Rubin, D.: Maximum likelihood from incomplete data via the EM algorithm. In: Proceedings of the Royal Statistical Society, pp. 1–38 (1976)
4. Gong, Y., et al.: Edgerec: recommendation system on edge in mobile Taobao. In: Proceedings of the 2020 ACM on Conference on Information and Knowledge Management. CIKM 2020 (2020)
5. Guo, H., Tang, R., Ye, Y., et al.: DeepFM: a factorization-machine based neural network for CTR prediction. In: Proceedings of the 26th International Joint Conference on Artificial Intelligence, pp. 1725–1731 (2017)
6. Hu, Z.: Initializing the EM Algorithm for Data Clustering and Sub-population Detection. Ph.D. thesis, The Ohio State University, Ohio, USA, December 2015
7. Li, J., Zhou, P., et al.: Prototypical contrastive learning of unsupervised representations. In: Proceedings of the 9th International Conference on Learning Representation. ICLR 2021 (2021)
8. Lin, S., et al.: Spatiotemporal-enhanced network for click-through rate prediction in location-based services. In: Proceedings of the 2022 ACM on Conference on Information and Knowledge Management. CIKM 2022 (2022)

9. Sheng, X.R., et al.: One model to serve all: Star topology adaptive recommender for multi-domain CTR prediction. In: Proceedings of the 30th ACM International CIKM (2021)
10. Sutton, R.S., Barto, A.G.: Reinforcement learning: an introduction. Robotica **17**(2), 229–235 (1999)
11. Wang, S., Hu, L., Wang, Y., Longbing, C., Sheng, Q.Z., Orgun, M.: Sequential recommender systems: challenges, progress and prospects. In: Proceedings of the Twenty-Eighth International Joint Conference on Artificial Intelligence IJCAI 2019 (2019)
12. Zhou, G., et al.: Deep interest network for click-through rate prediction. In: The 24th ACM SIGKDD Conference on Knowledge Discovery and Data Mining. KDD 2018 (2018)
13. Zhou, G., Zhu, X., et al.: Deep interest network for click-through rate prediction. In: Proceedings of the 24th ACM SIGKDD, pp. 1059–1068 (2018)

A Study on Hierarchical Text Classification as a Seq2seq Task

Fatos Torba[1,2](✉), Christophe Gravier[2], Charlotte Laclau[3], Abderrhammen Kammoun[1], and Julien Subercaze[1]

[1] AItenders, Saint-Etienne, France
{fatos.torba,abderrhammen.kammoun,julien.subercaze}@aitenders.com
[2] Laboratoire Hubert Curien, UMRCNRS, 5516 Saint-Etienne, France
christophe.gravier@univ-st-etienne.fr
[3] Télécom Paris, Institut Polytechnique de Paris, Paris, France
charlotte.laclau@telecom-paris.fr

Abstract. With the progress of generative neural models, Hierarchical Text Classification (HTC) can be cast as a generative task. In this case, given an input text, the model generates the sequence of predicted class labels taken from a label tree of arbitrary width and depth. Treating HTC as a generative task introduces multiple modeling choices. These choices vary from choosing the order for visiting the class tree and therefore defining the order of generating tokens, choosing either to constrain the decoding to labels that respect the previous level predictions, up to choosing the pre-trained Language Model itself. Each HTC model therefore differs from the others from an architectural standpoint, but also from the modeling choices that were made. Prior contributions lack transparent modeling choices and open implementations, hindering the assessment of whether model performance stems from architectural or modeling decisions. For these reasons, we propose with this paper an analysis of the impact of different modeling choices along with common model errors and successes for this task. This analysis is based on an open framework coming along this paper that can facilitate the development of future contributions in the field by providing datasets, metrics, error analysis toolkit and the capability to readily test various modeling choices for one given model.

Keywords: Hierarchical text classification · generative model · reproducibility

1 Introduction

Hierarchical Text classification (HTC) aims at tagging an input text with labels organized in an external label tree. It differs from standard text classification as an arbitrary number of labels are expected in the predictions, and also because the predictions have to be consistent with respect to the label tree [13]. HTC, which has long been challenging for machine learning and natural language

N. Goharian et al. (Eds.): ECIR 2024, LNCS 14610, pp. 287–296, 2024.
https://doi.org/10.1007/978-3-031-56063-7_20

processing, is also deemed mandatory in various industrial settings such as e-commerce applications [4], medical recording [2], and finance [8], to name a few.

Despite the advent of Large Language Models (LLMs), HTC is still considered a difficult task [1]. One initial reason is that this task involves classifying items into an arbitrary number of categories. Furthermore, this task is akin to choosing one or multiple, potentially incomplete, pathways from a class tree. Consequently, the hierarchical classifier must exhibit the ability to classify at a coarse-grain as well as a fine-grained level, which requires attention capacities proportional to the depth of the class tree. Moreover, the training data are imbalanced since the annotations for the leaves in the class tree are inevitably sparse due to the hierarchical structure [7], but also since the annotations can stop at any pre-leaf level by tagging design. Finally, there exists very few large-scale open datasets for hierarchical classification with respect to datasets available in standard text classification, due to the cost of such complex annotations and the straightforward industrial applications that can directly result from it. From this complexity, two main opportunities also arise to solve this task. The first one consists in leveraging the hierarchical information between classes to guide the model towards consistent and easier decisions [9,18,20,21]. The second subsists in exploiting the possible semantic cues between label tokens in the class tree path to predict and the text to classify, which are even sometimes expanded with label metadata [3,5,15,16].

Throughout the years, a diverse set of approaches and methodologies have been devised by researchers to address the challenge of HTC. These contributions are traditionally divided between global [3,9,18] and local approaches [17]. Local approaches consist of learning a set of localized classifiers that are learned for each node, for each parent node, or for each class tree level. The major shortcoming is the number of classifiers to learn as the class tree grows and the difficulty to accurately predict leaf nodes, which suffer from sparse annotations although they represent a fine-grain classification to perform. Contrary to local approaches, global approaches use one single model to capture the hierarchical information [21]. These approaches either flatten the class hierarchy as a single array of classes to predict or classify one node at a time, possibly constraining the next prediction to the ones already available for the previously predicted levels [6].

For this latter case, since the aim is to predict one or several paths from the class tree, the prediction can be seen as a sequence of label tokens, thus allowing to cast the problem as a sequence-to-sequence (seq2seq) task [1,15,19,20]. This is the setting that we are investigating in this paper because of the progress of seq2seq LLMs applied to HTC.

In what follows, SHTC will refer to "seq2seq HTC" models. Besides neural architectural choices, original and specific to each SHTC classifier, there exists a shared set of modeling choices to make when modeling the problem as a seq2seq task. While the literature lacks an in-depth exploration of the impact of these modeling choices, our paper addresses this gap by conducting a study on the three following key modeling choices for SHTC.

Modeling choice #1: First, there are different ways to traverse the nodes in the tree when producing the gold standard of sequences: 1) horizontally (BFS algorithm) or vertically (DFS algorithm), and 2) applying these algorithms from the root to the leaves as it is usually performed, but one can also consider traversing from the leaves to the root so that the hardest tokens (leaves are fine-grain classification levels) are at the beginning of the beam search at decoding time, trying to enforce more attention to them when training. The cartesian product of these two considerations for class tree traversal leads to four different options as illustrated in Fig. 1 ("Target Generator").
Modeling choice #2: SHTC can also lead to hierarchy incoherent prediction. One can choose to constrain or free the next token prediction to be based on the current predictions to enforce hierarchy consistency.
Modeling choice #3: Additionally the label name can encapsulate a semantic similarity with the input text. The model may easily predict the name of the labels rather than for instance their assigned acronym. The impact of such a modeling choice (name versus acronym) is yet unknown.

In this paper, we propose an open framework that allows testing the different variations for these three modeling choices and that we hope can facilitate the development of future contributions in the field by providing datasets, metrics, an error analysis toolkit, and the capability to readily test various modeling choices for one given model (Sec. 2). We also propose an evaluation of how the three main modeling choices for SHTC impact their performance (Sec. 3). The paper comes with an open source framework[1] which makes it possible to plug one's HTC model and run their own experiments on the provided datasets and perform error analysis, for all possible options of these three modeling choices.

2 Analysis Framework

Problem Definition Given an input text $X = \{x_1, x_2, .., x_n\}$ and a taxonomic hierarchy $T = \{V, E\}$ where n, V and E denote respectively the text length, the labels nodes set and the parent-child relationship, SHTC aims to predict $Y = \{y_1, y_2, \ldots, y_k\}$ where k is the number of true labels, y_i are the target nodes of X and Y a subset of V. We can model this problem as a seq2seq task by predicting the Y as a sequence conditioned on the given input token sequence X, so that $p(y_1, y_2, \ldots, y_k|X) = \prod_{t=1}^{k} p(y_t|y_{<t}, X)$, which is in turn approximated via beam search in practice.

2.1 Sequence Modeling

Since the target sequence can be constructed in various ways we investigate in this paper four options (Fig. 1 "Target Generator"). The Depth-First Search (DFS) starts at the root and explores as far as possible along each branch before backtracking ("Pop"). Breadth-First Search (BFS) explores all nodes at

[1] https://github.com/FatosTorba/SHTC.

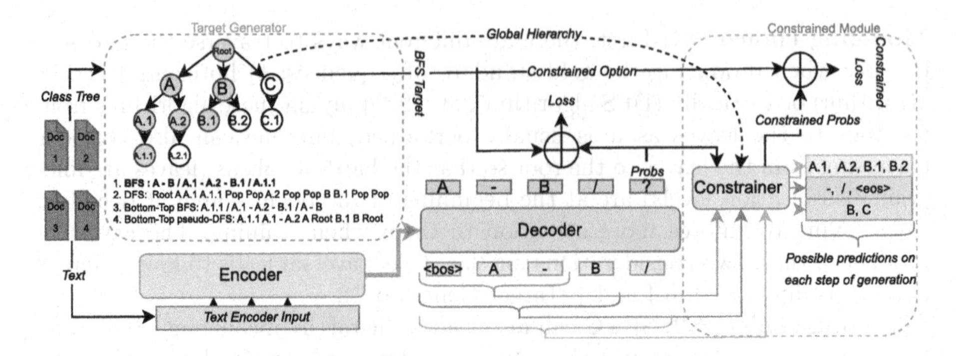

Fig. 1. SHTC illustration. The Target generator offers 4 target sequence options, green nodes denote correct document labels. The constrained module is optional, without it, SHTC operates like a standard T5 architecture. During auto-regressive decoding, this latter computes the probability distribution over the next possible tokens set based on prior predictions and global hierarchy, while nullifying others (constrained probs). To enhance clarity, we only consider a BFS target. For instance, if the model has already generated "A -" the constrained module will distribute the probabilities over {B, C} (indicated by the green arrow and green box). The purple and yellow arrows illustrate the next 2 constrained generation steps. (Color figure online)

the present level prior to moving on to the next one. The intra-level relationship is modeled here by "-", while "/" signifies an inter-level relationship. Bottom-Top BFS is the same process as BFS but starts from the deepest level. Bottom-Top pseudo-DFS starts from the left deepest node and moves either horizontally if it has a brother node (except for the "Root" parent) or upper if not and so on until all true nodes have been explored. Here "-" represents brother relationship.

Following the state of art [7,20] we choose T5 [14], a pre-trained Transformer-based architecture, as the backbone model. The presence of four target options results in the training of four distinct T5 models, each of which is given a specific name: BFS, BT-BFS, DFS, and BT-DFS, corresponding to BFS, Bottom-Top BFS, DFS, and Bottom-Top pseudo-DFS target sequences, respectively. The decoder can be constrained during both training and inference phases to generate a dynamic vocabulary as in [20]. We will provide a detailed description of this constrained decoding strategy in the following section.

2.2 Decoding Strategies

The decoder predicts a sequence of text tokens in an auto-regressive manner, one token at each step, starting from $<bos>$ (begin of sentence) token and ending when the $<eos>$ (end of sentence) token is predicted. At the ith generation step of a token y_i one can decide to constrain the latter to belong to a given set of tokens.

Let $T = \{V, E\}$ be the class taxonomic hierarchy and S a set of special tokens used to build the target sequence. Let $y_0, y_1, ..., y_{i-1} = Y_{i-1} \in V \cup S$ be the set of previous generated tokens. Suppose we can construct a function $f_{constrain}(Y_{i-1}, T) = C_T \in V \cup S$ where C_T is the possible next tokens set. Constraining y_i to belong to C_T is equivalent to computing the probability distribution over C_T tokens using $Softmax$ function as follow: $\forall k \in \{1, n\}, \quad p(y_i = t_k|Y_{i-1}, T) = \frac{exp(l_k)}{\sum\limits_{t_j \in C_T} exp(l_j)}$, if $t_k \in C_T$, 0 otherwise, where n, t_k and l_k are respectively the entire vocabulary length, the kth token and k^{th} logit value of the decoder output. Thus, during training, the model will focus more towards discriminating the possible tokens rather than learning not to predict those not included in C_T. It reduces the optimization space since only the C_T tokens will influence the $Loss$ function. During inference, it ensures a coherent prediction with respect to the hierarchy. The $f_{constrain}$ function depends on the choice of the target sequence. For the BFS target Fig. 1 shows an example of the output set of $f_{constrain}$ at the right of "Constrainer Module". For the DFS target $f_{constrain}$ will return the child's tokens set of the previous predicted parent or the backtracking token. For the "Root" node the <eos> is also a possible token to stop the generation where no more nodes of level one can be predicted.

2.3 Proposed Metrics

Inspired by existing works [20], we use four sets of metrics for SHTC evaluation:
F1 scores: F1-micro, F1-macro, per level and global.
C-F1 scores: (C-F1 micro, C-F1 macro, per level and global) corresponding respectively to F1-micro and F1-macro of constrained predictions obtained by post-processing, considering a node as `true` only if all its ancestor nodes have been predicted as `true`.
Hierarchical inconsistencies: Hierarchical Consistency Rate (HCR), True Positive Hierarchical Consistency Rate (TP-HCR) and False positive Hierarchical Consistency Rate (FP-HCR). HCR refers to the proportion of predicted labels that align with the parent-child relationships of labels within a predefined hierarchy. TP-HCR and FP-HCR refer to HCR of predicted labels where we remove respectively the false positive predicted labels and true positive predicted labels.
Depth of Prediction Rate: computed as DPR$= \frac{1}{|D|}\Sigma_{i=1}^{|D|} depth(y_{pred,i}) \geq depth(y_{true,i})$ where D is a set of documents, $y_{pred,i}$ and $y_{true,i}$ are respectively the predicted labels and the true labels of document i and $depth$ is a function that gives the deepest level number of a predicted label. A higher DPR suggests that the model tries to classify fine-grained labels and conversely a smaller one indicates that the model tends to focus more on higher-level categories and avoid taking the risk of predicting low-level label predictions.

3 Experiment and Datasets

3.1 Datasets and Implementation Details

Datasets We conduct experiments on three public datasets, including reuters corpus (RCV1V2[2] [11]), Blurb Genre Collection (BGC[3]) and Web-Of-Science (WOS[4] [10]). The labels of all this dataset are organized as a tree-like hierarchy. RCV1V2 and BGC are both multi-paths while WOS is single-path. More details for the datasets are shown in Table 1.

Table 1. Statistics of datasets: L_j is the level j, $|V|$ the total number of labels, *Depth* is the maximum level, $m|V|$, mD, mW_{max}, $mW(L_1/L_2/L_3/L_4)$ are the respective averages of the number, depth, maximum width and width of L_1, L_2, L_3, and L_4 of labels in each sample. $n(L_2/L_3/L_4)$ stands for the number of samples in the overall dataset for L_2, L_3, and L_4 (value for L_1 is the sum of train, val and test).

Dataset	$\|V\|$	*Depth*	$m\|V\|$	mD	mW_{max}	$mW(L_1/L_2/L3/L_4)$	$n(L_2/L_3/L_4)$	Train	Val	Test
RCV1V2	103	4	3.24	2.56	1.45	1.18/1.45/1.23/1	779702/452238/23211	20833	2316	781265
WOS	141	2	2	2	1	1/1/-/-	46798/0/0	30070	7518	9397
BGC	146	4	3.01	2.34	1.51	1.06/1.46/1.39/1.08	77438/45005/3213	58715	14785	18394

Implementation Details. For all datasets, we use the text description as input, ignoring all other metadata such as the author, title, or keywords for instance. Label names are kept pristine (no pre-processing). We trained T5 during 12 epochs with a batch size of 16 on a single GPU with a learning rate of $5e^{-5}$ and optimizing with AdamW [12]. During inference, we use the beam search strategy for sequence generation. For BGC we used the official split and partitioned the RCV1V2 training dataset into training and validation sets. For WOS, we made a random split into training, validation and test sets as no official split exists. We applied both constrained and not-constrained decoding strategies for BFS and DFS target sequences. For Bottom-Top BFS and Bottom-Top pseudo-DFS, no constrained decoding strategy is used since we want the model to learn the hierarchy. Constraining would reduce the possible next tokens set to parent tokens only. For RCV1V2 we conduct experiments using nodes designated by acronyms (only for the top two models, DFS and DFS-C), as presented by default in the dataset, as well as their corresponding full names in a dedicated experiment to assess the impact of semantic similarity between the input text and the label names.

[2] http://www.ai.mit.edu/projects/jmlr/papers/volume5/lewis04a/lyrl2004_rcv1v2_README.htm.

[3] https://www.inf.uni-hamburg.de/en/inst/ab/lt/resources/data/blurb-genre-collection.html.

[4] https://data.mendeley.com/datasets/9rw3vkcfy4/2.

3.2 Results

Experimental results are elaborated in Tables 2 and 3. The constrained decoding strategy consistently improves the performance at each level and globally while assuring, by definition, coherent prediction. The sole exception is RCV1V2, where BFS-C performs worse than BFS: as indicated by the DPR and F1 scores at Level 3, BFS-C terminates generation at higher levels. All sequence modeling choices encapsulate the hierarchical structure of the taxonomy, making it learnable for T5 and consequently, the C-F1 scores are almost equal to F1 ones. DFS modeling performs better but the fact that the class sub-trees have

Table 2. Global results, "-Acr" stands for models predicting the acronym nodes of RCV1V2.

Model	**WOS**					**RCV1V2**					**BGC**				
	Mi. F1	Ma. F1	C-Mi. F1	C-Ma. F1	DPR	Mi. F1	Ma. F1	C-Mi. F1	C-Ma. F1	DPR	Mi. F1	Ma. F1	C-Mi. F1	C-Ma. F1	DPR
BFS	86.12	80.69	**86.12**	**80.69**	99.89	85.59	64.92	**85.59**	64.92	**97.00**	75.76	59.63	75.76	59.63	92.92
BFS-C	86.61	80.96	-	-	**99.97**	85.47	62.20	-	-	94.15	76.84	60.89	-	-	88.36
BT-BFS	85.62	80.34	85.61	80.32	99.79	84.51	64.74	84.52	64.74	96.35	75.67	59.47	75.67	59.48	**93.53**
DFS	85.90	80.20	85.91	80.21	99.93	85.51	65.59	85.52	**65.59**	96.75	76.54	60.37	**76.54**	**60.37**	93.14
DFS-C	**86.42**	**81.67**	-	-	99.93	**85.61**	**68.01**	-	-	96.75	**77.79**	**62.41**	-	-	92.89
BT-DFS	85.87	80.56	85.88	80.57	99.70	84.91	64.91	84.92	64.91	95.99	75.59	58.68	75.59	58.68	93.31
DFS-Acr	-	-	-	-	-	85.48	61.90	85.48	60.91	95.26	-	-	-	-	-
DFS-C-Acr	-	-	-	-	-	85.56	62.54	-	-	96.37	-	-	-	-	-

Table 3. Per level results, for visibility, F1 scores are rounded to the nearest whole percentage, / is level separator, P-HCR, FP-HCR, HCR start from Level 2 since Level 1 can not violate the hierarchy.

Model	Mi. F1	Ma. F1	C-Mi. F1	C-Ma. F1	TP-HCR	FP-HCR	HCR
				WOS			
BFS	91/81	91/80	91/81	91/80	99.99	99.94	99.98
BFS-C	92/81	92/80	-	-	-	-	-
BT-BFS	91/80	91/80	91/80	91/80	99.95	99.89	99.94
DFS	91/81	91/80	91/81	91/80	100	99.89	99.98
DFS-C	91/82	91/81	-	-	-	-	-
BT-DFS	91/81	91/80	91/81	91/80	100	99.94	99.99
BGC							
BFS	91/71/64/56	81/64/58/51	91/71/64/57	81/64/58/51	100/100/99.55	100/99.91/99.37	100/99.97/99.46
BFS-C	92/72/65/64	83/65/60/59	-	-	-	-	-
BT-BFS	91/70/64/63	80/63/58/55	91/70/64/63	80/63/58/55	99.99/99.97/100	99.99/99.95/100	99.99/99.96/100
DFS	92/71/65/60	82/64/58/52	92/71/65/60	82/64/58/52	100/100/100	100/100/100	100/100/100
DFS-C	92/73/67/65	82/66/61/57	-	-	-	-	-
BT-DFS	91/70/65/62	80/62/57/54	91/70/65/62	80/62/57/54	100/100/100	99.99/100/100	100/100/100
RCV1V2							
BFS	94/80/84/13	92/65/62/13	94/80/84/13	92/65/62/13	100/99.99/100	100/99.75/100	100/99.95/100
BFS-C	94/80/83/30	92/65/56/30	-	-	-	-	-
BT-BFS	93/79/84/12	92/63/64/12	93/79/84/12	92/63/64/12	100/100/100	99.92/98.34/100	99.98/99.77/100
DFS	94/80/84/3	93/66/63/3	94/80/84/3	93/66/63/3	100/100/100	100/99.94/100	100/99.99/100
DFS-C	94/80/84/22	93/68/65/22	-	-	-	-	-
BT-DFS	93/79/84/22	92/65/63/22	93/79/84/22	92/65/63/22	100/100/100	99.83/99.95/100	99.97/99.99/100
DFS-Acr	94/79/83/03	92/62/57/03	94/79/83/03	92/62/57/03	100/99.99/100	100/99.93/100	100/99.98/100
DFS-C-Acr	94/79/83/03	92/62/60/03	-	-	-	-	-

bigger average depth than average width makes it unclear whether this improvement stems from the sequence modeling or the dataset properties. Bottom-top approaches decrease performance despite improvement in the last level, hierarchical coherency and depth rate. It is worth mentioning that the label names semantic information is crucial for the decoder. Predicting the RCV1V2 acronym taxonomy, as opposed to the full label names, significantly lowers the Macro F1 by approximately 4% for DFS and 5% for DFS-C.

4 Conclusion

In this work, we investigate three main design choices for HTC cast as a sequence-to-sequence task. First, we propose a complete framework, which can take any existing or future SHTC model and compare the impact of these different design choices on the architecture. This framework allows to validate that models performances stems from their intrinsic characteristics and not from different design choices from baselines. Second, through an empirical study of existing models, we devised interesting results for the practitioners: generating sequences from bottom to top yields more balanced predictions for each levels, generating sequence using DFS and use constrained generation is the best option overall, using label names instead of abstract tokens for classes have a significant impact on the performances of the system and invite to further investigate how to automatically select better sequence token for SHTC.

References

1. Bhambhoria, R., Chen, L., Zhu, X.: A simple and effective framework for strict zero-shot hierarchical classification. In: Proceedings of the 61st Annual Meeting of the Association for Computational Linguistics (Volume 2: Short Papers), pp. 1782–1792. Association for Computational Linguistics (2023)
2. Cao, P., Chen, Y., Liu, K., Zhao, J., Liu, S., Chong, W.: HyperCore: hyperbolic and co-graph representation for automatic ICD coding. In: Proceedings of the 58th Annual Meeting of the Association for Computational Linguistics, pp. 3105–3114. Association for Computational Linguistics (2020)
3. Chen, H., Ma, Q., Lin, Z., Yan, J.: Hierarchy-aware label semantics matching network for hierarchical text classification. In: Proceedings of the Annual Meeting of the Association for Computational Linguistics and the International Joint Conference on Natural Language Processing, ACL/IJCNLP, pp. 4370–4379. Association for Computational Linguistics (2021)
4. Chen, L., Chou, H.W., Zhu, X.: Developing prefix-tuning models for hierarchical text classification. In: Proceedings of Conference on Empirical Methods in Natural Language Processing: EMNLP, pp. 390–397. Association for Computational Linguistics (2022)
5. Deng, Z., Peng, H., He, D., Li, J., Yu, P.: HTCInfoMax: a global model for hierarchical text classification via information maximization. In: Proceedings of the Conference of the North American Chapter of the Association for Computational Linguistics: Human Language Technologies, pp. 3259–3265. Association for Computational Linguistics (2021)

6. Giunchiglia, E., Lukasiewicz, T.: Coherent hierarchical multi-label classification networks. In: Larochelle, H., Ranzato, M., Hadsell, R., Balcan, M., Lin, H. (eds.) NeurIPS 2020, December 6–12, 2020, virtual (2020)
7. Huang, W., Let al.: Exploring label hierarchy in a generative way for hierarchical text classification. In: Proceedings of the 29th International Conference on Computational Linguistics, pp. 1116–1127. International Committee on Computational Linguistics (2022)
8. Jiang, H., et al.: Financial news annotation by weakly-supervised hierarchical multi-label learning. In: Proceedings of the Second Workshop on Financial Technology and Natural Language Processing, pp. 1–7 (2020)
9. Jiang, T., Wang, D., Sun, L., Chen, Z., Zhuang, F., Yang, Q.: Exploiting global and local hierarchies for hierarchical text classification. In: Proceedings of EMNLP, pp. 4030–4039. Association for Computational Linguistics (2022)
10. Kowsari, K., Brown, D.E., Heidarysafa, M., Meimandi, K.J., Gerber, M.S., Barnes, L.E.: Hdltex: hierarchical deep learning for text classification. In: 2017 16th IEEE International Conference on Machine Learning and Applications (ICMLA), pp. 364–371, December 2017. https://doi.org/10.1109/ICMLA.2017.0-134
11. Lewis, D.D., Yang, Y., Russell-Rose, T., Li, F.: RCV1: a new benchmark collection for text categorization research. J. Mach. Learn. Res. **5**(Apr), 361–397 (2004)
12. Loshchilov, I., Hutter, F.: Decoupled weight decay regularization. In: 7th International Conference on Learning Representations, ICLR 2019, New Orleans, LA, USA, May 6–9, 2019. OpenReview.net (2019). https://openreview.net/forum?id=Bkg6RiCqY7
13. Mao, Y., Tian, J., Han, J., Ren, X.: Hierarchical text classification with reinforced label assignment. In: Proceedings of EMNLP-IJCNLP, pp. 445–455. Association for Computational Linguistics (2019)
14. Raffel, C., et al.: Exploring the limits of transfer learning with a unified text-to-text transformer. J. Mach. Learn. Res. **21**(140), 1–67 (2020). https://jmlr.org/papers/v21/20-074.html
15. Rivas Rojas, K., Bustamante, G., Oncevay, A., Sobrevilla Cabezudo, M.A.: Efficient strategies for hierarchical text classification: External knowledge and auxiliary tasks. In: Proceedings of the 58th Annual Meeting of the Association for Computational Linguistics, pp. 2252–2257. Association for Computational Linguistics (2020)
16. Shen, J., Qiu, W., Meng, Y., Shang, J., Ren, X., Han, J.: Taxoclass: hierarchical multi-label text classification using only class names. In: Proceedings of NAACL-HLT, pp. 4239–4249. Association for Computational Linguistics (2021)
17. Silla, C.N.J., Freitas, A.: A survey of hierarchical classification across different application domains. Data Mining Knowl. Disc. **22**(1–2), 31–72 (2011)
18. Wang, Z., Wang, P., Huang, L., Sun, X., Wang, H.: Incorporating hierarchy into text encoder: a contrastive learning approach for hierarchical text classification. In: Proceedings of ACL, pp. 7109–7119. Association for Computational Linguistics (2022)
19. Yang, P., Sun, X., Li, W., Ma, S., Wu, W., Wang, H.: SGM: sequence generation model for multi-label classification. In: Bender, E.M., Derczynski, L., Isabelle, P. (eds.) Proceedings of COLING 2018, Santa Fe, New Mexico, USA, August 20–26, 2018, pp. 3915–3926. Association for Computational Linguistics (2018)

20. Yu, C., Shen, Y., Mao, Y.: Constrained sequence-to-tree generation for hierarchical text classification. In: Proceedings of SIGIR, pp. 1865–1869. ACM (2022)
21. Zhou, J., et al.: Hierarchy-aware global model for hierarchical text classification. In: Proceedings of the 58th Annual Meeting of the Association for Computational Linguistics, pp. 1106–1117. Association for Computational Linguistics (2020)

Improving the Robustness of Dense Retrievers Against Typos via Multi-Positive Contrastive Learning

Georgios Sidiropoulos(✉) and Evangelos Kanoulas

University of Amsterdam, Amsterdam, The Netherlands
{g.sidiropoulos,e.kanoulas}@uva.nl

Abstract. Dense retrieval has become the new paradigm in passage retrieval. Despite its effectiveness on typo-free queries, it is not robust when dealing with queries that contain typos. Current works on improving the typo-robustness of dense retrievers combine (i) data augmentation to obtain the typoed queries during training time with (ii) additional robustifying subtasks that aim to align the original, typo-free queries with their typoed variants. Even though multiple typoed variants are available as positive samples per query, some methods assume a single positive sample and a set of negative ones per anchor and tackle the robustifying subtask with contrastive learning; therefore, making insufficient use of the multiple positives (typoed queries). In contrast, in this work, we argue that all available positives can be used at the same time and employ contrastive learning that supports multiple positives (multi-positive). Experimental results on two datasets show that our proposed approach of leveraging all positives simultaneously and employing multi-positive contrastive learning on the robustifying subtask yields improvements in robustness against using contrastive learning with a single positive.

Keywords: Dense retrieval · Typo-robustness · Contrastive learning

1 Introduction

Dense retrieval has become the new paradigm in passage retrieval. It has demonstrated higher effectiveness than traditional lexical-based methods due to its ability to tackle the vocabulary mismatch problem [5]. Even though dense retrievers are highly effective on typo-free queries, they can witness a dramatic performance decrease when dealing with queries that contain typos [11,12,14]. Recent works on robustifying dense retrievers against typos utilize data augmentation to obtain typoed versions of the original queries at training time. Moreover, they introduce additional robustifying subtasks to minimize the representation discrepancy between the original query and its typoed variants.

Sidiropoulos and Kanoulas [11] applied an additional contrastive loss to enforce the latent representations of the original, typo-free queries to be closer

N. Goharian et al. (Eds.): ECIR 2024, LNCS 14610, pp. 297–305, 2024.
https://doi.org/10.1007/978-3-031-56063-7_21

to their typoed variants. Zhuang and Zuccon [14] utilized a self-teaching training strategy to minimize the difference between the score distribution of the original query and its typoed variants. Alternatively, Tasawong *et al.*[13] employed dual learning in combination with self-teaching [14] and contrastively trained the dense retriever on the prime task of passage retrieval and the dual task of query retrieval (learns the query likelihood to retrieve queries for passages).

Despite the improvements in robustness, the existing typo-robust methods do not always make optimal use of the available typoed queries. In detail, they address the robustifying subtasks with contrastive learning, assuming a single positive sample (query) and a set of negative ones per anchor (depending on the approach, the anchor can be either a query or a passage). However, alongside the original query, its multiple typoed variants are available. Hence, there is more than one positive sample per anchor. As a result, we can leverage all the available positives simultaneously and apply multi-positive contrastive learning instead (i.e., contrastive learning that supports multiple positives). For instance, Tasawong *et al.*[13] computes the contrastive loss for the query retrieval subtask using only the original, typo-free query as relevant for a given passage. Given a passage, we argue that both the original query and its typoed variations can be considered as relevant and adopt a multi-positive contrastive loss instead.

Literature on contrastive learning has shown that including multiple positives can enhance the ability of the model to discriminate between signal and noise (negatives) [6,8]. Intuitively, multiple negatives focus on what makes the anchor and the negatives dissimilar, while multiple positives focus on what makes the anchor and the positives similar. To this end, contrasting among multiple positives and negatives can bring an anchor and all its positives closer together in the latent space while keeping them far from the negatives.

In this work, we revisit recent methods in typo-robust dense retrieval and unveil that, in many cases, they do not sufficiently utilize the multiple positives that are available. Specifically, when tackling the robustifying subtasks, they ignore that multiple positives are available per anchor and consider contrastive learning with a single positive. In contrast, we suggest leveraging all the available positives and adopting a multi-positive contrastive learning approach. We aim to answer the following research questions:

RQ1 Can our multi-positive contrastive learning approach increase the robustness of dense retrievers that use contrastive learning with a single positive?
RQ2 Does our multi-positive contrastive learning variant outperform its single-positive counterpart regardless of the number of positives?

Our experimental results on two datasets show that our proposed approach of employing multi-positive contrastive learning yields improvements in robustness compared to contrastive learning with a single positive.[1]

[1] https://github.com/GSidiropoulos/typo-robust-multi-positive-DR.

2 Methodology

Contrastive learning is a vital component for training an effective dense retriever. Current typo-robust dense retrievers use contrastive learning with a single positive sample and multiple negative ones for both the main task of passage retrieval and the robustifying subtasks. In detail, given an anchor x, a positive sample x^+, and a set of negative samples X^-, the contrastive prediction task aims to bring the positive sample closer to the anchor than any other negative sample:

$$\mathcal{L}_{CE}(x, x^+, X^-) = -\log \frac{e^{f(x,x^+)}}{e^{f(x,x^+)} + \sum_{x^- \in X^-} e^{f(x,x^-)}}, \quad (1)$$

where f is a similarity function (e.g., dot product).

However, in many cases, multiple positive samples are available per anchor and can be used simultaneously to increase the discriminative performance of the model. As opposed to the aforementioned contrastive loss that supports a single positive, we propose employing a multi-positive contrastive loss to benefit from all the available positives. Given an anchor x, multiple positive samples X^+, and multiple negatives X^-, a multi-positive contrastive loss [6] is computed as:

$$\mathcal{L}_{MCE}(x, X^+, X^-) = -\frac{1}{|X^+|} \sum_{x^+ \in X^+} \log \frac{e^{f(x,x^+)}}{e^{f(x,x^+)} + \sum_{x^- \in X^-} e^{f(x,x^-)}}. \quad (2)$$

This work aims to identify cases in typo-robust dense retrieval methods where the robustifying subtasks consider only a single positive sample, even though multiple ones are available, and optimize a contrastive loss. Next, we replace the contrastive loss with its multi-positive alternative to benefit from all the available positives. Below we present the typo-robust dense retrieval methods we build upon followed by our multi-positive variants. We focus on dense retrievers that follow the dual-encoder architecture [5]. A traditional dense retriever, **DR**, is optimized only with the passage retrieval task. Given a query q, a positive/relevant passage p^+, and a set of negative/irrelevant passages $P^- = \{p_i^-\}_{i=1}^N$, the learning task trains the query and passage encoders via minimizing the softmax cross-entropy: $\mathcal{L}_{CE}^p = \mathcal{L}_{CE}(q, p^+, P^-)$. Positive query-passage pairs are encouraged to have higher similarity scores and negative pairs to have lower scores.

2.1 Dense Retriever with Self-supervised Contrastive Learning

DR+CL alternates DR with an additional contrastive loss that maximizes the agreement between differently augmented views of the same query [11]. This loss enforces that a query q and its typoed variation q', sampled from a set of available typoed variations $Q' = \{q_i'\}_{i=1}^K$, are close together in the latent space and distant from other distinct queries $Q^- = \{q_i^-\}_{i=1}^M$: $\mathcal{L}_{CE}^t = \mathcal{L}_{CE}(q, q', Q^-)$. The final loss is computed as a weighted summation, $\mathcal{L} = w_1 \mathcal{L}_{CE}^p + w_2 \mathcal{L}_{CE}^t$.

DR+CL$_M$ is our multi-positive variant of DR+CL. Given a query q, instead of sampling a different typoed variant q' from a set Q' at each update, we propose simultaneously employing all typoed variants. To do so, we replace $\mathcal{L}^t_{CE}$ with the following multi-positive contrastive loss that accounts for multiple positives: $\mathcal{L}^t_{MCE} = \mathcal{L}_{MCE}(q, Q', Q^-)$. The final loss is: $\mathcal{L} = w_1\mathcal{L}^p_{CE} + w_2\mathcal{L}^t_{MCE}$.

2.2 Dense Retriever with Dual Learning

DR+DL trains a robust, dense retriever via a contrastive dual learning mechanism [7]. In contrast to classic DR, which is optimized for passage retrieval only ($\mathcal{L}^p_{CE}$), DR+DL is optimized for the prime task of passage retrieval (i.e., learns to retrieve relevant passages for queries) and the dual task of query retrieval (i.e., learns to retrieve relevant queries for passages). Therefore, given a passage p, a positive query q^+, and a set of negative queries $Q^- = \{q_i^-\}_{i=1}^M$, it further minimizes the loss for the dual task: $\mathcal{L}^q_{CE} = \mathcal{L}_{CE}(p, q^+, Q^-)$. The dual training loss is added to the prime training loss to conduct contrastive dual learning and train the dense retriever. Specifically, the final loss is computed as $\mathcal{L} = \mathcal{L}^p_{CE} + w\mathcal{L}^q_{CE}$, where w is used to weight the dual task loss.

DR+DL$_M$ is our multi-positive variant of DR+DL. Contrary to DR+DL, we propose that for the query retrieval task, given a passage p, we can have a set of relevant queries consisting of the typo-free query and its typoed variants, $\mathcal{Q} = \{q^+, q'_1, q'_2, \ldots, q'_K\}$. Thus, we replace the contrastive loss of $\mathcal{L}^q_{CE}$ with a multi-positive contrastive loss, which can account for multiple relevant queries at the same time. We define the multi-positive contrastive loss for the dual task as: $\mathcal{L}^q_{MCE} = \mathcal{L}_{MCE}(p, \mathcal{Q}, Q^-)$. The final loss is computed as $\mathcal{L} = \mathcal{L}^p_{CE} + w\mathcal{L}^q_{MCE}$.

2.3 Dense Retriever with Dual Learning and Self-Teaching

DR+ST+DL trains a dense retriever with dual learning and self-teaching [13]. Similar to DR+DL, it minimizes the $\mathcal{L}^p_{CE}$ and $\mathcal{L}^q_{CE}$ for the main task of passage retrieval and the subtask of query retrieval, respectively. The additional self-teaching mechanism distills knowledge from a typo-free query q into its typoed variants $Q' = \{q'_i\}_{i=1}^K$ by forcing the model to match score distributions of misspelled queries to the score distribution of the typo-free query for both the passage retrieval and query retrieval task. This is achieved by minimizing the KL-divergence losses: (i) $\mathcal{L}^p_{KL} = \frac{1}{K}\sum_{k=1}^K \mathcal{L}_{KL}(s'^k_p, s_p)$, where $\{s'^1_p, s'^2_p, \ldots, s'^K_p\}$ and s_p is the score distribution in a passage-to-queries direction (passage retrieval) for the typoed queries and the typo-free query, respectively and (ii) $\mathcal{L}^q_{KL} = \frac{1}{K}\sum_{k=1}^K \mathcal{L}_{KL}(s'^k_q, s_q)$, where $\{s'^1_q, s'^2_q, \ldots, s'^K_q\}$ and s_q is the score distribution in a query-to-passages direction (query retrieval) for the typoed queries and the typo-free query, respectively. The final loss is computed as the weighted summation of the four losses, $\mathcal{L} = (1-\beta)((1-\gamma)\mathcal{L}^p_{CE} + \gamma\mathcal{L}^q_{CE}) + \beta((1-\sigma)\mathcal{L}^p_{KL} + \sigma\mathcal{L}^q_{KL})$.

DR+ST+DL$_M$ is our multi-positive variant of DR+ST+DL. Even though DR+ST+DL simultaneously uses all the available typo variations of a query in order to calculate the KL divergence losses for the prime passage retrieval task and the dual query retrieval, it uses only the typo-free query to compute the contrastive loss for query retrieval ($\mathcal{L}^q_{CE}$). To fully benefit from the multiple available typoed queries per typo-free query, we replace the contrastive loss for query retrieval $\mathcal{L}^q_{CE}$ with a multi-positive variant that supports samples with multiple positives $\mathcal{L}^q_{MCE}$. The final loss is computed as the weighted summation, $\mathcal{L} = (1-\beta)((1-\gamma)\mathcal{L}^p_{CE} + \gamma\mathcal{L}^q_{MCE}) + \beta((1-\sigma)\mathcal{L}^p_{KL} + \sigma\mathcal{L}^q_{KL})$.

3 Experimental Setup

Query Augmentation. From the aforementioned methods, those employing queries with typos in their training scheme are augmentation-based. In detail, during training, the typoed queries are generated from the original, typo-free queries through a realistic typo generator [9]. The typo generator applies the following transformations that often occur in human-generated queries: random character insertion, deletion and substitution, swapping neighboring characters, and keyboard-based character swapping [4].

Datasets and Evaluation. We conduct our experiments on MS MARCO passage ranking [10] and DL-Typo [14] on their typo-free and typoed versions. Both datasets use the same underlying corpus of 8.8 million passages and $\sim 400K$ training queries but differ in evaluation queries. DL-Typo provides 60 real-world queries with typos alongside their manually corrected typo-free version. The development set of MS MARCO consists of $6,980$ queries (the test set is not publicly available). Following previous works [13,14], we obtain typo variations for each typo-free query via a synthetic typo generation model and repeat the typo generation process 10 times. To measure the retrieval performance, we report the official metrics of each dataset. For the evaluation on the typo version of MS MARCO, we report the metrics averaged for each repeated experiment since typoed queries are generated 10 times for each original query.

Implementation Details. We follow an in-batch negative training setting with 7 hard negative passages per query and a batch size of 16 to train the dense retrievers.[2] We use AdamW optimizer with a 10^{-5} learning rate, linear scheduling with $10K$ warm-up steps, and decay over the rest of the training steps. We train up to $150K$ steps. We implement the query and passage encoders with BERT [2]. When applicable, we set the query augmentation size to 40. For the remaining hyperparameters specific to each method (e.g., weight w in DR+CL), we use the values initially proposed by the creators of each method. We use the Tevatron toolkit [3] to train the models and the Ranx library [1] to evaluate the retrieval performance. Finally, we use the typo generators from the TextAttack toolkit [9] for all the methods we experiment with to augment the training queries.

[2] The original methods and our proposed counterparts employ the same number of original, typo-free query-passage pairs per batch. However, our method leverages multiple typoed variants for each query; therefore, the batch we need to fit in the GPU memory is larger.

4 Results

To answer **RQ1**, we compare the retrieval performance of our multi-positive contrastive learning approaches against the original models. From Table 1, we see that employing our multi-positive contrastive learning approach yields improvements in robustness against typos upon the original methods that use contrastive learning with a single positive.

As expected, the more dramatic improvement comes when applying multi-positive contrastive learning on DR+DL since the original work only considers the typo-free query as positive when computing the contrastive loss for query retrieval (see Sect. 2.2). In contrast, in our DR+DL$_M$, we consider the typo-free query and all its available typoed variants as positives and use a multi-positive contrastive loss for query retrieval. We also see improvements when comparing DR+CL vs. our DR+CL$_M$. In detail, employing all available positives (typoed queries) at once and using multi-positive contrastive loss outperforms sampling a different positive at each update and using a single positive contrastive loss (see Sect. 2.1). The improvements are held even when comparing our DR+DL+ST$_M$ against DR+DL+ST, a model that already uses multiple positives. As seen in Sect. 2.3, DR+DL+ST uses a contrastive loss with a single positive for the query retrieval dual task (i.e., $\mathcal{L}^q_{CE}$) while considering multiple positives simultaneously to compute the KL-divergence losses (i.e., $\mathcal{L}^p_{KL}$, $\mathcal{L}^q_{KL}$).

Table 1. Retrieval results for the settings of (i) clean queries (Clean), and (ii) queries with typos (Typo). Statistical significant gains (two-tailed paired t-test with Bonferroni correction, $p < 0.05$) obtained from models with multi-positive contrastive loss (ours) over their original version with standard contrastive loss are indicated by †.

Model	Multi-positive contrastive loss	MS MARCO				DL-Typo					
		Clean		Typo		Clean			Typo		
		MRR@10	R@1000	MRR@10	R@1000	nDCG@10	MRR	MAP	nDCG@10	MRR	MAP
DR	✗	.331	.953	.140	.698	.677	.850	.555	.264	.395	.180
DR+DL	✗	.332	.953	.140	.698	.679	.826	.557	.269	.411	.186
DR+DL$_M$	✓	.335	.958	.213†	.866†	.699	.864	.585	.347†	.452	.259†
DR+CL	✗	.321	.957	.170	.787	.659	.797	.535	.284	.411	.207
DR+CL$_M$	✓	.322	.956	.178†	.811†	.652	.847	.539	.290	.447	.215
DR+ST+DL	✗	.334	.951	.259	.893	.681	.868	.567	.412	.543	.315
DR+ST+DL$_M$	✓	.335	.955	.261	.902†	.687	.870	.579	.426†	.583	.342†

At this point, we want to explore how the different numbers of positives affect our multi-positive approach (**RQ2**). To do so, we compare our DR+DL+ST$_M$ against DR+DL+ST. In its training, the latter already employs multiple positives simultaneously to compute the KL-divergence losses. However, our multi-positive approach fully benefits from the multiple available positives by incorporating them when computing the contrastive loss for query retrieval ($\mathcal{L}^q_{CE} \rightarrow \mathcal{L}^q_{MCE}$). Table 2 unveils that our multi-positive variant consistently outperforms the original model for the different numbers of typoed variants per query.

Table 2. Retrieval results for different query augmentation sizes (K). We report the results in the format "R@1000 (MRR@10)" on MS MARCO with typos.

	Multi-positive contrastive loss	K 1	10	20	30	40
DR+ST+DL	✗	.884 (.251)	.892 (.258)	.894 (.258)	.893 (.259)	.893 (.259)
DR+ST+DL$_M$	✓	**.884 (.251)**	**.898 (.260)**	**.900 (.260)**	**.902 (.261)**	**.902 (.261)**

5 Conclusions

In this work, we revisit recent studies in typo-robust dense retrieval and showcase that they do not always make sufficient use of multiple positive samples. In detail, they assume a single positive sample and multiple negatives per anchor and use contrastive learning for the robustifying subtasks. Opposed to this, we propose to leverage all the available positives and employ multi-positive contrastive learning. Experimentation on two datasets shows that following a multi-positive contrastive learning approach yields improvements in the robustness of the underlying dense retriever upon contrastive learning with a single positive.

Acknowledgements. This research was supported by the NWO Innovational Research Incentives Scheme Vidi (016.Vidi.189.039). All content represents the opinion of the authors, which is not necessarily shared or endorsed by their respective employers and/or sponsors.

References

1. Bassani, Elias: `ranx`: a blazing-fast python library for ranking evaluation and comparison. In: Hagen, M., et al. (eds.) ECIR 2022. LNCS, vol. 13186, pp. 259–264. Springer, Cham (2022). https://doi.org/10.1007/978-3-030-99739-7_30
2. Devlin, J., Chang, M., Lee, K., Toutanova, K.: BERT: pre-training of deep bidirectional transformers for language understanding. In: Burstein, J., Doran, C., Solorio, T. (eds.) Proceedings of the 2019 Conference of the North American Chapter of the Association for Computational Linguistics: Human Language Technologies, NAACL-HLT 2019, Minneapolis, MN, USA, June 2–7, 2019, Volume 1 (Long and Short Papers), pp. 4171–4186. Association for Computational Linguistics (2019). https://doi.org/10.18653/v1/n19-1423
3. Gao, L., Ma, X., Lin, J., Callan, J.: Tevatron: an efficient and flexible toolkit for neural retrieval. In: Chen, H., Duh, W.E., Huang, H., Kato, M.P., Mothe, J., Poblete, B. (eds.) Proceedings of the 46th International ACM SIGIR Conference on Research and Development in Information Retrieval, SIGIR 2023, Taipei, Taiwan, July 23–27, 2023, pp. 3120–3124. ACM (2023). https://doi.org/10.1145/3539618.3591805

4. Hagen, M., Potthast, M., Gohsen, M., Rathgeber, A., Stein, B.: A large-scale query spelling correction corpus. In: Kando, N., Sakai, T., Joho, H., Li, H., de Vries, A.P., White, R.W. (eds.) Proceedings of the 40th International ACM SIGIR Conference on Research and Development in Information Retrieval, Shinjuku, Tokyo, Japan, August 7–11, 2017, pp. 1261–1264. ACM (2017). https://doi.org/10.1145/3077136.3080749
5. Karpukhin, V., et al.: Dense passage retrieval for open-domain question answering. In: Webber, B., Cohn, T., He, Y., Liu, Y. (eds.) Proceedings of the 2020 Conference on Empirical Methods in Natural Language Processing, EMNLP 2020, Online, November 16–20, 2020, pp. 6769–6781. Association for Computational Linguistics (2020). https://doi.org/10.18653/v1/2020.emnlp-main.550
6. Khosla, P., et al.: Supervised contrastive learning. In: Larochelle, H., Ranzato, M., Hadsell, R., Balcan, M., Lin, H. (eds.) Advances in Neural Information Processing Systems 33: Annual Conference on Neural Information Processing Systems 2020, NeurIPS 2020 (December), pp. 6–12, 2020. virtual (2020). https://proceedings.neurips.cc/paper/2020/hash/d89a66c7c80a29b1bdbab0f2a1a94af8-Abstract.html
7. Li, Y., Liu, Z., Xiong, C., Liu, Z.: More robust dense retrieval with contrastive dual learning. In: Hasibi, F., Fang, Y., Aizawa, A. (eds.) ICTIR 2021: The 2021 ACM SIGIR International Conference on the Theory of Information Retrieval, Virtual Event, Canada, July 11, 2021, pp. 287–296. ACM (2021). https://doi.org/10.1145/3471158.3472245
8. Małkiński, M., Mańdziuk, J.: Multi-label contrastive learning for abstract visual reasoning. IEEE Trans. Neural Netw. Learn. Syst. 1–13 (2022). https://doi.org/10.1109/TNNLS.2022.3185949
9. Morris, J.X., Lifland, E., Yoo, J.Y., Grigsby, J., Jin, D., Qi, Y.: Textattack: a framework for adversarial attacks, data augmentation, and adversarial training in NLP. In: Liu, Q., Schlangen, D. (eds.) Proceedings of the 2020 Conference on Empirical Methods in Natural Language Processing: System Demonstrations, EMNLP 2020 - Demos, Online, November 16–20, 2020, pp. 119–126. Association for Computational Linguistics (2020). https://doi.org/10.18653/v1/2020.emnlp-demos.16
10. Nguyen, T., et al.: MS MARCO: a human generated machine reading comprehension dataset. In: Besold, T.R., Bordes, A., d'Avila Garcez, A.S., Wayne, G. (eds.) Proceedings of the Workshop on Cognitive Computation: Integrating neural and symbolic approaches 2016 co-located with the 30th Annual Conference on Neural Information Processing Systems (NIPS 2016), Barcelona, Spain, December 9, 2016. CEUR Workshop Proceedings, vol. 1773. CEUR-WS.org (2016). http://ceur-ws.org/Vol-1773/CoCoNIPS_2016_paper9.pdf
11. Sidiropoulos, G., Kanoulas, E.: Analysing the robustness of dual encoders for dense retrieval against misspellings. In: Amigó, E., Castells, P., Gonzalo, J., Carterette, B., Culpepper, J.S., Kazai, G. (eds.) SIGIR 2022: The 45th International ACM SIGIR Conference on Research and Development in Information Retrieval, Madrid, Spain, July 11–15, 2022, pp. 2132–2136. ACM (2022). https://doi.org/10.1145/3477495.3531818
12. Sidiropoulos, G., Vakulenko, S., Kanoulas, E.: On the impact of speech recognition errors in passage retrieval for spoken question answering. In: Hasan, M.A., Xiong, L. (eds.) Proceedings of the 31st ACM International Conference on Information & Knowledge Management, Atlanta, GA, USA, October 17–21, 2022. pp. 4485–4489. ACM (2022). https://doi.org/10.1145/3511808.3557662
13. Tasawong, P., Ponwitayarat, W., Limkonchotiwat, P., Udomcharoenchaikit, C., Chuangsuwanich, E., Nutanong, S.: Typo-robust representation learning for dense

retrieval. In: Rogers, A., Boyd-Graber, J.L., Okazaki, N. (eds.) Proceedings of the 61st Annual Meeting of the Association for Computational Linguistics (Volume 2: Short Papers), ACL 2023, Toronto, Canada, July 9–14, 2023, pp. 1106–1115. Association for Computational Linguistics (2023). https://doi.org/10.18653/v1/2023.acl-short.95
14. Zhuang, S., Zuccon, G.: Characterbert and self-teaching for improving the robustness of dense retrievers on queries with typos. In: Amigó, E., Castells, P., Gonzalo, J., Carterette, B., Culpepper, J.S., Kazai, G. (eds.) SIGIR 2022: The 45th International ACM SIGIR Conference on Research and Development in Information Retrieval, Madrid, Spain, July 11–15, 2022, pp. 1444–1454. ACM (2022). https://doi.org/10.1145/3477495.3531951

Learning-to-Rank with Nested Feedback

Hitesh Sagtani(✉), Olivier Jeunen, and Aleksei Ustimenko

ShareChat, Bengaluru, India
{hiteshsagtani,jeunen,aleksei.ustimenko}@sharechat.co

Abstract. Many platforms on the web present ranked lists of content to users, typically optimized for engagement-, satisfaction- or retention-driven metrics. Advances in the Learning-to-Rank (LTR) research literature have enabled rapid growth in this application area. Several popular interfaces now include nested lists, where users can enter a 2nd-level feed via any given 1st-level item. Naturally, this has implications for evaluation metrics, objective functions, and the ranking policies we wish to learn. We propose a theoretically grounded method to incorporate 2nd-level feedback into any 1st-level ranking model. Online experiments on a large-scale recommendation system confirm our theoretical findings.

Keywords: Learning-to-Rank · Recommender systems · User feedback

1 Introduction and Related Work

Rankings are at the heart of how users interact with content on the web. This holds for applications and use-cases across web search and e-commerce recommendations to streaming and social media platforms. Indeed, platforms use ranked list interfaces to serve content to users. Machine learning models typically power these rankings, learned to optimize some metrics deemed relevant to the business or its users. Due to this wealth of application areas, the Learning-to-Rank (LTR) literature has seen vast industry adoption [13,14,18,23,33,39].

Nevertheless, real-world examples of ranking interfaces often deviate from the traditional setup where a single ranked list is shown. Famous examples here include the *gridwise* page layout that is now standardized in video streaming platforms [11], and the research literature has looked into more general interfaces as well [32,40]. These works keep a broad focus, making few assumptions about the setting to provide effective methods with universal appeal.

Recently, *nested* ranking lists have started to gain in popularity, particularly seeing adoption in short-video feeds on social media platforms such as Reddit, Instagram, and ShareChat [16]. Here, users are presented with a scrollable 1st-level feed, where they can enter a full-screen 2nd-level feed via any given 1st-level item. This differs from the grid layout discussed above, as only a single level is presented to the user at any given time. Figure 1 visualizes such a layout of *nested* feeds, which are the core topic of this paper.

N. Goharian et al. (Eds.): ECIR 2024, LNCS 14610, pp. 306–315, 2024.
https://doi.org/10.1007/978-3-031-56063-7_22

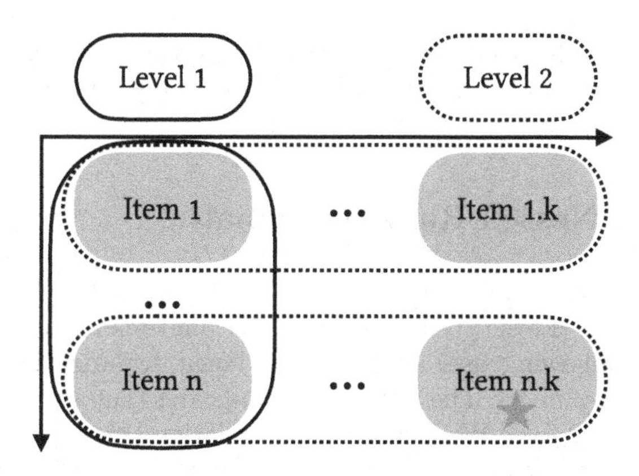

Fig. 1. A nested-feed interface. When receiving feedback on item $n.k$ in the 2nd-level feed (denoted ★), the 1st-level item n should be attributed as well.

A naïve modeling approach would instantiate both levels with independent ranking policies that optimize level-specific objectives. This implicitly assumes that rewards across feeds are independent, which ignores the hierarchically nested structure. Indeed, as shown in Fig. 1, positive feedback on 2nd-level items should be (partially) attributed to the relevant 1st-level item, as to allow the 1st-level ranking policy to learn from this *nested* feedback.

Existing methods in the literature that go beyond single lists either deal with complex settings where no assumptions are placed on the reward or examination dependencies in the interface [32], or page-level re-ranking strategies when multiple independent lists are present [40]. Even though nested ranking interfaces are prevalent in practical applications, they are currently under-explored in the research literature. Our work formally introduces this problem setting. Under the commonly assumed position-based model [8], we theoretically derive the optimal objective for the 1st-level ranking model when 2nd-level nested feedback is available.

Related Work. Learning-to-Rank (LTR) is a classical problem in information retrieval that has received extensive research attention [27,29]. It has found widespread adoption in various application areas, including web search [35], question answering [43], e-commerce recommendations [24], streaming platforms [23], and recommendation systems [9,18,22]. LTR approaches can be categorized into different types, such as pointwise [28], pairwise [4,10,26], and listwise [6,7,41,42] methods. Recent advancements in LTR include using online learning algorithms with non-linear models [31] and the proposal of stochastic bandit algorithms like BatchRank [45]. In the context of social media platforms, LTR algorithms are usually trained on implicit signals derived from user interactions, such as user clicks [19], dwell time [44], and various other engagement signals [25], that are then treated as indicators of relevance. However, these signals are subject to biases, such as positional bias, where the ranking position influences user clicks. To address this issue, many studies have explored methods to mitigate bias in ranking using counterfactual inference frameworks through empirical risk minimization [1,20]. Qingyao *et al.* [2] evaluated and compared the performance of such state-of-the-art unbiased LTR algorithms.

2 Modeling Nested Ranking Signals

In the rest of the work, we refer to the 1st-level feed as L1 (Level 1), and the 2nd-level feed as L2 (Level 2). To calculate the final relevance label for a feed, we combine the following user signals using linear scalarization: *likes*, *shares*, *favourites*, and *video clicks*. The latter signal indicates that a user clicks on an L1 video to enter the L2 feed. We observe the correlation among these signals to be significantly different for L1 and L2 feed for our platform, highlighting variations in user behaviours across feeds, resulting in different trade-offs between signals.

We consider some distribution of users $u \sim \mathcal{U}$ that interact with our platform. Classical LTR deals with the problem of finding and evaluating a ranking for a list of items $A = \{a_1, \ldots, a_n\}$ with per-item relevance signals $y_i \forall i \in [1, n]$, e.g., a click or crowd-sourced label. Typically, ranking quality is measured by Discounted Cumulative Gain (DCG):

$$\text{DCG} = \sum_{i=1}^{n} \frac{y_i}{\log_2(1+i)}. \tag{1}$$

Given some model $f(u, a)$ we can sort items $a \in A$ in decreasing order to obtain a ranking. LTR is then: finding a model $f(u, a)$ such that $\mathbb{E}_{u \sim \mathcal{U}}\text{DCG}$ is maximized. DCG objectives can be effectively optimized by state-of-the-art LTR approaches like LambdaRank [5], StochasticRank [38] or YetiRank [12].

Denote $R(u, a)$ as the online relevance signal that we observe for the ranked list. Note that DCG adheres to the Position-Based Model (PBM) [8], when we assume that the probability of user viewing position i is equal to $\frac{1}{\log_2(1+i)}$. As a result, $y_i = \frac{R(u,a_i)}{\mathbb{P}(u \text{ viewed position } i)}$ is an unbiased estimator of $\mathbb{E}\big(R(u, a_i)|u, a, \text{viewed}\big)$. Under these reasonable assumptions, DCG can be seen as an offline estimator of an online metric Q [15,17]:

$$Q = \mathbb{E}_{u \sim \mathcal{U}} \sum_{i=1}^{n} R(u, a_i) = \tag{2}$$

$$\mathbb{E}_{u \sim \mathcal{U}} \sum_{i=1}^{n} \mathbb{E}\big(R(u, a_i)|u, a, \text{viewed}\big)\mathbb{P}(u \text{ viewed position } i) = \mathbb{E}_{u \sim \mathcal{U}}\text{DCG}. \tag{3}$$

In our case, when the user clicks on the item a_i, our system retrieves another *nested* list of items $B_i = \{b_{i1}, \ldots, b_{im}\}$, ranks them, and presents them to the user. For an item $a \in A$ we denote as $R_A(u, a)$ a *relevance* signal (e.g., an indicator of a positive interaction) that indicates user preference on the L1 feed, and we analogously define $R_B(u, b)$ for the L2 feed. The goal of the LTR model is to maximize the following online metric:

$$Q = \mathbb{E}_{u \sim \mathcal{U}} \sum_{i=1}^{n} \left(R_A(u, a_i) + \sum_{j=1}^{m} R_B(u, b_{ij}) \right). \tag{4}$$

Suppose we ignore the contribution from R_B and assume a classic setup with only R_A: items with lower L1 relevance but better L2 feeds are penalised and thus, a model trained only on L1 signals will have *worse* online performance for metric Q. This leads to a degradation of the recommendation system overall, while the "quality" of ranking on the L1 feed appears high.

In our setup, we consider only how to train a ranker model to rank list A so that Q is maximized while assuming the ranking for B is fixed. From the very definition of Q it is seen that we should define $\widetilde{R}(u, a_i) = R_A(u, a_i) + \sum_{j=1}^{m} R_B(u, b_{ij})$ as the new relevance signal for the item a_i. That is, the relevance signal for the L1 feed should account for the relevance signal on the L2 feed if we want to find a ranker that maximizes overall quality measured by Q.

Thus, to train a model we consider historical logs consisting of (u, A, B, Y) where $Y = (y_{ij})_{i=1,j=0}^{n,m}$ denotes observed values for $y_{i0} \sim R_A(u, a_i)$ and $y_{ij} \sim R_B(u, b_{ij}), j > 0$. These logs are transformed into triplets $(u, A, \widetilde{Y})$, where we define $\widetilde{y}_i = \sum_{j=0}^{m} y_{ij} \sim \widetilde{R}(u, a_i)$. The dataset $\{(u, A, \widetilde{Y})\}$ now resembles a classic LTR dataset and, therefore, can serve as input to well-established LTR methods.

We also note that because the ranking on the L2 feed (i.e. the ranking of list B) is fixed, that means that we should not measure the contribution from the L2 feed using DCG. This is because our observed reward in the logs $R_B(u, b_{ij})$ is already positionally biased. To see that, we can formally write:

$$\mathbb{E}_{u \sim \mathcal{U}} \sum_{j=1}^{m} R_B(u, b_{ij}) = \mathbb{E}_{u \sim \mathcal{U}} \frac{R_B(u, b_{ij})}{\mathbb{P}(u \text{ viewed position } j)} \mathbb{P}(u \text{ viewed position } j) \tag{5}$$

This means that we can get DCG if we consider debiased labels $\frac{R_B(u, b_{ij})}{\mathbb{P}(u \text{ viewed position } j)}$ but since the ranking of B is fixed, those debiased labels have *the same* denominator as the positional bias and, thus, it cancels out. Therefore, we should not discount observed relevance signals on B and include them just as sums.

Our final feed-level relevance signals, denoted as $R_A(u, a_i)$ for the L1 feed and $R_B(u, b_{ij})$ for the L2 feed, are obtained by linearly combining individual user signals. These signals are fine-tuned internally to optimize user retention. However, since our objective is to evaluate the incorporation of user feedback from the L2 feed into the L1 feed, we do not delve into the specifics of how we tune the weights for each individual feed. In the subsequent section, we assign the following synthetic labels as the relevance signals for evaluation: S_1 as $R_A(u, a_i)$, S_2 as $\widetilde{R}(u, a_i)$ with DCG of $R_B(u, b_{ij})$, and S_3 as $\widetilde{R}(u, a_i)$ with sum of $R_B(u, b_{ij})$.

3 Experimental Validation

To the best of our knowledge, no datasets containing logged user feedback for nested feed interfaces are publicly available. For this reason, we resort to a proprietary dataset obtained from ShareChat, a widely popular social media application with over 180 million monthly active users across 18 regional languages in India, to substantiate our hypothesis empirically.

We leverage a range of dynamic user and post attributes based on embeddings, user sign-up date, genre, tags, fatigue score [37] and language, along with several interaction-based attributes. We find strong correlations between the relevance label with certain interaction features: the dot product between user and post embeddings is one such example. We applied negative sampling to the dataset based on a union of all individual signals, i.e. *likes*, *shares*, *favs* and *clicks*. We collected a random sample of 100 million instances, where less than 5% of the dataset had a positive signal, while the remaining instances were labeled as '0'. To ensure representative data for training, validation, and testing, we used stratified sampling, dividing the dataset in a 70:15:15 ratio, respectively.

3.1 Offline Training and Experiment Results

Gradient-Boosted Decision Tree (GBDT) methods such as LambdaMart have long remained the state-of-the-art approach [36] for tabular LTR datasets. Recent empirical evaluations have shown YetiRank to outperform LambdaMart and other competing algorithms in most cases [30]. To evaluate the synthetic labels, we use the YetiRank algorithm implemented in the Catboost [34] library. For hyperparameter tuning, we perform around 100 iterations of Bayesian optimization using the hyperopt [3] library for each synthetic label. To address the class imbalance problem in the dataset during training, we scale the loss of positive class by the ratio of negative to the positive examples using `scale_pos_weight` parameter. We use an overfitting detector on the validation dataset and stop the training if there is no improvement after 50 iterations on *DCG@10* and use the best-performing model on the validation set for each signal.

We evaluated the predictors (S_1, S_2, S_3) discussed in the previous section, considering *likes*, *shares*, *favs*, *clicks*, S_1, S_2 and S_3 as the true relevance labels and evaluated on the *DCG@k* metric. Table 1 shows the % loss in DCG metric values of predictor models compared to the model trained on the true relevance signal for ranking. For example, when using *likes* as true relevance, we observe a 20.7% decrease in *DCG@3* metric for model trained on S_3 synthetic label, compared to the model trained on *likes* signal.

From Table 1, we observe that models trained on S_2 and S_3 labels exhibit lower DCG loss compared to S_1 for all the individual user signals: *likes*, *shares*, *favs* and *clicks*, with S_3 having minimum loss across most of the signals. This validates our hypothesis that incorporating L2 feedback leads to improved overall ranking, with a sum-aggregation of L2 feedback to be optimal over discounting, as outlined in Sect. 2.

We also treat each synthetic label as the *true* relevance signal to compare them relatively. We note that the % loss in DCG is much lower when using S_3 as predictor for other synthetic labels as the true relevance label (S_1, S_2), followed by S_2 and finally S_1 exhibiting the highest loss. Since, S_3 and S_2 predictors have

Table 1. Offline % loss in DCG of predictor label compared to a model trained and ranked only on various "true" relevance signals.

DCG @k	Label	True Relevance Signal						
		likes	shares	favs	clicks	S_1	S_2	S_3
3	S_1	27.8	26.8	23.7	17.3	**0**	17.3	18.1
	S_2	24.1	25.2	**20.5**	14.8	10.2	**0**	15.1
	S_3	**20.7**	**24.7**	20.8	**13.4**	12.4	8.4	**0**
5	S_1	25.9	25.8	20.3	15.6	**0**	16.5	17.7
	S_2	22.9	**24.1**	**17.5**	14.4	10.1	**0**	15.2
	S_3	**19.2**	24.3	18.2	**12.6**	10.5	8.1	**0**
10	S_1	24.5	24.2	19.6	15.1	**0**	16.1	17.3
	S_2	20.8	23.6	**16.7**	11.6	9.3	**0**	14.4
	S_3	**17.8**	**23.2**	17.4	**10.8**	10.4	7.7	**0**

both the L1 and L2 feedback information in their synthetic label, they are able to capture user behavior on both the feeds, compared to the S_1 predictor which has feedback for only L1 feed. In addition, the loss is much lower when using the S_3 predictor on S_2 as true relevance compared to the S_2 predictor with S_3 as true relevance, the S_3 label is better able to capture the information on L2 feed than S_2.

3.2 Online A/B Experimentation

Based on the offline results, we conducted an A/B experiment to evaluate various synthetic labels. Table 2 tabulates the online metrics for ranking based on S_2 and S_3 labels in *variant-1* and *variant-2* respectively, with S_1 being the control.

From Table 2a, we observe that overall engagements (likes, shares, favs) have increased for both variants, indicating better posts on the L1 feed that lead to more engagements on the L2 feed. We also observe increased clicks on the L1 feed, indicating higher user convergence to the L2 feed compared to the model trained solely on the S_1 label. While there is a decrease in dwell time on the L1 feed, there is an increase for the L2 feed, suggesting that users spend more time overall in the L2 feed than in the L1 feed. The increase in clicks on the L1 feed also supports this insight.

To further validate our hypothesis of user convergence to the L2 feed, Table 2b shows a decrease in L1 depth (indicating the count of subsequent feed fetches) and an increase in L2 depth. Increment in L2 transition (#times users switch from the L1 feed to the L2 feed) and decrease in S2L2 (Second to L2: time in seconds users take to open any post on L2 feed after session start) confirms higher and early user convergence to L2 feed. Including the feedback from the L2 feed improves user experience on the overall platform as indicated by the platform-level metrics in Table 2b, with metrics showing increased user retention, engagements, and interactions with the platform.

Table 2. Online % gain in metrics when compared to the model trained only on L1 signal. All results are statistically significant according to a 2-tailed t-test at $p < 0.05$ after Bonferroni correction, except for those marked by^{+}.

(a) Engagement signals for L1 and L2 feed

Feed	variant	likes	shares	favs	dwell time	clicks
L1	variant-1	5.9	4.3	5.7	-1.0^{+}	5.1
	variant-2	6.5	5.6	5.2	-1.2^{+}	7.1
L2	variant-1	3.6	1.1^{+}	3.1	1.5	-
	variant-2	4.2	2.7	2.5	3.3	-

(b) Platform & transition metrics

	L1 to L2 transition metrics				Platform level metrics		
variant	S2L2	L1 depth	L2 depth	L2 transition	Engagements	#Session	Retention
variant-1	-0.9	-1.43^{+}	2.53	8.15	0.43	0.3^{+}	0.12
variant-2	-2.1	-4.14	5.42	11.23	0.71	0.5	0.17

We also examined the composition of the suggested posts between the variants and observed that including the L2 feedback results in more suggestions from personalized candidate generators like field-aware factorization machines [21] compared to non-personalized candidate generators (e.g. popularity). However, we did not observe any statistically significant difference in the genre distribution of the posts. Finally, we found that variant-2 performs better in most of the metrics compared to variant-1, which is consistent with the offline numbers we observed and our hypothesis of using the sum of feedback on the L2 feed being optimal compared to when discounting is used.

4 Discussion and Future Work

In this work, we have proposed a modeling framework for the nested feed structure that is prevalent in modern applications. We provided a theoretical explanation for why optimizing for combined signals on both feeds is more effective than optimizing them independently. We considered different user feedback signals as the proper relevance signal. We used the DCG metric to demonstrate that incorporating L2 feedback in the L1 predictor reduces the loss in DCG compared to using the L1 predictor alone. Furthermore, we observed consistency between our online metric and the offline metric and observed improvement in both short-term engagement and user retention at the platform level. Our current work focused on optimizing the ranking policy for the L1 feed while keeping the L2 feed ranking policy fixed. We envision future extensions of this work where we can jointly optimize the L1, and L2 feeds using a single policy trained end-to-end.

References

1. Ai, Q., Bi, K., Luo, C., Guo, J., Croft, W.B.: Unbiased learning to rank with unbiased propensity estimation. In: The 41st International ACM SIGIR Conference on Research and Development in Information Retrieval, pp. 385–394 (2018)
2. Ai, Q., Yang, T., Wang, H., Mao, J.: Unbiased learning to rank: online or offline? ACM Trans. Inf. Syst. (TOIS) **39**(2), 1–29 (2021)
3. Bergstra, J., Yamins, D., Cox, D.D., et al.: Hyperopt: a python library for optimizing the hyperparameters of machine learning algorithms. In: Proceedings of the 12th Python in Science Conference, vol. 13, p. 20. Citeseer (2013)
4. Burges, C., Shaked, T., Renshaw, E., Lazier, A., Deeds, M., Hamilton, N., Hullender, G.: Learning to rank using gradient descent. In: Proceedings of the 22nd International Conference on Machine Learning, pp. 89–96 (2005)
5. Burges, C., Ragno, R., Le, Q.: Learning to rank with nonsmooth cost functions. In: Advances in Neural Information Processing Systems, vol. 19 (2006)
6. Burges, C.J.: From ranknet to lambdarank to lambdamart: an overview. Learning **11**(23–581), 81 (2010)
7. Cao, Z., Qin, T., Liu, T.Y., Tsai, M.F., Li, H.: Learning to rank: from pairwise approach to listwise approach. In: Proceedings of the 24th International Conference on Machine Learning, pp. 129–136 (2007)
8. Chuklin, A., Markov, I., de Rijke, M.: Click Models for Web Search. Morgan & Claypool (2015). https://doi.org/10.2200/S00654ED1V01Y201507ICR043
9. Duan, Y., Jiang, L., Qin, T., Zhou, M., Shum, H.Y.: An empirical study on learning to rank of tweets. In: Proceedings of the 23rd International Conference on Computational Linguistics (Coling 2010), pp. 295–303 (2010)
10. Freund, Y., Iyer, R., Schapire, R.E., Singer, Y.: An efficient boosting algorithm for combining preferences. J. Mach. Learn. Res. **4**(Nov), 933–969 (2003)
11. Gomez-Uribe, C.A., Hunt, N.: The Netflix recommender system: algorithms, business value, and innovation. ACM Trans. Manage. Inf. Syst. **6**(4), 1–9 (2016). https://doi.org/10.1145/2843948
12. Gulin, A., Kuralenok, I., Pavlov, D.: Winning the transfer learning track of yahoo learning to rank challenge with yetirank. In: Proceedings of the Learning to Rank Challenge, pp. 63–76. PMLR (2011)
13. Haldar, M., et al.: Improving deep learning for airBNB search. In: Proceedings of the 26th ACM SIGKDD International Conference on Knowledge Discovery and Data Mining, pp. 2822–2830. KDD 2020, ACM (2020). https://doi.org/10.1145/3394486.3403333
14. Jagerman, R., Wang, X., Zhuang, H., Qin, Z., Bendersky, M., Najork, M.: Rax: composable learning-to-rank using Jax. In: Proceedings of the 28th ACM SIGKDD Conference on Knowledge Discovery and Data Mining, pp. 3051–3060. KDD 2022, ACM (2022). https://doi.org/10.1145/3534678.3539065
15. Jeunen, O.: Offline approaches to recommendation with online success. Ph.D. thesis, University of Antwerp (2021)
16. Jeunen, O.: A probabilistic position bias model for short-video recommendation feeds. In: Proceedings of the 17th ACM Conference on Recommender Systems, pp. 675–681. RecSys 2023, ACM (2023). https://doi.org/10.1145/3604915.3608777
17. Jeunen, O., Potapov, I., Ustimenko, A.: On (normalised) discounted cumulative gain as an offline evaluation metric for top-n recommendation (2023)

18. Jeunen, O., et al.: On gradient boosted decision trees and neural rankers: a case-study on short-video recommendations at ShareChat. In: Proceedings of the 15th Annual Meeting of the Forum for Information Retrieval Evaluation. FIRE 2023, ACM (2023)
19. Joachims, T., Granka, L., Pan, B., Hembrooke, H., Gay, G.: Accurately interpreting clickthrough data as implicit feedback. In: ACM SIGIR Forum, vol. 51, pp. 4–11. ACM New York, NY, USA (2017)
20. Joachims, T., Swaminathan, A., Schnabel, T.: Unbiased learning-to-rank with biased feedback. In: Proceedings of the tenth ACM International Conference on Web Search and Data Mining, pp. 781–789 (2017)
21. Juan, Y., Zhuang, Y., Chin, W.S., Lin, C.J.: Field-aware factorization machines for CTR prediction. In: Proceedings of the 10th ACM Conference on Recommender Systems, pp. 43–50 (2016)
22. Karatzoglou, A., Baltrunas, L., Shi, Y.: Learning to rank for recommender systems. In: Proceedings of the 7th ACM Conference on Recommender Systems, pp. 493–494 (2013)
23. Karmaker Santu, S.K., Sondhi, P., Zhai, C.: On application of learning to rank for e-commerce search. In: Proceedings of the 40th International ACM SIGIR Conference on Research and Development in Information Retrieval, pp. 475–484. SIGIR 2017, ACM (2017). https://doi.org/10.1145/3077136.3080838
24. Karmaker Santu, S.K., Sondhi, P., Zhai, C.: On application of learning to rank for e-commerce search. In: Proceedings of the 40th International ACM SIGIR Conference on Research and Development in Information Retrieval, pp. 475–484 (2017)
25. Lalmas, M., Hong, L.: Tutorial on metrics of user engagement: applications to news, search and e-commerce. In: Proceedings of the Eleventh ACM International Conference on Web Search and Data Mining, pp. 781–782 (2018)
26. Leaman, R., Islamaj Doğan, R., Lu, Z.: DNorm: disease name normalization with pairwise learning to rank. Bioinformatics **29**(22), 2909–2917 (2013)
27. Li, H.: A short introduction to learning to rank. IEICE Trans. Inf. Syst. **94**(10), 1854–1862 (2011)
28. Li, P., Wu, Q., Burges, C.: McRank: learning to rank using multiple classification and gradient boosting. In: Advances in Neural Information Processing Systems, vol. 20 (2007)
29. Liu, T.Y., et al.: Learning to rank for information retrieval. Found. Trends® Inf. Retr. **3**(3), 225–331 (2009)
30. Lyzhin, I., Ustimenko, A., Gulin, A., Prokhorenkova, L.: Which tricks are important for learning to rank? In: International Conference on Machine Learning, pp. 23264–23278. PMLR (2023)
31. Oosterhuis, H., de Rijke, M.: Differentiable unbiased online learning to rank. In: Proceedings of the 27th ACM International Conference on Information and Knowledge Management, pp. 1293–1302 (2018)
32. Oosterhuis, H., de Rijke, M.: Ranking for relevance and display preferences in complex presentation layouts. In: The 41st International ACM SIGIR Conference on Research and Development in Information Retrieval, pp. 845–854. SIGIR 2018, ACM (2018). https://doi.org/10.1145/3209978.3209992
33. Pasumarthi, R.K., et al.: TF-ranking: scalable tensorflow library for learning-to-rank. In: Proceedings of the 25th ACM SIGKDD International Conference on Knowledge Discovery and Data Mining, pp. 2970–2978. KDD 2019, ACM (2019). https://doi.org/10.1145/3292500.3330677

34. Prokhorenkova, L., Gusev, G., Vorobev, A., Dorogush, A.V., Gulin, A.: Catboost: unbiased boosting with categorical features. In: Advances in Neural Information Processing Systems, vol. 31 (2018)
35. Qin, T., Liu, T.Y., Zhang, X.D., Wang, D.S., Xiong, W.Y., Li, H.: Learning to rank relational objects and its application to web search. In: Proceedings of the 17th International Conference on World Wide Web, pp. 407–416 (2008)
36. Qin, Z., et al.: Are neural rankers still outperformed by gradient boosted decision trees? (2021)
37. Sagtani, H., Jhanwar, M.G., Gupta, A., Mehrotra, R.: Quantifying and leveraging user fatigue for interventions in recommender systems. In: Proceedings of the 46th International ACM SIGIR Conference on Research and Development in Information Retrieval (2023)
38. Ustimenko, A., Prokhorenkova, L.: Stochasticrank: global optimization of scale-free discrete functions. In: International Conference on Machine Learning, pp. 9669–9679. PMLR (2020)
39. Wang, R., et al.: Dcn v2: Improved deep and cross network and practical lessons for web-scale learning to rank systems. In: Proceedings of the Web Conference 2021, pp. 1785–1797. WWW 2021, ACM (2021). https://doi.org/10.1145/3442381.3450078
40. Xi, Y., et al.: A bird's-eye view of reranking: from list level to page level. In: Proceedings of the Sixteenth ACM International Conference on Web Search and Data Mining, pp. 1075–1083. WSDM 2023, ACM (2023). https://doi.org/10.1145/3539597.3570399
41. Xia, F., Liu, T.Y., Wang, J., Zhang, W., Li, H.: Listwise approach to learning to rank: theory and algorithm. In: Proceedings of the 25th International Conference on Machine Learning, pp. 1192–1199 (2008)
42. Xu, J., Li, H.: Adarank: a boosting algorithm for information retrieval. In: Proceedings of the 30th Annual International ACM SIGIR Conference on Research and Development in Information Retrieval, pp. 391–398 (2007)
43. Yang, L., et al.: Beyond factoid QA: effective methods for non-factoid answer sentence retrieval. In: Ferro, N., et al. (eds.) ECIR 2016. LNCS, vol. 9626, pp. 115–128. Springer, Cham (2016). https://doi.org/10.1007/978-3-319-30671-1_9
44. Yi, X., Hong, L., Zhong, E., Liu, N.N., Rajan, S.: Beyond clicks: dwell time for personalization. In: Proceedings of the 8th ACM Conference on Recommender Systems, pp. 113–120 (2014)
45. Zoghi, M., Tunys, T., Ghavamzadeh, M., Kveton, B., Szepesvari, C., Wen, Z.: Online learning to rank in stochastic click models. In: International Conference on Machine Learning, pp. 4199–4208. PMLR (2017)

MMCRec: Towards Multi-modal Generative AI in Conversational Recommendation

Tendai Mukande[1,3](✉), Esraa Ali[1,4], Annalina Caputo[1,4], Ruihai Dong[2,5], and Noel E. O'Connor[1,3,5]

[1] Dublin City University, Dublin 9, Dublin, Ireland
tendai.mukande2@mail.dcu.ie, {annalina.caputo,noel.oconnor}@dcu.ie
[2] University College Dublin, Dublin 4, Dublin, Ireland
ruihai.dong@ucd.ie
[3] SFI ML-LABS, Dublin, Ireland
[4] ADAPT Centre, Dublin, Ireland
esraa.ali@adaptcentre.ie
[5] Insight SFI Research Centre for Data Analytics, Dublin, Ireland
https://www.ml-labs.ie/ , https://www.adaptcentre.ie/ , https://www.dcu.ie/ , https://www.ucd.ie/ , https://www.insight-centre.org/

Abstract. Personalized recommendation systems have become integral in this digital age by facilitating content discovery to users and products tailored to their preferences. Since the Generative Artificial Intelligence (GAI) boom, research into GAI-enhanced Conversational Recommender Systems (CRSs) has sparked great interest. Most existing methods, however, mainly rely on one mode of input such as text, thereby limiting their ability to capture content diversity. This is also inconsistent with real-world scenarios, which involve multi-modal input data and output data. To address these limitations, we propose the Multi-Modal Conversational Recommender System (MMCRec) model which harnesses multiple modalities, including text, images, voice and video to enhance the recommendation performance and experience. Our model is capable of not only accepting multi-mode input, but also generating multi-modal output in conversational recommendation. Experimental evaluations demonstrate the effectiveness of our model in real-world conversational recommendation scenarios.

Keywords: Generative AI · Large Language Model · Conversational Recommendation · Graph Neural Network · Diffusion Model

1 Introduction

The data generated in the increasingly digital world is a valuable source of information to create personalized recommendations. This data comes in multiple forms such as text, audio, videos, and images. Recommender Systems (RSs) have become essential components of e-commerce platforms, content streaming

N. Goharian et al. (Eds.): ECIR 2024, LNCS 14610, pp. 316–325, 2024.
https://doi.org/10.1007/978-3-031-56063-7_23

services, and social media networks [21]. Most traditional RSs rely on historical offline user-item interaction data, and this constrains their ability to fully capture dynamic user preferences [43]. To resolve this issue, CRSs have been proposed to enable proactive dialogue with users as well as generate more accurate and explainable recommendations [18]. Most state-of-the-art CRSs however, rely on unimodal data such as text, which limits their ability to capture the complexity of real-world user intent [5]. This limitation has spurred research into CRSs that are multi-modal and interactive [6,17]. The motivation behind multi-modal CRSs is to harness the power of multiple modalities [10] in order to enrich user experiences and offer explainable recommendations that align more closely with user preferences [22]. In this way, users would be able to express their preferences through several ways which include textual queries, images and voice commands. Moreover, CRSs incorporating diverse modalities would provide richer and more nuanced representations of user intent, thereby comprehending and responding to user needs more effectively [29]. Several methods have been proposed for multi-modal CRSs but they do not sufficiently address the current CRS challenges [27] such as incorporating multi-modal input/output combinations of the data forms [17,20,39] as well as limited explainability [7], thereby limiting the transparency of the recommendations [14] (Fig. 1).

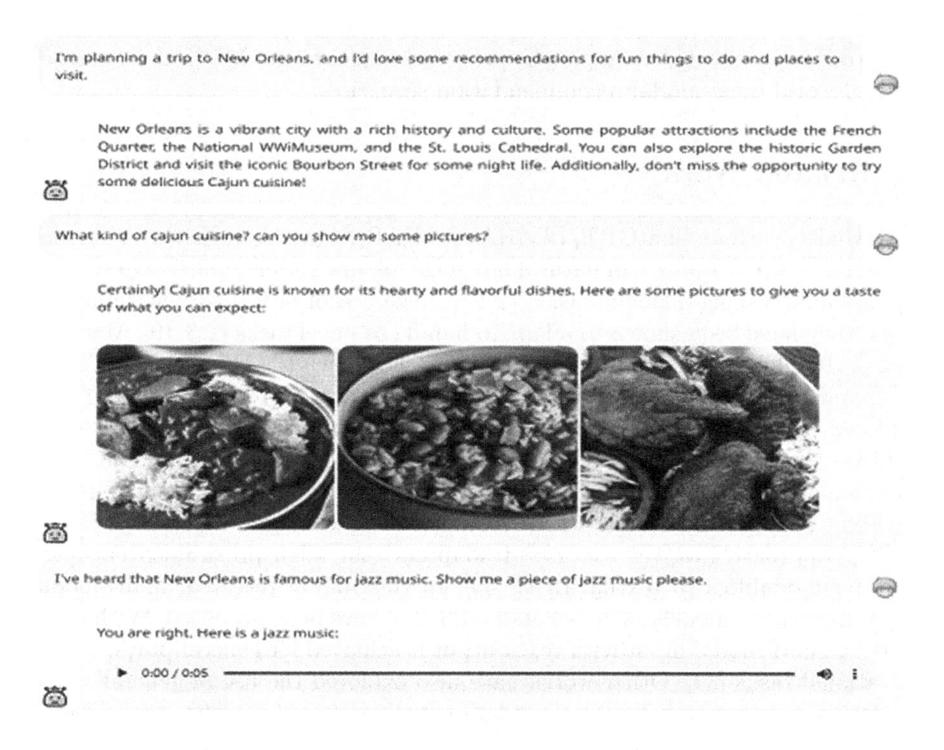

Fig. 1. An example of multi-modal Conversational Recommendation. Source: [1]

To address the highlighted challenges, we propose the Multi-Modal Conversational Recommender System (MMCRec) model capable of handling input and output combinations of text, images, videos and audio data. Using a Large Language Model (LLM) as the backbone of the model, multi-modal input encoders and output diffusion decoder modules enable a context-aware recommendation experience. Users can interact with the model through text, share images, audio and videos of their desired items. This multi-modal approach does not only enrich user engagement but also facilitates better understanding of user intent and context, thereby improving the quality of the recommendations [15]. Our **contributions** are as follows:

- We extend the NExT-GPT [1] model to the conversational recommendation scenario. Leveraging on the multi-modal input encoding as well as output image, audio and video diffusion modules, the resultant framework is a *multi-modal input-multi-modal output* LLM-based CRS which has understanding, reasoning and explanation abilities.
- We apply a Graph-of-Thought (GoT) [44] prompting model to enhance the LLM reasoning process which leverages on cross-attention among the text, video, image and video modalities [36]. We hypothesize that this approach adds additional context and semantic understanding of the queries by the model [37].

In our experiments, we apply the proposed model to the M5Product dataset for real world cross-modal recommendation scenarios.

2 Related Work

GAI Models such as ChatGPT, LLaMA and BARD have been shown to demonstrate remarkable abilities in natural language understanding, multi-step reasoning and undertaking dialogue tasks [12,28]. In zero-shot or few-shot learning settings, they have been shown to adapt to handle to novel tasks [2,3,19]. Methods such as Reinforcement Learning from Human Feedback (RLHF) [11,16], Chain-of-Thought Prompting [36], self-consistency and boosting techniques have been explored by prior research to enhance contextual understanding and reasoning in GAIs [31]. To improve perception in real-world scenarios, the integration of multi-modal capabilities into GAI models has also been studied, such as understanding of other modalities which include image video, audio, etc. [9,13]. A notable approach involves the use of adapters that align pre-trained encoders in other modalities to textual LLM [40]. In this line of research, multi-modal Language-Vision models such as PandaGPT [26] have been proposed. With these models, multi-modal input with text output is achieved for general-purpose conversational tasks [30]. Other works have also explored the use of generative AI models in conversational recommendation tasks [19,23,38]. The main limitation with these models is that they do not generate multi-modal recommendations, which is inconsistent with practical scenarios. Wu et al. proposed the Next-GPT LLM in which they added multi-modal output generation with multi-modal input content [1]. To address the limitations of state-of-the-art methods, we build upon this approach in conversational recommendation settings, which is consistent with real-world scenarios.

3 Methodology

In this section, we introduce our MM-MMCRec model. The novelty of our approach lies in the multi-modal graph-based derivation of potential user-item interactions. We hypothesize that the generated multi-modal recommendations enhance the user experience. Input instructions to the LLM module are passed in the form of audio, video, text or image. We leverage on the multi-modal joint embedding capabilities of ImageBIND [24] to enable multi-modal input queries to the CRS. We adopt a GNN cross-attention mechanism [8] to capture semantic relationships among the text, image, video and audio modalities. In this way multi-modal context is added to the queries, thereby complementing the reasoning abilities of the model. We also adopt Multi-modal Alignment Learning and Modality-switching Instruction Tuning as proposed by Wu et al. [1] to understand user input queries and generate the requested multi-modal recommendation outputs (Fig. 2).

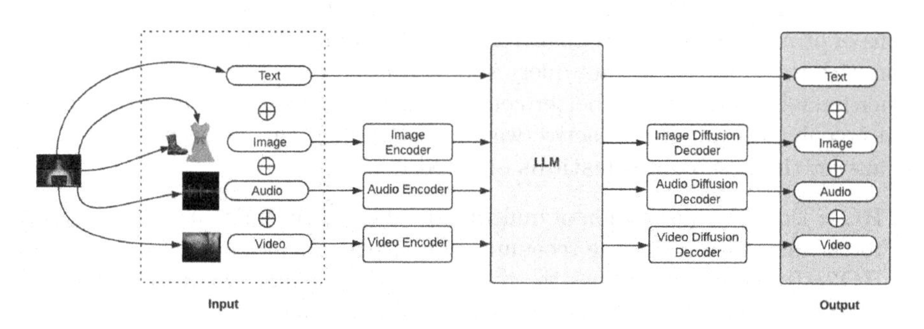

Fig. 2. MMCRec model overview. Input instructions to the LLM module can be passed in the form of audio, video, text or image. Graph-engineered prompts enable cross-attention between the multi-modal input data. Audio, Image and Video diffusion models decode and synthesize the required output as one or a combination of the required modalities.

We adopt the Vicuna LLM [25] which has understanding abilities to multi-modal input representations [26]. Following the NextGPT model [1], we adopt the transformer-based multi-modal output generation whereby multi-modal signal token representations from the LLM are mapped to Image Diffusion, Audio Diffusion and Video Diffusion decoders. We also adopt the following latent diffusion models to generate the multi-modal outputs: Stable Diffusion [33] for image synthesis, Zeroscope[1] for video synthesis, and AudioLDM5 [34] for audio

[1] https://huggingface.co/cerspense.

synthesis. For our model instruction tuning, we leverage on the Low-Rank Adaptation (LoRA) approach to minimize the number of trainable parameters for the conversational recommendation task [35].

In adopting the GoT prompting model, multi-modal inputs to the LLM are represented as nodes and edges to capture the non-sequential complex scenarios in real-world scenarios. In this way, the intermediate LLM reasoning model is enhanced to mimic the human-like thought process in generating the recommendations. Following the GoT framework [44], the graph input (representation of the user-item interactions) to the LLM models the reasoning stage to improve the model's contextual understanding of complex real-world heterogeneous recommendation scenarios. This reasoning process is modelled as a heterogeneous graph G = (V, E) where V is a set of nodes representing the users/items and E is a set of edges representing the node relationships. The recommendation task, involves the graph-enhanced reasoning multi-modal model which maps node relationships based on the respective user queries.

4 Experiments

We consider the multi-modal product recommendation task for our experiments. The task aims to find the most relevant target products using one or a combinations of or more modalities, enhanced by the LLM and the multi-modal interaction with the CRS. User input query is in the form of a combination of text and other modalities to provide better context whereas the recommendation output can be enhanced in an interactive way. In conducting our experiments, we aim to answer the **research questions** outlined below:

- **RQ1:** Does the integration of multi-modal data in the CRS model contribute to the improvement of the recommendation performance?
- **RQ2:** To what extend does the multi-modal input modules in the proposed MMCRec model impact the recomendation performance?

Dataset:We use the publicly available M5Product real world e-commerce dataset [32] which contains 6,313,067 samples of multi-modal information (images, text, table, video, and audio) consisting of 6,232 product categories and 5,679 attributes such as **appearance, usage, specification, selling point, production, material** and **category descriptions** for products which include clothes, cosmetics and instruments.

Training: The model is trained using a combination of positive and negative samples. The training objective is to minimize the ranking loss function, which measures the difference between predicted scores and true interactions.

Metrics:We use the Recall (R) and Normalized Discounted Cumulative Gain (NDCG) metrics to evaluate the recommendation performance. A scoring function is applied to the recommendations for each item. The model returns Top-N items with the highest scores which are recommended to the user. We adopt the GoT ranking model to evaluate the recommendation scores [44].

Baselines: We compare our approach to the following methods: **P5**: A unified "Pretrain, Personalized Prompt & Predict Paradigm" which learns related recommendation tasks through a unified sequence-to-sequence framework [39] as well as **Macaw-LLM**: A multi-modal model which integrates image, video, audio, and text information features into the LLM input sequence [41]. We also include **CHATRec**: A ChatGPT-augmented model for conversational recommendation which uses prompts based on historical user profiles and interactions [4]. In addition, **CLIP**: A Contrastive Language-Image Pre-Training model which jointly learns image-text pairs and can be instructed in natural language to predict the most relevant text snippets, given an image [42].

5 Results and Discussion

Performance Evaluation (RQ1): From the recommendation performance comparison results in Table 1, multi-modal recommendation models Macaw-LLM and MMCRec achieve better performance compared to the uni-modal approaches P5 and CHATRec, which shows the effectiveness of incorporating multi-modal inference into the CRS. The CLIP model, which combines text and image input, performs better than the unimodal approaches, but worse than MMCRec and Macaw-LLM approaches which combine more modalities. Our model, which incorporates graph-based reasoning into the LLM, outperforms the other approaches. This is mainly attributed to the additional context into the recommendation scenarios.

Table 1. Comparing the recommendation performance against baselines. Bold indicates the performance improvement against the second-best baseline at 0.05 significance.

Model	Modality	R@1	R@5	R@10	NDCG@1	NDCG@5	NDCG@10
P5	T	0.398	0.286	0.384	0.312	0.249	0.298
CHATRec	T	0.387	0.415	0.482	0.423	0.344	0.465
CLIP	T+I	0.486	0.547	0.576	0.408	0.442	0.451
Macaw-LLM	V+A+I+T	0.545	0.597	0.628	0.496	0.537	0.522
MMCRec	T+V+A+I	**0.625***	**0.688***	**0.734***	**0.584***	**0.607***	**0.692***

Effect of the Input Module Components (RQ2): We study the effect of each multi-modal input modules to our model performance. Experimental results conducted by deactivating the model components indicated in Table 2 show degraded performance, which is consistent with the earlier observation that the incorporation of the multi-modal components provides additional context to the model, resulting in better recommendation performance. In these experiments, the MMCRec model outperforms the variants in which the audio, video and image input modules are removed, which indicates their positive influence on the overall recommendation performance.

Table 2. Comparing the influence of module components. Removal of the audio, image and video input modules from the MMCRec model shows degraded performance.

Model	R@1	R@5	NDCG@1	NDCG@5
MMCRec	**0.625**	**0.688**	**0.584**	**0.607**
w/o Audio	0.581	0.640	0.552	0.56
w/o Video	0.38	0.471	0.340	0.459
w/o Image	0.32	0.356	0.307	0.338

6 Ethical Considerations

As the development of multi-modal CRSs involves the use of data forms such as images, audio and videos, several ethical issues need to be addressed. Paramount among these is user data privacy and security to ensure that personal information is not misused or exposed without consent. To achieve this, robust security measures should be implemented to protect user data from unauthorized access or breaches. Informed user consent should be emphasized before data collection or usage and users should be aware of how their data is being used and have the option to opt in or out. CRSs should be inclusive, for instance, provide alternative interaction modalities to accommodate different users with diverse abilities and needs. Content moderation mechanisms should be implemented to filter out inappropriate or harmful content. Accountability mechanisms are essential to ensure compliance with relevant data privacy and protection regulations such as the General Data Protection Regulation (GDPR)[2] and other applicable laws. In conclusion, collaboration with stakeholders which include users, researchers, and policymakers is necessary to address any ethical lapses or unintended consequences.

7 Conclusion

In this work, we propose a multi-modal GAI-enhanced Conversational Recommender System MMCRec. By integration of an LLM module with multi-modal encoders and diffusion decoders, MMCRec is capable of accepting GNN enhanced multi-modal input and generating multi-modal output in a combination of text, audio, video and images based on the input request. This approach offers users a more dynamic and engaging way to explore content and products tailored to their preferences and contexts. In future we plan to extend our model to include other modalities such as tabular and web data as well as take into account user satisfaction metrics.

Acknowledgements. This publication has emanated from research supported by Science Foundation Ireland under Grant number 18/CRT/6183.

[2] https://gdpr-info.eu/.

References

1. Wu, S., Fei, H., Qu, L., Ji, W., Chua, T.: NExT-GPT: any-to-any multimodal LLM. ArXiv Preprint: ArXiv:2309.05519 (2023)
2. Cui, Z., Ma, J., Zhou, C., Zhou, J., Yang, H.: M6-Rec: generative pre-trained language models are open-ended recommender systems. ArXiv Preprint: ArXiv:2205.08084 (2022)
3. Hou, Y., et al.: Large language models are zero-shot rankers for recommender systems. ArXiv Preprint: ArXiv:2305.08845 (2023)
4. Gao, Y., Sheng, T., Xiang, Y., Xiong, Y., Wang, H., Zhang, J.: Chat-REC: towards interactive and explainable LLMs-augmented recommender system. ArXiv Preprint: ArXiv:2303.14524 (2023)
5. Salah, A., Truong, Q., Lauw, H.: Cornac: a comparative framework for multimodal recommender systems. J. Mach. Learn. Res. **21**, 3803–3807 (2020)
6. Liu, Q., Hu, J., Xiao, Y., Gao, J., Zhao, X.: Multimodal recommender systems: a survey. ArXiv Preprint ArXiv:2302.03883 (2023)
7. Chen, X., et al.: Personalized fashion recommendation with visual explanations based on multimodal attention network: towards visually explainable recommendation. In: Proceedings of the 42nd International ACM SIGIR Conference on Research and Development in Information Retrieval, pp. 765–774 (2019)
8. Gu, R., Wang, X., Yang, Q.: Multimodal cross-attention graph network for desire detection. In: International Conference on Artificial Neural Networks, pp. 512–523 (2023)
9. Yao, Y., Liu, Z., Lin, Y., Sun, M.: Cross-modal representation learning. In: Representation Learning for Natural Language Processing, pp. 211–240 (2023)
10. Zhu, L., Wang, T., Li, F., Li, J., Zhang, Z., Shen, H.: Cross-modal retrieval: a systematic review of methods and future directions. ArXiv Preprint: ArXiv:2308.14263 (2023)
11. Tao, S., Qiu, R., Ping, Y., Ma, H.: Multi-modal knowledge-aware reinforcement learning network for explainable recommendation. Knowl.-Based Syst. **227**, 107217 (2021)
12. Huang, H., et al.: ChatGPT for shaping the future of dentistry: the potential of multi-modal large language model. Int. J. Oral Sci. **15**, 29 (2023)
13. Hu, Z., Cai, S., Wang, J., Zhou, T.: Collaborative recommendation model based on multi-modal multi-view attention network: movie and literature cases. Appl. Soft Comput., 110518 (2023)
14. Yan, A., He, Z., Li, J., Zhang, T., McAuley, J.: Personalized showcases: generating multi-modal explanations for recommendations. In: Proceedings of the 46th International ACM SIGIR Conference on Research and Development in Information Retrieval, pp. 2251–2255 (2023)
15. Wu, Y., Macdonald, C., Ounis, I.: Goal-oriented multi-modal interactive recommendation with verbal and non-verbal relevance feedback. In: Proceedings of the 17th ACM Conference on Recommender Systems, pp. 362–373 (2023)
16. Xin, X., Pimentel, T., Karatzoglou, A., Ren, P., Christakopoulou, K., Ren, Z.: Rethinking reinforcement learning for recommendation: a prompt perspective. In: Proceedings of the 45th International ACM SIGIR Conference on Research and Development in Information Retrieval, pp. 1347–1357 (2022)
17. Chen, X., Lu, Y., Wang, Y., Yang, J.C.M.B.F.: Cross-modal-based fusion recommendation algorithm. Sensors **21**, 5275 (2021)

18. Friedman, L., et al.: Leveraging large language models in conversational recommender systems. ArXiv Preprint: ArXiv:2305.07961 (2023)
19. Dai, S., et al.: Uncovering ChatGPT's capabilities in recommender systems. ArXiv Preprint: ArXiv:2305.02182 (2023)
20. Bao, K., Zhang, J., Zhang, Y., Wang, W., Feng, F., He, X.: TALLRec: an effective and efficient tuning framework to align large language model with recommendation. ArXiv Preprint: ArXiv:2305.00447 (2023)
21. Wang, W., Lin, X., Feng, F., He, X., Chua, T.: Generative recommendation: towards next-generation recommender paradigm. ArXiv Preprint: ArXiv:2304.03516 (2023)
22. Li, J., Zhang, W., Wang, T., Xiong, G., Lu, A., Medioni, G.: GPT4Rec: a generative framework for personalized recommendation and user interests interpretation. ArXiv Preprint: ArXiv:2304.03879 (2023)
23. Wang, X., Tang, X., Zhao, W., Wang, J., Wen, J.: Rethinking the evaluation for conversational recommendation in the era of large language models. ArXiv Preprint: ArXiv:2305.13112 (2023)
24. Girdhar, R., et al.: ImageBind: one embedding space to bind them all. In: Proceedings of the IEEE/CVF Conference on Computer Vision and Pattern Recognition, pp. 15180–15190 (2023)
25. Chiang, W., et al.: Vicuna: an open-source chatbot impressing GPT-4 with 90%* ChatGPT quality (2023). See https://vicuna.Lmsys.Org. Accessed 14 Apr 2023
26. Su, Y., Lan, T., Li, H., Xu, J., Wang, Y., Cai, D.: PandaGPT: one model to instruction-follow them all. ArXiv Preprint: ArXiv:2305.16355 (2023)
27. Wang, X., Qin, J.: Multimodal recommendation algorithm based on dempster-Shafer evidence theory. Multimedia Tools Appl., 1–16 (2023)
28. Luo, L., Ju, J., Xiong, B., Li, Y., Haffari, G., Pan, S.: ChatRule: mining logical rules with large language models for knowledge graph reasoning. ArXiv Preprint: ArXiv:2309.01538 (2023)
29. Wu, Y., et al.: State graph reasoning for multimodal conversational recommendation. IEEE Trans. Multimedia (2022)
30. Liao, L., Long, L., Zhang, Z., Huang, M., Chua, T.: MMConv: an environment for multimodal conversational search across multiple domains. In: Proceedings of the 44th International ACM SIGIR Conference on Research and Development in Information Retrieval, pp. 675–684 (2021)
31. Viswanathan, S., Guillot, F., Grasso, A.: What is natural? Challenges and opportunities for conversational recommender systems. In: Proceedings of the 2nd Conference on Conversational User Interfaces, pp. 1–4 (2020)
32. Dong, X., et al.: M5product: self-harmonized contrastive learning for e-commercial multi-modal pretraining. In: Proceedings of the IEEE/CVF Conference on Computer Vision and Pattern Recognition, pp. 21252–21262 (2022)
33. Rombach, R., Blattmann, A., Lorenz, D., Esser, P., Ommer, B.: High-resolution image synthesis with latent diffusion models. In: Proceedings of the IEEE/CVF Conference on Computer Vision and Pattern Recognition, pp. 10684–10695 (2022)
34. Liu, H., et al.: AudioLDM: text-to-audio generation with latent diffusion models. ArXiv Preprint: ArXiv:2301.12503 (2023)
35. Hu, E., et al.: Lora: low-rank adaptation of large language models. ArXiv Preprint: ArXiv:2106.09685 (2021)
36. Zhang, Z., Zhang, A., Li, M., Zhao, H., Karypis, G., Smola, A.: Multimodal chain-of-thought reasoning in language models. ArXiv Preprint: ArXiv:2302.00923 (2023)

37. Liu, Z., Yu, X., Fang, Y., Zhang, X.: GraphPrompt: unifying pre-training and downstream tasks for graph neural networks. In: Proceedings of the ACM Web Conference 2023, 417–428 (2023)
38. Wu, L., et al.: A survey on large language models for recommendation. ArXiv Preprint: ArXiv:2305.19860 (2023)
39. Geng, S., Liu, S., Fu, Z., Ge, Y., Zhang, Y.: Recommendation as language processing (RLP): a unified pretrain, personalized prompt & predict paradigm (p5). In: Proceedings of the 16th ACM Conference on Recommender Systems, pp. 299–315 (2022)
40. Lin, J., et al.: How can recommender systems benefit from large language models: a survey. ArXiv Preprint: ArXiv:2306.05817 (2023)
41. Lyu, C., et al.: Macaw-LLM: multi-modal language modeling with image, audio, video, and text integration. ArXiv Preprint: ArXiv:2306.09093 (2023)
42. Radford, A., et al.: Learning transferable visual models from natural language supervision. In: International Conference on Machine Learning, pp. 8748–8763 (2021)
43. He, M., Wang, J., Ding, T., Shen, T.: Conversation and recommendation: knowledge-enhanced personalized dialog system. Knowl. Inf. Syst. **65**, 261–279 (2023)
44. Besta, M., et al.: Graph of thoughts: solving elaborate problems with large language models. ArXiv Preprint: ArXiv:2308.09687 (2023)

GenQREnsemble: Zero-Shot LLM Ensemble Prompting for Generative Query Reformulation

Kaustubh D. Dhole(✉) and Eugene Agichtein

Department of Computer Science, Emory University, Atlanta, USA
{kaustubh.dhole,eugene.agichtein}@emory.edu

Abstract. Query Reformulation(QR) is a set of techniques used to transform a user's original search query to a text that better aligns with the user's intent and improves their search experience. Recently, zero-shot QR has been shown to be a promising approach due to its ability to exploit knowledge inherent in large language models. By taking inspiration from the success of ensemble prompting strategies which have benefited many tasks, we investigate if they can help improve query reformulation. In this context, we propose an ensemble based prompting technique, GenQREnsemble which leverages paraphrases of a zero-shot instruction to generate multiple sets of keywords ultimately improving retrieval performance. We further introduce its post-retrieval variant, GenQREnsembleRF to incorporate pseudo relevant feedback. On evaluations over four IR benchmarks, we find that GenQREnsemble generates better reformulations with relative nDCG@10 improvements up to 18% and MAP improvements upto 24% over the previous zero-shot state-of-art. On the MSMarco Passage Ranking task, GenQREnsembleRF shows relative gains of 5% MRR using pseudo-relevance feedback, and 9% nDCG@10 using relevant feedback documents.

Keywords: Query Reformulation · Zero-Shot · Prompting · Relevance Feedback

1 Introduction

Users searching for relevant documents might not always be able to accurately express their information needs in their initial queries. This could result in queries being vague or ambiguous or lacking the necessary domain vocabulary. Query Reformulation (QR) is a set of techniques used to transform a user's original search query to a text that better aligns with the user's intent and improves their search experience. Such reformulation alleviates the vocabulary mismatch problem by expanding the query with related terms or paraphrasing it into a suitable form by incorporating additional context.

Recently, with the success of large language models (LLMs) [5,8], a plethora of QR approaches have been developed. The generative capabilities of LLMs have been exploited to produce novel queries [17], as well as useful keywords

N. Goharian et al. (Eds.): ECIR 2024, LNCS 14610, pp. 326–335, 2024.
https://doi.org/10.1007/978-3-031-56063-7_24

to be appended to the users' original queries [14]. By gaining exposure to enormous amounts of text during pre-training, prompting has become a promising avenue for utilizing knowledge inherent in an LLM for the benefit of subsequent downstream tasks [18] especially QR [16,43].

Unlike training or few-shot learning, zero-shot prompting does not rely on any labelled examples. The advantage of a zero-shot approach is the ease with which a standalone generative model can be used to reformulate queries by prompting a templated piece of instruction along with the original query. Particularly, zero-shot QR can be used to generate keywords by prompting the user's original query alongwith an instruction that defines the task of query reformulation in natural language like `Generate useful search terms for the given query:'List all the breweries in Austin'`.

However, such a zero-shot prompting approach is still contingent on the exact instruction appearing in the prompt providing plenty of avenues of improvement. While LLMs have been known to vary significantly in performance across different prompts [10,11] and generation settings [19], many natural language tasks have benefited by exploiting such variation via ensembling multiple prompts or generating diverse reasoning paths [23–25]. Whether such improvements also transfer to tasks like QR are yet to be determined. In Fig. 1, a vast difference is noticed in the keywords generated when the input instruction is altered to a semantically similar variant. We hypothesize that QR might naturally benefit from such variation – An ensemble of zero-shot reformulators with paraphrastic instructions can be tasked to look at the input query in diverse ways so as to elicit different expansions. This work proposes the following contributions:

Instruction	Expansions Generated
Increase the search efficacy by offering beneficial expansion keywords for the query	age goldfish grow outsmart outlive ageing species...
Enhance search outcomes by recommending beneficial expansion terms to supplement the query	Goldfish breed sizes What kind of goldfish grows the fastest...

Fig. 1. Keywords generated for the query ("do goldfish grow") differ drastically when generated from two paraphrastic instructions prompted to `flan-t5-xxl` [12].

- We propose a novel method,**GenQREnsemble** – a zero-shot **Ensemble** based **Gen**erative **Q**uery **R**eformulator which exploits multiple zero-shot instructions for QR to generate a more effective query reformulation than possible with an individual instruction. (Section 3)
- We further introduce an extension **GenQREnsembleRF** to incorporate **R**elevance **F**eedback into the process. (Section 3)
- We evaluate the proposed methods over four standard IR benchmarks, demonstrating significant relative improvements vs. recent state of the art, of up to 18% on nDCG@10 in pre-retrieval settings, and of up to 9% nDCG@10 on post-retrieval (feedback) settings, demonstrating increased generalizability of our approach.

Next, we summarize the prior work to place our contributions in context.

2 Related Work

Query reformulation has been shown to be effective in many settings [1]. It can be done pre-retrieval, or post-retrieval, via incorporating evidence from feedback, obtained either from a user or from top-ranked results in the sparse retrieval setting [2], and in the dense retrieval setting [3,4].

Recently, zero-shot approaches to query reformulation have received considerable attention. Wang et al. [14] design a query reformulator by fine-tuning a sequence-to-sequence transformer, T5 [13] on pairs of raw and transformed queries. Their zero-shot prompting approach uses an instruction tuned model, FlanT5 [12] to generate keywords for query expansion and incorporating PRF. Jagerman et al. [16] demonstrate LLMs can be more powerful than traditional methods for query expansion. Mo et al. [21] propose a framework to reformulate conversational search queries using LLMs. Gao et al. [35]'s framework performs retrieval through fake documents generated by prompting LLMs with user queries. Alaofi et al. [44] prompt LLMs with information descriptions to generate query variants.

However, using a single query reformulation can sometimes degrade performance compared to the original query. To address this drawback, prior efforts have incorporated ensemble strategies via keywords from numerous sources or fusing documents from different queries. Gao et al. [39], combine features derived from various translation models to generate better query rewrites. Si et al. [40] perform QR by utilizing multiple external biomedical resources. Hsu and Taksa [41] present a data fusion framework suggesting that diverse query formulations represent distinct evidence sources for inferring document relevance. Later, Mohankumar et al. [38] generated diverse queries by introducing a diversity driven RL algorithm. For other tasks, recent works demonstrated the benefits of ensemble strategies for prompting LLMs, including self-consistency [23] for arithmetic and common sense tasks, Chain of Verification [22] for improving factuality and Diverse [24] for question answering. However, zero-shot based ensemble methods for LLM have not been explored for the Query Reformulation task, as we propose in this paper.

3 Proposed Approach: GenQREnsemble

In this section, we describe two variations of our proposed approach, for the pre- and post-retrieval settings. In the pre-retrieval setting, a Query Reformulation R transforms a user's expressed query q_0 into a novel reformulated version q_r to improve retrieval effectiveness for a given search task (e.g., passage or document retrieval). We also consider the post-retrieval setting, wherein the reformulator can incorporate additional contextual information like document or passage-level feedback.

Pre-retrieval: We propose **GenQREnsemble** – an ensemble prompting based query reformulator which uses N diverse paraphrases of a QR instruction to enhance retrieval. Specifically, we first use an LLM to paraphrase the instruction

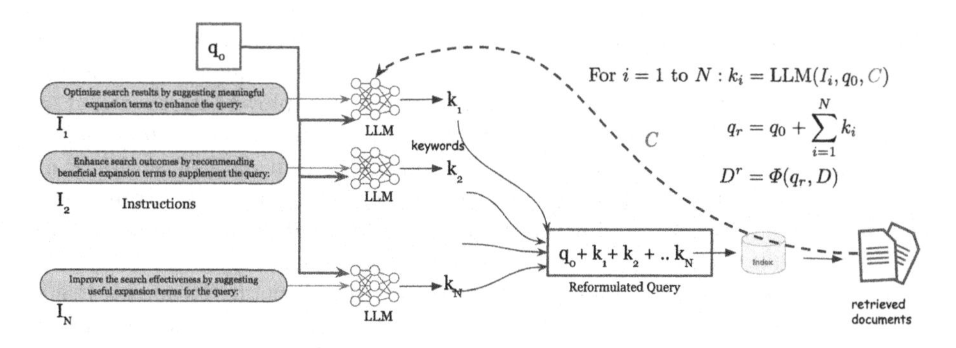

Fig. 2. The complete flow and algorithm shown on the top right.

I_1 to create N instructions with different surface forms viz. I_1 to I_N. This is required to be done once. Each instruction is then prompted alongwith the user's query q_0 to generate instruction-specific keywords. All the keywords are then appended to the original query, resulting in a reformulated query, which is then executed against a document index D to retrieve relevant documents D^r. The complete process and algorithm are shown in Fig. 2.

Post-retrieval: To assess how well our method can incorporate additional context like document feedback, we introduce **GenQREnsembleRF**. Here, we prepend the N instructions described earlier with a fixed context capturing string "`Based on the given context information {C},`" used[1] in [14] to create their PRF counterparts – where `C` is a space (' ') delimited concatenation of feedback documents $\texttt{C} = d_1 + \ldots + d_m$, obtained either as pseudo-relevance feedback from initial retrieval, or manually chosen by the user (Fig. 4).

4 Experiments

We now describe the experiments and analysis performed for different retrieval settings.

To instruct the LLM to generate query reformulations, we start with the instruction empirically chosen by Wang et al. [14] – as our base QR instruction I_1. We use this instruction to generate N paraphrases of the instruction ($N = 10$). To this aim, we invoke GPT-3.5 API with the paraphrase generating prompt, namely, I_p=`Generate 10 paraphrases for`

#	Instruction
1	Improve the search effectiveness by suggesting expansion terms for the query
2	Recommend expansion terms for the query to improve search results
3	Improve the search effectiveness by suggesting useful expansion terms for the query
4	Maximize search utility by suggesting relevant expansion phrases for the query
5	Enhance search efficiency by proposing valuable terms to expand the query
6	Elevate search performance by recommending relevant expansion phrases for the query
7	Boost the search accuracy by providing helpful expansion terms to enrich the query
8	Increase the search efficacy by offering beneficial expansion keywords for the query
9	Optimize search results by suggesting meaningful expansion terms to enhance the query
10	Enhance search outcomes by recommending beneficial expansion terms to supplement the query

Fig. 3. Reformulation instructions generated (N=10).

[1] We found prepending the string in the prompt performs better than appending it at the end.

`the following instruction:`– and the base QR instruction I_1 to obtain I_2 to I_{10}. These paraphrases serve as our instruction set for subsequent experiments.

For generating the actual query reformulations, we employ **`flan-t5-xxl`** [12], an instruction tuned model. The FlanT5 set of models are created by fine-tuning the text-to-text transformer model, `T5` [13] on instruction data of a variety of NL tasks. We use the checkpoint[2] provided through HuggingFace's Transformers library [6]. Nucleus sampling is performed with a cutoff probability of 0.92 keeping top 200 tokens (top_k) and a repetition penalty of 1.2.

For evaluation, we use four popular benchmarks through IRDataset [15]'s interface: 1)**TP19**: TREC 19 Passage Ranking which uses the MSMarco dataset [16,20] consisting of search engine queries. 2)**TR04**: TREC Robust 2004 Track, a task intended for testing poorly performing topics. In our experiments, we use the Title as our choice of query. And two tasks from the BEIR [26] benchmark 3)**WT**: Webis Touche [29] for argument retrieval 4)**DE**: DBPedia Entity Retrieval [27].

4.1 Baselines:

We compare our work against the following using the Pyterrier [7] framework. For all the post-retrieval experiments, we use 5 documents as feedback.

With BM25 Retriever:

- BM25: Here, we retrieve using raw queries without any reformulation
- FlanQR [14]: We implement Wang et al's single-instruction zero-shot QR [14] which is also a specific case of GenQREnsemble when N=1
- BM25+RM3 [32]: BM25 retrieval with RM3 expanded queries (#feedback terms=10)
- BM25+FlanPRF [14]: BM25 retrieval with FlanPRF expanded queries

With Neural Reranking: Here, we re-evaluate the above settings in conjunction with a MonoT5 neural reranker [42] with all other parameters constant.

- BM25+MonoT5: BM25 retrieval using raw queries, re-ranked with MonoT5 model [42]
- FlanQR+MonoT5: BM25 retrieval with FlanQR reformulations, re-ranked with MonoT5 model
- BM25+RM3+MonoT5: BM25 retrieval with RM3 expanded queries, re-ranked with MonoT5 model
- BM25+FlanPRF+MonoT5: BM25 retrieval with FlanPRF expanded queries, re-ranked with MonoT5 model

5 Results and Analysis

We now report the results of query reformulation for pre- and post-retrieval settings.

[2] https://huggingface.co/google/flan-t5-xxl.

5.1 Pre-retrieval Performance

We first compare the retrieval performances of raw queries and reformulations from FlanQR, and GenQREnsemble in Table 1. GenQREnsemble outperforms the raw queries as well as generates better reformulations than FlanQR's reformulated queries across all the four benchmarks over a BM25 retriever, indicating the usefulness of paraphrasing initial instructions. On TP19, nDCG@10 and MAP improve significantly with relative improvements of 18% and 24% respectively. This is further validated through the querywise analysis shown in Fig. 4 – Relative to BM25, nDCG@10 scores of GenQREnsemble (shown in green) are overall better than FlanQR (shown in blue).GenQREnsemble seems more robust too as it avoids drastic degradation in at least 6 queries on which FlanQR fails.

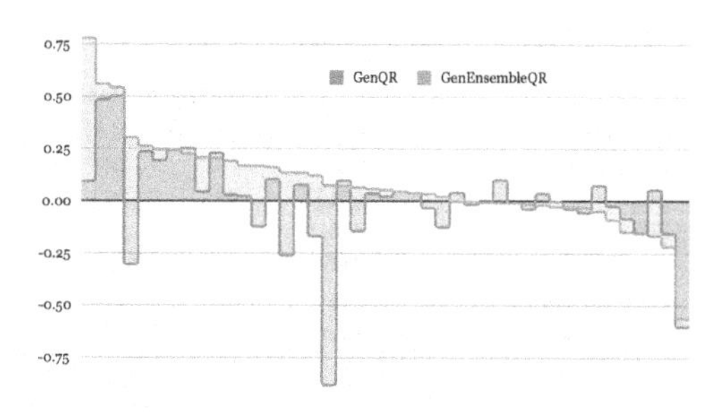

Fig. 4. nDCG@10 Scores of GenQREnsemble and FlanQR relative to BM25

Table 1. Performance of GenQREnsemble on the four benchmarks. α denotes significant improvements (paired t-test with Holm-Bonferroni correction, $p < 0.05$) over FlanQR. +% indicates % improvements relative to FlanQR (as whole numbers).

Evaluation Set	TREC Passage 19			TREC Robust 04			Webis Touche			DBpedia Entity		
Setting	nDCG@10	MAP	MRR	P@10	nDCG@10	MRR	nDCG@10	MAP	MRR	nDCG@10	MAP	MRR
BM25	.480	.286	.642	.426	.434	.154	.260	.206	.454	.321	.168	.297
FlanQR	.477	.302	.593	.473	.483	.151	.315	.241	.511	.342	.196	.345
FlanQR$_{\beta=.05}$	.511	.323	.621	.469	.477	.150	.276	.221	.476	.353	.188	.339
GenQREnsemble	**.564**$^{\alpha}$+18%	**.375**$^{\alpha}$+24%	**.706**+19%	**.500**$^{\alpha}$+6%	**.513**$^{\alpha}$+6%	**.159**+6%	**.317**+1%	**.257**+6%	**.555**+9%	**.374**$^{\alpha}$+9%	**.212**$^{\alpha}$+8%	**.376**$^{\alpha}$+9%
GenQREnsemble$_{\beta=.05}$	**.575**$^{\alpha}$	**.377**$^{\alpha}$	**.714**	**.502**$^{\alpha}$	**.512**$^{\alpha}$	**.159**	.292	**.242**	.489	**.377**$^{\alpha}$	**.212**$^{\alpha}$	**.380**$^{\alpha}$
BM25+MonoT5	.718	.477	**.881**	**.492**	**.513**	**.173**	**.299**	.216	.525	.414	.249	.444
FlanQR+MonoT5	.707	.486	.847	.490	.510	.170	.292	.215	.530	.415	.255	.446
GenQREnsemble+MonoT5	**.722**+2%	**.503**+3%	.862+2%	.484-1%	.506-1%	.170	.298+3%	**.219**+2%	**.548**+3%	**.420**+1%	**.258**+1%	**.450**+1%

We further look at GenQREnsemble under the neural reranker setting shown at the bottom half of Table 1. In three of the four settings, viz., TP19, WT, and DE, GenQREnsemble is preferable to its zero-shot variant, FlanQR. Evidently, the gains of both the zero-shot approaches in the traditional setting are stronger vis-à-vis the neural setting. We hypothesize this could be due to GenQREnsemble and FlanQR both expanding the query via incorporating semantically similar but lexically different keywords. Comparatively, neural models are adept at capturing notions of semantic similarity and might benefit less with query expansion. This also is in line with Weller et al.'s [43] recent analysis on the non-ensemble variant.

5.2 Post-retrieval Performance

Table 2. Comparison of PRF performance on the TREC 19 Passage Ranking Task

	With BM25 Retriever				With Neural Reranking			
Setting	nDCG@10	nDCG@20	MAP	MRR	nDCG@10	nDCG@20	MAP	MRR
BM25	.480	.473	.286	.642	.718	.696	.477	.881
RM3	.504	.496	.311	.595	.716	.699	.480	.858
FlanPRF	.576	.553	.363	.715	.722	.703	.486	.874
GenQREnsembleRF	**.585**+2%	**.560**+1%	**.373**+3%	**.753**+5%	**.729**+1%	**.706**+1%	**.501**+3%	**.894**+2%
FlanPRF (Oracle)	.753	.728	.501	.936	.742	.734	.545	.881
GenQREnsembleRF (Oracle)	**.820**$^{\alpha}$+9%	**.773**+6%	**.545**+9%	**.977**+4%	**.756**+2%	**.751**+2%	**.545**	**.897**+2%

We now investigate if GenQREnsembleRF can effectively incorporate PRF in Table 2. We find that GenQREnsembleRF improves retrieval performance as compared to other PRF approaches and is able to incorporate feedback from a BM25 retriever better than RM3 as well as its zero-shot counterpart. To assess if GenQREnsembleRF and FlanPRF can at all benefit from incorporating relevant documents, we perform oracle testing by providing the highest relevant gold documents as context. We find that GenQREnsembleRF is able to improve over GenQREnsemble (without feedback) showing that it is able to capture context effectively as well as benefit from it. Further, it can incorporate relevant feedback better than its single-instruction counterpart FlanPRF. We notice improvements even under the neural reranker setting as GenQREnsembleRF outperforms RM3 and FlanPRF. Besides, the oracle improvements are higher with only a BM25 retriever as compared to when a neural reranker is introduced (Fig. 5).

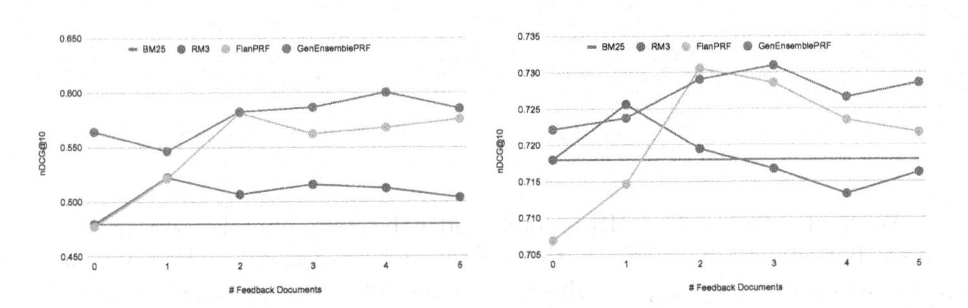

Fig. 5. Effect of feedback documents under sparse (BM25) and neural (MonoT5) rankers

We further evaluate the effect of varying the number of feedback documents from 0 to 5. We notice that resorting to an ensemble approach is highly beneficial. In the BM25 setting, ensemble approach seems always preferable. Under the neural reranker setting too,GenQREnsembleRF almost always outperforms FlanPRF.

6 Conclusions

Zero-shot QR is advantageous since it does not rely on any labelled relevance judgements and allows eliciting pre-trained knowledge in the form of keywords by prompting the model with the original query and an appropriate instruction. By introducing GenQREnsemble, we show that zero-shot performance can be further enhanced by using multiple views of the initial instruction. We also show that the extension GenQREnsembleRF is able to effectively incorporate relevance feedback, either automated or from users. While generative QR greatly benefits from our ensemble approach, the proposed methods come at a cost of potentially increased latency, but this is becoming less problematic with increased availability of batch inference for LLMs. The proposed ensemble approach could also be applied to other settings, for example to address different aspects of queries or metrics to optimize, or to better control the generated reformulations.

References

1. Carpineto, C., Romano, G.: A survey of automatic query expansion in information retrieval. ACM Comput. Surv. 44(1), 50 (2012). https://doi.org/10.1145/2071389.2071390. Article 1
2. Li, H., Mourad, A., Zhuang, S., Koopman, B., Zuccon, G.: Pseudo relevance feedback with deep language models and dense retrievers: successes and pitfalls. ACM Trans. Inf. Syst. 41(3), 40 (2023). https://doi.org/10.1145/3570724. Article 62
3. Wang, X., Macdonald, C., Tonellotto, N., Ounis, I.: ColBERT-PRF: semantic pseudo-relevance feedback for dense passage and document retrieval. ACM Trans. Web 17(1), 39 (2023). https://doi.org/10.1145/3572405. Article 3
4. Yu, H., Xiong, C., Callan, J.: Improving query representations for dense retrieval with pseudo relevance feedback. In Proceedings of the 30th ACM International Conference on Information & Knowledge Management (CIKM '21), pp. 3592–3596. Association for Computing Machinery, New York (2021). https://doi.org/10.1145/3459637.3482124
5. Brown, T., et al.: Language models are few-shot learners. In: Advances in Neural Information Processing Systems, vol. 33, pp. 1877–1901 (2020)
6. Wolf, T., et al.: Transformers: State-of-the-art natural language processing. In: Proceedings of the 2020 Conference on Empirical Methods in Natural Language Processing: System Demonstrations, pp. 38–45 (2020)
7. Macdonald, C., Tonellotto, N., MacAvaney, S., Ounis, I.: PyTerrier: declarative experimentation in Python from BM25 to dense retrieval. In: Proceedings of the 30th ACM International Conference on Information and Knowledge Management, pp. 4526–4533 (2021)
8. Peng, B., Li, C., He, P., Galley, M., Gao, J.: Instruction tuning with GPT-4. arXiv preprint: arXiv:2304.03277 (2023)
9. Craswell, N., Mitra, B., Yilmaz, E., Campos, D., Voorhees, E.M.: Overview of the TREC 2019 deep learning track (2020)
10. Zhao, Z., Wallace, E., Feng, S., Klein, D., Singh, S.: Calibrate before use: improving few-shot performance of language models. In: International Conference on Machine Learning, pp. 12697–12706. PMLR (2021)

11. Dhole, K., et al.: NL-augmenter: a framework for task-sensitive natural language augmentation. Northern Eur. J. Lang. Technol. **9**(1) (2023)
12. Chung, H.W., et al.: Scaling instruction-finetuned language models. arXiv preprint: arXiv:2210.11416 (2022)
13. Raffel, C., et al.: Exploring the limits of transfer learning with a unified text-to-text transformer. J. Mach. Learn. Res. **21**(1), 5485–5551 (2020)
14. Wang, X., MacAvaney, S., Macdonald, C., Ounis, I.: Generative query reformulation for effective Adhoc search. In: The First Workshop on Generative Information Retrieval, SIGIR (2023)
15. MacAvaney, S., Yates, A., Feldman, S., Downey, D., Cohan, A., Goharian, N.: Simplified data Wrangling with ir_datasets. In: Proceedings of the 44th International ACM SIGIR Conference on Research and Development in Information Retrieval (SIGIR '21), pp. 2429–2436. Association for Computing Machinery, New York (2021). https://doi.org/10.1145/3404835.3463254
16. Jagerman, R., Zhuang, H., Qin, Z., Wang, X., Bendersky, M.: Query expansion by prompting large language models. arXiv preprint: arXiv:2305.03653 (2023)
17. Nogueira, R., Lin, J., Epistemic, A.I.: From doc2query to docTTTTTquery. Online Prepr. **6**, 2 (2019)
18. Srivastava, A., et al.: Beyond the imitation game: quantifying and extrapolating the capabilities of language models. Trans. Mach. Learn. Res. (2023)
19. Wiher, G., Meister, C., Cotterell, R.: On decoding strategies for neural text generators. Trans. Assoc. Comput. Linguist. **10**, 997–1012 (2022)
20. Nguyen, T., et al.: Ms marco: a human-generated machine reading comprehension dataset (2016)
21. Mo, F., Mao, K., Zhu, Y., Wu, Y., Huang, K., Nie, J.Y.: ConvGQR: generative query reformulation for conversational search. In: Proceedings of the 61st Annual Meeting of the Association for Computational Linguistics (Volume 1: Long Papers), pp. 4998–5012, Toronto, Canada. Association for Computational Linguistics (2023)
22. Dhuliawala, S., Komeili, M., Xu, J., Raileanu, R., Li, X., Celikyilmaz, A., Weston, J.: Chain-of-verification reduces hallucination in large language models. arXiv preprint: arXiv:2309.11495 (2023)
23. Wang, X., et al.: Self-consistency improves chain of thought reasoning in language models (2023)
24. Li, Y., et al.: Making language models better reasoners with step-aware verifier. In: Proceedings of the 61st Annual Meeting of the Association for Computational Linguistics (Volume 1: Long Papers), pp. 5315–5333 (2023)
25. Arora, S., et al.: Ask me anything: a simple strategy for prompting language models. In: The Eleventh International Conference on Learning Representations (2022)
26. Thakur, N., Reimers, N., Rücklé, A., Srivastava, A., Gurevych, I.: BEIR: a heterogeneous benchmark for zero-shot evaluation of information retrieval models (2021)
27. Hasibi, F., et al.: DBpedia-entity v2: a test collection for entity search. In: Proceedings of the 40th International ACM SIGIR Conference on Research and Development in Information Retrieval, pp. 1265–1268 (2017)
28. Voorhees, E.M.: The TREC robust retrieval track. In: ACM SIGIR Forum, vol. 39, no. 1, pp. 11–20. ACM, New York (2005)
29. Bondarenko, A., et al.: Overview of Touché 2020: argument retrieval. In: Arampatzis, A., et al. (eds.) Experimental IR Meets Multilinguality, Multimodality, and Interaction. Lecture Notes in Computer Science(), vol. 12260, pp. 384–395. Springer, Cham (2020). https://doi.org/10.1007/978-3-030-58219-7_26

30. Voorhees, E., et al.:TREC-COVID: constructing a pandemic information retrieval test collection. In: ACM SIGIR Forum, vol. 54, no. 1, pp. 1–12. ACM, New York (2021)
31. Amati, G., Van Rijsbergen, C.J.: Probabilistic models of information retrieval based on measuring the divergence from randomness. ACM Trans. Inf. Syst. (TOIS) **20**(4), 357–389 (2002)
32. Abdul-Jaleel, N., et al.: UMass at TREC 2004: novelty and HARD. Comput. Sci. Dept. Fac. Publ. Ser., 189 (2004)
33. Harman, D.: Evaluation issues in information retrieval. Inf. Process. Manage. **28**(4), 439–40 (1992)
34. Järvelin, K., Kekäläinen, J.: Cumulated gain-based evaluation of IR techniques. ACM Trans. Inf. Syst. (TOIS) **20**(4), 422–446 (2002)
35. Gao, L., Ma, X., Lin, J., Callan, J.: Precise zero-shot dense retrieval without relevance labels. In: Proceedings of the 61st Annual Meeting of the Association for Computational Linguistics (Volume 1: Long Papers), pp. 1762–1777, Toronto, Canada. Association for Computational Linguistics (2023)
36. Paulus, R., Xiong, C., Socher, R.: A deep reinforced model for abstractive summarization. In: International Conference on Learning Representations (2018)
37. Klein, G., Kim, Y., Deng, Y., Senellart, J., Rush, A.M.: OpenNMT: open-source toolkit for neural machine translation. In: Proceedings of ACL 2017, System Demonstrations, pp. 67–72 (2017)
38. Mohankumar, A.K., Begwani, N., Singh, A.: Diversity driven query rewriting in search advertising. In: Proceedings of the 27th ACM SIGKDD Conference on Knowledge Discovery and Data Mining, pp. 3423–3431 (2021)
39. Gao, J., Xie, S., He, X., Ali, A.: Learning lexicon models from search logs for query expansion. In: Proceedings of EMNLP (2012)
40. Si, L., Lu, J., Callan, J.: Combining multiple resources, evidences and criteria for genomic information retrieval. In: TREC (2006)
41. Frank Hsu, D., Taksa, I.: Comparing rank and score combination methods for data fusion in information retrieval. Inf. Retrieval **8**(3), 449–480 (2005)
42. Pradeep, R., Nogueira, R., Lin, J.: The expando-mono-duo design pattern for text ranking with pretrained sequence-to-sequence models. arXiv preprint: arXiv:2101.05667 (2021)
43. Weller, O., et al.: When do generative query and document expansions fail? A comprehensive study across methods, retrievers, and datasets. arXiv preprint: arXiv:2309.08541 (2023)
44. Alaofi, M., Gallagher, L., Sanderson, M., Scholer, F., Thomas, P.: Can generative LLMs create query variants for test collections? An exploratory study. In: Proceedings of the 46th International ACM SIGIR Conference on Research and Development in Information Retrieval (SIGIR '23), pp. 1869–1873. Association for Computing Machinery, New York (2023). https://doi.org/10.1145/3539618.3591960

Towards Reliable and Factual Response Generation: Detecting Unanswerable Questions in Information-Seeking Conversations

Weronika Łajewska(✉) and Krisztian Balog

University of Stavanger, Stavanger, Norway
{weronika.lajewska,krisztian.balog}@uis.no

Abstract. Generative AI models face the challenge of hallucinations that can undermine users' trust in such systems. We approach the problem of conversational information seeking as a two-step process, where relevant passages in a corpus are identified first and then summarized into a final system response. This way we can automatically assess if the answer to the user's question is present in the corpus. Specifically, our proposed method employs a sentence-level classifier to detect if the answer is present, then aggregates these predictions on the passage level, and eventually across the top-ranked passages to arrive at a final answerability estimate. For training and evaluation, we develop a dataset based on the TREC CAsT benchmark that includes answerability labels on the sentence, passage, and ranking levels. We demonstrate that our proposed method represents a strong baseline and outperforms a state-of-the-art LLM on the answerability prediction task.

Keywords: Conversational search · Conversational response generation · Unanswerability detection

1 Introduction

Conversational information seeking (CIS) systems allow users to fulfill their complex information needs via a sequence of interactions. This problem is often approached as a passage retrieval task [5,14], rather than employing generative AI techniques, to allow for the grounding of responses in supporting documents and to avoid hallucinations. However, the ultimate goal is to return informative, concise, and reliable answers instead of top-ranked passages. In an ideal scenario, when the passages from the top of the ranking answer the question, the task of response generation boils down to summarization [15]. However, it is often the case that the answer to the user's question is not contained in the top retrieved passages. In such cases, summaries generated from those passages would result in hallucinations [3,22].

In this paper, we make the first step towards reliable and factual conversational response generation. We propose a mechanism for detecting unanswerable

N. Goharian et al. (Eds.): ECIR 2024, LNCS 14610, pp. 336–344, 2024.
https://doi.org/10.1007/978-3-031-56063-7_25

questions for which the correct answer is not present in the corpus or could not be retrieved. More specifically, given a set of top-ranked passages that have been identified as most relevant to the given question, we predict if the question can be answered (at least partially) based on information contained in those passages. This enables us to move beyond the notion of passage relevance and focus more on the actual presence of the information that answers the question. Introducting this additional step of answerability prediction in the CIS pipeline, to be performed after the passage retrieval and before the response generation steps, could help mitigate hallucinations and factual errors. It would enable the system to transparently communicate to the user if the answer to the query could not be found, instead of generating a response from only marginally relevant passages.

Unanswerability detection has been addressed in the context of machine reading comprehension [9,10] and question answering [4,18,19,21], both of which differ significantly from the conversational search setup. Information-seeking dialogues pose additional challenges, such as open-ended questions requiring descriptive answers, indirect answers requiring an inference or background/context knowledge, and complex queries with partial answers spread across passages. Therefore, unanswerability detection is a novel, still unsolved task in CIS, and, to the best of our knowledge, no public dataset exists for this problem.

As our first main contribution, we develop a dataset, based on the TREC CAsT benchmark, to train and evaluate methods for question answerability prediction. Utilizing an existing resource of snippet-level answer annotations [12], our dataset provides answerability labels on three levels: (1) sentences, (2) passages, and (3) rankings (i.e., top-ranked passages retrieved by a CIS system). Notably, we generate input passage rankings with various degrees of difficulty in answerability prediction, mixing passages that contain answers with those with no answers, in a controlled way. As a result, passage rankings range from all passages containing an answer to "no answer found in the corpus."

As our second main contribution, we provide a baseline approach for predicting answerability based on an input ranking. Our proposed approach predicts which sentences from the top-ranked passages contribute to the answer and aggregates the obtained answerability scores on the passage and ranking levels. We demonstrate that this simple method provides a strong baseline that outperforms ChatGPT-3.5 on the same task. Further, we show that augmenting our dataset with additional training samples for unanswerable question detection (from the SQuAD 2.0 dataset [18]) does not improve ranking-level answerability prediction in conversational search, underscoring the distinct character of this task. Our benchmark dataset (CAsT-answerability) as well as the implementation of our proposed method are made publicly available at https://github.com/iai-group/answerability-prediction.

2 Related Work

Research on information-seeking conversations is largely driven by test collections developed as part of the TREC Conversational Assistance Track

Table 1. Statistics for the CAsT-answerability dataset.

	Answerable?	
	Yes	**No**
#question-sentence pairs (train+test)	6,395	19,043
#question-passage pairs (train+test)	1,778	1,932
#question-ranking pairs (test)	4,035	504

(CAsT) [5–7,15]. Unlike generative AI approaches, answers in this benchmark are grounded in passages, hence the problem boils down to that of conversational passage retrieval [14,17,23]. Aggregating results from top-ranked passages into a single answer is an open problem [2] that has been first piloted in the 2022 edition, where a subtask of generating summaries from retrieved results was introduced [15]. Ren et al. [20] propose an approach for response generation divided into three stages: (optional) query rewriting, finding supporting sentences in results displayed on a SERP, and summarizing them into a short conversational response. While the authors acknowledge the problem of unanswerability in conversational search, they do not address it in their proposed approach. In this paper, we aim to fill that gap.

The problem of unanswerability has been addressed in the context of machine reading comprehension (MRC) [9,10] and extractive question-answering (QA) [1,8,13]. Solutions proposed include answerability prediction using prompt-tuning [13], modeling high-level semantic relationships between objects from question and context [10], and combining the output of reading and verification modules in MRC systems [9,24]. Our proposed solution is based on a sentence-level classifier that is learned on CIS-specific training data, and can further be augmented with QA answerability data.

3 Dataset

This paper builds upon the CAsT-snippets dataset [12],[1] which extends the TREC CAsT'20 and '22 datasets with snippet-level annotations for the top-retrieved results. Specifically, it contains annotations of information nuggets defined as "minimal, atomic units of relevant information" [16], representing key pieces of information required to answer the given question. Snippets in the dataset are identified for every question in the 5 most relevant passages according to ground truth judgments. To balance the collection, we also include 5 randomly selected non-relevant passages to each question. The resulting dataset, named *CAsT-answerability*, contains around 1.8k answerable and 1.9k unanswerable question-passage pairs. We further consider answerability on the level of sentences and on the level of rankings, as follows. For sentence-level answerability, we leverage annotations of information nuggets from the CAsT-snippets dataset

[1] https://github.com/iai-group/CAsT-snippets.

as follows: each sentence that overlaps with an information nugget, as per annotations in the originating CAsT-snippets dataset, is labeled as 1 (answer in the sentence), otherwise as 0 (no answer in the sentence).

For ranking-level answerability, which is the ultimate task we are addressing, we consider different input rankings, i.e., sets of $n = 3$ passages, for the same input question. Specifically, for each unique input test question (38), we generate all possible n-element subsets of passages available for this question (both containing and not containing an answer), thereby simulating passage rankings of varying quality. These rankings represent inputs with various degrees of difficulty for the same question, ranging from all passages containing an answer to a single passage with an answer to "no answer found in the corpus." This yields a total of 4.5k question-ranking pairs, of which 0.5k are unanswerable.[2]

Overall, our CAsT-answerability dataset contains binary answerability labels on three levels: sentence, passage, and ranking. Sentence- and passage-level answerability is divided into train (90%), and test (10%) portions; the splitting is done on the question level to avoid information leakage. Ranking-level answerability has only a test set. See Table 1 for a summary.

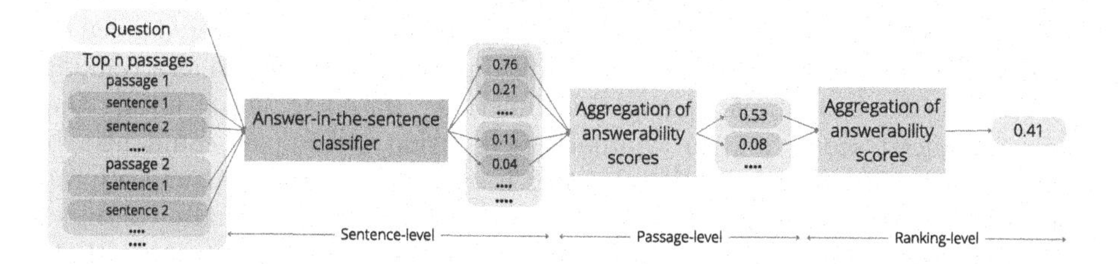

Fig. 1. Overview of our answerability detection approach.

4 Answerability Detection

The challenge of answerability in CIS arises from the fact that the answer is typically not confined to a single entity or text snippet, but rather spans across multiple sentences or even multiple passages. Note that answerability extends beyond the general notion of relevance and asks for the presence of a specific answer. At the core of our approach is a sentence-level classifier that can distinguish sentences that contribute to the answer from ones that do not. These sentence-level estimates are then aggregated on the passage level and then further on the ranking level (i.e., set of top-n passages) to determine whether the question is answerable; see Fig. 1. Operating on the sentence level is a design decision that has the added benefit that a future summary generation step may take a filtered set of sentences that contribute to the final answer as input.

[2] Examples of data samples with annotated information nuggets and answerability scores on various levels are provided in the repository accompanying the paper.

4.1 Answer-in-the-Sentence Classifier

The answer-in-the-sentence classifier is trained on sentence-level data from the train portion of the CAsT-answerability dataset. In some of the experiments, this data is augmented by data from the SQuAD 2.0 dataset [18] to provide the classifier with additional training material and thus guidance in terms of questions that can be answered with a short snippet contained in a single sentence. Data from SQuAD 2.0 is downsampled to be balanced in terms of the number of answerable and unanswerable question-sentence pairs. The classifier is built using a BERT transformer model with a sequence classification head on top (BertForSequenceClassification provided by HuggingFace[3]). Each data sample contains `question [SEP] sentence` as input and a binary answerability label. The output of the classifier is the probability that the sentence contains (part of) the answer to the question.

4.2 Aggregation of Sentence-Level Answerability Scores

In reality, answers are not confined to a single sentence but can be spread across several passages. We thus need a method to aggregate results obtained from the sentence-level classifier to decide whether the question can be answered given (1) a particular passage or (2) a set of top-ranked passages, referred to as a ranking.

We consider two simple aggregation functions, *max* and *mean*, noting that more advanced score- and/or content-based fusion techniques could also be applied in the future [11]. Intuitively, *max* is expected to work particularly well for factoid questions where the answer is relatively short and usually contained in a single sentence, while *mean* should capture the cases where pieces of the answer are spread over several sentences within the passage or across passages. The aggregated answerability score for a given passage is compared against a fixed threshold; passages with an aggregated score exceeding this threshold are identified as containing the answer. We set the threshold values on a validation partition (10% of the dataset, sampled from the training partition); 0.5 for max and 0.25 for mean.

An analogous procedure is repeated for the top $n = 3$ passages in the ranking to decide on ranking-level answerability. Here, the aggregation methods take the passage-level answerability scores as input (obtained using max or mean aggregation of sentence-level probabilities). The resulting values are compared against a fixed threshold (using the same values as for passage-level aggregation) to yield a final ranking-level answerability prediction.

5 Results

Table 2 presents the answerability results on the sentence-, passage-, and ranking-levels on the test partition of CAsT-answerability in terms of accuracy.

[3] https://huggingface.co/docs/transformers/model_doc/bert#transformers.BertForSequenceClassification.

Table 2. Answerability detection results in terms of classification accuracy. The best scores for each level are in boldface. For the augmented classifier (rows 5–8), significant differences against the respective method (rows 1–4) are indicated by *. ChatGPT results are tested against the best classifier in rows 1–8. We use McNemar's test with $p < 0.05$.

Classifier	**Sentence**	**Passage**		**Ranking**	
	Acc.	**Aggr.**	**Acc.**	**Aggr.**	**Acc.**
CAsT-answerability	0.752	Max	0.634	Max	0.790
				Mean	**0.891**
		Mean	0.589	Max	0.332
				Mean	0.829
CAsT-answerability augmented with SQuAD 2.0	**0.779***	Max	0.676*	Max	0.810*
				Mean	0.848*
		Mean	0.639*	Max	0.468*
				Mean	0.672*
ChatGPT passage-level (zero-shot)			**0.787***	T=0.33	0.839*
				T=0.66	0.623*
ChatGPT ranking-level (zero-shot)					0.669*
ChatGPT ranking-level (two-shot)					0.601*

Does Data Augmentation Help Answerability Detection?. On the sentence level, we find that augmenting the CAsT-answerability dataset with additional training examples from SQuAD 2.0 improves performance. These improvements also carry over to the first aggregation step on the passage level. However, the best ranking-level results are obtained by aggregating results obtained from the classifier trained only on CAsT-answerability. It may result from the fact that SQuAD 2.0 training data focuses on questions with short-span answers (like entities or numbers) confined to a single sentence. This could mislead the classifier to overlook answers spanning multiple sentences or passages. Thus, while sentence-level answerability prediction benefits from augmented data, this does not translate to effective passage or ranking-level answerability prediction.

Which of the Two Aggregation Methods Performs Better?. In all cases, max aggregation on the passage level followed by mean aggregation on the ranking level gives the best results. Intuitively, this configuration captures single sentences with high answerability scores in individual passages (max aggregation on passage level) that give a high average score for the whole ranking (mean aggregation on ranking level) for answerable questions.

How Competitive are these Baselines in Absolute Terms?. Ours is a novel task, with no established baselines to compare against. However, using a large language model (LLM) for generating the final response from the top retrieved passages is a natural choice. Therefore, for reference, we compare against a state-of-the-art LLM, using the most recent snapshot of GPT-3.5 (gpt-3.5-turbo-0301) via the ChatGPT API. We consider two settings: given a passage (analogous to the passage-level setup) and given a set of passages as input (analogous to the ranking-level setup). We prompt the model to verify whether the question is answerable in the provided passage(s) and return 0 or 1 accordingly.[4] In the passage-level setup, the passage-level predictions returned by ChatGPT are aggregated using fixed thresholds to obtain a ranking-level prediction. The max aggregation boils down to checking whether any of the passages is predicted to contain the answer. In the case of mean aggregation, a threshold of 0.33 or 0.66 (based on the fact that binary values are returned for passage-level answerability predictions) would mean that 1 or 2 out of 3 passages, respectively, contain the answer. In the ranking-level setup, we experiment with both a zero-shot setting, where neither examples nor context is given to the model, and a two-shot setting containing a question followed by two sentences (one positive and one negative example) extracted from the passage. We observe that the passage-level answerability scores of ChatGPT are higher than ours, but after ranking-level aggregation, it is no longer the case. Further, performing the ranking-level task directly results in significantly lower performance. These results indicate that LLMs have a limited ability to detect answerability without additional guidance. Our baseline methods trained on small datasets and based on simple classifiers with multi-step results aggregation turn out to be more effective for answerability prediction and thus represent a strong baseline.

6 Conclusion

Unanswerable questions pose a challenge in conversational information seeking. To study this problem, we have developed a test collection, based on two editions of the TREC CAsT benchmark, with sentence-, passage-, and ranking-level answerability labels. We have also presented a baseline approach based on the idea of sentence-level answerability classification and multi-step results aggregation, and evaluated multiple instantiations of this approach with different configurations. Despite their simplicity, our baselines have been shown to outperform a state-of-the-art LLM on the task of answerability prediction.

In this paper, we have simplified the scenario by treating answerability as binary concept: a question is answerable if any sentence in the returned passages contains the answer. In practice, answerability is more nuanced, with some pieces of the information found but not all. A more realistic future approach would involve an ordinal scale (e.g., unanswerable, partially answerable, fully answerable), which would necessitate ground truth assessments with an explicit specification of the different facets/aspects of the answer. We are not aware of

[4] The prompts are available in the repository accompanying the paper.

any dataset for information-seeking tasks (conversational or not) that would provide this information.

Acknowledgments. This research was supported by the Norwegian Research Center for AI Innovation, NorwAI (Research Council of Norway, project number 309834).

References

1. Asai, A., Choi, E.: Challenges in information-seeking QA: unanswerable questions and paragraph retrieval. In: Proceedings of the 59th Annual Meeting of the Association for Computational Linguistics and the 11th International Joint Conference on Natural Language Processing. ACL-IJNLP '21, vol. 1, pp. 1492–1504 (2021)
2. Bolotova-Baranova, V., Blinov, V., Filippova, S., Scholer, F., Sanderson, M.: WikiHowQA: A comprehensive benchmark for multi-document non-factoid question answering. In: Proceedings of the 61st Annual Meeting of the Association for Computational Linguistics. ACL '23 (2023)
3. Cao, Z., Li, W., Li, S., Wei, F., Li, Y.: AttSum: joint learning of focusing and summarization with neural attention. In: Proceedings of COLING 2016, the 26th International Conference on Computational Linguistics. COLING '16, pp. 547–556 (2016)
4. Choi, E., et al.: QuAC: question answering in context. In: Findings of the Association for Computational Linguistics: EMNLP 2018. EMNLP '18, pp. 2174–2184 (2018)
5. Dalton, J., Xiong, C., Callan, J.: TREC CAsT 2019: the conversational assistance track overview. In: The Twenty-Eighth Text Retrieval Conference Proceedings. TREC '19 (2019)
6. Dalton, J., Xiong, C., Callan, J.: CAsT 2020: the conversational assistance track overview. In: The Twenty-Ninth Text Retrieval Conference Proceedings. TREC '20 (2020)
7. Dalton, J., Xiong, C., Callan, J.: TREC CAsT 2021: the conversational assistance track overview. In: The Thirtieth Text Retrieval Conference Proceedings. TREC '21 (2021)
8. Godin, F., Kumar, A., Mittal, A.: Learning when not to answer: a ternary reward structure for reinforcement learning based question answering. In: Proceedings of the 2019 Conference of the North American Chapter of the Association for Computational Linguistics: Human Language Technologies. NAACL-HLT '19, pp. 122–129 (2019)
9. Hu, M., Wei, F., Peng, Y., Huang, Z., Yang, N., Li, D.: Read + verify: machine reading comprehension with unanswerable questions. arXiv cs.CL/1808.05759 (2018)
10. Huang, K., Tang, Y., Huang, J., He, X., Zhou, B.: Relation module for non-answerable predictions on reading comprehension. In: Proceedings of the 23rd Conference on Computational Natural Language Learning. CoNLL '19, pp. 747–756 (2019)
11. Kurland, O., Culpepper, J.S.: Fusion in information retrieval: SIGIR 2018 half-day tutorial. In: Proceedings of the 41st International ACM SIGIR Conference on Research and Development in Information Retrieval. SIGIR '18, pp. 1383–1386 (2018)
12. Łajewska, W., Balog, K.: Towards filling the gap in conversational search: from passage retrieval to conversational response generation. In: Proceedings of the

32nd ACM International Conference on Information and Knowledge Management. CIKM '23, pp. 5326–5330 (2023)
13. Liao, J., Zhao, X., Zheng, J., Li, X., Cai, F., Tang, J.: PTAU: prompt tuning for attributing unanswerable questions. In: Proceedings of the 45th International ACM SIGIR Conference on Research and Development in Information Retrieval. SIGIR '22, pp. 1219–1229 (2022)
14. Luan, Y., Eisenstein, J., Toutanova, K., Collins, M.: Sparse, dense, and attentional representations for text retrieval. Trans. Assoc. Comput. Linguist. **9**, 329–345 (2021)
15. Owoicho, P., Dalton, J., Aliannejadi, M., Azzopardi, L., Trippas, J.R., Vakulenko, S.: TREC CAsT 2022: going beyond user ask and system retrieve with initiative and response generation. In: The Thirty-First Text REtrieval Conference Proceedings. TREC '22 (2022)
16. Pavlu, V., Rajput, S., Golbus, P.B., Aslam, J.A.: IR system evaluation using nugget-based test collections. In: Proceedings of the fifth ACM International Conference on Web Search and Data Mining. WSDM '12, pp. 393–402 (2012)
17. Pradeep, R., Nogueira, R., Lin, J.: The expando-mono-duo design pattern for text ranking with pretrained sequence-to-sequence models. arXiv cs.IR/2101.05667 (2021)
18. Rajpurkar, P., Jia, R., Liang, P.: Know what you don't know: Unanswerable questions for SQuAD. In: Proceedings of the 56th Annual Meeting of the Association for Computational Linguistics. ACL '18, pp. 784–789 (2018)
19. Reddy, S., Chen, D., Manning, C.D.: CoQA: a conversational question answering challenge. Trans. Assoc. Comput. Linguist. **7**, 249–266 (2019)
20. Ren, P., Chen, Z., Ren, Z., Kanoulas, E., Monz, C., De Rijke, M.: Conversations with search engines: SERP-based conversational response generation. ACM Trans. Inf. Syst. **39**(4), 1–29 (2021)
21. Sulem, E., Hay, J., Roth, D.: Yes, no or idk: the challenge of unanswerable yes/no questions. In: Proceedings of the 2022 Conference of the North American Chapter of the Association for Computational Linguistics: Human Language Technologies. NAACL-HLT '22, pp. 1075–1085 (2022)
22. Tang, L., et al.: Understanding factual errors in summarization: errors, summarizers, datasets, error detectors. arXiv cs.CL/2205.12854 (2023)
23. Vakulenko, S., Longpre, S., Tu, Z., Anantha, R.: Question rewriting for conversational question answering. In: Proceedings of the 14th ACM International Conference on Web Search and Data Mining. WSDM '21, pp. 355–363 (2021)
24. Zhang, J., Zhao, Y., Saleh, M., Liu, P.J.: PEGASUS: pre-training with extracted gap-sentences for abstractive summarization. arXiv cs.CL/1912.08777 (2020)

MFVIEW: Multi-modal Fake News Detection with View-Specific Information Extraction

Marium Malik(✉), Jiaojiao Jiang, Yang Song, and Sanjay Jha

School of Computer Science and Engineering, University of New South Wales, Sydney, Australia
{marium.malik,jiaojiao.jiang,yang.song1,sanjay.jha}@unsw.edu.au

Abstract. The spread of fake news on social media is a rapidly growing problem that is impacting both the general public and the government. Current methods for detecting false news often fail to take full advantage of the multi-modal information that is available, which can lead to inconsistent decisions due to modality ambiguity. Moreover, existing methods often overlook the unique information pertaining to view-specific details that could significantly boost their discriminative power and overall performance. To this end, we introduce a novel model, MFVIEW (Multi-Modal Fake News Detection with View-Specific Information Extraction), that unifies the modeling of multi-modal and view-specific information within a single framework. Specifically, the proposed model consists of a View-Specific Information Extractor that incorporates an orthogonal constraint within the shared subspace, enabling the utilization of discriminative information unique to each modality, and an Ambiguity Cross-Training Module that detects inherent ambiguity across different modalities by capturing their correlation. Extensive experiments on two publicly available datasets show that MFVIEW outperforms state-of-the-art fake news detection approaches with an accuracy of 91.0% on the Twitter dataset and 93.3% on the Weibo dataset.

Keywords: Multi-Modal Fake News Detection · Multi-Modal Fusion · View-Specific Information Extraction

1 Introduction

The widespread availability of social media networks has shifted news sharing from traditional to online platforms, making it easier to share information on sites like Twitter (rebranded as X) and Weibo. However, these platforms also serve as hotbeds for the spread of misinformation [21], posing serious threats such as panic and social discord. For example, multimodal fake posts about the massive earthquake in Nepal in 2015 circulated on Twitter, as shown in Fig. 1. Traditional manual review struggles against the flood of fake news on social media, pushing researchers to innovate detection methods [9,13,18,24,26].

Early research mainly focused on analyzing a single modality, such as text [2,4,14,31], image (spliced or tampered) [3], or distribution pattern [10,12],

N. Goharian et al. (Eds.): ECIR 2024, LNCS 14610, pp. 345–353, 2024.
https://doi.org/10.1007/978-3-031-56063-7_26

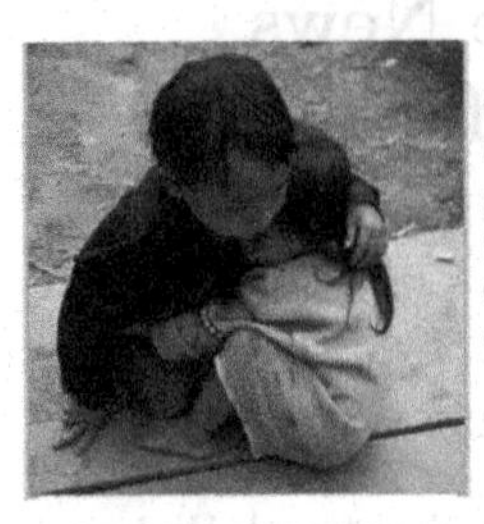

(a)From Nepal. , Two and a half year old sister protected by four year old brother in Nepal. Perhaps one the most divine picture of the century !!!

(b)#NepalQuakeRelief- Heartbreaking Pic from Nepal. Please help. Details of charities/Org at – nyti.ms/1EWcUGo

Fig. 1. Fake news on Twitter.

neglecting the crucial dimension of multiple modalities in modern news. However, research on multi-modal data is increasing now aiming to capture complex patterns by incorporating diverse data types [1,5,6,8,23,30,33]. Zhang et al. [32] trained SceneFND, on FakeNewsNet [16], outperforming baselines like SpotFake [20]. However, this work lacks a deeper analysis of semantic correlations. Most other generic frameworks either explore the cross-domain challenges [17], semantic information [6], or cross-modal feature correlation [25]. Silva et al. [17] showcase improvements in rarely-appearing domains using independent embedding spaces. Moreover, our research draws inspiration from the utility of view-specific information extraction in multi-view multi-label learning [27,29], where strengths in certain views enhance the performance in multi-label tasks. View represents the distinct characteristics of each modality. These are the features that are often ignored in multimodal fake news detection models that only rely on shared subspace exploitation by independently mapping original uni-modalities to a shared subspace, whereas incorporation of view-specific information can help further enrich the feature representation as they enhance heterogeneous communication between the modalities. Thus, improving performance in the fake news classification task.

To address the aforementioned limitations we propose a Multi-Modal Fake News Detection with VIEW-Specific Information Extraction (MFVIEW) framework. It integrates both shared multi-modal and view-specific information within a unified framework as shown in Fig. 2. We begin by extracting information from text and images using the modality-specific encoder, leveraging an orthogonal constraint for distinctive details. In parallel, ambiguity analysis is performed to enhance fake news detection. The primary contributions of our work include incorporating innovative orthogonal constraints for extracting view-specific insights within a shared subspace for fake news detection. Experimenting with Twitter [3] and Weibo [8] datasets, our method demonstrates a substantial improvement over state-of-the-art models.

2 Methodology

Problem Definition. Given a social media post $P = \{P_t, P_v\} \in \mathbb{D}$, P_t and P_v are the textual and visual embeddings from dataset $\mathbb{D}$ respectively. We denote their representations as $\{R_t\} \in \mathbb{D}$ and $\{R_v\} \in \mathbb{D}$. The main goal of this work is to predict whether P is fake news, $\hat{y} = 0$, or real news, $\hat{y} = 1$, by learning parameters θ. For each post, we extract uni-modal features using the Modality-Specific Encoder, then utilize two components of our framework, Ambiguity Cross-Training for aligning the features and analyzing their correlations, and the View-Specific Feature Extractor to extract distinct features from the modalities to enhance multi-modal fake news detection.

Modal-Specific Encoder. Pretrained techniques BERT [11] and ResNet-50 [7] are utilized to encode the text P_t and image P_v into unimodal embedding e_t and e_v respectively. For text, we feed a sequence of words to obtain a rich textual embedding representation which is further refined by passing it through a fully connected layer. For the image, the regional features are transformed using a fully connected layer.

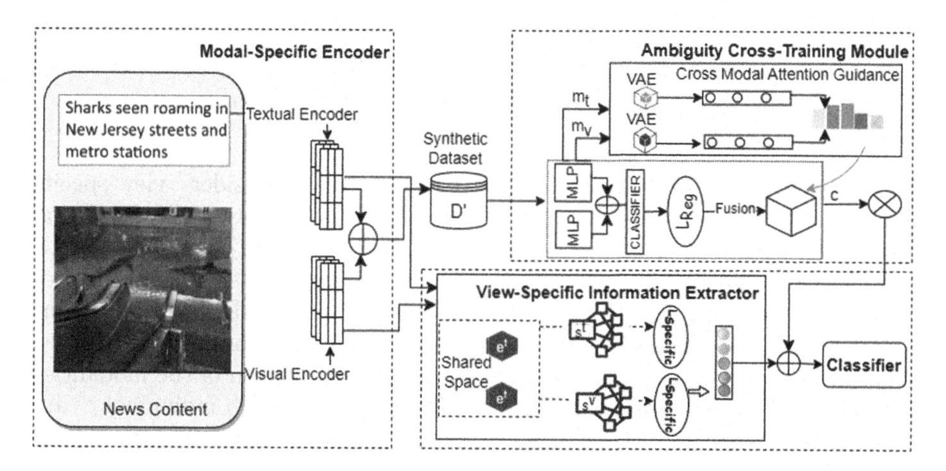

Fig. 2. Overview of MFVIEW model using Modal-Specific Encoder, View-Specific Information Extractor, and Ambiguity Cross-Training modules for utilizing shared and discriminative features.

Ambiguity Cross-Training Module. Existing fake posts frequently experience semantic discrepancies between images and text. To address the disparity, an ambiguity-aware method is employed, adapted from [5] aligning features by transforming the unimodal information into a shared subspace followed by an auxiliary correlation learning function to assess their similarity. Specifically, a synthetic dataset $\mathbb{D}'$ dataset is generated labeling the text-image pairs as positive ($y' = 1$) if they are from the same real news, and negative ($y' = 0$) otherwise. The shared subspace exploitation task is facilitated by a modality-specific multilayer perceptron (MLP) to acquire shared semantics. Subsequently, the shared embeddings are used in binary classification under the supervision of cosine embedding loss with a margin d set to 0.2 based on observational studies, as shown in Eq. 1:

$$L_{\text{Reg}} = (1 - \cos(e'_t, e'_v))\text{if } y' = 1,;\ L_{\text{Reg}} = \max(0, \cos(e'_t, e'_v) - d) \quad \text{if } y' = 0 \tag{1}$$

where e'_t and e'_v are shared embeddings in $\mathbb{D}'$. The obtained semantically aligned unimodal embeddings are then passed through the attention weight based on the ambiguity score, followed by a fusion module that learns cross-modal correlation based on the attention weight. First, the inter-modal weights are obtained based on the correlation between the unimodality a'_t and a'_v that accumulates text features to visual features by normalizing it by the square root of the dimension and applying a softmax function as shown in the Eq. 2:

$$M = \left\{ \text{softmax}\left(\frac{[a'_v][a'_t]^T}{\sqrt{\dim}}\right) \quad \text{for } t \rightarrow v; \quad \text{softmax}\left(\frac{[a'_t][a'_v]^T}{\sqrt{\dim}}\right) \quad \text{for } v \rightarrow t \right. \tag{2}$$

Next, we obtain the weighted sum of the obtained textual feature set and one visual feature to obtain a correlation map c for explicit features as indicated in Eq. 3:

$$\hat{m}_v = M_{t \rightarrow v} \times a'_v;\ \hat{m}_t = M_{v \rightarrow t} \times a'_t;\ c = \hat{m}_v \otimes \hat{m}_t \tag{3}$$

Ultimately, the interaction matrix M_I, which fuses the multimodal features is defined by taking the outer product of $\hat{m}_v$ and $\hat{m}_t$.

View-Specific Information Extractor. MFVIEW considers view-specific information, incorporating discriminative features previously ignored by most multi-modal methods that predominantly emphasize shared subspace exploitation. The view-specific information is derived by removing shared information from the original features through the enforcement of an orthogonal constraint. Let s^t and s^v be the specific feature vectors of the extraction layer E. Considering the feature vector c', containing the shared information of the modalities. Given that c' is obtained using element-wise addition of each individual c^t and c^v as below : $c' = \sum_{v=1}^{k} c^v \oplus c^t$. L_{Specific} loss is applied to encourage the orthogonality between the shared vector feature c' and specific feature vectors s^t and s^v represented by the Eq. 4, where $\| \cdot \|_2^2$ is the squared L2-norm :

$$L_{\text{Specific}} = \| \ (s)^T c' \ \|_2^2 \tag{4}$$

Fake News Detection. The final representation $\tilde{f}$ from the multimodal data is obtained by concatenating the representations of each modality, their correlations, and the attention weights. Next, the final representations $\tilde{f}$ are fed into a fully connected layer to learn intricate patterns and representations to predict a label. To ensure the prediction is accurate, cross-entropy loss, L_{ce} is incorporated. Finally, we define the final loss function for MFVIEW by combining the loss functions L_{Reg} in Eq. 1, L_{Specific} in Eq. 4 and L_{ce}, as shown in Eq. 5:

$$L_{\text{combined}} = L_{\text{Reg}} + \lambda L_{\text{Specific}} + L_{\text{ce}} \tag{5}$$

Table 1. Performance comparison between MFVIEW and other baselines.

Dataset	Model	Acc.	Fake News			Real News		
			Prec.	Rec.	F1.	Prec.	Rec.	F1.
Twitter	EANN	0.648	0.810	0.498	0.617	0.584	0.759	0.660
	SAFE	0.762	0.831	0.724	0.774	0.695	0.811	0.748
	HMCAN	0.897	**0.971**	0.801	0.878	0.853	**0.979**	0.912
	MCAN	0.809	0.889	0.765	0.822	0.732	0.871	0.795
	CAFE	0.806	0.807	0.799	0.803	0.805	0.813	0.809
	LIIMR	0.831	0.836	0.832	0.830	–	–	–
	COOLANT	0.900	0.879	0.922	0.900	**0.923**	0.880	0.901
	MFVIEW	**0.910**	0.875	**0.933**	**0.903**	0.889	0.941	**0.915**
Weibo	EANN	0.827	0.847	0.812	0.829	0.807	0.843	0.825
	SAFE	0.816	0.818	0.815	0.817	0.816	0.818	0.817
	HMCAN	0.885	0.920	0.845	0.881	0.856	0.926	0.890
	MCAN	0.899	0.913	0.899	0.901	0.884	0.909	0.897
	CAFE	0.840	0.855	0.830	0.842	0.825	0.851	0.837
	LIIMR	0.900	0.882	0.823	0.847	–	–	–
	COOLANT	0.923	**0.927**	0.923	0.925	0.919	**0.922**	0.920
	MFVIEW	**0.933**	0.910	**0.961**	**0.935**	**0.957**	0.904	**0.930**

3 Experiments

Dataset. The model is evaluated on two datasets from the real world: Twitter [3] and Weibo [8]. The Twitter dataset consists of around 6,840 real and 5,007 fake tweets in the training set, and 1,406 posts in the test set. The Weibo dataset comprises 3,749 fake and 3,783 real news in the training set, and 1,000 fake and 996 real news in the testing set respectively. Our experiments follow established procedures [8] to ensure dataset quality by removing duplicated and low-quality images.

Baselines. Our model, MFVIEW, was compared against established robust multimodal fake news detection models as baselines. EANN [23] eliminating event-specific characteristics using a GAN-based approach, SAFE [33] converting images into textual data and calculating similarity, HMCAN [15] integrating multi-modal context information and hierarchical text semantics, MCAN [28] using co-attention layers, CAFE [5] adaptively aggregating cross-modal correlations, LIIMR [19] leveraging intra and inter-modality relationships, and COOLANT [22] enhancing image-text alignment through a contrastive learning framework with attention and guidance mechanisms.

Implementation Details. For textual encoding using 256-dimensional BERT, the text input size is 200 words. For visual encoding, the size is 224×224 and we use the ResNet-50 pretrained model. The hyperparameter λ for orthogonal loss is set to 0.01. The data split is consistent across comparisons. We employ the Adam optimizer with an initial learning rate set at 10^{-4} for a duration of 50 epochs, implementing early stopping as a prevention measure against overfitting.

Overall Performance. Table 1 presents the performance of MFVIEW compared with seven baseline methods in terms of Accuracy, Precision, Recall, and F1-Score. MFVIEW exhibits superior performance across both datasets, surpassing all compared models with an accuracy of 91.0% on the Twitter dataset and 93.3% on the Weibo dataset, accompanied by notably the highest F1-Score. MFVIEW excels due to its comprehensive approach, combining shared subspace utilization and view-specific information. Additionally, the Ambiguity Cross-Training Module aligns modalities, and learning ambiguity to adaptively mitigate noise.

Ablation Study. To further investigate the effectiveness of each component of MFVIEW, we perform a quantitative analysis by removing each component and comparing it with the following variants: M-Text and M-Image: The view-specific information extractor only considers textual and visual information respectively. M-Amb: The view-specific information extractor module is entirely removed, utilizing cross-modal ambiguity analysis similar to [5].

Table 2. Ablation study on the architecture of the proposed model MFVIEW on Twitter and Weibo datasets.

Model	Twitter				Weibo			
	Acc.	Prec.	Rec.	F1.	Acc.	Prec.	Rec.	F1.
M-Text	0.76	0.58	0.80	0.67	0.90	0.90	0.89	0.89
M-Image	0.79	0.88	0.54	0.67	0.87	0.89	0.83	0.86
M-Amb	0.82	**0.92**	0.71	0.80	0.89	0.90	0.87	0.89
M-View	**0.91**	0.88	**0.93**	**0.90**	**0.93**	**0.91**	**0.96**	**0.94**

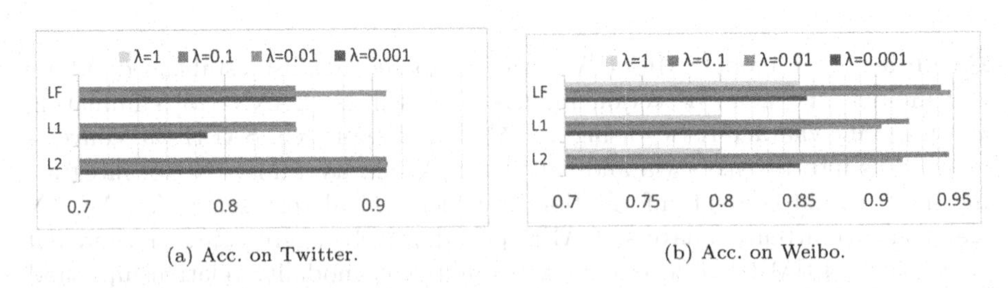

(a) Acc. on Twitter.

(b) Acc. on Weibo.

Fig. 3. Parameter sensitivity analysis on the View-Specific Information Extractor of MFVIEW

The results shown in Table 2 demonstrate that M-Text and M-Image perform poorly due to limited textual and visual information utilization, highlighting the importance of leveraging both modalities for view-specific information extraction. Moreover, M-Amb yields poor performance advocating the usage of the View-Specific information along with the shared subspace exploitation as in M-View that refers to MFVIEW.

Parameter Sensitivity Analysis. Further, we have conducted a parameter sensitivity analysis on the View-Specific Information Extractor to observe the performance of the model on the Weibo and Twitter datasets. As shown in Figs. 3a and 3b the value of λ at 0.01 balances the regularization impact, preventing overemphasis on orthogonality or sparsity, leading to improved model performance; deviations below 0.01 reduce the regularization effect, while values above 0.1 excessively constrain the model.

4 Conclusion

In this paper, we propose a novel framework, MFVIEW, that uses a View-Specific Information Extractor to leverage the utilization of multimodal data for fake news detection alongside the cross-modal correlation analysis. Incorporating an orthogonal constraint for view-specific information extraction with respect to the shared subspace enhances discriminative modeling, overcoming challenges associated with inconsistent decision-making. Whereas, the Ambiguity Cross-Training Module enforces the utilization of relevant features adaptively. Experiments on two datasets show that MFVIEW outperforms other methods, with 91.0% accuracy on Twitter and 93.3% on Weibo, respectively. In the future, this approach can be extended by exploring dynamic multi-label learning methods. This will help to adapt to the changing nature of misinformation, making our framework more resilient and effective against evolving deceptive tactics in the realm of fake news.

References

1. Agrawal, T., Gupta, R., Narayanan, S.: Multimodal detection of fake social media use through a fusion of classification and pairwise ranking systems. In: 2017 25th European Signal Processing Conference (EUSIPCO), pp. 1045–1049. IEEE (2017)
2. Bian, T., et al.: Rumor detection on social media with bi-directional graph convolutional networks. In: Proceedings of the AAAI Conference on Artificial Intelligence, vol. 34, pp. 549–556 (2020)
3. Boididou, C., Papadopoulos, S., Zampoglou, M., Apostolidis, L., Papadopoulou, O., Kompatsiaris, Y.: Detection and visualization of misleading content on twitter. Int. J. Multimedia Inf. Retrieval **7**(1), 71–86 (2018)
4. Chen, T., Li, X., Yin, H., Zhang, J.: Call attention to rumors: deep attention based recurrent neural networks for early rumor detection. In: Ganji, M., Rashidi, L., Fung, B., Wang, C. (eds.) Trends and Applications in Knowledge Discovery and Data Mining. Lecture Notes in Computer Science(), vol. 11154, pp. 40–52. Springer, Cham (2018). https://doi.org/10.1007/978-3-030-04503-6_4
5. Chen, Y., et al.: Cross-modal ambiguity learning for multimodal fake news detection. In: Proceedings of the ACM Web Conference 2022, pp. 2897–2905 (2022)

6. Giachanou, A., Zhang, G., Rosso, P.: Multimodal fake news detection with textual, visual and semantic information. In: Sojka, P., Kopecek, I., Pala, K., Horak, A. (eds.) Text, Speech, and Dialogue. Lecture Notes in Computer Science(), vol. 12284, pp. 30–38. Springer, Cham (2020). https://doi.org/10.1007/978-3-030-58323-1_3
7. He, K., Zhang, X., Ren, S., Sun, J.: Deep residual learning for image recognition (2015)
8. Jin, Z., Cao, J., Guo, H., Zhang, Y., Luo, J.: Multimodal fusion with recurrent neural networks for rumor detection on microblogs. In: Proceedings of the 25th ACM International Conference on Multimedia, pp. 795–816 (2017)
9. Jin, Z., Cao, J., Guo, H., Zhang, Y., Wang, Y., Luo, J.: Detection and analysis of 2016 us presidential election related rumors on twitter. In: Lee, D., Lin, Y.R., Osgood, N., Thomson, R. (eds.) Social, Cultural, and Behavioral Modeling. Lecture Notes in Computer Science(), vol. 10354, pp. 14–24. Springer, Cham (2017). https://doi.org/10.1007/978-3-319-60240-0_2
10. Jin, Z., Cao, J., Zhang, Y., Zhou, J., Tian, Q.: Novel visual and statistical image features for microblogs news verification. IEEE Trans. Multimedia **19**(3), 598–608 (2016)
11. Kenton, J.D.M.W.C., Toutanova, L.K.: BERT: pre-training of deep bidirectional transformers for language understanding. In: Proceedings of NAACL-HLT, vol. 1, p. 2 (2019)
12. Liu, Y., Wu, Y.F.: Early detection of fake news on social media through propagation path classification with recurrent and convolutional networks. In: Proceedings of the AAAI Conference on Artificial Intelligence, vol. 32 (2018)
13. Nguyen, V.H., Sugiyama, K., Nakov, P., Kan, M.Y.: FANG: leveraging social context for fake news detection using graph representation. In: Proceedings of the 29th ACM International Conference on Information & Knowledge Management, pp. 1165–1174 (2020)
14. Qi, P., Cao, J., Yang, T., Guo, J., Li, J.: Exploiting multi-domain visual information for fake news detection. In: 2019 IEEE International Conference on Data Mining (ICDM), pp. 518–527. IEEE (2019)
15. Qian, S., Wang, J., Hu, J., Fang, Q., Xu, C.: Hierarchical multi-modal contextual attention network for fake news detection. In: Proceedings of the 44th International ACM SIGIR Conference on Research and Development in Information Retrieval, pp. 153–162 (2021)
16. Shu, K., Mahudeswaran, D., Wang, S., Lee, D., Liu, H.: FakeNewsNet: a data repository with news content, social context, and spatiotemporal information for studying fake news on social media. Big Data **8**(3), 171–188 (2020)
17. Silva, A., Luo, L., Karunasekera, S., Leckie, C.: Embracing domain differences in fake news: Cross-domain fake news detection using multi-modal data. In: Proceedings of the AAAI Conference on Artificial Intelligence, vol. 35, pp. 557–565 (2021)
18. Singh, B., Sharma, D.K.: Predicting image credibility in fake news over social media using multi-modal approach. Neural Comput. Appl. **34**(24), 21503–21517 (2022)
19. Singhal, S., Pandey, T., Mrig, S., Shah, R.R., Kumaraguru, P.: Leveraging intra and inter modality relationship for multimodal fake news detection. In: Companion Proceedings of the Web Conference 2022, pp. 726–734 (2022)
20. Singhal, S., Shah, R.R., Chakraborty, T., Kumaraguru, P., Satoh, S.: SpotFake: a multi-modal framework for fake news detection. In: 2019 IEEE Fifth International Conference on Multimedia Big Data (BigMM), pp. 39–47. IEEE (2019)

21. Wang, J., Makowski, S., Cieślik, A., Lv, H., Lv, Z.: Fake news in virtual community, virtual society, and metaverse: a survey. IEEE Trans. Comput. Soc. Syst. (2023)
22. Wang, L., Zhang, C., Xu, H., Zhang, S., Xu, X., Wang, S.: Cross-modal contrastive learning for multimodal fake news detection. arXiv preprint: arXiv:2302.14057 (2023)
23. Wang, Y., et al.: EANN: event adversarial neural networks for multi-modal fake news detection. In: Proceedings of the 24th ACM SIGKDD International Conference on Knowledge Discovery & Data Mining, pp. 849–857 (2018)
24. Wang, Y., et al.: Weak supervision for fake news detection via reinforcement learning. In: Proceedings of the AAAI Conference on Artificial Intelligence, vol. 34, pp. 516–523 (2020)
25. Wei, Z., Pan, H., Qiao, L., Niu, X., Dong, P., Li, D.: Cross-modal knowledge distillation in multi-modal fake news detection. In: ICASSP 2022–2022 IEEE International Conference on Acoustics, Speech and Signal Processing (ICASSP), pp. 4733–4737. IEEE (2022)
26. Wu, L., Long, Y., Gao, C., Wang, Z., Zhang, Y.: MFIR: multimodal fusion and inconsistency reasoning for explainable fake news detection. Inf. Fusion **100**, 101944 (2023)
27. Wu, X., et al.: Multi-view multi-label learning with view-specific information extraction. In: IJCAI, pp. 3884–3890 (2019)
28. Wu, Y., Zhan, P., Zhang, Y., Wang, L., Xu, Z.: Multimodal fusion with co-attention networks for fake news detection. In: Findings of the Association for Computational Linguistics: ACL-IJCNLP 2021, pp. 2560–2569 (2021)
29. Xing, Y., Yu, G., Domeniconi, C., Wang, J., Zhang, Z.: Multi-label co-training. In: Proceedings of the 27th International Joint Conference on Artificial Intelligence, pp. 2882–2888 (2018)
30. Xue, J., Wang, Y., Tian, Y., Li, Y., Shi, L., Wei, L.: Detecting fake news by exploring the consistency of multimodal data. Inf. Proc. Manage. **58**(5), 102610 (2021)
31. Yu, F., et al.: A convolutional approach for misinformation identification. In: IJCAI, pp. 3901–3907 (2017)
32. Zhang, G., Giachanou, A., Rosso, P.: SceneFND: multimodal fake news detection by modelling scene context information. J. Inf. Sci., 01655515221087683 (2022)
33. Zhou, X., Wu, J., Zafarani, R.: Safe: similarity-aware multi-modal fake news detection, vol. 2 . Preprint: arXiv:2003.04981 (2020)

BertPE: A BERT-Based Pre-retrieval Estimator for Query Performance Prediction

Maryam Khodabakhsh[1](✉), Fattane Zarrinkalam[2], and Negar Arabzadeh[3]

[1] Shahrood University of Technology, Shahrood, Iran
m_khodabakhsh@shahroodut.ac.ir
[2] University of Guelph, Guelph, ON, Canada
fzarrink@uoguelph.ca
[3] University of Waterloo, Waterloo, ON, Canada
narabzad@uwaterloo.ca

Abstract. Query Performance Prediction (QPP) aims to estimate the effectiveness of a query in addressing the underlying information need without any relevance judgments. More recent works in this area have employed the pre-trained neural embedding representations of the query to go beyond the corpus statistics of query terms and capture the semantics of the query. In this paper, we propose a supervised QPP method by adopting contextualized neural embeddings to directly learn the performance through fine-tuning. To address the challenges arising from disparities in the evaluation of retrieval models through sparse and comprehensive labels, we introduce an innovative strategy for creating synthetic relevance judgments to enable effective performance prediction for queries, irrespective of whether they are evaluated with sparse or more comprehensive labels. Through our experiments on four different query sets accompanied by MS MARCO V1 collection, we show that our approach shows significantly improved performance compared to the state-of-the-art Pre-retrieval QPP methods.

Keywords: Pre-retrieval query performance prediction · Bert · Synthetic relevance judgments

1 Introduction

Detecting poorly performing queries, often referred to as "difficult" queries, in real-time of a search engine is a crucial concern in information retrieval (IR) systems which facilitates addressing such queries using techniques like query reformulation, query suggestion, and seeking clarifications to enhance search results [4,21,33]. Query Performance Prediction (QPP) is the task of estimating query difficulty by assessing a retrieval system's effectiveness for a given query, all without relying on relevance judgments [1,2,10,18,19,22,27,35,36]. QPP methods are categorized into two groups: pre-retrieval and post-retrieval methods [9]. Pre-retrieval QPP methods exclusively consider query content and corpus statistics gathered during indexing to predict query difficulty [6,7]. Since they

N. Goharian et al. (Eds.): ECIR 2024, LNCS 14610, pp. 354–363, 2024.
https://doi.org/10.1007/978-3-031-56063-7_27

make predictions prior to retrieval, they find more practical applications, avoiding unnecessary latency associated with retrieval operations [9]. Therefore, this paper focuses on pre-retrieval QPP methods to enhance the real-time detection of difficult queries in IR systems.

The majority of existing pre-retrieval QPP methods fall into *linguistic* and *statistical* predictors [28]. However, previous research has shown that these linguistic predictors are less competitive compared to statistical predictors such as Inverse Document Frequency (IDF) [9] and Term Weight Variance (VAR) [41], which rely on analyzing the distribution of query term frequencies within the collection [28]. With the increasing success of neural networks in IR, recent studies have introduced pre-retrieval predictors based on *pre-trained neural embeddings*. These approaches have demonstrated significantly improved retrieval effectiveness and performance prediction accuracy when compared to their non-neural counterparts [6,8,34]. For instance, [6] proposed a set of pre-retrieval QPP methods that leverage neural embedding representations of query terms to build an ego network of query terms and consider the graph connectivity metrics as indicators of query performance. However, we contend that static pre-trained neural embeddings representing query terms may not adequately distinguish between potentially different semantics associated with the same surface form of a term, as they are context-independent [20].

Motivated by recent advancements in employing contextualized embeddings for post-retrieval QPP [3], our study extends this concept to pre-retrieval QPP by moving beyond the single-term representation of queries. Our approach involves a supervised method that directly learns query performance by fine-tuning pre-trained language models. In contrast to neural-based models used in ad hoc retrieval, which primarily focus on learning relevance, our proposed approach, known as `BertPE` (Bert-based Pre-retrieval Estimator), specifically targets the quality of search results as an indicator of query performance. Further, recognizing the challenges of limited labeled data for learning and predicting query performance, we introduce a novel strategy of leveraging *synthetic relevance judgments* as an enhancement for pre-retrieval QPP [11,25,40]. Our experiments reveal that learning the performance from incomplete relevance judgments could be quite noisy. Therefore, we suggest first expanding the relevance judgments for a given query through a reliable multi-stage ranking stack and using the top-retrieved documents as synthetic relevance documents. Further, we propose to learn the actual performance of a given query with the synthetic relevance judgments. We show that the incorporation of synthetic relevance judgments contributes to more stable and robust performance predictions, particularly in datasets with sparse labels like MS MARCO dev set, as well as query sets with relatively more comprehensive labels, such as TREC deep learning tracks and DL-Hard. To demonstrate the effectiveness of BertPE, we conduct experiments on various datasets, including the MS-MARCO passage retrieval dataset, TREC Deep Learning Track 2019, 2020,2021 and 2022 queries, and Deep Learning Hard (DL-Hard) dataset. Our results show that `BertPE` outperforms state-of-the-art pre-retrieval QPP methods in terms of metrics such as Pearson's ρ, Kendall's τ, and Spearman's ρ, illustrating its superior predictive capability.

2 Proposed Approach

Given a collection of documents C and a user query q, a pre-retrieval query performance predictor μ needs to estimate the performance of q with respect to an IR evaluation metric M, as $\widehat{M_q} \leftarrow \mu(q, C)$. Where $\widehat{M_q}$ is the estimation of M for query q and the association between the actual value of M for query q, i.e., M_q, and $\widehat{M_q}$ is often considered as the usefulness of the predictor μ.

In this paper, our primary objective is to introduce a predictive model denoted as μ, which harnesses the capabilities of *pre-trained transformer language models* [20,37]. These language models excel in capturing the structural, contextual, and linguistic attributes of textual inputs. They are founded on the transformer encoder architecture, which processes a textual sequence by embedding it and passing it through a stack of L self-attentive transformer modules [38]. Additionally, they utilize special tokens, including $[CLS]$ tokens, which often encapsulate the semantics of a sequence of terms [20]. In a formal sense, when presented with an input text query q, we employ transformer language models to process the query as follows:

$$[h^0_{[CLS]} : h^0_q] = Emb([CLS]q[SEP]) \tag{1}$$

$$[h^l_{[CLS]} : h^l_q] = Transformer(h^{l-1}_{[CLS]} : h^{l-1}_q) \tag{2}$$

where $Emb()$ and $Transformer()$ are the embedding layer and self-attentive transformer module. $[h^l_{[CLS]} : h^l_q]$ is the representation of the tokens of $[CLS]$ and the queries at the layer l. To predict M_q for a given input query q, we feed the contextual representation of $[CLS]$ token in the last layer as input to a fully connected layer to predict the performance for the query q as follows:

$$\widehat{M_q} = \sigma(W_2.\phi(W_1.h^L_{[CLS]} + b_1) + b_2) \tag{3}$$

where $h^L_{[CLS]}$ is the final hidden states of the $[CLS]$ token and denotes the BERT encoding of the query q. $\sigma()$ and $\phi(.)$ are the sigmoid and linear activation functions respectively, W_1, and W_2 are weighted matrices for the first and second fully-connected layers respectively, b_1, and b_2 are biases.

Considering that $h^L_{[CLS]}$ is proficient in encapsulating latent information within the query $q's$ structural, semantic, and contextual aspects, we consider $h^L_{[CLS]}$ can potentially be an effective representation of the query and useful in predicting the performance of the query. Given that the pre-retrieval QPP task can be framed as a regression problem, as previously explored in [3], and with access to a training data tuple (q, M_q), where M_q denotes the value of an evaluation metric such as MRR (Mean Reciprocal Rank) for query q, our pre-retrieval QPP task aims to minimize the squared error between the predicted QPP score, denoted as $\widehat{M_q}$, and the actual M_q. Formally, the loss function for our pre-retrieval QPP method can be expressed as Mean squared error as $(M_q - \widehat{M_q})^2$.

Our research is centered on exclusive training to acquire a novel concept: query performance. We distinguish our work from BERT-QPP [3] by emphasizing our sole reliance on query input, with no incorporation of information from retrieved documents. Furthermore, our study underscores the significant contribution made by the utilization of synthetic relevance judgments and synthetic labels during the training of `BertPE`, resulting in a substantial enhancement in its predictive performance. It is noted that the MS MARCO dataset typically includes an average of one relevant document per query [29] and it left no information on the unjudged documents [5]. These documents could potentially contain multiple passages that hold varying degrees of relevance and different aspects for the same query as shown in previous studies [5,32]. Consequently, QPP models may struggle to accurately estimate the performance of each query based on an inaccurate quantified actual performance. This motivates us to focus on generating ***Synthetic relevance judgments*** for the queries through document re-ranking algorithms as a preliminary step, to generate pseudo-performance labels just before training `BertPE`. We believe that augmenting and extending the relevance judgments with the high-quality retrieved documents from a more expensive multi-stage stack of rankers can lead to a more precise estimation of the IR evaluation metric M and enhance the effectiveness of the predictor μ especially when the ground truth labels are quite sparse.

3 Experiments

Dataset. The MS MARCO collection [29] is a well-known dataset that includes over 8.8 million passages in its V1 variation and over 138 million passages in its v2 edition. MS MARCO V1 collection comes with over $500k$ search queries, each annotated with at least one relevant passage. We used the train set of MS MARCO V1 collection to train our proposed method `BertPE` and assess it on 4 different query sets including 1) MS MARCO Development set which consists of 6,980 queries 2) TREC Deep Learning (DL) Track 2019 [15] and 2020 [13] which include 97 queries that were judgment comprehensively on a 4-scale relevance grade on MS MARCO V1, 3) Deep Learning Hard (DL-Hard) [26] query sets with 50 of the most challenging queries from judged and adjudged queries of DL2019 and DL2020 and 4) Queries in Deep Learning tracks of 2021 and 2022 which have been annotated on MS MARCO V2 edition [12,14,16,17].

Baselines. We compared our proposed method to statistical-based predictors including *Inverse Document Frequency (IDF)* [9], *Term weight Variance (VAR)* [41], *Collection Query Similarity (SCQ)* [41] and *Simplified Clarity Score (SCS)* [23] which are formalized in Table 1. We also considered a set of **pre-trained neural embedding-based** metrics, proposed in [6] in which they build an ego network for each query terms and its neighbors. Then, they measure the specificity of the ego node through the use of graph connectivity metrics such as *Edge Count (EC)*, *Edge Weight Sum (EWS)*, *Inverse Edge Frequency (IEF)*,

Table 1. Statistical-based pre-retrieval predictors. t, C, and N are a term of the query, the corpus, and the total number of documents in corpus C respectively. N_t is the total number of documents that contain the term t. $tf(t, C)$ is the occurrence frequency of term t in C. $avgICTF(q) = \frac{1}{|q|} \sum_{t \in q} log \frac{N}{tf(t,C)}$.

Pre-retrieval metrics	Formula
IDF [9]	$IDF(t) = log \frac{N}{N_t}$
VAR [41]	$VAR(t)$
SCQ [41]	$SCQ(t) = (1 + log(tf(t, C))).IDF(t)$
SCS [23]	$SCS(q) = log \frac{1}{\lvert q \rvert} + avgICTF(q)$

and *Degree Centrality (DC)*. Recently, Zamani et al. [39] proposed a new representation learning framework named *MRL* that learns a multivariate distribution and uses negative multivariate KL divergence to compute the similarity between distributions and the norm of the covariance matrix for each query. The authors in this work have shown that their representation could be considered as a signal to compute the query performance.

Implementation Details. Our proposed method was trained for one epoch in an end-to-end fashion on MS MARCO train set and the model parameters were estimated using stochastic gradient descent with Adam [24] for optimization. In our experiments, we used the pre-trained BERT [20] from HuggingFace with twelve layers, 768 hidden layers, 12 heads, and $110M$ parameters trained on lower-cased English text, often known as BERT-Base Uncased. The dimension of fully connected layers was 100 (thus W1 and W2 should have a shape of 768×100). The learning rate was fixed at $2e-5$.

Toward reaching the best synthetic relevance judgments, we adapt MonoBERT for reranking top-1000 retrieved documents by BM25 [30] followed by duoBERT pairwise reranking [30]. Following the multi-stage douBERT reranker, in this paper, we used the top-1000 hits of BM25 (a common setting proposed by Nogueira and Cho [30]) and fed them to the monoBERT in order to re-rank the initial retrieved list. Given that duoBERT used time-consuming pairwise loss, only top-100 re-ranked passages from monoBERT are considered as input to duoBERT. Finally, as suggested in [30] and due to the high computational expenses of running pairwise rerankers, we selected the top-50 of duoBERT as synthetic relevance judgments and augmented the query relevance judgments with them. We note that due to pairwise comparisons that need to be made, having a higher number of documents ranked would heavily hurt the query latency.

Evaluation Metrics. We measure the correlation between the set of queries that are ranked based on their predicted performance against their actual performance with the official evaluation metric of each dataset i.e., $MRR@10$ for MS MARCO dev set and $NDCG@10$ for the other query sets, in order to evaluate our method for the task of pre-retrieval QPP. To this end, we used Kendall's

Table 2. The performance on MS MARCO Dev set in terms of correlation with MRR@10 and the rest of the query sets in terms of correlations with NDCG@10. * indicates a statistically significant correlation with NDCG@10 with p-value <0.05.

	2019/2020			2021/2022		
	Pearson's ρ	Kendall's τ	Spearman's ρ	Pearson's ρ	Kendall's τ	Spearman's ρ
IDF	0.347*	0.249*	0.372*	−0.044	0.161*	0.252*
VAR	0.347*	**0.263***	**0.39***	0.098	0.121*	0.195*
SQC	−0.1	−0.058	0.086	−0.05	−0.042	−0.056
SCS	0.273*	0.145	0.212	−0.032	−0.019	−0.027
EC	0.121	0.075	0.116	0.113	0.044	0.069
EWS	0.125	0.074	0.113	0.121	0.047	0.071
IEF	0.134	0.094	0.146	0.125	0.068	0.1
DC	0.137	0.086	0.127	0.03	0.005	0.01
MRL	0.172	0.1	0.144	0.066	0.041	0.066
BertPE	**0.366***	0.246*	0.353*	**0.238***	**0.173***	**0.254***
	MS MARCO Dev Set			DL-Hard		
	Pearson's ρ	Kendall's τ	Spearman's ρ	Pearson's ρ	Kendall's τ	Spearman's ρ
IDF	0.104*	0.214*	0.285*	0.2	0.319*	0.432*
VAR	−0.011	0.181*	0.241*	0.26	0.26*	0.363*
SQC	0.072*	0.071*	0.094*	0.154	0.133	0.2
SCS	0.168*	0.162*	0.215*	0.286*	0.235*	0.342*
EC	0.074*	0.059*	0.079*	0.221	0.084	0.119
EWS	0.075*	0.059*	0.079*	0.218	0.089	0.123
IEF	0.073*	0.055*	0.073*	0.228	0.107	0.155
DC	0.069*	0.064*	0.085*	0.166	0.056	0.075
MRL	0.18	0.024*	0.032*	−0.013	0.049	0.057
BertPE	**0.375***	**0.308***	**0.397***	**0.421***	**0.329***	**0.478***

τ and Spearman's ρ coefficient as a ranking correlation metric and Pearson's ρ coefficient as a linear correlation metric to report the quality of our method in predicting the difficulty of the queries. The higher the correlation value is, the more accurate the predicted performance is.

3.1 Experimental Results and Findings

The results of our proposed method compared to the state-of-the-art baselines are reported in Table 2 in terms of Pearson's ρ, Kendall's τ, and Spearman's ρ. Predictors that are calculated on individual term levels, are aggregated toward $avg()$ function as suggested in [9,23].

Based on the results reported in Table 2, one can observe that our proposed method, `BertPE`, has significantly outperformed statistical-based pre-retrieval baselines (i.e., IDF, VAR, SQC, and SCS) in terms of all the three correlation metrics over Trec DL 2021/2022, Dev set, and DL Hard. However, on

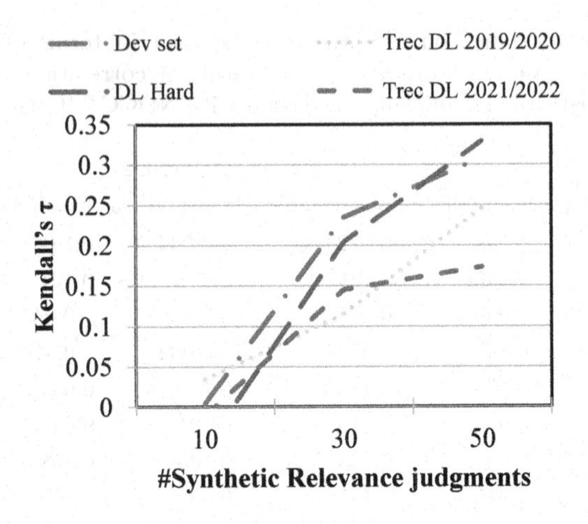

Fig. 1. The impact of the #synthetic relevance judgments.

DL2019/2020 datasets VAR statistical-based predictor had slight improvement over BertPE in terms of Kendal's τ and Spearman's ρ. On the other hand, BertPE shows a higher linear correlation compared to this baseline. Our method also outperforms neural embedding-based specificity metrics (i.e., EC, EWS, IEF, and DC). Although these metrics benefit from the similarity between the terms of queries and collection in embedding space, they suffer from context-independent problems and are not able to recognize the various contexts of a term. Further, the results on the DL-Hard dataset and MS MARCO Dev set confirm our observation that BertPE is significantly better than other baselines, as shown in Table 2. We also observe that there is no single baseline that consistently outperforms in terms of linear correlation (Pearson ρ) and rank-based correlations (Kendall τ and Spearman ρ). For example, among the baselines on the DL-HARD dataset, while SCS shows the highest linear correlation, IDF exhibits the highest rank-based correlation. Consequently, we empirically conclude that the baselines exhibit inconsistency across linear and rank-based correlations. However, our proposed method demonstrates a relatively higher degree of consistency between linear and rank-based methods, as shown in Table 2.

To further analyze the impact of synthetic relevance judgments, we investigated the ability of BertPE when trained with varying numbers of synthetic relevance judgments in Fig. 1. We choose to limit the number of synthetic relevance judgments to 50 because higher numbers would entail increased computational expenses and query latency times. Additionally, prior studies have indicated that pairwise reranking of the top 50 documents already yields promising performance [30,31]. As illustrated in the Figure, the model's performance consistently improves as the number of synthetic relevance judgments increases across four evaluation datasets. We conclude that having more synthetic relevance documents contributes to learning more accurate representations of query performance and, consequently, more precise QPP. However, even with 50 additional relevance-judged documents, we are able to outperform the baselines.

4 Concluding Remarks

We have proposed a pre-retrieval QPP method to predict the performance of a given query before the retrieval process by fine-tuning a contextualized pre-trained transformer, i.e., BERT. We leveraged synthetic relevance labels to add synthetic relevance judgments when training our proposed method on sparse labels. Our experiments showed that `BertPE` outperformed state-of-the-art pre-retrieval QPP baselines on four different query sets.

References

1. Arabzadeh, N., Bigdeli, A., Zihayat, M., Bagheri, E.: Query performance prediction through retrieval coherency. In: Hiemstra, D., Moens, M.F., Mothe, J., Perego, R., Potthast, M., Sebastiani, F. (eds.) ECIR 2021, Part II. LNCS, vol. 12657, pp. 193–200. Springer, Cham (2021). https://doi.org/10.1007/978-3-030-72240-1_15
2. Arabzadeh, N., Hamidi Rad, R., Khodabakhsh, M., Bagheri, E.: Noisy perturbations for estimating query difficulty in dense retrievers. In: Proceedings of the 32nd ACM International Conference on Information and Knowledge Management, pp. 3722–3727 (2023)
3. Arabzadeh, N., Khodabakhsh, M., Bagheri, E.: Bert-QPP: contextualized pre-trained transformers for query performance prediction. In: Proceedings of the 30th ACM International Conference on Information & Knowledge Management, pp. 2857–2861 (2021)
4. Arabzadeh, N., Seifikar, M., Clarke, C.L.A.: Unsupervised question clarity prediction through retrieved item coherency (2022)
5. Arabzadeh, N., Vtyurina, A., Yan, X., Clarke, C.L.A.: Shallow pooling for sparse labels. CoRR abs/2109.00062 (2021). https://arxiv.org/abs/2109.00062
6. Arabzadeh, N., Zarrinkalam, F., Jovanovic, J., Al-Obeidat, F., Bagheri, E.: Neural embedding-based specificity metrics for pre-retrieval query performance prediction. Inf. Process. Manag. **57**(4), 102248 (2020)
7. Arabzadeh, N., Zarrinkalam, F., Jovanovic, J., Bagheri, E.: Neural embedding-based metrics for pre-retrieval query performance prediction. In: Jose, J., et al. (eds.) ECIR 2020, Part II. LNCS, vol. 12036, pp. 78–85. Springer, Cham (2020). https://doi.org/10.1007/978-3-030-45442-5_10
8. Arabzadeh, N., Zarrinkalam, F., Jovanovic, J., Bagheri, E.: Geometric estimation of specificity within embedding spaces. In: Proceedings of the 28th ACM International Conference on Information and Knowledge Management, pp. 2109–2112 (2019)
9. Carmel, D., Yom-Tov, E.: Estimating the query difficulty for information retrieval. Synthesis Lectures on Information Concepts, Retrieval, and Services, vol. 2, no. 1, pp. 1–89 (2010)
10. Chen, X., He, B., Sun, L.: Groupwise query performance prediction with BERT. In: Hagen, M., et al. (eds.) ECIR 2022. LNCS, vol. 13186, pp. 64–74. Springer, Cham (2022). https://doi.org/10.1007/978-3-030-99739-7_8

11. Cormack, G.V., Palmer, C.R., Clarke, C.L.: Efficient construction of large test collections. In: Proceedings of the 21st Annual International ACM SIGIR Conference on Research and Development in Information Retrieval, pp. 282–289 (1998)
12. Craswell, N., Mitra, B., Yilmaz, E., Campos, D.: Overview of the TREC 2020 deep learning track. In: TREC (2020)
13. Craswell, N., Mitra, B., Yilmaz, E., Campos, D.: Overview of the TREC 2020 deep learning track. CoRR abs/2102.07662 (2021). https://arxiv.org/abs/2102.07662
14. Craswell, N., Mitra, B., Yilmaz, E., Campos, D., Voorhees, E.M.: Overview of the TREC 2019 deep learning track. In: TREC (2019)
15. Craswell, N., Mitra, B., Yilmaz, E., Campos, D., Voorhees, E.M.: Overview of the TREC 2019 deep learning track. arXiv preprint arXiv:2003.07820 (2020)
16. Craswell, N., Mitra, B., Yilmaz, E., Campos, D.F., Lin, J.: Overview of the TREC 2021 deep learning track. In: TREC (2021)
17. Craswell, N., et al.: Overview of the TREC 2022 deep learning track. In: TREC (2022)
18. Datta, S., Ganguly, D., Greene, D., Mitra, M.: Deep-QPP: a pairwise interaction-based deep learning model for supervised query performance prediction. In: Proceedings of the Fifteenth ACM International Conference on Web Search and Data Mining, pp. 201–209 (2022)
19. Datta, S., Ganguly, D., Mitra, M., Greene, D.: A relative information gain-based query performance prediction framework with generated query variants. ACM Trans. Inf. Syst. **41**(2), 1–31 (2022)
20. Devlin, J., Chang, M., Lee, K., Toutanova, K.: BERT: pre-training of deep bidirectional transformers for language understanding. In: Proceedings of the 2019 Conference of the North American Chapter of the Association for Computational Linguistics: Human Language Technologies, NAACL-HLT 2019, Minneapolis, MN, USA, 2–7 June 2019, Volume 1 (Long and Short Papers), pp. 4171–4186. Association for Computational Linguistics (2019). https://doi.org/10.18653/v1/n19-1423
21. Diaz, F.: Pseudo-query reformulation. In: Ferro, N., et al. (eds.) ECIR 2016. LNCS, vol. 9626, pp. 521–532. Springer, Cham (2016). https://doi.org/10.1007/978-3-319-30671-1_38
22. Faggioli, G., et al.: Towards query performance prediction for neural information retrieval: challenges and opportunities. In: Proceedings of the 2023 ACM SIGIR International Conference on Theory of Information Retrieval, pp. 51–63 (2023)
23. He, B., Ounis, I.: Inferring query performance using pre-retrieval predictors. In: Apostolico, A., Melucci, M. (eds.) SPIRE 2004. LNCS, vol. 3246, pp. 43–54. Springer, Heidelberg (2004). https://doi.org/10.1007/978-3-540-30213-1_5
24. Kingma, D.P., Ba, J.: Adam: a method for stochastic optimization. In: International Conference on Learning Representations (ICLR) (2015)
25. Losada, D.E., Parapar, J., Barreiro, A.: Multi-armed bandits for adjudicating documents in pooling-based evaluation of information retrieval systems. Inf. Process. Manag. **53**(5), 1005–1025 (2017)
26. Mackie, I., Dalton, J., Yates, A.: How deep is your learning: the DL-hard annotated deep learning dataset. In: Proceedings of the 44th International ACM SIGIR Conference on Research and Development in Information Retrieval, pp. 2335–2341 (2021)
27. Meng, C., Arabzadeh, N., Aliannejadi, M., de Rijke, M.: Query performance prediction: from ad-hoc to conversational search. arXiv preprint arXiv:2305.10923 (2023)

28. Mothe, J., Tanguy, L.: Linguistic features to predict query difficulty. In: ACM SIGIR 2005 Workshop on Predicting Query Difficulty - Methods and Applications (2005)
29. Nguyen, T., Rosenberg, M., Song, X., Gao, J., Tiwary, S., Majumder, R., Deng, L.: MS MARCO: a human generated machine reading comprehension dataset. In: CoCo@ NIPS (2016)
30. Nogueira, R., Yang, W., Cho, K., Lin, J.: Multi-stage document ranking with bert (2019)
31. Pradeep, R., Nogueira, R., Lin, J.: The expando-mono-duo design pattern for text ranking with pretrained sequence-to-sequence models. arXiv preprint arXiv:2101.05667 (2021)
32. Qu, Y., et al.: RocketQA: an optimized training approach to dense passage retrieval for open-domain question answering. In: Proceedings of the 2021 Conference of the North American Chapter of the Association for Computational Linguistics: Human Language Technologies, pp. 5835–5847. Association for Computational Linguistics, Online (2021). https://doi.org/10.18653/v1/2021.naacl-main.466
33. Roitman, H., Erera, S., Feigenblat, G.: A study of query performance prediction for answer quality determination. In: Proceedings of the 2019 ACM SIGIR International Conference on Theory of Information Retrieval, pp. 43–46 (2019)
34. Roy, D., Ganguly, D., Mitra, M., Jones, G.J.: Estimating gaussian mixture models in the local neighbourhood of embedded word vectors for query performance prediction. Inf. Process. Manag. **56**(3), 1026–1045 (2019)
35. Salamat, S., Arabzadeh, N., Seyedsalehi, S., Bigdeli, A., Zihayat, M., Bagheri, E.: Neural disentanglement of query difficulty and semantics. In: Proceedings of the 32nd ACM International Conference on Information and Knowledge Management, pp. 4264–4268 (2023)
36. Singh, A., Ganguly, D., Datta, S., Macdonald, C.: Unsupervised query performance prediction for neural models utilising pairwise rank preferences. def **1**, 2 (2023)
37. Vaswani, A., et al.: Attention is all you need. CoRR abs/1706.03762 (2017). http://arxiv.org/abs/1706.03762
38. Vaswani, A., et al.: Attention is all you need. In: Advances in Neural Information Processing Systems, vol. 30 (2017)
39. Zamani, H., Bendersky, M.: Multivariate representation learning for information retrieval. arXiv preprint arXiv:2304.14522 (2023)
40. Zerveas, G., Rekabsaz, N., Eickhoff, C.: Enhancing the ranking context of dense retrieval methods through reciprocal nearest neighbors. arXiv preprint arXiv:2305.15720 (2023)
41. Zhao, Y., Scholer, F., Tsegay, Y.: Effective pre-retrieval query performance prediction using similarity and variability evidence. In: Macdonald, C., Ounis, I., Plachouras, V., Ruthven, I., White, R.W. (eds.) ECIR 2008. LNCS, vol. 4956, pp. 52–64. Springer, Heidelberg (2008). https://doi.org/10.1007/978-3-540-78646-7_8

On the Influence of Reading Sequences on Knowledge Gain During Web Search

Wolfgang Gritz[1(✉)], Anett Hoppe[2], and Ralph Ewerth[2]

[1] TIB – Leibniz Information Centre for Science and Technology, Welfengarten 1B, Hannover 30167, Germany
wolfgang.gritz@tib.eu

[2] L3S Research Center, Leibniz University Hannover, Appelstraße 9A, Hannover 30167, Germany

Abstract. Nowadays, learning increasingly involves the usage of search engines and web resources. The related interdisciplinary research field *search as learning* aims to understand how people learn on the web. Previous work has investigated several feature classes to predict, for instance, the expected knowledge gain during web search. Therein, eye-tracking features have not been extensively studied so far. In this paper, we extend a previously used line-based reading model to one that can detect reading sequences across multiple lines. We use publicly available study data from a web-based learning task to examine the relationship between our feature set and the participants' test scores. Our findings demonstrate that learners with higher knowledge gain spent significantly more time reading, and processing more words in total. We also find evidence that faster reading at the expense of more backward regressions, i.e., re-reading previous portions of text, may be an indicator of better web-based learning. We make our code publicly available at https://github.com/TIBHannover/reading_web_search.

Keywords: Search as Learning · Web Search · Eye-Tracking · Reading · Knowledge Gain

1 Introduction

The research field *Search as Learning* (SAL) examines knowledge acquisition processes during and with the help of web search. Whereas conventional information retrieval systems focus on satisfying an information need, the objective here is to determine how individuals can be best supported to perform an engaging and successful learning session. The learning success is often measured by *knowledge gain*, that is, the change in the learner's knowledge state achieved through the search [14,16].

Past SAL research on knowledge gain has followed various directions, e.g., the time frames in which learning happens [22,29], the characteristics of a web resource indicating suitability for learning [11,18,23,26,27], and the *user*

N. Goharian et al. (Eds.): ECIR 2024, LNCS 14610, pp. 364–373, 2024.
https://doi.org/10.1007/978-3-031-56063-7_28

behavior. The research to *user behavior* is based on, e.g., the analysis of input queries [8], navigation logs [9], or other behavioral features [10,23,26,27]. One branch of user behavior research attempts to associate reading behavior with learning using eye-tracking data. Bhattacharya and Gwizdka [2] conducted a web search study to compare the reading behavior of more and less successful learners. For this, they relied on the line-based reading model from previous work [6,7]. However, this model has limitations in accurately capturing reading behavior, particularly as it ignores the association of fixations with actual words on web pages and the handling of line breaks.

In this paper, we suggest to extend their reading model in two regards: (1) to verify that the fixations actually apply to words on the web page and (2) to also cover line breaks and paragraph breaks. This extension of the reading model allows for the computation of new and more in-depth features such as the number of words read and reading speed, while known features such as the number of backward regressions can become more informative. Overall, the reading behavior can be represented more accurately and, perspectively, a more sophisticated analysis of the influence on learning success can be carried out. For the evaluation, we use the publicly available *SaL-Lightning dataset* [17], which consists of data from an explorative web search study. We aim to answer the following questions: (1) Can we reproduce the results for content pages in Bhattacharya and Gwizdka [2] on other study data? (2) Can we model a learner's prior knowledge by behavioral reading features? (3) Can we use behavioral reading features to determine learning success? Contrary to Bhattacharya and Gwizdka [2], our experimental results show that learners, who spent more time reading, processed more words, and exhibited faster reading with more backward regressions, achieved higher knowledge gain and a better web-based learning performance.

The remainder of the paper is structured as follows: In Sect. 2, we summarize related work that investigates indicators for successful web search. In Sect. 3, we describe how reading sequences are identified from individual fixations of the eye-tracking data. In Sect. 4, we analyze the mean differences between more and less successful learners. Finally, we summarize the results in Sect. 5 and discuss the limitations of the approach.

2 Related Work

Central research questions in SAL concern the description and detection of typical SAL-related user behavior and the design of optimized retrieval and ranking algorithms for learning [5,21,28]. For determining characteristics that can indicate learning success, previous work has examined both user behavior and visited resources (i.e., web pages, their appearance, and content). For example, Vakkari [24] conducted a survey that captured features indicating potential knowledge gain during the search process. In the context of resource-centric features, Otto et al. [18] examined the number and type of graphics on websites. In contrast, Gritz et al. [11] studied the influence of the textual complexity of the page. In this context, Pardi et al. [20] found evidence that text-based web pages seem to have a more substantial influence on a user's knowledge gain. Yu et al. [26,27] and Tang et al. [23] also used textual and HTML features in order to predict knowledge gain. But besides resource features, they additionally suggest behavioral features of the user. Similar to Eickhoff et al. [9], they also explore the entered queries, as well as the click and scroll behavior. Other approaches aim to predict the perceived webpage relevance [3,4] for the user, or even of single paragraphs [1]. Cole et al. [6,7] and Bhattacharya and Gwizdka [2] determined the influence of reading sequences on learning by analyzing eye-tracking data. Therefore, they used a line-based reading model to determine, i.e., temporally successive fixations that are associated with a reading process of the user. It is based on the assumption that reading sequences can be determined by successive fixations in a roughly horizontal trajectory. However, the reading model ignores the underlying text and treats sequences as single lines.

3 Detection of Reading Sequences

In this section, we present our method for the detection of reading sequences that rely on actual words and can deal with line breaks, as opposed to the model of Cole et al. [6,7] and Bhattacharya and Gwizdka [2]. This gives us the opportunity to enhance features, e.g., backward regressions that can reach back over several lines, and to develop deeper features such as the number of read words or the reading speed. Our method is based on the reading model of Cole et al. [6] as it is well-established in the SAL domain and has been widely used in other studies [2, 6,7]. The model relies only on the coordinates of fixations as input without considering the actual web page. We extend the approach by incorporating data from the rendered web pages provided by the *reading protocol* [13]. It receives a web page and resolution as input and then renders the web page. As an output, we get the position of the individual words as bounding boxes with a continuous index I_w that also crosses paragraph boundaries. This allows us to enhance the features describing the reading behavior, including backward regressions and reading sequences spanning multiple lines, and to compute deeper features such as the number of read words and reading speed.

(Para)foveal Region Definition: Crucial for the detection of reading sequences are the values for the foveal and parafoveal regions on the screen plane. The foveal region is responsible for clear central focus during reading, while the parafoveal region detects surrounding words. In the algorithm, they determine the maximum distance between two fixations that can potentially be considered as a reading sequence. Since the values for the foveal and parafoveal region radius r_{foveal} and $r_{parafoveal}$ are not contained in the used dataset [17], we calculate them in a preprocessing step. We use the following values from the literature [12]: $d_{foveal}{=}2^\circ$, $d_{parafoveal}{=}7^\circ$, and $d_{display}{=}65cm$, i.e., the diameter of the foveal or parafoveal region on the screen and the average distance from participant to display. The dataset description also gives a screen diagonal of 24" and 1920×1080 resolution. This yields the values $r_{foveal}{\approx}41$ and $r_{parafoveal}{\approx}185$ in pixels. These

values are crucial for the accuracy of reading sequence detection. Too small values could result in too few candidates for the grouping of fixations into reading sequences, whereas too large values could introduce noise and contradict the actual human perception.

Definitions: Let: F_t denote the fixation at time t; W represents the set of words in the text; L be a reading line; $d(F_t, w)$ be the Euclidean distance between the fixation F_t and word $w \in W$; $PR(F_t)$ be the parafoveal region for a fixation F_t. We define the parafoveal region to the right as $r_{parafoveal}$ and to the left, top, and bottom as r_{foveal}.

Candidate Selection: Eye-tracking data are noisy and may have errors due to measurement inaccuracies and aggregation of measurement points to fixations. Therefore, we do not necessarily assign the word with the smallest distance to a fixation, but determine a set of the most likely candidates. We consider only those words that fall within a certain radius r around the fixation point: $C(F_t) = \{w \mid d(F_t, w) \leq r\}$. We assume a maximum error of one complete foveal region in each direction, resulting in $r = 2 \cdot r_{foveal}$. We sort the candidate words in ascending order based on their distances to the fixation. If no candidates exist, we assume the fixation was not on text and cannot be part of a reading sequence.

Sequential Line Processing. We first check, independent of the words, whether successive fixations are approximately in the same line. For each fixation F_t, we check if the next fixation F_{t+1} falls within the parafoveal region $PR(F_t)$: (1) If F_{t+1} is within $PR(F_t)$, we add F_{t+1} to the current reading line L, or (2) if F_{t+1} is not within $PR(F_t)$, we conclude that the current reading line L has ended. For each reading line L, we apply the Viterbi algorithm to find the most likely words for the fixations. The cost function is defined as:

$$\mathcal{L} = C_R(F_t)^2 + (1 - C_I(F_{t+1}) - C_I(F_t))^2 \tag{1}$$

where $C_R(F_t)$ represents the rank of the candidate words $C(F_t)$ for the fixation F_t, and $C_I(F_{t+1})$ the index of the candidate words on the web page. The two summands are intended to represent a tradeoff between the spacing of (1) fixations and words and (2) words in the text.

Reading Sequence Definition: After the assignment of words to fixations, we only consider the indices to define reading sequences. We define that a fixation F_t belongs to a *reading sequence* RS, if: (1) $C_I(F_{t-1}) \leq C_I(F_t) \leq C_I(F_{t-1}) + K$, indicating that the next fixation is at most K words away, or (2) $\min(RS) \leq C_I(F_t) \leq \max(RS)$, indicating that the fixation lies within the previously read *reading sequence*. We refer to the latter case as *regression*. We decided to set the value for K=4 based on the assumption that the maximum distance between two fixations should be within the parafoveal region. As the words are of different lengths, we use the average word length in pixels, i.e., $K = \left\lfloor \frac{r_{parafoveal}}{w_{avg}} \right\rfloor = \left\lfloor \frac{185}{44} \right\rfloor = 4$, where w_{avg} is the average width of all words in the dataset. The selection of this value can have a large influence, as it determines how the data is grouped into reading sequences. In the future, the impact of alternative calculation approaches on the accuracy needs to be examined.

Feature Calculation: From the preprocessed data, we calculated features to encapsulate various aspects of the reading behavior. They include fixation behavior consisting of total and average fixation durations and their count. Furthermore, the reading behavior is examined by the maximum and mean reading durations per content page, and the average duration and count of reading sequences. To describe backward regressions, the features include their durations, counts, coordinates, and ratios concerning fixations and reading sequences. Additionally, we calculate reading sequence characteristics as the length in both pixels and word indices and the average duration until the first regression per reading sequence. Finally, to enable further insights into language processing, we calculate word and unique word counts, as well as rates of reading unique words and total words per second.

4 Experimental Results

Dataset: For the evaluation, we rely on the publicly available *SaL-Lightning dataset* [17] of an exploratory web search with N=106 participants. Since the study language was German, the reading direction was from left to right. The participants were instructed to gather knowledge about the formation of thunder and lightning. In addition to the tracking data, the authors also provided the original web page data. Since we primarily aim to investigate the influence of reading sequences on (text-based) content pages, we filter out fixations on search engine result pages and video platforms. Before (*Pre*) and after (*Post*) the web search, (1) a multiple choice questionnaire and (2) a written essay have been performed, further denoted as *MCQ* and *Essay*, respectively. For *MCQ*, the correct answers have been counted; for *Essay*, the authors counted the correctly identified concepts, allowing us to define the knowledge gain as *KG*=*Post*−*Pre*.

The key differences between the study in *SaL-Lightning dataset* [17] and the one described by Bhattacharya and Gwizdka [2] include: in the *SaL-Lightning dataset* the participants (1) had a time limit of 30 min ([2]: unrestricted), (2) had to learn about one search task about the formation of lightning and thunder ([2]: two search tasks from the medical domain), and (3) have been tested by a multiple choice questionnaire and a written essay ([2]: similarity between a free item response test and an expert set of words and phrases).

Metric: We divide the participants into two groups per setting: a *Low* group if they scored lower than the average score across participants or *High* if higher. The distributions for *MCQ* and *Essay* for *Pre*, *Post*, and *KG* are as follows: For *Pre*, 63 participants are in *Low* for *MCQ*, 57 in *Essay*, and 42 assigned *Low* for both, respectively 43, 49 and 42 for *High*; for *Post*, 54 participants are in *Low* for *MCQ*, 54 in *Essay*, and 30 assigned *Low* for both, respectively 52, 52 and 28 for *High*; finally, for *KG*, 66 participants are in *Low* for *MCQ*, 46 in *Essay*, and 32 assigned *Low* for both, respectively 40, 60 and 26 for *High*.

Since the overlaps between *MCQ* and *Essay* are not very strong, we report the results for both test forms. As the feature values are not normally distributed, we use the parameter-free Mann-Whitney U test to determine significance.

Table 1. Mean and p-value based on a Mann Whitney U test in multiple choice questionnaire (MCQ) and essay for knowledge gain and prior knowledge (after double line), respectively. Differences with $0.05 \leq p < 0.1$ are underlined, significant differences with $0.01 < p \leq 0.05$ **bold**, and very significant differences with $p < 0.01$ **both**.

	MCQ			Essay			
Feature	**Low**	**High**	**p**	**Low**	**High**	**p**	**Description**
sum_RFix_dur	186.9	177.0	.747	162.1	199.3	**.040**	*sum of reading fixation durations (in s) on content pages*
avg_RFix_dur	40.90	34.83	.722	33.00	42.91	.130	*sum of reading fixation durations (in s), averaged across content pages*
avg_n_RFix	84.58	79.44	.582	65.35	95.89	**.045**	*count of reading fixations, averaged across content pages*
n_CP_visited	6.59	6.42	.768	6.15	6.82	.170	*number of visited content pages*
avg_Fix_dur	471.1	432.9	**.015**	476.3	441.7	.059	*average duration of any fixations (in ms)*
dur_per_RSeq	1941	1774	.079	1969	1808	**.046**	*average duration of reading sequences (in ms)*
n_RSeq	99.05	101.4	.611	82.83	113.1	**.006**	*number of reading sequences*
avg_RFix_dur	489.8	447.3	**.024**	497.0	455.9	**.046**	*average duration of reading fixations (in ms)*
n_RFix	392.4	406.0	.560	326.9	451.7	**.009**	*number of reading fixations*
n_Reg	25.76	26.32	.351	20.48	30.18	**.021**	*number of backward regressions*
n_Reg_per_sec	0.13	0.15	**.026**	0.13	0.14	.099	*number of regressions per second*
n_unique_word	528.1	533.4	.553	454.7	588.0	**.014**	*total number of unique words read*
n_words	692.6	727.6	.510	587.3	796.7	**.010**	*total count of read words (including duplicates, e.g., by regressions)*
words_per_sec	3.82	4.12	**.023**	3.84	4.00	.151	*read words per second*
max_y_of_RFix	2889	3644	**.016**	2790	3667	.057	*maximum y-position of reading fixations*
avg_y_of_RFix	1160	1272	.073	1123	1301	**.005**	*average y-position of reading fixations*

Results: To answer the first research question (1), we analyze the variables regarding the total reading duration on content pages, the average reading duration per content page, and finally the number of reading fixations per content page (see first part of Table 1). We observe that learners from the *Low* group read for significantly shorter durations and had fewer fixations per content page (for *Essay*) compared to the *High* group. In contrast to Bhattacharya and Gwizdka [2], our results do not reveal that learners with low knowledge gain have read longer in total or on average, nor do they have more reading fixations. Possible reasons could be the time limit imposed in the source study or characteristics of the task domain. This highlights the need for future studies across multiple domains.

For the second research question (2), we examine the reading features in relation to the pretest results. Only few of them differ significantly when comparing *Low* and *High*. This suggests that reading behavior features are of limited potential for modeling prior knowledge. However, significant differences in average and maximum y-coordinate of reading fixations exist for both *MCQ* and *Essay* (see Table 1 after double line). As there is no significant difference in the number of visited content pages, this suggests that readers with more prior knowledge engage more deeply with web pages than those with less prior knowledge. According to previous research, prior knowledge on a topic may lead to different user behavior [15,25]. A possible hypothesis could be that learners with higher prior knowledge read web pages more extensively to deepen their knowledge. However, this behavior does not seem to imply a significant difference in *KG*.

To answer the last research question (3), we analyze a subset of eye-tracking features regarding *KG* (see second part of Table 1). Our results indicate that more successful learners read longer in total on content pages (in terms of the number of fixations and reading sequences). Furthermore, learners who processed more words and unique words achieved noticeably better results in the *Essay* scores. Furthermore, a difference in the time learners read until regressions occur can be seen for both scores. Combined with the higher reading speed for successful learners (for *MCQ*), this may indicate that it can be more efficient for learning to skim web pages for new information and regress as required, as opposed to slow and thorough reading. However, these observations are based on a single search task and need further validation in different search domains.

5 Conclusions

Summary: In this paper, we have presented the extension of an existing reading model [2,6,7]. While prior work solely relied on fixations, we incorporated additional information on the positions of text on web pages and sequences across line breaks. To determine the impact on learning outcomes, we calculated eye-tracking features for textual content pages, such as the total number of read words and the read words per second. We evaluated the impact of the features on pretest, posttest, and knowledge gain scores from multiple choice and essay assessments. Our findings revealed that learners with higher knowledge gain had spent more time reading and had an increased number of fixations, contradicting the opposite observation of Bhattacharya and Gwizdka [2]. We also found that learners who had higher prior knowledge, read more deeply on content pages, without necessarily experiencing higher knowledge gains. Finally, we observed that learners with high knowledge gains read more intensively in terms of the number of words read and unique words read. In addition, reading speed and time to backward regressions were also increased for more successful learners.

Limitations: Currently, our model only works for languages with a left-to-right reading direction, but an adaptation is possible. Furthermore, while we calculated the influence of the computed features on knowledge gain, we did not evaluate the reading model itself, i.e., how accurately reading sequences are recognized by our model. System parameters, such as the minimum number of words for reading sequences, were chosen based on commonsense reasoning. In the future, empirical evidence should be provided to support these choices. Finally, the scope of this study was limited to a single task and to the textual content of web pages. As prior work underlined, learning processes are often supported by other visual formats, such as images and videos [18,19]. Their impact on gaze direction and knowledge gain provide interesting directions for future work.

Acknowledgments. Part of this work was financially supported by the Leibniz Association, Germany (Leibniz Competition 2023, funding line "Collaborative Excellence", project VideoSRS [K441/2022]).

References

1. Barz, M., Bhatti, O.S., Sonntag, D.: Implicit estimation of paragraph relevance from eye movements. Front. Comp. Sci. **3**, 808507 (2021). https://doi.org/10.3389/fcomp.2021.808507
2. Bhattacharya, N., Gwizdka, J.: Measuring learning during search: differences in interactions, eye-gaze, and semantic similarity to expert knowledge. In: Conference on Human Information Interaction and Retrieval, CHIIR 2019, Glasgow, Scotland, UK, March 10–14, 2019, pp. 63–71. ACM (2019). https://doi.org/10.1145/3295750.3298926
3. Bhattacharya, N., Rakshit, S., Gwizdka, J.: Towards real-time webpage relevance prediction using convex hull based eye-tracking features. In: ACM Symposium on Eye Tracking Research and Applications. ETRA 2020 Adjunct, Association for Computing Machinery, New York, USA (2020). https://doi.org/10.1145/3379157.3391302
4. Bhattacharya, N., Rakshit, S., Gwizdka, J., Kogut, P.: Relevance prediction from eye-movements using semi-interpretable convolutional neural networks. In: Conference on Human Information Interaction and Retrieval, CHIIR 2020, Vancouver, BC, Canada, March 14–18, 2020, pp. 223–233. ACM (2020). https://doi.org/10.1145/3343413.3377960
5. Câmara, A., Zein, D.E., da Costa Pereira, C.: RULK: a framework for representing user knowledge in search-as-learning. In: International Conference on Design of Experimental Search and Information REtrieval Systems, DESIRES 2022, San Jose, CA, USA, August 30–31, 2022, pp. 1–13. CEUR-WS.org (2022). https://ceur-ws.org/Vol-3480/paper-01.pdf
6. Cole, M.J., Gwizdka, J., Liu, C., Belkin, N.J.: Dynamic assessment of information acquisition effort during interactive search. In: Bridging the Gulf: Communication and Information in Society, Technology, and Work - Proceedings of the 74th ASIS&T Annual Meeting, ASIST 2011, New Orleans, USA, pp. 1–10. Wiley (2011). https://doi.org/10.1002/meet.2011.14504801149

7. Cole, M.J., Gwizdka, J., Liu, C., Belkin, N.J., Zhang, X.: Inferring user knowledge level from eye movement patterns. Inf. Process. Manage. **49**(5), 1075–1091 (2013). https://doi.org/10.1016/j.ipm.2012.08.004
8. Collins-Thompson, K., Rieh, S.Y., Haynes, C.C., Syed, R.: Assessing learning outcomes in web search: a comparison of tasks and query strategies. In: Conference on Human Information Interaction and Retrieval, CHIIR 2016, Carrboro, North Carolina, USA, March 13–17, 2016, pp. 163–172. ACM (2016). https://doi.org/10.1145/2854946.2854972
9. Eickhoff, C., Teevan, J., White, R., Dumais, S.T.: Lessons from the journey: a query log analysis of within-session learning. In: International Conference on Web Search and Data Mining, WSDM 2014, New York, NY, USA, February 24–28, 2014, pp. 223–232. ACM (2014). https://doi.org/10.1145/2556195.2556217
10. Gadiraju, U., Yu, R., Dietze, S., Holtz, P.: Analyzing knowledge gain of users in informational search sessions on the web. In: Conference on Human Information Interaction and Retrieval, CHIIR 2018, New Brunswick, New Jersey, USA, March 11–15, 2018, pp. 2–11. ACM (2018). https://doi.org/10.1145/3176349.3176381
11. Gritz, W., Hoppe, A., Ewerth, R.: On the impact of features and classifiers for measuring knowledge gain during web search - a case study. In: Workshops co-located with the International Conference on Information and Knowledge Management, CIKM 2021, Gold Coast, Australia, November 1–5, 2021. CEUR-WS.org (2021), http://ceur-ws.org/Vol-3052/paper6.pdf
12. Gwizdka, J., Zhang, Y., Dillon, A.: Using the eye-tracking method to study consumer online health information search behaviour. Aslib J. Inf. Manage. **71**(6), 739–754 (2019). https://doi.org/10.1108/AJIM-02-2019-0050
13. Hienert, D., Kern, D., Mitsui, M., Shah, C., Belkin, N.J.: Reading protocol: understanding what has been read in interactive information retrieval tasks. In: Conference on Human Information Interaction and Retrieval, CHIIR 2019, Glasgow, Scotland, UK, March 10–14, 2019, pp. 73–81. ACM (2019). https://doi.org/10.1145/3295750.3298921
14. Hoppe, A., Holtz, P., Kammerer, Y., Yu, R., Dietze, S., Ewerth, R.: Current challenges for studying search as learning processes. In: 7th Workshop on Learning and Education with Web Data (LILE2018), in Conjunction with ACM Web Science (2018)
15. von Hoyer, J., et al.: The search as learning spaceship: toward a comprehensive model of psychological and technological facets of search as learning. Front. Psychol. **13**, 827748 (2022). https://doi.org/10.3389/fpsyg.2022.827748, https://www.frontiersin.org/articles/10.3389/fpsyg.2022.827748
16. Machado, M., Gimenez, P., Siqueira, S.: Raising the dimensions and variables for searching as a learning process: a systematic mapping of the literature. In: Anais do XXXI Simpósio Brasileiro de Informática na Educação, pp. 1393–1402. SBC, Porto Alegre, RS, Brasil (2020). https://sol.sbc.org.br/index.php/sbie/article/view/12895
17. Otto, C., et al.: Sal-lightning dataset: search and eye gaze behavior, resource interactions and knowledge gain during web search. In: Conference on Human Information Interaction and Retrieval, CHIIR 2022, Regensburg, Germany, March 14–18, 2022, pp. 347–352. ACM (2022). https://doi.org/10.1145/3498366.3505835
18. Otto, C., et al.: Predicting knowledge gain during web search based on multimedia resource consumption. In: Roll, I., McNamara, D., Sosnovsky, S., Luckin, R., Dimitrova, V. (eds.) AIED 2021. LNCS (LNAI), vol. 12748, pp. 318–330. Springer, Cham (2021). https://doi.org/10.1007/978-3-030-78292-4_26

19. Pardi, G., Gottschling, S., Gerjets, P., Kammerer, Y.: The moderating effect of knowledge type on search result modality preferences in web search scenarios. Comput. Educ. Open. **4**, 100126 (2023). https://www.sciencedirect.com/science/article/pii/S2666557323000058
20. Pardi, G., von Hoyer, J., Holtz, P., Kammerer, Y.: The role of cognitive abilities and time spent on texts and videos in a multimodal searching as learning task. In: Conference on Human Information Interaction and Retrieval, CHIIR 2020, Vancouver, BC, Canada, March 14–18, 2020, pp. 378–382. ACM (2020). https://doi.org/10.1145/3343413.3378001
21. Rokicki, M., Yu, R., Hienert, D.: Learning to rank for knowledge gain. In: Joint Proceedings of the 10th International Workshop on News Recommendation and Analytics (INRA 2022) and the 3rd International Workshop on Investigating Learning During Web Search (IWILDS 2022) co-located with 45th International ACM SIGIR Conference on Research and Development in Information Retrieval (SIGIR 2022), Madrid, Spain, July 15, 2022, pp. 60–68. CEUR-WS.org (2022). https://ceur-ws.org/Vol-3411/IWILDS-paper2.pdf
22. Roy, N., Moraes, F., Hauff, C.: Exploring users' learning gains within search sessions. In: Conference on Human Information Interaction and Retrieval, CHIIR 2020, Vancouver, BC, Canada, March 14–18, 2020, pp. 432–436. ACM (2020). https://doi.org/10.1145/3343413.3378012
23. Tang, R., Yu, R., Rokicki, M., Ewerth, R., Dietze, S.: Domain-specific modeling of user knowledge in informational search sessions. In: Workshops co-located with the International Conference on Information and Knowledge Management, CIKM 2021, Gold Coast, Australia, November 1–5, 2021. CEUR-WS.org (2021). https://ceur-ws.org/Vol-3052/paper8.pdf
24. Vakkari, P.: Searching as learning: a systematization based on literature. J. Inf. Sci. **42**(1), 7–18 (2016). https://doi.org/10.1177/0165551515615833
25. White, R.W., Dumais, S.T., Teevan, J.: Characterizing the influence of domain expertise on web search behavior. In: Baeza-Yates, R., Boldi, P., Ribeiro-Neto, B.A., Cambazoglu, B.B. (eds.) Proceedings of the Second International Conference on Web Search and Web Data Mining, WSDM 2009, Barcelona, Spain, February 9–11, 2009, pp. 132–141. ACM (2009). https://doi.org/10.1145/1498759.1498819
26. Yu, R., Gadiraju, U., Holtz, P., Rokicki, M., Kemkes, P., Dietze, S.: Predicting user knowledge gain in informational search sessions. In: International Conference on Research and Development in Information Retrieval, SIGIR 2018, Ann Arbor, Michigan, USA, July 8–12, 2018, pp. 75–84. ACM (2018). https://doi.org/10.1145/3209978.3210064
27. Yu, R., Tang, R., Rokicki, M., Gadiraju, U., Dietze, S.: Topic-independent modeling of user knowledge in informational search sessions. Inf. Retr. J. **24**(3), 240–268 (2021). https://doi.org/10.1007/s10791-021-09391-7
28. Zein, D.E., Câmara, A., da Costa Pereira, C., Tettamanzi, A.: RULKNE: representing user knowledge state in search-as-learning with named entities. In: Conference on Human Information Interaction and Retrieval, CHIIR 2023, Austin, TX, USA, March 19–23, 2023, pp. 388–393. ACM (2023). https://doi.org/10.1145/3576840.3578330
29. Zein, D.E., da Costa Pereira, C.: The evolution of user knowledge during search-as-learning sessions: a benchmark and baseline. In: Conference on Human Information Interaction and Retrieval, CHIIR 2023, Austin, TX, USA, March 19–23, 2023, pp. 454–458. ACM (2023). https://doi.org/10.1145/3576840.3578273

Bias Detection and Mitigation in Textual Data: A Study on Fake News and Hate Speech Detection

Apostolos Kasampalis, Despoina Chatzakou(✉), Theodora Tsikrika, Stefanos Vrochidis, and Ioannis Kompatsiaris

Information Technologies Institute, Centre for Research and Technology Hellas, Thessaloniki, Greece
{apkas,dchatzakou,theodora.tsikrika,stefanos,ikom}@iti.gr

Abstract. Addressing bias in NLP-based solutions is crucial to promoting fairness, avoiding discrimination, building trust, upholding ethical standards, and ultimately improving their performance and reliability. On the topic of bias detection and mitigation in textual data, this work examines the effect of different bias detection models along with standard debiasing methods on the effectiveness of fake news and hate speech detection tasks. Extensive discussion of the results draws useful conclusions, highlighting the inherent difficulties in effectively managing bias.

Keywords: Bias · NLP · Fake news detection · Hate speech detection

1 Introduction

Despite the undeniable benefits of Natural Language Processing (NLP) tools, the possible presence of bias (e.g., gender, race, cultural, and ideological bias) in such tools is a major issue with potentially negative impact on society, as it can lead to discrimination against certain social groups. Among others, such biases often emerge due to the inherent biases in the data that are used for training and developing the respective models. Therefore, bias detection in training data and bias mitigation in the derived NLP models are crucial to ensure their fairness, equity, and reliability.

Focusing on textual data, *bias detection* methods range from rule-based heuristics [35] to neural network-based models that identify subtle biases. For instance, Leavy [11] delves into "gender bias" observed in the English language, identifying instances of bias in naming, ordering, descriptions, metaphors, and the presence of gender-specific terms, shedding light on the deeply embedded nature of gender bias in language. Moreover, for the detection of "media bias", commonly referring to the unfair favoritism and reporting of certain ideas or viewpoints, a BERT-based text classifier has been developed in addition to a dataset for the same purpose [25]. From a more theoretical perspective, Doughman et al. [6] have proposed a taxonomy of gender bias in textual data, focusing on its origins and societal implications. In particular, the proposed taxonomy

N. Goharian et al. (Eds.): ECIR 2024, LNCS 14610, pp. 374–383, 2024.
https://doi.org/10.1007/978-3-031-56063-7_29

can act as a guide for the technical and research communities, providing insights into various forms of gender bias through a comprehensive classification system.

Regarding *bias mitigation*, Sun et al. [28] provide a comprehensive review on gender bias mitigation in NLP, using e.g., gender swapping, whereby each word defined as male is swapped with its female equivalent and vice versa [12], while also discussing the trade-offs and challenges associated with debiasing methods. In addition, data augmentation based on gender swapping has been proposed to reduce gender bias, focusing in particular on coreference resolution tasks, with promising results [35]. Finally, concerning the neural relation extraction (NRE) and its vulnerability to gender bias, Gaut et al. [7] employ Piecewise Convolutional Neural Networks (PCNN) with selective attention and debiasing techniques, such as data augmentation and word embedding adjustments, for bias mitigation. Based on the analysis conducted, they concluded that existing bias mitigation approaches have a negative effect on NRE.

In this context, this work performs a comprehensive evaluation and analysis of commonly considered solutions for detecting as well as mitigating bias in textual data, with particular interest in media and gender bias. The focus is on two case studies of NLP models for *fake news detection* and *hate speech detection*, given their significance in our society [2], as also illustrated by the various NLP models developed thus far for tackling these tasks, e.g., [31,32].

In particular, this work examines four classification models for bias detection and two simple yet common methods for bias mitigation (gender swapping and data augmentation) on two datasets for fake news detection and one dataset for hate speech detection. The results indicate that bias detection models perform well when trained on data from the same source as the test data (e.g. train and test on Twitter data), but their performance drops significantly otherwise, even when built to detect the same type of bias (e.g. gender bias). The results also show that there is some bias in the employed data and its mitigation leads to better performance for both the fake news and hate speech detection models, with data augmentation appearing to perform better compared to gender swapping.

2 Methodology

This section presents: (i) the models developed to detect bias, with an emphasis on media and gender bias, (ii) the employed bias mitigation methods, and (iii) the case studies considered to evaluate the bias detection and mitigation methods.

2.1 Bias Detection

The bias detection models were developed based on the following datasets:

- **MBIC** [26]: pertains to media bias; it contains 2,036 biased and 1,066 unbiased samples from online sources (e.g. Reuters, Fox News, and HuffPost).
- **MBIB** [34]: constitutes a comprehensive benchmark dataset that groups different types of media bias, such as cognitive, political, and gender bias. Here, we focused on gender bias that can negatively affect, for instance, perceptions of professions, role models, and voting decisions. In particular, the following three subsets of MBIB data were considered:

- *MBIB_workplace*: focuses on identifying gender bias in workplace-related content; it contains 624 biased and 513 unbiased samples.
- *MBIB_reddit*: addresses gender bias on Reddit; it consists of 2,033 biased and 943 unbiased samples.
- *MBIB_twitter*: consists of Twitter data that are potentially sexist; it contains 1,809 biased and 11,822 unbiased samples.

All four datasets (the MBIC dataset and the three MBIB subsets) were divided into a train set (90%) (with 10% kept for validation) and a test set (10%).

Preprocessing. All the aforementioned datasets were subjected to preprocessing to remove non-informative pieces of text, such as stopwords and special characters. In addition, tokenization and lowercasing were carried out.

Deep Neural Network-Based Architectures. Five neural-based architectures were examined to develop effective classification models for bias detection based on the dataset at hand. The first layer of the models' architecture is the embedding layer; its purpose is to map each word in a sequence to a layer of higher dimension. In the case of the first four architectures (listed below), we opted for GloVe [17] pretrained embeddings of 100 dimensions, while for the BERT-based architecture, the DistilBERT (uncased) pretrained model was used [5].

These architectures are (in all cases *sigmoid* is used as activation function):

1. **Bidirectional Long Short-Term Memory [biLSTM]**: one bi-LSTM layer of 64, spatial dropout of $p = 0.4$ before, and a dropout of $p = 0.2$ after the LSTM layer to reduce/avoid overfitting.
2. **Convolutional Neural Network [CNN]**: three 1D convolutional (Conv1D) layers of 128 filters and kernel size of 3, 4, 5 respectively, one spatial dropout layer with $p = 0.2$ and one 1D average pool layer before and after each of the first two Conv1D, respectively, and one 1D global average pooling.
3. **Combined biLSTM and CNN [CNN_biLSTM]**: one Conv1D layer of 16 filters and kernel size of 2, one 1D max pooling layer of pool size of 4, a spatial dropout layer of $p = 0.4$, and lastly a biLSTM layer of 8 units.
4. **Gated Recurrent Unit [GRU]**: one GRU layer of 8 units, spatial dropout of $p = 0.5$ before and a dropout layer of $p = 0.5$ after the GRU.
5. **Bidirectional Encoder Representations from Transformers [BERT]**: the DistilBERT is followed by a dense layer of 128 units.

2.2 Bias Mitigation

Overall, the following two methods were considered for bias mitigation:

Gender Swapping. This method constitutes a rule-based approach where gendered words in a sentence are swapped with the opposite gender. Inspired by [13,16,19,24,33], we applied two rules: (i) **[Rule#1]** *Gendered Words*: The swapping was performed based on compiled lists of gendered words [1,4,35], such as he-she, male-female, etc.; (ii) **[Rule#2]** *Gender-Neutral Words* [6,12,33]: Similarly to above, a compiled list of gender-neutral words was used [1,4], where an example of swapping would be, for instance, "fireman" to "firefighter". We then applied these rules in two different ways: (a) **[M2F_swap]** male-to-female swapping, and (b) **[F2M_swap]** female-to-male swapping.

Data Augmentation [DA_swap]. This method essentially uses gender swapping to balance the dataset with regards to gender. Particularly, the original sentences are first gender swapped and then added to the dataset, effectively creating a larger dataset with equal numbers of male- and female-gendered sentences. To avoid gender-swapped sentences becoming meaningless, but to also make the method more robust, names were anonymized; names were detected based on examples from [10,14,30] and the Gender-guesser [8] library.

2.3 Case Studies

For an in-depth evaluation of the bias detection and mitigation methods considered in this work, we focused on two general tasks: *fake news detection* and *hate speech detection*. In both cases, dedicated classification models were developed for fake news and hate speech detection by following a similar process to the one described in Sect. 2.1; the same preprocessing steps were followed and the same five neural-based architectures were examined to arrive at the one with the best performance based on the data available. In particular:

Fake News Detection. Two popular datasets were considered: (i) WELFake [23] consisting of 72,134 news articles with around $35k$ real and $37k$ fake news, and (ii) ISOT [3] consisting of around $21k$ real and $25k$ fake news articles. For each dataset, the best performing neural-based architecture among the five aforementioned ones corresponds to the fake news detection model based on that dataset.

Hate Speech Detection. A set of data collected from the 4chan website and in particular from the Politically Incorrect Board [20] was considered. Each entry in the dataset is characterized by its toxicity level on a [0, 1] scale; we considered as hate speech those with a toxicity level above 0.4 and randomly selected 1,592 toxic and 6,069 non-toxic samples. Similarly to before, the best performing neural-based architecture among the five aforementioned ones corresponds to the hate speech detection model based on this dataset.

Overall, although more sophisticated models for both fake news and hate speech detection have been recently developed (e.g. [18,21,22]), particularly due to the emergence of a wide variety of large language models, we proceeded with the well-established models presented above, as the focus of this work is on the evaluation of common bias detection and mitigation methods.

3 Experiments

3.1 Experimental Setup

For our experiments, we used Keras [9] with TensorFlow [29] on a server equipped with one NVIDIA GeForce RTX 3080 of 10 GB memory. For training, we used binary cross-entropy as loss function and AdamW [15] as optimizer. A maximum of 200 epochs was allowed, while also a validation set was used for early stopping; training was interrupted if the validation loss did not drop in 5 consecutive epochs. For evaluation, standard metrics are considered: accuracy (Acc), precision (Prec), recall (Rec), and weighted area under the ROC curve (AUC). We repeated each experiment 5 times and report the average (AVG) values. Finally, for fake news and hate speech detection, the best results are highlighted in bold.

3.2 Experimental Results

Table 1 presents the results obtained with the best performing model in the test set of each dataset described in Sect. 2.1; these are BERT for MBIC, biLSTM for MBIB_workplace, CNN for MBIB_reddit, and biLSTM for MBIB_twitter. Overall, we observe a quite satisfactory performance in all cases, with the best results achieved for MBIC that focuses on the more general type of media bias.

We then examine how well a model built to detect the more general media bias can detect a more specific type of bias, namely gender bias; in particular we applied the BERT model built on MBIC on the three MBIB datasets. The results in Table 2, reporting also the standard deviation (STD), are particularly poor. An exception is the recall value of the true class (Rec (T)), with scores above 82% in all cases, indicating that the general model can quite successfully identify gender-bias; the precision though of both the true and false classes is in most cases quite low, thus significantly limiting the discriminative ability of the model. Similar results were obtained when a gender-biased model trained on data from platform A was tested on data from platform B (e.g., trained on Reddit data and tested on Twitter data); the results are omitted due to

Table 1. Performance of the Bias detection models (average).

	MBIC	MBIB_workplace	MBIB_reddit	MBIB_twitter
Prec	91.26	83.56	82.23	85.41
Rec	89.06	83.41	79.28	88.61
Acc	91.00	83.33	83.56	93.65
AUC	89.06	83.41	79.28	88.61

Table 2. Performance of the media bias detection MBIC model on the MBIB datasets.

	MBIB_workplace		MBIB_reddit		MBIB_twitter	
	AVG	STD	AVG	STD	AVG	STD
Prec (T)	55.45	1.00	67.54	0.59	13.04	0.36
Rec (T)	82.50	12.09	90.92	7.57	95.72	6.44
Prec (F)	49.04	6.25	19.57	5.68	83.92	7.48
Rec (F)	19.30	12.30	5.90	5.91	2.42	3.60
Accuracy	53.99	1.95	63.98	3.38	16.87	5.04
AUC	50.90	1.63	48.41	1.14	49.07	1.42

space limits. This observation highlights the difficulty of this task, i.e., it is not always possible to create models to detect similar types or the exact same type of bias on data collected from different sources (possibly with different inherent characteristics, such as text length, formal vs. informal language, use of slang).

Next, we evaluate the effect of the bias detection and mitigation process on the *fake news detection* and *hate speech detection* tasks. In particular, we follow these steps: (1) Apply classification models to detect fake news and hate speech in textual data (see Sect. 2.3): among the neural-based architectures presented in Sect. 2.1, the BERT model resulted in the best AUC for both tasks and was thus used for the experiments presented next; (2) Apply the four best performing bias detection models (Table 1) on the datasets used by the fake news and hate speech classification models; (3) For texts identified as biased, apply the two debiasing methods (Sect. 2.2); and (4) Re-build the fake news and hate speech classification models on the updated sets of data: when gender swapping is used as the debiasing method, the updated dataset consists of the original texts detected as non-biased, along with the debiased texts, while for data augmentation the dataset includes both original and debiased texts.

For *fake news detection*, Table 3 presents the results for both the WELfake and ISOT datasets. Due to space limitations, we only present results for two bias detection models: BERT trained on MBIC and CNN trained on MBIB_reddit; a similar behavior is observed for the MBIB_workplace and MBIB_twitter models. The results indicate that the overall performance improves, even slightly, in most cases when debiasing methods are applied, with the best results observed when data augmentation is used as the debiasing method. Regarding gender swapping, we observe that M2F swapping yields better performance compared to F2M in the case of the WELfake dataset, indicating that this particular dataset is probably more female-biased. The opposite behavior is observed in the case of the ISOT dataset, indicating that its data are probably more biased towards males. Table 4 presents how the performance of the *hate speech detection model*

Table 3. MBIC & MBIB_reddit on the fake news detection task.

	Original	M2F_swap	F2M_swap	DA_swap	Original	M2F_swap	F2M_swap	DA_swap
	BERT bias detection model on MBIC							
	WELfake				ISOT			
Prec	96.23	96.38	96.10	**97.20**	98.72	97.64	98.73	**99.25**
Rec	94.27	94.57	94.39	**95.65**	98.77	98.30	98.89	**99.55**
Acc	97.85	97.98	97.72	**99.10**	98.69	98.95	98.82	**99.09**
AUC	98.50	98.79	98.37	**99.44**	99.12	98.65	99.31	**99.75**
	CNN bias detection model on MBIB_reddit							
	WELfake				ISOT			
Prec	96.23	96.23	96.19	**97.06**	98.72	97.64	99.04	**99.36**
Rec	94.27	94.77	94.51	**95.36**	98.77	98.30	98.43	**99.47**
Acc	97.85	98.18	97.96	**98.65**	98.69	98.95	98.82	**99.10**
AUC	98.50	98.99	98.21	**99.04**	99.12	98.65	98.78	**99.13**

Table 4. MBIC & MBIB_reddit on the hate speech detection task.

	Original	M2F_swap	F2M_swap	DA_swap
	BERT bias detection model on MBIC			
Prec	88.41	88.22	88.41	**89.10**
Rec	93.65	93.57	93.65	**94.58**
Acc	93.09	92.96	93.09	**93.68**
AUC	93.66	93.57	93.66	**94.24**
	CNN bias detection model on MBIB_reddit			
Prec	88.41	88.56	88.56	**89.38**
Rec	93.65	93.95	93.95	**94.84**
Acc	93.09	93.22	93.22	**94.07**
AUC	93.66	93.95	93.95	**94.45**

changes after data debiasing. Overall, results similar to the fake news detection task are observed, with data augmentation achieving improved performance.

Discussion. Overall, classification models are widely used to detect bias in textual data as they constitute an easy-to-understand approach, requiring though a sufficient amount of properly annotated data that reflect the bias we wish to detect. Thus, although they appear to be a promising solution, they can only help towards effective bias detection, if the annotated datasets used are sufficiently large, trustworthy, and of good quality in terms of syntax/content. But even in such a case, their generalization is not always possible as highlighted above.

The application of bias detection models along with the data debiasing methods leads overall to improved results for both fake news and hate speech detection tasks. According to Park et al. [16] and Zhao et al. [35], *gender-swapping* has been found to be a quite effective solution, leading to improved classification results after the models' retraining. The better performance of the *data augmentation* in our case could be attributed to different factors. First, as the dataset is augmented by including data presenting the same content as a function of both genders, the retrained model is forced to focus on more gender-neutral features and avoid making predictions based on gender-related information. Moreover, the mere fact of increasing the size of the dataset has been shown to lead to improved performance [27], particularly for neural network-based solutions. Finally, we should mention that even if bias may be addressed at the data level, it may remain to some extent in the classification models, due to the fact that pretrained embeddings were used, where bias is inherent. Therefore, more holistic solutions are required to address bias more effectively in NLP-based models.

4 Conclusions

This work performed a comprehensive evaluation of (media/gender) bias detection and mitigation in textual data on NLP models for fake news and hate speech detection. Overall, the proposed bias detection and mitigation process appears to have a positive effect on the effectiveness of fake news and hate speech detection models, although the generalization of specific bias detection models to other types of bias is rather poor. These findings could act as a stepping stone for future bias management studies towards improving the understanding, detection, and removal of bias. Next steps could involve using larger datasets, further optimizing the models, and debiasing the word embeddings themselves.

Acknowledgements. This project has received funding from the European Union's H2020 research and innovation programme as part of the STARLIGHT (GA No 101021797) project.

References

1. Biased words. https://github.com/gregology/biased-words. Accessed 2023
2. Blanco-Herrero, D., Sánchez-Holgado, P.: Fake news and hate speech: who is to blame? Study of the perceptions of Spanish citizens about the actors responsible for the production and spread of fake news and hate speech. In: Ninth International Conference on Technological Ecosystems for Enhancing Multiculturality (TEEM 2021), pp. 448–451 (2021)
3. Clement Bisaillon. https://www.kaggle.com/datasets/clmentbisaillon/fake-and-real-news-dataset Accessed 2023
4. Debiaswe: try to make word embeddings less sexist. https://github.com/tolga-b/debiaswe/tree/master/data. Accessed 2023
5. DistilBERT base model (uncased): https://huggingface.co/distilbert-base-uncased. Accessed 2023
6. Doughman, J., Khreich, W., El Gharib, M., Wiss, M., Berjawi, Z.: Gender bias in text: origin, taxonomy, and implications. In: Proceedings of the 3rd Workshop on Gender Bias in Natural Language Processing, pp. 34–44 (2021)
7. Gaut, A., et al.: Towards understanding gender bias in relation extraction. In: Proceedings of the 58th Annual Meeting of the Association for Computational Linguistics, pp. 2943–2953 (2020)
8. Gender Guesser. https://github.com/lead-ratings/gender-guesser. Accessed 2023
9. Keras (2020). https://keras.io/
10. Lample, G., Ballesteros, M., Subramanian, S., Kawakami, K., Dyer, C.: Neural architectures for named entity recognition. In: Proceedings of the 2016 Conference of the North American Chapter of the Association for Computational Linguistics: Human Language Technologies, pp. 260–270 (2016)
11. Leavy, S.: Gender bias in artificial intelligence: the need for diversity and gender theory in machine learning. In: Proceedings of the 1st International Workshop on Gender Equality in Software Engineering, pp. 14–16, May 2018
12. Lu, K., Mardziel, P., Wu, F., Amancharla, P., Datta, A.: Gender bias in neural natural language processing. In: Nigam, V., et al. (eds.) Logic, Language, and Security. LNCS, vol. 12300, pp. 189–202. Springer, Cham (2020). https://doi.org/10.1007/978-3-030-62077-6_14
13. Manzini, T., Lim, Y.C., Tsvetkov, Y., Black, A.W.: Black is to criminal as caucasian is to police: detecting and removing multiclass bias in word embeddings. In: Proceedings of NAACL-HLT, pp. 615–621 (2019)

14. NER Tagger. https://github.com/glample/tagger. Accessed 2023
15. OverLordGoldDragon: Keras adamW. GitHub. Note (2019). https://github.com/OverLordGoldDragon/keras-adamw/
16. Park, J.H., Shin, J., Fung, P.: Reducing gender bias in abusive language detection. In: Proceedings of the 2018 Conference on Empirical Methods in Natural Language Processing, pp. 2799–2804 (2018)
17. Pennington, J., Socher, R., Manning, C.D.: Glove: global vectors for word representation. In: Proceedings of the 2014 Conference on Empirical Methods in Natural Language Processing (EMNLP), pp. 1532–1543 (2014)
18. Plaza-Del-Arco, F.M., Molina-González, M.D., Ureña-López, L.A., Martín-Valdivia, M.T.: A multi-task learning approach to hate speech detection leveraging sentiment analysis. IEEE Access **9**, 112478–112489 (2021)
19. Prates, M.O., Avelar, P.H., Lamb, L.C.: Assessing gender bias in machine translation: a case study with Google Translate. Neural Comput. Appl. **32**, 6363–6381 (2020)
20. Raiders of the Lost Kek. https://zenodo.org/records/3606810#.YH2TYCXivIU. Accessed 2023
21. Raza, S., Ding, C.: Fake news detection based on news content and social contexts: a transformer-based approach. Int. J. Data Sci. Anal. **13**(4), 335–362 (2022)
22. Samadi, M., Mousavian, M., Momtazi, S.: Deep contextualized text representation and learning for fake news detection. Inf. Process. Manag. **58**(6), 102723 (2021)
23. Saurabh Shahane. https://www.kaggle.com/datasets/saurabhshahane/fake-news-classification. Accessed 2023
24. Seaborn, K., Chandra, S., Fabre, T.: Transcending the "male code": implicit masculine biases in NLP contexts. In: Proceedings of the 2023 CHI Conference on Human Factors in Computing Systems, pp. 1–19 (2023)
25. Spinde, T., Plank, M., Krieger, J.D., Ruas, T., Gipp, B., Aizawa, A.: Neural media bias detection using distant supervision with babe-bias annotations by experts. In: Findings of the Association for Computational Linguistics: EMNLP 2021, pp. 1166–1177 (2021)
26. Spinde, T., Rudnitckaia, L., Sinha, K., Hamborg, F., Gipp, B., Donnay, K.: MBIC - a media bias annotation dataset including annotator characteristics. arXiv preprint arXiv:2105.11910 (2021)
27. Stylianou, N., Chatzakou, D., Tsikrika, T., Vrochidis, S., Kompatsiaris, I.: Domain-aligned data augmentation for low-resource and imbalanced text classification. In: Kamps, J., et al. (eds.) European Conference on Information Retrieval, vol. 13981, pp. 172–187. Springer, Cham (2023). https://doi.org/10.1007/978-3-031-28238-6_12
28. Sun, T., et al.: Mitigating gender bias in natural language processing: literature review. In: Proceedings of the 57th Annual Meeting of the Association for Computational Linguistics, pp. 1630–1640, July 2019
29. TensorFlow. https://www.tensorflow.org/. Accessed 2023
30. Transition-based NER system. https://github.com/clab/stack-lstm-ner. Accessed 2023
31. del Valle-Cano, G., Quijano-Sánchez, L., Liberatore, F., Gómez, J.: SocialHaterBERT: a dichotomous approach for automatically detecting hate speech on twitter through textual analysis and user profiles. Expert Syst. Appl. **216**, 119446 (2023)
32. Verma, P.K., Agrawal, P., Amorim, I., Prodan, R.: WELFake: word embedding over linguistic features for fake news detection. IEEE Trans. Comput. Soc. Syst. **8**(4), 881–893 (2021)

33. Vig, J., et al.: Causal mediation analysis for interpreting neural NLP: the case of gender bias (2020). CoRR arXiv (2004)
34. Wessel, M., Horych, T., Ruas, T., Aizawa, A., Gipp, B., Spinde, T.: Introducing MBIB - the first media bias identification benchmark task and dataset collection. In: Proceedings of the 46th International ACM SIGIR Conference on Research and Development in Information Retrieval, pp. 2765–2774 (2023)
35. Zhao, J., Wang, T., Yatskar, M., Ordonez, V., Chang, K.W.: Gender bias in coreference resolution: evaluation and debiasing methods. In: Proceedings of the 2018 Conference of the North American Chapter of the Association for Computational Linguistics: Human Language Technologies, Volume 2 (Short Papers), pp. 15–20 (2018)

Estimating the Usefulness of Clarifying Questions and Answers for Conversational Search

Ivan Sekulić[1(✉)], Weronika Łajewska[2], Krisztian Balog[2], and Fabio Crestani[1]

[1] Università della Svizzera italiana, Lugano, Switzerland
{ivan.sekulic,fabio.crestani}@usi.ch
[2] University of Stavanger, Stavanger, Norway
{weronika.lajewska,krisztian.balog}@uis.ch

Abstract. While the body of research directed towards constructing and generating clarifying questions in mixed-initiative conversational search systems is vast, research aimed at processing and comprehending users' answers to such questions is scarce. To this end, we present a simple yet effective method for processing answers to clarifying questions, moving away from previous work that simply appends answers to the original query and thus potentially degrades retrieval performance. Specifically, we propose a classifier for assessing usefulness of the prompted clarifying question and an answer given by the user. Useful questions or answers are further appended to the conversation history and passed to a transformer-based query rewriting module. Results demonstrate significant improvements over strong non-mixed-initiative baselines. Furthermore, the proposed approach mitigates the performance drops when non useful questions and answers are utilized.

Keywords: Conversational search · Mixed initiative · Clarifying questions

1 Introduction

The goal of a conversational search (CS) system is to satisfy the user's information need. To this end, several aspects of CS systems have enjoyed significant advancements, including conversational passage retrieval [22], query rewriting [21], and intent detection [13]. In recent years, the mixed-initiative paradigm of conversational search has attracted significant research attention [2,16,18]. Under the mixed-initiative paradigm, the CS system can at any point in a conversation offer suggestions or ask clarifying questions.

Asking clarifying questions has been identified as an invaluable component of modern-day CS systems [14]. Such questions are directed at users and aim to elucidate their underlying information needs. While there is a growing body of research on constructing and generating clarifying questions [2,19,23], work

N. Goharian et al. (Eds.): ECIR 2024, LNCS 14610, pp. 384–392, 2024.
https://doi.org/10.1007/978-3-031-56063-7_30

aimed at processing and comprehending users' answers to such questions is scarce. Nonetheless, recent research suggests their usefulness by demonstrating improvements in passage retrieval performance after asking a clarifying question and receiving an answer [1].

To bridge the aforementioned research gap, we make a first step towards processing the answers given to clarifying questions. We hypothesize that not all information acquired through such interactions with the user would benefit the CS system, i.e., yield improvements in retrieval effectiveness. Thus, the main novelty of our approach is that we do not blindly utilize the questions and the answers, but only when they are deemed to be useful. Specifically, we focus on the task of conversational passage retrieval and design a classifier aimed at assessing usefulness of the asked clarifying question and the provided answer. We utilize the question or the answer only if they are deemed useful, by appending them to the conversational history and employing a query rewriting method to attain a more information-dense query. Results on the TREC 2022 Conversational Assistance Track (CAsT'22) [11] demonstrate significant improvements in passage retrieval performance with the use of enhanced queries, as opposed to a non-mixed-initiative retrieval system (12% and 3% relative improvement in terms of Recall@1000 and nDCG, respectively). Further, when contrasting our approach to an established method that simply appends the prompted clarifying question and its answer to the original query [1], we observe differences in performance. Specifically, if neither the question nor the answer are deemed useful, but still used, there is a relative performance decrease of 13% in terms of nDCG@3, compared to non-mixed-initiative baselines. In other words, it is better not to use any information provided by such questions and answers, than to use it wrongly. Our contributions can be summarized as follows:

- We propose a simple, yet effective, method for processing answers to clarifying questions. The method is based on classifying usefulness of the prompted question and the given answer.
- We identify scenarios where asking clarifying questions resulted in improved passage retrieval, and where it decreased the retrieval performance.

2 Related Work

To facilitate further research in conversational search (CS), the TREC Conversational Assistance Track (CAsT) [3] aims to provide a reusable benchmark composed of several pre-defined conversational trajectories over a variety of topics. The most recent edition, CAsT'22, offers a subtask oriented towards mixed-initiative (MI) interactions [11]. Under the MI paradigm of CS, the system can at any point of a conversation take initiative and prompt the user with various suggestions or questions [14]. One of the most prominent usages of mixed-initiative is asking clarifying questions with a goal of elucidating users underlying information need [2]. Recent research demonstrates a positive impact of clarifying questions both on user experience [5,24] and retrieval performance [1]. Although other collections with clarifying questions and answers exist, most notably ClariQ [1], in this work, we focus on the aforementioned CAsT'22 as we additionally have control over which kind of question is being asked. In general, two streams of approaches to constructing clarifying questions exist: (1) select an appropriate question from a pre-defined pool of questions [1,2,11,16,18]; (2) generate the question [9,19,23]. However, despite the abundance of research on clarifying question construction, researched aimed at processing users' answers to such questions is scarce. To bridge this gap, Krasakis et al. [6] conduct an analysis of users' answers and find that they vary in polarity and length, as well as that retrieval effectiveness is often hurt. Thus, in this work, we aim to automatically assess their usefulness, with a goal of mitigating this undesired effect.

3 Methodology

In this section, we formally define the task of conversational passage retrieval under the mixed-initiative (MI) paradigm and present our methods for each of the components of the task, i.e., query rewriting, clarifying question selection, answer processing, and passage retrieval.

3.1 Task Formulation

At a current conversational turn t, given a user utterance u^t and a conversation history $H = [(u^1, s^1), \ldots, (u^{t-1}, s^{t-1})]$, the task is to generate a system response s^t. For clarity, we omit the superscript t from the subsequent definitions. In MI CS systems, the system's response s can either be a clarifying question s_{cq} or a ranked list of passages s_D, $D = [d_1, d_2, \ldots, d_N]$, where N is the number of passages retrieved and d_i is the i-th passage in the list. Similarly, user utterance u can take form of a query u_q or an answer u_a to system's question s_{cq}. Modeling other types of user utterances, such as explicit feedback, falls out of the scope of this study. Following prior work [21], the first task, i.e., query rewriting, is aimed towards resolution of the user query u_q in the context of the conversation history, resulting in $u'_q = \gamma(u_q|H)$, where γ is a query rewriting method.

Following the MI setting introduced at TREC CAsT'22 [11], we address the problem of conversational passage retrieval through the following three components: (i) Produce system utterance s_{cq} by selecting an appropriate clarifying question cq from a given pool of questions PQ; (ii) Process the given answer u_a and incorporate relevant information to the current query, resulting in $u''_q = \theta(u'_q, s_{cq}, u_a)$; (iii) Return a ranked list of passages s_D. Next, we define our approaches to the described components. We note that a clarifying question might be needed only for ambiguous, faceted, or unclear user requests. Thus, for queries not requiring clarification, the system might opt to return a ranked list of passages without asking further questions. However, following the setup enabled by CAsT'22 track, we do not explicitly model clarification need and thus design a system that prompts the user with a clarifying question at each turn.

Table 1. Examples of annotated subset of ClariQ, indicating cases when clarifying question, answer, both, or neither are useful.

Query (initial request)	Clarifying Question	Answer	Useful	Prevalence
I'm looking for information on hobby stores	Do you want to know hours of operation?	No	*None*	32%
Tell me information about computer programming	Are you interested in a coding bootcamp?	No, I want to know what career options programmers have	*Answer*	53%
Find me map of USA	Do you want to see a map of US territories?	Yes	*Question*	11%
All men are created equal	Would you like to know more about the declaration of independence?	Yes, I'd like to know who wrote it	*Question* and *answer*	6%

3.2 Clarifying Question Selection

For each query u'_q, we rank the potential candidates cq_i based on their semantic similarity to u'_q. Specifically, we utilize a T5 model fine-tuned on the CANARD dataset [4], available at HuggingFace,[1] as our γ rewriting function, which yields a resolved utterance u'_q. To rank the potential candidates, we use MPNet [20] from SentenceTransformers [17], trained for semantic matching. We select cq_i with the highest score, as predicted by the MPNet: $s_{cq} = \text{argmax}_{cq_i \in PQ_f} MPNet(u'_q, cq_i)$, where PQ_f is a pool of clarifying questions with potentially misleading, unreliable, and faulty questions automatically filtered from the pool [8].

3.3 Answer Processing

In this subsection, we describe our novel usefulness-based approach to processing answers given to the asked clarifying questions. To address the issue, we move away from previous approaches that simply append the question and the answer to the original query [1,2], regardless of the actual information gain. In fact, a recent study by Krasakis et al. [6] demonstrated that such practice can cause a decrease in retrieval effectiveness. Moreover, they show that multi-word answers are informative (e.g., "yes, I'm looking for info on spiders in Europe"), thus improving retrieval performance. Similarly, short negative answers are not informative (e.g., "no"), while multi-word negative answers are (e.g., "no, I'm interested in buying aquarium cleaner"). Thus, we define four possible actions, based on the current resolved utterance u'_q, the clarifying question asked s_{cq}, and the answer u_a:

1. In case the answer is affirmative (e.g., "yes" or "Yes, that is what I'm looking for"), we expand the u'_q by appending the clarifying question asked.
2. In case the answer is deemed useful, i.e., the underlying information need is explained in greater detail, we expand u'_q by appending the answer.
3. In case the answer is affirmative and it provides additional details, we expand u'_q with both the clarifying question and the answer.

[1] https://huggingface.co/castorini/t5-base-canard.

4. If neither (1), (2), nor (3) is the case, we do not expand the utterance.

Examples of the described cases are presented in Table 1 and are all aimed at incorporating additional useful information to the current utterance. Formally:

$$u''_q = \begin{cases} u'_q, & \psi(u'_q, s_{cq}, u_a) = 0 \\ \phi(u'_q, s_{cq}), & \psi(u'_q, s_{cq}, u_a) = 1 \\ \phi(u'_q, u_a), & \psi(u'_q, s_{cq}, u_a) = 2 \\ \phi(u'_q, u_a, s_{cq}), & \psi(u'_q, s_{cq}, u_a) = 3 \end{cases} \quad (1)$$

where $\psi(u'_q, s_{cq}, u_a)$ is a usefulness classifier, which aims to predict which of the above described actions to take. The labels 0, 1, 2, and 3 correspond to neither s_{cq} or u_a were deemed useful, s_{cq} was deemed useful, u_a was deemed useful, and both were useful, respectively. Similarly to Sect. 3.1, the function ϕ rewrites the original query given the context, in this case s_{cq} or u_a.

Specifically, to model ψ, we fine-tune a large transformer-based model, namely T5 [15], for multi-class classification. To fine-tune the classifier, we manually annotate a portion of ClariQ (150 samples) for the specific aforementioned cases. The annotations were performed by two authors of the paper with an inter-annotator agreement Cohen's kappa of 0.89. The differences in annotations were discussed and resolved consensually. Examples of annotations are presented in Table 1 and classification performance is reported in Sect. 4. We dub our novel mixed-initiative classifier-based method *MI-Clf.* Moreover, we assess the prevalence of each of the cases, and find, as presented in Table 1, that 68% of interactions contain new, useful information. In the other 32% of the cases, the answer simply negates the prompted clarifying question. Although this interaction also provides valuable insights into the user's information need, current approaches would expand the query by appending the prompted clarifying question and the answer. However, such an expanded query contains terms the user is not interested in, which can potentially degrade retrieval performance. We compare our proposed method to such a baseline, which always extends the query as: $u''_q = \phi(u'_q, scq, u_a)$. This method is dubbed *MI-All.*

Passage Retrieval. Finally, the rewritten utterance u''_q is fed into a standard two-stage retrieve-and-rerank pipeline [7]. We utilize BM25 ($k1 = 0.95$, $b = 0.45$) with RM3, where the initial query is extended with the highest-weighting terms from top-k scoring passages ($k = 10$ and number of terms $m = 10$). Next, we use a point-wise monoT5 re-ranker [10] to re-rank the top 1000, followed by a pair-wise duoT5 re-ranker [12] to additionally re-rank the top 50 passages. The non-mixed-initiative baseline, dubbed *DuoT5*, uses the same retrieval pipeline.

4 Results

Usefulness Classifier. The proposed usefulness classifier, described in Sect. 3.3, achieves an average macro-F_1 score of 0.75 and accuracy of 89% in a stratified 5-fold evaluation on the aforementioned annotated subset of ClariQ. Next, we

Table 2. Performance of baselines and our mixed-initiative approaches on CAsT'22.

Approach/RunID	R@1000	MAP	MRR	NDCG	NDCG@3	NDCG@5
BM25_T5_automatic	0.3244	0.1498	0.5272	0.2987	0.3619	0.3443
BM25_T5_manual	0.4651	0.2309	0.7155	0.4228	0.5031	0.4831
our_baseline (DuoT5)	0.3846	0.1680	0.4990	0.3392	0.3593	0.3502
+MI-All	**0.4441**	0.1741	**0.5297**	0.3594	**0.3722**	0.3508
MI-Clf	0.4302	**0.1776**	0.5144	**0.3613**	0.3697	**0.3581**

employ the trained classifier to predict the usefulness of (u'_q, s_{cq}, u_a) at each turn in the CAsT'22 dataset. The question s_{cq} was classified as useful in 28% of turns, while the answer u_a in 37%. In the rest 35% of the cases, neither was predicted to be useful. While the distribution of the predictions is similar to the prevalence in human-annotated data reported in Table 1, some differences can be observed. For example, in CAsT'22 28% of the clarifying questions were deemed useful, as opposed to the 13% in ClariQ.

Retrieval Performance. Results of the end-to-end conversational passage retrieval task, after the applied mixed-initiative answer processing methods (*MI-All* and *MI-Clf*) are presented in Table 2. For reference, we also include the organizers' baselines in the table. We make several observations from the presented results. First, both methods that utilize mixed-initiative show improvements over the DuoT5 method. This confirms previous findings on the positive impact of clarifications in conversational search. Second, differences between *MI-All* and *MI-Clf* are not statistically significant, across all metrics. However, we note that our classifier-based method utilizes clarifying question or the answer only when deemed useful, which is in about 70% of the cases in CAsT'22. On the contrary, *MI-All* always utilizes both the clarifying question and the answer. The equal performance of the two methods suggests that our usefulness classifier successfully includes only relevant information.

Analysis. In cases where the usefulness classifier predicted that neither the clarifying question s_{cq} nor the answer u_a is useful, we observe a drop of the *MI-All* method's retrieval performance, in terms of nDCG@3 (−13%). Recall, however, is not impacted by incorporating potentially not useful information and even shows a slight increase (+3.3%). As this method always appends both s_{cq} and u_a to the query u'_q, the performance drop is expected, especially in the re-ranking stage, as the re-ranker might be confused by the additional non-relevant information. Moreover, for both MI methods, we observe higher performance gains when the answer is useful (+19.8% for *MI-All* and +23.4% for *MI-Clf* in recall), compared to cases when the question is useful (+12.5% for *MI-All* and +8.5% *MI-Clf* in recall). This could be explained by the fact that users' answers are deemed useful when they are longer and thus provide more detail on the underlying information need [6]. On the contrary, a clarifying question can be deemed useful even when tangibly addressing the user's need. In other words, a good clarifying question can make a small step towards elucidating the user's information need. However, the user's answer can contain detailed expression of their information need, thus making further gains.

5 Conclusion

In this study, we proposed a classifier-based method, *MI-Clf*, for processing answers to clarifying questions in conversational search, which extends the original query only when either is deemed useful. Results on the TREC CAsT'22 dataset demonstrate clear improvements of the *MI-Clf* method over non-mixed-initiative baselines (+12% and +3% relative improvement in terms of Recall@1000 and nDCG). Moreover, we observed a performance drop for established methods that always use both the clarifying question and the answer, in cases where neither is useful (−13% in terms of nDCG@3), thus incorporating noisy information. This study makes the first steps towards improved answer processing methods.

Acknowledgments. This research was partially supported by the Norwegian Research Center for AI Innovation, NorwAI (Research Council of Norway, project number 309834).

References

1. Aliannejadi, M., Kiseleva, J., Chuklin, A., Dalton, J., Burtsev, M.: Building and evaluating open-domain dialogue corpora with clarifying questions. In: Proceedings of the 2021 Conference on Empirical Methods in Natural Language Processing, EMNLP 2020, pp. 4473–4484 (2021)
2. Aliannejadi, M., Zamani, H., Crestani, F., Croft, W.B.: Asking clarifying questions in open-domain information-seeking conversations. In: Proceedings of the 42nd International ACM SIGIR Conference on Research and Development in Information Retrieval, SIGIR 2019, pp. 475–484 (2019)
3. Dalton, J., Xiong, C., Callan, J.: TREC CAsT 2019: the conversational assistance track overview. In: The Twenty-Eighth Text REtrieval Conference Proceedings, TREC 2019 (2019)
4. Elgohary, A., Peskov, D., Boyd-Graber, J.: Can you unpack that? Learning to rewrite questions-in-context. In: Empirical Methods in Natural Language Processing, EMNLP 2019, pp. 5918–5924 (2019)
5. Kiesel, J., Bahrami, A., Stein, B., Anand, A., Hagen, M.: Toward voice query clarification. In: The 41st International ACM SIGIR Conference on Research & Development in Information Retrieval, SIGIR 2018, pp. 1257–1260 (2018)
6. Krasakis, A.M., Aliannejadi, M., Voskarides, N., Kanoulas, E.: Analysing the effect of clarifying questions on document ranking in conversational search. In: Proceedings of the 2020 ACM SIGIR on International Conference on Theory of Information Retrieval, ICTIR 2020, pp. 129–132 (2020)

7. Lajewska, W., Balog, K.: From baseline to top performer: a reproducibility study of approaches at the TREC 2021 conversational assistance track. In: Kamps, J., et al. (eds.) Proceedings of the 45th European Conference on Information Retrieval, ECIR 2023, vol. 13982, pp. 177–191. Springer, Cham (2023). https://doi.org/10.1007/978-3-031-28241-6_12
8. Lajewska, W., Bernard, N., Kostric, I., Sekulić, I., Balog, K.: The University of Stavanger (IAI) at the TREC 2022 conversational assistance track. In: Proceedings of the Thirty-First Text REtrieval Conference, TREC 2022 (2023)
9. Majumder, B.P., Rao, S., Galley, M., McAuley, J.: Ask what's missing and what's useful: improving clarification question generation using global knowledge. In: Proceedings of the 2021 Conference of the North American Chapter of the Association for Computational Linguistics: Human Language Technologies, NAACL 2021, pp. 4300–4312 (2021)
10. Nogueira, R., Jiang, Z., Lin, J.: Document ranking with a pretrained sequence-to-sequence model. In: Findings of the Association for Computational Linguistics, EMNLP 2020, pp. 708–718 (2020)
11. Owoicho, P., Dalton, J., Aliannejad, M., Azzopardi, L., Trippas, J.R., Vakulenko, S.: TREC CAsT 2022: going beyond user ask and system retrieve with initiative and response generation. In: The Thirty-First Text REtrieval Conference Proceedings, TREC 2022 (2022)
12. Pradeep, R., Nogueira, R., Lin, J.: The expando-mono-duo design pattern for text ranking with pretrained sequence-to-sequence models. arXiv cs.IR/2101.05667 (2021)
13. Qu, C., Yang, L., Croft, W.B., Zhang, Y., Trippas, J.R., Qiu, M.: User intent prediction in information-seeking conversations. In: Proceedings of the 2019 Conference on Human Information Interaction and Retrieval, CHIIR 2019, pp. 25–33 (2019)
14. Radlinski, F., Craswell, N.: A theoretical framework for conversational search. In: Proceedings of the 2017 Conference on Human Information Interaction and Retrieval, CHIIR 2017, pp. 117–126 (2017)
15. Raffel, C., et al.: Exploring the limits of transfer learning with a unified text-to-text transformer. J. Mach. Learn. Res. **21**(1), 5485–5551 (2020)
16. Rao, S., Daumé III, H.: Learning to ask good questions: ranking clarification questions using neural expected value of perfect information. In: Proceedings of the 56th Annual Meeting of the Association for Computational Linguistics (Volume 1: Long Papers), ACL 2018, pp. 2737–2746 (2018)
17. Reimers, N., Gurevych, I.: Sentence-BERT: sentence embeddings using Siamese BERT-networks. In: Proceedings of the 2019 Conference on Empirical Methods in Natural Language Processing and the 9th International Joint Conference on Natural Language Processing, EMNLP-IJCNLP 2019, pp. 3982–3992 (2019)
18. Rosset, C., et al.: Leading conversational search by suggesting useful questions. In: Proceedings of The Web Conference 2020, WWW 2020, pp. 1160–1170 (2020)
19. Sekulić, I., Aliannejadi, M., Crestani, F.: Towards facet-driven generation of clarifying questions for conversational search. In: Proceedings of the 2021 ACM SIGIR International Conference on Theory of Information Retrieval, ICTIR 2021, pp. 167–175 (2021)
20. Song, K., Tan, X., Qin, T., Lu, J., Liu, T.Y.: MPNet: masked and permuted pre-training for language understanding. arXiv preprint arXiv:2004.09297 (2020)

21. Vakulenko, S., Voskarides, N., Tu, Z., Longpre, S.: A comparison of question rewriting methods for conversational passage retrieval. In: Hiemstra, D., Moens, M.-F., Mothe, J., Perego, R., Potthast, M., Sebastiani, F. (eds.) ECIR 2021. LNCS, vol. 12657, pp. 418–424. Springer, Cham (2021). https://doi.org/10.1007/978-3-030-72240-1_43
22. Yu, S., Liu, Z., Xiong, C., Feng, T., Liu, Z.: Few-shot conversational dense retrieval. In: Proceedings of the 44th International ACM SIGIR Conference on Research and Development in Information Retrieval, SIGIR 2021, pp. 829–838 (2021)
23. Zamani, H., Dumais, S., Craswell, N., Bennett, P., Lueck, G.: Generating clarifying questions for information retrieval. In: Proceedings of The Web Conference 2020, WWW 2020, pp. 418–428 (2020)
24. Zamani, H., et al.: Analyzing and learning from user interactions for search clarification. In: Proceedings of the 43rd International ACM SIGIR Conference on Research and Development in Information Retrieval, SIGIR 2020, pp. 1181–1190 (2020)

Navigating Uncertainty: Optimizing API Dependency for Hallucination Reduction in Closed-Book QA

Pierre Erbacher[1](✉), Louis Falissard[2,3], Vincent Guigue[4], and Laure Soulier[1]

[1] Sorbonne Université, Paris, France
{pierre.erbacher,laure.soulier}@isir.upmc.fr
[2] Sorbonne Center of Artificial Intelligence, Paris, France
[3] Université Paris 8, Saint Denis, France
[4] AgroParisTech, Paris-Saclay, France
vincent.guigue@agroparistech.fr

Abstract. While Large Language Models (LLM) are able to accumulate and restore knowledge, they are still prone to hallucination. Especially when faced with factual questions, LLM cannot only rely on knowledge stored in parameters to guarantee truthful and correct answers. Augmenting these models with the ability to search on external information sources, such as the web, is a promising approach to ground knowledge to retrieve information. However, searching in a large collection of documents introduces additional computational/time costs. An optimal behavior would be to query external resources only when the LLM is not confident about answers. In this paper, we propose a new LLM able to self-estimate if it is able to answer directly or needs to request an external tool. We investigate a supervised approach by introducing a hallucination masking mechanism in which labels are generated using a close book question-answering task. In addition, we propose to leverage parameter-efficient fine-tuning techniques to train our model on a small amount of data. Our model directly provides answers for 78.2% of the known queries and opts to search for 77.2% of the unknown ones. This results in the API being utilized only 62% of the time.

Keywords: budgeted search · hallucination

1 Introduction

Language models have demonstrated remarkable performances in a wide range of Natural Language Processing (NLP) tasks, including conversational agents, summarization, translation, and question-answering [2,3,28]. As scaling these models increases their ability to incorporate more and more knowledge [2,4], instead of relying on traditional search engines, Metzler et al. [19] suggest using

P. Erbacher and L. Falissard—Authors contributed equally.

N. Goharian et al. (Eds.): ECIR 2024, LNCS 14610, pp. 393–402, 2024.
https://doi.org/10.1007/978-3-031-56063-7_31

LLM as a unified knowledge base able to perform question answering as well as document retrieval. However, even the larger models [2] are prone to producing inaccurate or false responses, commonly known as hallucinations [8]. These have been extensively explored in various NLP tasks, including summarization and translation [5,8,15,17,29]. Numerous approaches have been suggested to tackle this problem, with all of them employing external techniques to detect and mitigate hallucinations. In question answering, retrieval augmented methods such as REALM [6], RAG [16], or RETRO [1,13], were proposed to reduce LLM's hallucinations. These approaches consist in grounding LLM with a retriever model to add context from a large corpus of documents and to generate answers. These architectures are effective as they both improve factualness and reduce hallucinations for specific knowledge-intensive tasks such as Open-domain Question Answering [13]. However, retrieved documents are always considered without consideration of their helpfulness in solving the task. In a second line of work, models, such as LaMDA, BlenderBot, WebGPT, Toolformer [20,24,25,27] are specifically trained to generate a query and rely on a search engine when confronted with questions. While these LLMs accumulated a lot of knowledge during pre-training, they are fine-tuned to rely on external databases for each question, without considering the model's inherent ability to answer the question. **Toolformer calls the web API for almost all the questions,** 99.3% **with no real discernment between directly answerable questions and the real need for external knowledge.** LLMs have accumulated a lot of information and may be able to answer directly when confronted with widely known facts [2,3,28]. Several works proposed methods to evaluate LLMs knowledge uncertainty. Kadavath et al. [10], suggest that LLMs are able to self-evaluate their knowledge by prompting them to estimate the probability of their predicted answer being true.

Additionally, Kuhn et al. [11] introduced the notion of semantic entropy. This entropy-inspired, measurement which incorporates linguistic invariances derived from an external BART model was found more reliable than standard likelihood estimations to assess model uncertainty. However, all these proposed methods still require models to first generate an answer before any uncertainty estimation can be performed. As a consequence, whether a LLM can be taught to identify known from unknown answers before prediction remains unclear.

In this paper, we study a more nuanced approach that leverages external knowledge while also incorporating LLMs' intrinsic knowledge. We, therefore, propose a model that either generates a natural language answer or an API call (e.g. $\langle search \rangle$) only when the model is not self-confident about the answer, minimizing the dependency on external resources helps to save inference time and computational costs. We focus on closed-book question-answering (CBQA) tasks and carried out on two datasets (Natural Questions (NQ) [12] and TriviaQA (TQA) [9]). We study how LLMs perform at self-estimating their ability to correctly answer factual questions.

2 Learning When to Search with LLMs

Problem Formalization. We consider the set of factual questions Q and a set of all possible answers Ω. Each question $q \in Q$ is associated with a set A of correct answers with $A \subset \Omega$. In CBQA, models are able to answer factual questions without supporting documents, this means relying only on the knowledge stored in parameters. Formally, we define an LLM that maps answers with questions: $LM_\theta : Q \mapsto \Omega$. Let's consider an LLM that generates an answer $\hat{a}$ given a question q, with $\hat{a}$ that might be in the set A of ground truth or not (in this last case the answer is hallucinated). Our objective is to train an LLM to query external resources instead of generating hallucinations or to generate directly the answers, otherwise:

$$LM_{\theta'} : Q \mapsto \Omega \cup \{\langle search \rangle\} \tag{1}$$

where $\langle search \rangle$ is a call to an external tool such as a search engine. Training this LLM of parameter θ' can be seen as a budgeted open QA task in which we minimize the probability of generating a wrong answer $P(\hat{a} \notin A|\theta', q)$ and the probability to call an external tool $P(\hat{a} = \langle search \rangle|\theta', q)$ with a budget λ:

$$\underset{\theta}{\text{argmin}} \prod_{q \in Q} [P(\hat{a} = \langle search \rangle|\theta', q) + \lambda P(\hat{a} \notin A|\theta', q)] \tag{2}$$

where $\lambda \geq 1$ is a hyper-parameter controlling the relative importance between accessing the external resource and a hallucination behavior. The objective of this formulation is to encourage the model to provide direct answers whenever possible, therefore minimizing the cost of searching in an external resource. In this paper, we limit this analysis to $\lambda = 1.0$ (Eq. 2). A natural solution to label the model's hallucinations is to verify whether the model outputs are correct and factual. However, the ability to answer correctly to a question is inherent to the model's size and training. There is therefore no fixed dataset with supervision labels identifying when to call an API. We, therefore, propose to leverage a language model fine-tuned on a QA dataset to infer pseudo labels from language model performance during the training. More particularly, we aim to teach LLMs to generate a special sequence of tokens ($\langle search \rangle$), instead of answering incorrectly, without deteriorating the ability of the model to answer questions thanks to a "Hallucination Masking Mechanism".

Hallucination Masking Mechanism. Our objective here is to update the model LM_θ to display similar performances on CBQA tasks while also detecting hallucinations. Given LM_θ a model able to perform a CQBA task, we learn parameters θ' such that $LM_{\theta'} : Q \mapsto \Omega \cup \{\langle search \rangle\}$ where the LLM can still perform CBQA but predict the $\langle search \rangle$ token instead of hallucinating an answer. We introduce ψ a Hallucination Masking Mechanism (HalM) allowing to mask wrong answers with $\langle search \rangle$ tokens. Formally, $\psi \circ LM_\theta : Q \mapsto \Omega \cup \{\langle search \rangle\}$:

$$\psi(LM_\theta(q)) = \begin{cases} \mathbb{1}(\hat{a}), & \text{if } \hat{a} \in A \\ \langle search \rangle, & \text{otherwise} \end{cases} \tag{3}$$

Table 1. Table showing how predictions are considered. The LM_θ is the language model after the fine-tuning on the CQBA datasets and $LM_{\theta'}$ after the second one, namely including the HalM (Sect. 2).

		$LM_{\theta'}$		
		C	H	S
LM_θ	C	TP	FP	FN
	H	TP	FP	TN

With $\mathbb{1}$ the identity function and $\langle search \rangle$ the sequence of tokens used to query an external knowledge base. ψ enables to generate labels for data where the identity function is applied for questions answered correctly and hallucinations are masked using $\langle search \rangle$ tokens. This mechanism is composed on top of the original LM_θ to conserve the ability of the model to answer directly when the answer is correct. To avoid additional biases in the experiment, we limit ourselves to in-domain hallucination detection, where questions used for QA fine-tuning and hallucination detection come from the same distribution. This means that we use a single dataset for both steps, avoiding distribution shifts and shared example problems.

3 Evaluation Protocol

3.1 Datasets

We consider two open-domain CBQA datasets to perform our experiments.
Natural Question Open (NQ) [14]: an open domain question answering benchmark derived from the Natural Question dataset [12] which consists of questions from web queries accompanied by a list of appropriate answers, but without the original context provided in Natural Question.
TriviaQA (TQA) [9]: a dataset including questions gathered from quiz league websites and also accompanied by a list of appropriate answers.

3.2 Metrics

The standard approach for assessing generative CQBA model performances is based on the consideration that a generated answer is correct if and only if it constitutes an exact match or correct answer (noted **C**) with at least one element in a list of admissible answers. This metric alone, however, is insufficient to paint a comprehensive picture of the model's behavior, and we propose to extend it based on a comparison between the ground truth and both LM_θ and $LM_{\theta'}$'s predictions. Model output can be associated with three distinct events. A model prediction is either Correct (noted **C**), incorrect (noted **H** for Hallucinated), or a query to an external tool, namely $\langle search \rangle$, (noted S). As aforementioned, LM_θ predictions can only correspond to **C** or **H** events, while $LM_{\theta'}$ predictions can also correspond to S type events. Following these considerations, we define true positive (TP), false positive (FP), true negative (TN), and false negative (FN) events as shown in Table 1, and use them to report F1-scores in the results.

3.3 Model Architectures and Fine-Tuning

We consider sequence-to-sequence (encoder/decoder) models [26] with different sizes to assess how scale might affect the generated data, and therefore performances. All experiments utilize both the large and XXL T5-SSM models [22] (770M and 11B parameters, respectively) specifically trained for CBQA using Salient Span Masking (SSM) [6]. In addition, these models have official checkpoints that were fine-tuned on NQ and TQA, saving us the computational cost of training them ourselves. Large models are used FP32, however, 11B parameters models are quantized into int8 to fit on GPUs.

Models are finetuned on the train of each dataset to perform traditional CBQA. The dev set is then used to perform the second HalM-based finetuning step. This ensures high CBQA performances. Specifically, we focus on detecting hallucinations within the domain of interest, using questions from the same distribution as those used for QA fine-tuning. By utilizing a single dataset for both steps, we mean to avoid issues related to distribution shifts and shared example problems.

3.4 Baselines and Model Variants

For our models based on T5-Large and T5-XXL models, we consider two strategies to fine-tune with HalM: 1) the standard fine-tuning (**FT**) and 2) using Low-Rank Adaptation (**LoRA**) [7]. Due to computational constraints, the XXL (11B) T5 model is only fine-tuned with LoRA.

We compare our model variants to different baselines:

T5-Large and T5-XXL: the models fine-tuned on the train set of the CBQA task. Note that these models have not been trained to call external API, solely to generate answers.

T5-Large+PPL-t and T5-XXL+PPL-t: which is the strongest exogenous hallucination detection method known in the literature [5,15], based on a perplexity threshold. This heuristic assesses the model's output's perplexity score and classifies it as a hallucination if it exceeds a predefined, data-derived threshold.

Mistral-7B[1]**:** an in-context learning strong model with 16 examples randomly extracted from the train set. Wrong answers are masked with the 'search' sequence. These are examples of how the model should behave and be used for in-context learning. We observed that if the number of masked hallucinations and direct answers is unbalanced in the prompt, this also leads to very unbalanced prediction. Thus, we used a balanced set of examples.

[1] https://huggingface.co/mistralai/Mistral-7B-v0.1.

Mistral-7B-instruct[2]: a strong instruction-based LLM prompted to follow this instruction: *Answer to the question only if you know the answer, otherwise answer "I don't know"* followed by the question.

Table 2. Table showing the distribution of Exact match, Hallucination, and search sequence for NQ and TQA dataset. (%) are showing the remaining fraction of the same behavior (C, H, S) regarding predictions of the base model.

	NQ				TQA			
	C (↑)	H (↓)	Search	F1	C (↑)	H (↓)	Search	F1
T5-Large	27.3	72.7	0.0	–	19.4	80.6	0.0	–
T5-Large + PPL-t	18.6 (68.1%)	8.9 (12.2%)	72.5	**67.7**	12.7 (65.4%)	12.7 (15.7%)	74.6	**56.4**
T5-Large-HalM (FT)	15.7 (57.5%)	7.1(9.7%)	77.2	62.4	14.4 (74.2%)	29.9 (37.0%)	55.8	45.0
T5-Large-HalM (LoRA)	21.3 (78%)	16.6(22.8%)	62.0	65.0	13.6 (70.1%)	27.2 (34.8%)	59.2	45.1
T5-XXL	35.2	64.8	0.0	–	51.9	48.1	0.0	–
T5-XXL + PPL-t	21.7 (61.6%)	12.1 (18.6%)	66.3	65.4	27.7 (53.3%)	24.6 (51.1%)	47.7	63.1
T5-XXL-HalM (LoRA)	23.4 (66.5%)	15.9 (24.5%)	60.7	**66.1**	28.0 (53.9%)	24.4 (50.7%)	47.6	**63.5**
Mistral-7B (16 shots)	28.8 (91.7%)	49.9 (72.5%)	21.4	50.6	34.1 (51.8%)	14.7 (42.9%)	51.2	60.0
Mistral-7B-instruct	2.40	2.95	94.65	–	25.9	13.7	60.3	

Table 3. A table illustrating the fluctuation in hallucination and correct answer rates as the accuracy of Search S varies

Ratio of correct S	0.0	0.1	0.2	0.3	0.4	0.5	0.6	0.7	0.8	0.9	1.0
C	21.3	27.5	33.7	39.9	46.1	52.3	58.5	64.7	70.9	77.1	83.3
H	78.6	72.4	66.2	60.0	53.8	47.6	41.4	35.2	29.0	22.8	16.6

3.5 Implementation Details

We used existing checkpoints trained for closed-book QA [23]. These checkpoints are available on hugging face hub[3]. To infer our label, we follow [23] and use greedy decoding. To classify if a prediction is correct in generative QA, predictions are compared using Exact Match against a list of ground truths (GT). Because the list of GT is not exhaustive, a relative amount of predictions are False Negative, introducing noise in the training data for the second fine-tuning. For example, if the model generates "Napoleon I" but the GT only contains "Napoleon", the answer is considered False. To mitigate this, the model predictions are compared with the list of normalized ground truth: all values are lowercased and stopwords and punctuations are removed. We used PEFT [18] and Adapter-transformers [21] libraries to plug the parameters efficient method to LLMs. Large models are used FP32, however, 11B parameters models are quantized into int8 to fit on GPUs. For training LoRA, we used a warmup with a ratio of 0.1, $r = 16$, $alpha = 32$, and a learning rate of $1e-4$ and $7e-5$ for

[2] https://huggingface.co/mistralai/Mistral-7B-Instruct-v0.1.
[3] https://huggingface.co/.

large and XXL models. Regarding TQA, a checkpoint is available for the XXL model, and the T5 Large SSM was fine-tuned by our care using the following hyperparameters used in [23]: constant learning rate of $1e-3$ for 10000 steps, dropout of 0.1 and batch size of 128 and gradient accumulation of 8. However, contrary to what is reported by authors in [23], we encounter overfitting after a few thousand steps.

4 Results

Table 2 shows the results of all model variants and baselines for Natural Question (NQ) and TriviaQA (TQA) datasets, according to the F1-score, and proportions of correct answers C, Hallucination H and search S as defined in Sect. 3.2. Every rate on adapted models is accompanied by the remaining fractions for each behavior (C, H, S) regarding predictions of the first model.

From a general point of view, all hallucination reduction strategies (PPL-t and HalM) are able to reduce hallucinations regarding the models fined-tuned without consideration of searching on an external API. We notice that for the T5-Large variants the PPL-t (Perplexity-threshold) strategy outperforms the HalM with a high search rate; that might be costly. Our variant T5-Large-HalM fine-tuned with LoRA seems to have a better balance between accurate answer generation and the search rate. By focusing on the T5-XXL architecture, we show that LoRA consistently outperforms PPL-T on both NQ and TQA datasets for most of the metrics. Indeed, the LoRA strategy retains a higher fraction of correct answers on the NQ dataset. For the TQA dataset, both approaches exhibit similar behaviors, with a slight advantage for LoRA, and manage to filter out around half the Hallucinations while retaining a similar amount of Correct answers. Altogether, these results support the claim that our proposed approach enables LLMs to endogenously identify their potential for hallucination better than perplexity-based methods.

To understand the benefit of this mechanism, we consider different correct answer ratios for the search. If the ratio is set to 0.0, all API calls return only incorrect answers, while a ratio of 1.0 returns only correct answers. Table 3 shows the variation of this rate for the T5-Large-HalM (LoRA) on NQ. We can see that the user experience heavily relies on the boost of correct answers provided by the search. With a ratio of 0.1, this has similar performances as the T5-Large on NQ. While a ratio of 1.0 provides 83.3% of correct answer while the model only searches for 62% of the questions.

Focusing on in-context learning models, one can see that the Mistral-7b does not perform very well with an F1-score of 50.6% on NQ. Additionally, the model is very sensitive to the balance of examples in the prompt. Regarding the instructed capabilities with Mistral-7b-instruct, we observed that the model catastrophically failed to perform the given task as the model outputs 94.65% 'I don't know' while only having 2.4% correct answers on NQ. This suggests that abilities outlined in LLM, namely instructions and in-context learning, are not consistent with the specific behavior to identify uncertainties without devoted fine-tuning.

5 Conclusion

We introduced a new model to teach an LLM to internally assess its ability to answer properly a given query, without using anything more than data used for its training. The resulting model can directly identify its ability to answer a given question, with performances comparable -if not superior- to widely accepted hallucination detection baselines such as perplexity-based approaches which are strong exogenous baselines. In addition, this approach enables large language models to condition their generation on their ability to answer appropriately on a given query, a crucially important feature in the Toolformer approach that can learn to search only when needed. In future work, we plan to assess the impact of the λ hyperparameter in the Hallucinations risk/Search trade-off.

Acknowledgements. We would also like to thank the ANR JCJC project SESAMS (Project-ANR18-CE23-0001) and the Sorbonne Center for Artificial Intelligence for supporting this work.

References

1. Borgeaud, S., et al.: Improving language models by retrieving from trillions of tokens. In: Chaudhuri, K., Jegelka, S., Song, L., Szepesvari, C., Niu, G., Sabato, S. (eds.) Proceedings of the 39th International Conference on Machine Learning. Proceedings of Machine Learning Research, vol. 162, pp. 2206–2240. PMLR, 17–23 July 2022. https://proceedings.mlr.press/v162/borgeaud22a.html
2. Brown, T., et al.: Language models are few-shot learners. In: Larochelle, H., Ranzato, M., Hadsell, R., Balcan, M., Lin, H. (eds.) Advances in Neural Information Processing Systems, vol. 33, pp. 1877–1901. Curran Associates, Inc. (2020). https://proceedings.neurips.cc/paper_files/paper/2020/file/1457c0d6bfcb4967418bfb8ac142f64a-Paper.pdf
3. Bubeck, S., et al.: Sparks of artificial general intelligence: Early experiments with GPT-4 (2023)
4. Chowdhery, A., et al.: PaLM: Scaling language modeling with pathways (2022)
5. Guerreiro, N.M., Voita, E., Martins, A.: Looking for a needle in a haystack: a comprehensive study of hallucinations in neural machine translation. In: Proceedings of the 17th Conference of the European Chapter of the Association for Computational Linguistics, Dubrovnik, Croatia, pp. 1059–1075. Association for Computational Linguistics, May 2023. https://aclanthology.org/2023.eacl-main.75
6. Guu, K., Lee, K., Tung, Z., Pasupat, P., Chang, M.W.: REALM: retrieval-augmented language model pre-training. In: Proceedings of the 37th International Conference on Machine Learning, ICML 2020. JMLR.org (2020)
7. Hu, E.J., et al.: LoRA: low-rank adaptation of large language models. In: International Conference on Learning Representations (2022). https://openreview.net/forum?id=nZeVKeeFYf9
8. Ji, Z., et al.: Survey of hallucination in natural language generation. ACM Comput. Surv. **55**(12) (2023). https://doi.org/10.1145/3571730

9. Joshi, M., Choi, E., Weld, D., Zettlemoyer, L.: TriviaQA: a large scale distantly supervised challenge dataset for reading comprehension. In: Proceedings of the 55th Annual Meeting of the Association for Computational Linguistics (Volume 1: Long Papers), Vancouver, Canada, pp. 1601–1611. Association for Computational Linguistics, July 2017. https://doi.org/10.18653/v1/P17-1147. https://aclanthology.org/P17-1147
10. Kadavath, S., et al.: Language models (mostly) know what they know (2022)
11. Kuhn, L., Gal, Y., Farquhar, S.: Semantic uncertainty: linguistic invariances for uncertainty estimation in natural language generation (2023)
12. Kwiatkowski, T., et al.: Natural questions: a benchmark for question answering research. Trans. Assoc. Comput. Linguist. **7**, 452–466 (2019). https://doi.org/10.1162/tacl_a_00276. https://aclanthology.org/Q19-1026
13. Lee, K., Chang, M.W., Toutanova, K.: Latent retrieval for weakly supervised open domain question answering. In: Proceedings of the 57th Annual Meeting of the Association for Computational Linguistics, Florence, Italy, pp. 6086–6096. Association for Computational Linguistics, July 2019. https://doi.org/10.18653/v1/P19-1612. https://aclanthology.org/P19-1612
14. Lee, K., Chang, M.W., Toutanova, K.: Latent retrieval for weakly supervised open domain question answering. In: Proceedings of the 57th Annual Meeting of the Association for Computational Linguistics, Florence, Italy, pp. 6086–6096. Association for Computational Linguistics, July 2019. https://doi.org/10.18653/v1/P19-1612. https://www.aclweb.org/anthology/P19-1612
15. Lee, N., Bang, Y., Madotto, A., Fung, P.: Towards few-shot fact-checking via perplexity. In: Proceedings of the 2021 Conference of the North American Chapter of the Association for Computational Linguistics: Human Language Technologies, pp. 1971–1981. Association for Computational Linguistics, Online, June 2021. https://doi.org/10.18653/v1/2021.naacl-main.158. https://aclanthology.org/2021.naacl-main.158
16. Lewis, P., et al.: Retrieval-augmented generation for knowledge-intensive NLP tasks. In: Proceedings of the 34th International Conference on Neural Information Processing Systems, NIPS 2020. Curran Associates Inc., Red Hook (2020)
17. Manakul, P., Liusie, A., Gales, M.J.F.: SelfCheckGPT: zero-resource black-box hallucination detection for generative large language models (2023)
18. Mangrulkar, S., Gugger, S., Debut, L., Belkada, Y., Paul, S.: PEFT: state-of-the-art parameter-efficient fine-tuning methods (2022). https://github.com/huggingface/peft
19. Metzler, D., Tay, Y., Bahri, D., Najork, M.: Rethinking search: making domain experts out of dilettantes. SIGIR Forum **55**(1) (2021). https://doi.org/10.1145/3476415.3476428
20. Nakano, R., et al.: WebGPT: browser-assisted question-answering with human feedback (2022)
21. Pfeiffer, J., et al.: AdapterHub: a framework for adapting transformers. In: Proceedings of the 2020 Conference on Empirical Methods in Natural Language Processing: System Demonstrations, pp. 46–54 (2020)
22. Raffel, C., et al.: Exploring the limits of transfer learning with a unified text-to-text transformer (2020)
23. Roberts, A., Raffel, C., Shazeer, N.: How much knowledge can you pack into the parameters of a language model? In: Proceedings of the 2020 Conference on Empirical Methods in Natural Language Processing (EMNLP), pp. 5418–5426. Association for Computational Linguistics, Online, November 2020. https://doi.org/10.18653/v1/2020.emnlp-main.437. https://aclanthology.org/2020.emnlp-main.437

24. Schick, T., et al.: Toolformer: language models can teach themselves to use tools (2023)
25. Shuster, K., et al.: BlenderBot 3: a deployed conversational agent that continually learns to responsibly engage (2022)
26. Sutskever, I., Vinyals, O., Le, Q.V.: Sequence to sequence learning with neural networks. In: Ghahramani, Z., Welling, M., Cortes, C., Lawrence, N., Weinberger, K. (eds.) Advances in Neural Information Processing Systems, vol. 27. Curran Associates, Inc. (2014). https://proceedings.neurips.cc/paper_files/paper/2014/file/a14ac55a4f27472c5d894ec1c3c743d2-Paper.pdf
27. Thoppilan, R., et al.: LaMDA: language models for dialog applications (2022)
28. Wei, J., et al.: Emergent abilities of large language models (2022)
29. Yuan, W., Neubig, G., Liu, P.: BARTscore: evaluating generated text as text generation. In: Ranzato, M., Beygelzimer, A., Dauphin, Y., Liang, P., Vaughan, J.W. (eds.) Advances in Neural Information Processing Systems, vol. 34, pp. 27263–27277. Curran Associates, Inc. (2021). https://proceedings.neurips.cc/paper_files/paper/2021/file/e4d2b6e6fdeca3e60e0f1a62fee3d9dd-Paper.pdf

Simple Domain Adaptation for Sparse Retrievers

Mathias Vast[1,2](✉), Yuxuan Zong[2], Benjamin Piwowarski[2], and Laure Soulier[2]

[1] Sinequa, Paris, France
[2] Sorbonne Université, CNRS, ISIR, 75005 Paris, France
{mathias.vast,yuxuan.zong,benjamin.piwowarski,laure.soulier}@isir.upmc.fr

Abstract. In Information Retrieval, and more generally in Natural Language Processing, adapting models to specific domains is conducted through fine-tuning. Despite the successes achieved by this method and its versatility, the need for human-curated and labeled data makes it impractical to transfer to new tasks, domains, and/or languages when training data doesn't exist. Using the model without training (zero-shot) is another option that however suffers an effectiveness cost, especially in the case of first-stage retrievers. Numerous research directions have emerged to tackle these issues, most of them in the context of adapting to a task or a language. However, the literature is scarcer for domain (or topic) adaptation. In this paper, we address this issue of cross-topic discrepancy for a sparse first-stage retriever by transposing a method initially designed for language adaptation. By leveraging pre-training on the target data to learn domain-specific knowledge, this technique alleviates the need for annotated data and expands the scope of domain adaptation. Despite their relatively good generalization ability, we show that even sparse retrievers can benefit from our simple domain adaptation method.

Keywords: Pretrained Language Models · Cross-Topic Adaptation · Zero-Shot

1 Introduction

Nowadays, most of the models that achieve state-of-the-art results in Natural Language Processing (NLP) rely on the "*pre-train then fine-tune*" pipeline [4,22]. In this setup, a model is first trained with self-supervised objectives, such as Masked Language Modeling, on a large unlabeled corpus, before being fine-tuned on a labeled dataset for a specific task. In Information Retrieval (IR), this method has delivered huge improvements over the previous "pre-BERT" models and has given birth to a large variety of frameworks including, but not limited to, dense retrievers [13], learned sparse retrievers [5] and cross-encoders [22]. For a given task, when enough good-quality labeled data is available, fine-tuning an already pre-trained model is the best option. Thanks to datasets such as MS MARCO [21], this condition might be met for the majority of the downstream

N. Goharian et al. (Eds.): ECIR 2024, LNCS 14610, pp. 403–412, 2024.
https://doi.org/10.1007/978-3-031-56063-7_32

tasks in the generic domain for the English language, but in other languages and/or specific domains, these resources may not exist.

This poses a problem since a fine-tuned model evaluated outside its training domain might perform poorly [29]. When the volume of training data is insufficient for fine-tuning, it can even degrade the model's performance [32]. Besides, the increasing size of these models coupled with the difficulty of collecting a satisfying amount of high-quality labeled data has led to a growing interest in zero-shot or few-shot methods.

It is possible to tackle this issue by leveraging unlabeled data and pursuing pre-training over task or domain-related data [7,14] before fine-tuning. Providing a better initialization point alleviates the dependency on the amount of in-domain labeled data available for fine-tuning. Differently, Artetxe et al. [2] leverage pre-training as a means to adapt a model *subpart* for a specific language, different from the source language used for pre-training and fine-tuning the base model. In this paper, we consider a setup of cross-domain adaptation where the model to be transferred can be trained on a source domain D_{source}, where an annotated corpus containing enough documents for both pre-training and fine-tuning is available, whereas the target domain D_{target} contains enough data for pre-training but no annotation. We extend the work of [2] in two ways to deal with ad-hoc IR. First, inspired by [3], we study which subparts of the model should be pre-trained or fine-tuned – we do not only consider the embeddings to be domain-dependent. Second, we propose a new pre-training procedure on both the source and target domains, and not only on the target one. This allows a model to learn task-specific parameters (on the source domain) when (and only when) its domain-specific ones are well set. As domain shift impacts more strongly first-stage retrievers, we study a first-stage sparse retriever, namely SPLADE [5]. Our approach allows one to share the fine-tuning resources across multiple domains with less drop in effectiveness due to domain discrepancies.

2 Related Works

Transfer learning and adaptation to a specific context are long-standing research topics. With the recent arrival of large language models, research shifted to transferring these models at the lowest computational and annotation cost possible [1]. To get rid of human annotation, works inspired by Doc2Query [23] propose to generate queries for which a document in the target domain would be relevant [16,30]. Inversely, generative models can also be used to produce pseudo-relevant documents [6] before fine-tuning a model on it. [9] even proposed a method that generates a complete collection of both documents and queries from a simple description of the target domain. For each of these methods, one can apply the traditional "*pretrain then fine-tune*" framework. Besides the fact that this type of approach can be computationally costly, studying, as in our work, pre-training techniques, is complementary. Indeed, as highlighted in [15], better pre-training, either by using different self-supervised objectives [12] or pursuing it longer [7,14], often provides better results compared to fine-tuning for first-stage methods.

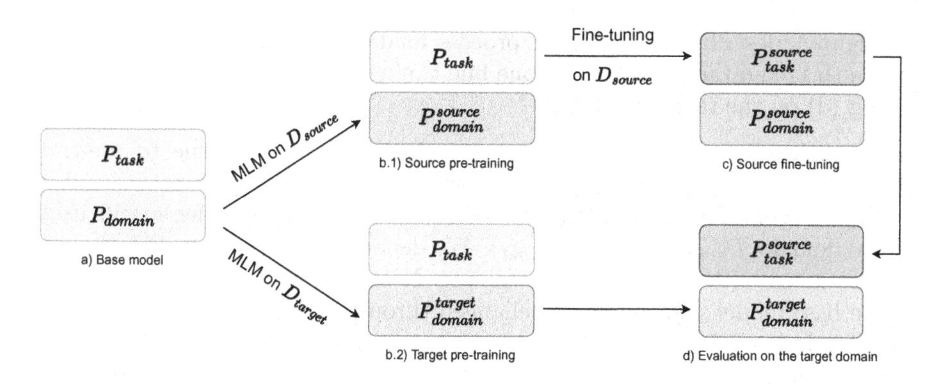

Fig. 1. Illustration of the cross-domain adaptation process

Other works have tried to ease the transfer of (large) pre-trained models. We can separate them into two categories, even though they share the same underlying principle of distinguishing different sets of parameters during training. The first family, referred to as Parameter-Efficient Fine-Tuning (or PEFT) [10,11,17,19], achieves this objective by modifying the original model with an *adapter* whose parameters are fine-tuned. Its success relies on the quality of the underlying pre-trained model, whose generality is worth being conserved while tuning only a small portion of parameters on the target task. In IR, these methods have started to be explored recently and, so far, have proven to be successful [18,24,28]. Orthogonal to these works, some papers do not use freezing to preserve generality or to spare computations, but to adapt specific parts of the model for a given language or task. In particular, [31] applies PEFT methods to Dense Retrievers to disentangle domain and relevance learning. For each domain, a Transformer backbone is trained to learn its linguistic features (MLM). Separately, a distinct module is trained once and then shared, to predict relevance based on these extracted features. Artetxe et al. [2] adapt a similar reasoning but apply the disentangling within the Transformer backbone's layers, thus introducing a zero inference overhead. In this work, we extend their approach to domain adaptation in the context of IR. Note that using PEFT following [31] is an alternative that we will explore in the future.

3 Cross-Domain Adaptation of a Neural Network Model

We describe here how we adapted the pipeline introduced by Artetxe et al. [2] to cross-domain adaptation. This approach can be extended to any model in any setting by establishing a distinction between the model's parameters dedicated to learning the domain/language model, denoted P_{domain}, and the ones dedicated to learning the task denoted P_{task}, which are supposed to be distinct, i.e. $P_{domain} \cap P_{task} = \emptyset$. Given the IR setting, we define P_{domain} to be the embeddings together with the first k layers of the transformer architecture – whereas in the language adaptation setting, [2] only considered P_{domain} to be the embeddings.

Figure 1 describes the training process made of two distinct pre-trainings; source (b.1) and target (b.2); and one fine-tuning on the source (c), before evaluating (d) on the target domain:

Pre-training for D_{source} (resp. D_{target}): As in [2], we continue to pre-train BERT over D_{source} (resp. D_{target}) while keeping the parameters' subset P_{task} frozen. This step adapts the parameters P_{domain} to the specificity of the domain D_{source} (resp. D_{target}). We denote P^{source}_{domain} (resp. P^{target}_{domain}) the parameters after the MLM pre-training. Note that contrary to [2], we keep the base model's vocabulary unchanged throughout the pipeline.

Fine-tuning on D_{source}: We leverage the previously pre-trained subset P^{source}_{domain} but keep it frozen while only fine-tuning the P_{task} parameters. The objective of this step is to specialize the subset of parameters P_{task} to the IR task. We denote as P^{source}_{task} the fine-tuned parameters.

Contrary to [2], we add an additional pre-training step over the source corpus (b.1 in Fig. 1). We suppose that thanks to this step, the model can better learn the downstream task as it is not fine-tuned on the same dataset (MS MARCO) that it was originally pre-trained on (Wikipedia).

Finally, for inference on the D_{target}'s dataset, we combine the two subsets P^{target}_{domain} and P^{source}_{task}. The underlying intuition is that the subset P^{source}_{task} is task-specific while P^{target}_{domain} is domain-specific. The model should be able to generalize better when used forIR in the target domain.

4 Experiments

We use $BERT_{base}$ pre-trained on the English version of Wikipedia[1] as our base model and share its base vocabulary through every domain. We focus on a sparse first-stage ranker, SPLADE [5], as first-stage rankers particularly suffer from domain shifts and SPLADE is known to be at the state-of-the-art among them.

4.1 Datasets, Baselines, and Ablations

Two datasets in IR might differ in 1) domain/vocabulary, 2) document types, and/or 3) topic types [25]. For a neural model, any of these distinctions can impact the generalization power. The source dataset we consider across all our experiments is MS MARCO [21] as it is the defacto dataset for fine-tuning pre-trained models in IR and its domain can be regarded as "generic Web search". We assemble a collection of target datasets from a subset of the BEIR benchmark (*) [29], the LoTTE's collection (**) [27], as well as other sources. The decision about using or not some datasets from the BEIR benchmark is based on multiple criteria, including the existence of distinct train and test sets and the task affiliated with the dataset as we are only interested in pure IR in this case. We processed them with the *ir-datasets* library [20]. Table 1 gives details about the characteristics of each dataset and shows how we covered various domains, ranging from bio-medical to literature.

[1] The model is made available by Google on the HuggingFace' Hub: bert-base-uncased.

Table 1. Datasets considered in the study. Statistics on the average topic and document length for each dataset are computed using BERT tokens.

Dataset	Domain	Dataset Statistics		
		Avg top./doc. len	# Top	# Doc
MSM-Passage	Generic	7.5/74.9	808K	8.8M
TREC-COVID (*)	Bio-Medical	16.0/243.5	50	171.3K
NFCorpus (*)	Bio-Medical	5.0/338.1	325	5.4K
BioASQ (*)	Bio-Medical	13.1/320.3	500	14.9M
FiQA-2018 (*)	Finance	13.6/175.1	648	57.6K
TREC-NEWS (*)	News	18.2/652.5	50	595K
Robust04 (*)	News	18.7/638.5	250	528.1K
ANTIQUE [8]	Web	11.9/52.3	200	403.7K
LoTTe-Wri. (**)	Literature	8.7/165.2	1071	200K
LoTTe-Tec. (**)	Technology	9.7/216.6	596	638.5K
LoTTe-Rec. (**)	Entertainment	9.3/181.6	924	167K

To verify the benefits of our adaptation method, we experimented with multiple variants, differing on the P_{domain} / P_{task} split by the number of layers $k \in \{0, 1, 2, 3, 4, 6, 8, 10\}$ after the embedding belonging to P_{domain}. We also included two baselines and two ablations that we describe in what follows:

- **BM25** [26], is a very strong baseline in domain adaptation [29] and is also a sparse first-stage retriever;
- **Zero-Shot Learning** evaluation of the SPLADE model fine-tuned on MS MARCO to quantify the benefits we can obtain with our adaptation method;
- Ablation of the source pre-training, referred to as **w/o source**, as this corresponds to the original approach described in Artetxe et al. (with $k = 0$);
- Ablation of the source *and* target pre-trainings, referred to as **w/o pre-training** (again with $k = 0$). Note that this is a zero-shot model. As we are starting from a model pre-trained on Wikipedia, this ablation evaluates whether our proposition has an impact.

All the project code, based on the library *experimaestro-ir* [33], including the experimental details, is freely accessible[1].

4.2 Results

Table 2 contains the results of these methods from which we can draw the following conclusions: (i) Compared to the Zero-Shot Learning baseline, our adaptation approach allows us to gain on average between 0.7 and 1.4 points in nDCG@10, and more than 1 point for some datasets; (ii) Comparing "w/o source" to our approach, it seems that the additional pre-training step over the source domain helps the model to generalize better, with an average improvement between 0.1

[1] https://git.isir.upmc.fr/mat_vast/cross_domain_adaptation.

Table 2. Performance in nDCG@10 of our approach versus BM25 and Zero-shot (SPLADE). Bold values are strictly superior to the Zero-shot baseline. † and ‡: Improves upon BM25 baseline and Zero-Shot baseline respectively with statistical significance ($p \leq 0.05$) under the two-tailed Student's t-test.

Method (→)	Baselines		Proposed pipeline ($k = \ldots$)				Ablations	
Dataset (↓)	BM25	0-shot	0 layer	1 layer	2 layers	4 layers	w/o pre-training	w/o source
TREC-COVID	58.1	71.3	68.8†	70.5†	**72.1**†	70.9†	67.9†	67.7†
NFCorpus	24.5	24.9	**25.6**	**25.4**	**25.1**	24.1	**25.1**	**25.6**
BioASQ	**52.3**	43.1	**45.4**‡	**44.2**	**45.2**‡	**46.2**‡	**44.8**‡	**45.3**‡
FiQA-2018	23.6	27.3	27.2†	**27.5**†	**29.5**†‡	**29.0**†‡	25.2†	**28.3**†
TREC-NEWS	31.4	32.1	**33.3**	**32.8**	**33.0**	**36.5**‡	31.9	**33.1**
Robust04	40.8	39.6	**42.9**‡	**42.9**‡	**43.5**‡	**42.6**‡	**41.2**	**42.1**‡
ANTIQUE	**45.4**	43.3	42.9	**44.0**	**43.7**	**44.1**	**45.0**‡	**45.0**‡
LoTTE-Rec.	39.3	46.2	**47.6**†‡	**47.7**†‡	**47.8**†‡	**46.9**†	**46.8**†	**46.7**†
LoTTE-Wri.	41.3	52.7	**53.0**†	**53.8**†‡	**53.4**†	**54.1**†‡	52.0†	**52.9**†
LoTTE-Tec.	25.3	36.8	**38.2**†	**37.6**†	**37.4**†	36.8†	36.8†	36.8†
Avg.	38.2	41.7	**42.5**	**42.6**	**43.1**	**43.1**	**41.9**	**42.4**

Table 3. Evolution of the performance in nDCG@10 and of the sparsity of SPLADE when the number of layers dedicated to the learning of the target domain increases. Best result is highlighted in bold.

Method (→)	0-shot	k = 0	k = 1	k = 2	k = 3	k = 4	k = 6	k = 8	k = 10
Average Performance (nDCG@10)	41.7	42.5	42.6	43.1	42.6	**43.1**	43.1	43.0	42.6
Sparsity (x10^3)	9.3	7.3	9.2	11.5	8.4	10.3	10	9.8	9.3

and 0.8 points (for nDCG@10). This is sensible since the model learns to adapt the task-specific parameters from domain-specific parameters; (iii) Similarly, no pre-training at all can hurt the performance up to 1.2 points in nDCG@10 ("w/o pre-training"), showing the benefit of our pre-training procedure. However, in a zero-shot setting, we observe that fine-tuning only the Transformer's layers performs ("w/o pre-training") on par with fine-tuning the whole model ("0-shot"). Both setups are evaluated without using the target dataset, and can thus be considered as Zero-Shot Learning. It indicates that we could save some computations during fine-tuning without hurting performance. (iv) Pre-training over additional layers can provide additional gains of up to 0.5 points in nDCG@10 (columns "1 layer", "2 layers" and "4 layers" compared to the column "0 layer"). Results however seemed to plateau beyond $k = 2$ as illustrated by the minor differences between the columns $k = 4$, with the best result in average, and $k = 2$. Table 3 summarizes the evolution of the average performance along with the number of additional layers reserved for learning the target domain. These results seem to comfort recent findings on the role of the first layers' attention head in a Transformer model [3]. It also highlights that even though SPLADE-base models highly rely on the Masked Language Modeling task, fine-tuning nevertheless remains an important part of the process, and increasing the share

of the parameters dedicated to the former can infringe the model's performance on the IR task.

Table 3 also gives additional insights into the sparsity achieved by the document encoder of our SPLADE variants. Query encoder sparsity isn't specified as it says consistent along the variants and the baseline. We note that the best models usually have a higher sparsity level, which tends to indicate that the associated model is able to build a more appropriate representation of each target domain. However, more work is needed to understand why some other variants happen to have lower sparsity values than the baseline while performing better.

Discussion and Limitations. This work is a preliminary study of how pre-training can be used to ease domain adaptation in IR. We led the same series of experiments with a second-stage ranker, namely monoBERT [22]. Interestingly, the results showed that our approach didn't provide any improvements compared to the Zero-Shot Learning approach in this case. Previous work [15] already mentioned the fact that second-stage rankers benefit more from larger collections compared to pre-training on specific datasets. Our conclusion is that our approach is too naive to provide any improvement at all, given the generalization power of monoBERT, and suggests that a better understanding of which parameters are domain or task-specific, as well as their interplay, is necessary. Further work is also needed to explore the impact of pre-training time given corpus length on the final results. We leave both of these directions for future works as well as the understanding of why second-stage rankers cannot draw any benefits from specific pre-training. Finally, we did not compare with generative methods (i.e., generating a query matching a document in the target domain) such as [16] – these methods are costly, and we also need to investigate whether the improvements are complementary with our (more cost-efficient) approach.

5 Conclusion

We presented a simple approach to domain adaptation for IR, based on a language adaptation approach [2]. Results from our experiments highlight the potential of our method with learned sparse retrievers and could be particularly helpful when transferring fine-tuned models in contexts where high-quality annotated data isn't available or impossible to collect because it is either too expensive or complex. Another benefit of our approach is its re-usability. Once the expensive fine-tuning has been performed, the subset P_{task} can be re-used with different subsets P_{domain} pre-trained over different domains, which eliminates the need to perform a fine-tuning over all the model for every new domain. In the future, we would like to include dense retrievers as well as more costly methods in our study. In addition, we also want to extend it to more sophisticated pre-training approaches.

Ackowledgements. We wish to thank Basile Van Cooten, from Sinequa, for his support and supervision on Mathias Vast's PhD and particularly on this paper. This work is supported by the ANR project ANR-23-IAS1-0003.

References

1. Althammer, S., Zuccon, G., Hofstätter, S., Verberne, S., Hanbury, A.: Annotating data for fine-tuning a neural ranker? Current active learning strategies are not better than random election. In: Proceedings of the Annual International ACM SIGIR Conference on Research and Development in Information Retrieval in the Asia Pacific Region, SIGIR-AP 2023, pp. 139–149. Association for Computing Machinery, New York (2023). https://doi.org/10.1145/3624918.3625333
2. Artetxe, M., Ruder, S., Yogatama, D.: On the cross-lingual transferability of monolingual representations. In: Proceedings of the 58th Annual Meeting of the Association for Computational Linguistics, pp. 4623–4637 (2020). https://doi.org/10.18653/v1/2020.acl-main.421. arXiv:1910.11856 [cs]
3. Clark, K., Khandelwal, U., Levy, O., Manning, C.D.: What does BERT look at? An analysis of BERT's attention. In: Proceedings of the 2019 ACL Workshop BlackboxNLP: Analyzing and Interpreting Neural Networks for NLP (2019). https://doi.org/10.18653/v1/w19-4828. http://dx.doi.org/10.18653/v1/W19-4828
4. Devlin, J., Chang, M.W., Lee, K., Toutanov, K.: BERT: pre-training of deep bidirectional transformers for language understanding. In: 2019 Conference of the North American Chapter of the Association for Computational Linguistics: Human Language Technologies. Association for Computational Linguistic (2018). https://doi.org/10.48550/arXiv.1810.04805
5. Formal, T., Lassance, C., Piwowarski, B., Clinchant, S.: SPLADE v2: sparse lexical and expansion model for information retrieval (2021). https://doi.org/10.48550/ARXIV.2109.10086. https://arxiv.org/abs/2109.10086
6. Gao, L., Ma, X., Lin, J., Callan, J.: Precise zero-shot dense retrieval without relevance labels (2022)
7. Gururangan, S., et al.: Don't stop pretraining: adapt language models to domains and tasks. In: Jurafsky, D., Chai, J., Schluter, N., Tetreault, J. (eds.) Proceedings of the 58th Annual Meeting of the Association for Computational Linguistics, pp. 8342–8360. Association for Computational Linguistics, Online, July 2020. https://doi.org/10.18653/v1/2020.acl-main.740. https://aclanthology.org/2020.acl-main.740
8. Hashemi, H., Aliannejadi, M., Zamani, H., Croft, W.B.: ANTIQUE: a non-factoid question answering benchmark. In: Jose, J.M., et al. (eds.) ECIR 2020. LNCS, vol. 12036, pp. 166–173. Springer, Cham (2020). https://doi.org/10.1007/978-3-030-45442-5_21
9. Hashemi, H., Zhuang, Y., Kothur, S.S.R., Prasad, S., Meij, E., Croft, W.B.: Dense retrieval adaptation using target domain description. In: Proceedings of the 2023 ACM SIGIR International Conference on Theory of Information Retrieval, ICTIR 2023, pp. 95–104. Association for Computing Machinery, New York (2023). https://doi.org/10.1145/3578337.3605127
10. Houlsby, N., et al.: Parameter-efficient transfer learning for NLP, June 2019. http://arxiv.org/abs/1902.00751
11. Hu, E.J., et al.: LoRA: low-rank adaptation of large language models, October 2021. arXiv:2106.09685 [cs]
12. Izacard, G., et al.: Unsupervised dense information retrieval with contrastive learning. Trans. Mach. Learn. Res. (2022). https://openreview.net/forum?id=jKN1pXi7b0
13. Karpukhin, V., et al.: Dense passage retrieval for open-domain question answering. In: Proceedings of the 2020 Conference on Empirical Methods in Natural Lan-

guage Processing (EMNLP), pp. 6769–6781. Association for Computational Linguistics, Online, November 2020. https://doi.org/10.18653/v1/2020.emnlp-main.550. https://aclanthology.org/2020.emnlp-main.550
14. Krishna, K., Garg, S., Bigham, J., Lipton, Z.: Downstream datasets make surprisingly good pretraining corpora. In: Rogers, A., Boyd-Graber, J., Okazaki, N. (eds.) Proceedings of the 61st Annual Meeting of the Association for Computational Linguistics (Volume 1: Long Papers), pp. 12207–12222. Association for Computational Linguistics, Toronto, July 2023. https://doi.org/10.18653/v1/2023.acl-long.682. https://aclanthology.org/2023.acl-long.682
15. Lassance, C., Dejean, H., Clinchant, S.: An experimental study on pretraining transformers from scratch for IR. In: Kamps, J., et al. (eds.) Advances in Information Retrieval. LNCS, vol. 13980, pp. 504–520. Springer, Cham (2023). https://doi.org/10.1007/978-3-031-28244-7_32. https://link.springer.com/10.1007/978-3-031-28244-7_32
16. Li, M., Gaussier, E.: Domain adaptation for dense retrieval through self-supervision by pseudo-relevance labeling (2022)
17. Li, X.L., Liang, P.: Prefix-tuning: optimizing continuous prompts for generation, January 2021. http://arxiv.org/abs/2101.00190, arXiv:2101.00190 [cs]
18. Litschko, R., Vulić, I., Glavaš, G.: Parameter-efficient neural reranking for cross-lingual and multilingual retrieval. In: Calzolari, N., et al. (eds.) Proceedings of the 29th International Conference on Computational Linguistics, Gyeongju, Republic of Korea, pp. 1071–1082. International Committee on Computational Linguistics, October 2022. https://aclanthology.org/2022.coling-1.90
19. Liu, X., et al.: P-Tuning v2: prompt tuning can be comparable to fine-tuning universally across scales and tasks, March 2022. arXiv:2110.07602 [cs]
20. MacAvaney, S., Yates, A., Feldman, S., Downey, D., Cohan, A., Goharian, N.: Simplified data wrangling with ir_datasets. In: SIGIR (2021)
21. Nguyen, T., et al.: MS MARCO: a human generated machine reading comprehension dataset. CoRR **abs/1611.09268** (2016). http://dblp.uni-trier.de/db/journals/corr/corr1611.html#NguyenRSGTMD16
22. Nogueira, R., Cho, K.: Passage re-ranking with BERT (2019). http://arxiv.org/abs/1901.04085
23. Nogueira, R., Yang, W., Lin, J.J., Cho, K.: Document expansion by query prediction. ArXiv **abs/1904.08375** (2019). https://api.semanticscholar.org/CorpusID:119314259
24. Pal, V., Lassance, C., Déjean, H., Clinchant, S.: Parameter-efficient sparse retrievers and rerankers using adapters. In: Kamps, J., et al. (eds.) Advances in Information Retrieval. LNCS, pp. 16–31. Springer, Cham (2023). https://doi.org/10.1007/978-3-031-28238-6_2
25. Pan, S.J., Yang, Q.: A survey on transfer learning. IEEE Trans. Knowl. Data Eng. **22**(10), 1345–1359 (2010). https://doi.org/10.1109/TKDE.2009.191
26. Robertson, S.E., Walker, S., Jones, S., Hancock-Beaulieu, M., Gatford, M.: Okapi at TREC-3. In: Text Retrieval Conference (1994). https://api.semanticscholar.org/CorpusID:3946054
27. Santhanam, K., Khattab, O., Saad-Falcon, J., Potts, C., Zaharia, M.: ColBERTv2: Effective and efficient retrieval via lightweight late interaction. In: Carpuat, M., de Marneffe, M.C., Meza Ruiz, I.V. (eds.) Proceedings of the 2022 Conference of the North American Chapter of the Association for Computational Linguistics: Human Language Technologies), pp. 3715–3734. Association for Computational Linguistics, July 2022. https://doi.org/10.18653/v1/2022.naacl-main.272. https://aclanthology.org/2022.naacl-main.272

28. Tam, W.L., et al.: Parameter-efficient prompt tuning makes generalized and calibrated neural text retrievers, July 2022. https://doi.org/10.48550/arXiv.2207.07087, http://arxiv.org/abs/2207.07087
29. Thakur, N., Reimers, N., Rücklé, A., Srivastava, A., Gurevych, I.: BEIR: a heterogenous benchmark for zero-shot evaluation of information retrieval models, October 2021. https://doi.org/10.48550/arXiv.2104.08663. http://arxiv.org/abs/2104.08663
30. Wang, K., Thakur, N., Reimers, N., Gurevych, I.: GPL: generative pseudo labeling for unsupervised domain adaptation of dense retrieval. In: Proceedings of the 2022 Conference of the North American Chapter of the Association for Computational Linguistics: Human Language Technologies (2022). https://doi.org/10.18653/v1/2022.naacl-main.168. http://dx.doi.org/10.18653/v1/2022.naacl-main.168
31. Zhan, J., et al.: Disentangled modeling of domain and relevance for adaptable dense retrieval, August 2022. arXiv:2208.05753 [cs]
32. Zhang, X., Yates, A., Lin, J.: A little bit is worse than none: ranking with limited training data. In: Moosavi, N.S., et al. (eds.) Proceedings of SustaiNLP: Workshop on Simple and Efficient Natural Language Processing, pp. 107–112. Association for Computational Linguistics, Online, November 2020. https://doi.org/10.18653/v1/2020.sustainlp-1.14, https://aclanthology.org/2020.sustainlp-1.14
33. Zong, Y., Piwowarski, B.: XpmIR: a modular library for learning to rank and neural IR experiments. In: Proceedings of the 46th International ACM SIGIR Conference on Research and Development in Information Retrieval, SIGIR 2023, pp. 3185–3189. Association for Computing Machinery, New York (2023). https://doi.org/10.1145/3539618.3591818

SPARe: Supercharged Lexical Retrievers on GPU with Sparse Kernels

Tiago Almeida(✉) and Sérgio Matos

IEETA/DETI, LASI, University of Aveiro, 3810-193 Aveiro, Portugal
{tiagomeloalmeida,aleixomatos}@ua.pt

Abstract. Lexical sparse retrievers, rely on efficient searching algorithms that operate over inverted index structures, tailored specifically for CPU. This CPU-centric design poses a challenge when adapting these algorithms for highly parallel accelerators, such as GPUs, thus deterring potential performance gains. To address this, we propose to leverage the recent advances in sparse computations offered by deep learning frameworks to directly implementing sparse retrievals on these accelerators.

This paper presents the SPARe (SPArse Retrievers) Python package, which provides a high-level API to deal with sparse retrievers on (single or multi)-accelerators, by leveraging deep learning frameworks at its core.

Experimental results show that SPARe, running on an accessible GPU (RTX 2070), can calculate the BM25 scores for close to 9 million MSMARCO documents at a rate of 800 questions per second with our specialized algorithm. Notably, SPARe proves highly effective for denser LSR indexes, significantly surpassing the performance of established systems such as PISA, Pyserini and PyTerrier.

SPARe is publicly available at https://github.com/ieeta-pt/SPARe.

Keywords: Sparse Retrieval · CSR/CSC Sparse Matrices · Information Retrieval · Sparse Kernels · Python Package

1 Introduction

Information retrieval (IR) is predominantly shaped by transformer-based solutions that harness the recent advancements in deep learning to learn strong supervised retrieval models. Nonetheless, many of these cutting-edge approaches continue to depend on lexical retrievals, commonly referred to as sparse retrievers. These sparse retrievers serve various purposes, from acting as a benchmark for validation to playing integral roles such as the initial retrieval stage for rerankers or for negative mining during the training of retrieval and reranking models. More recently, owing to the superior efficiency of lexical retrievals, works such as DeepCT [3], DeepImpact [15], and SPLADE [6] proposed to build more robust sparse representations that are compatible with the current searching algorithms employed by lexical retrievals, highlighting the still paramount importance of these models in the field.

N. Goharian et al. (Eds.): ECIR 2024, LNCS 14610, pp. 413–421, 2024.
https://doi.org/10.1007/978-3-031-56063-7_33

Traditionally, sparse retrievers have heavily leaned on specialized algorithms, based on inverted-indexes and designed for general-purpose processors (CPUs). But with the rapid evolution of technology, GPUs and other accelerators present a promising avenue to challenge this and redefine the horizons of retrieval efficiency for sparse retrievers. Unlike previous works [4,7,8], our objective is not to transfer the current existent CPU algorithms, like WAND [2], to GPU, but rather embrace the unified conceptual retrieval framework [9] and the rapidly evolving general-purpose GPU (GPGPU) ecosystem.

In this work, we aim to challenge the established strategy, by introducing a framework designed to seamlessly execute sparse retrievers on single or multiple accelerators, including GPUs. Specifically, we harness the latest advancements in sparse computational capabilities [1] to directly implement sparse retrievers using highly-optimized kernels. Our motivation is that by leveraging these accelerators, we can achieve higher question throughput when compared to the traditional searching approach. Furthermore, these accelerators are becoming an essential commodity for any researcher currently working on neural IR.

Our endeavours have culminated in an open-sourced Python package named SPARe (SPArse Retrievers). This package features a user-friendly, high-level interface that facilitates direct interactions with the sparse retrievers on accelerators, shielding users from unnecessary complexity. Notably, while our package introduces new capabilities, it is designed to be compatible with existing searching systems, specifically with Terrier and Anserini indexes. This design choice reflects our intention not to supplant but rather to complement current retrieval systems, enhancing them with accelerator-based search capabilities.

To assess the computational efficiency of SPARe, we benchmark its performance against well established retrieval packages, namely PyTerrier [13], Pyserini [10] and PISA [16], using datasets ranging from 8k to 9M documents. Additionally, we conduct further experiments with SPARe to delve deeper into its performance nuances.

Our contributions can be summarized as follows: (1) an innovative method for implementing sparse retrievals on accelerators; (2) an iterative algorithm for computing inner product by using a CSC sparse matrix; (3) a practical connection with the unified conceptual framework for implementing sparse retrievals; (4) Python package publicly available at https://github.com/ieeta-pt/SPARe.

2 SPARe Framework

The SPARe framework is built on top of the conceptual framework for representational approach proposed by Lin [9]. This approach offers a unified formulation to describe any retrieval model, whether dense or sparse, from a representational perspective. With more detail, let's define the top-k ranking order of any retrieval function as f. Let q represent a question and D represent a collection of documents. Then, according to the conceptual framework, f can be expressed as a function of an arbitrary scoring operation, denoted by ϕ, over a transformation of each query and document, represented by η_q and η_d, respectively, as illustrated in Eq. 1.

$$f(q, D) = \arg \underset{d \in D}{\text{top-k}}\ \phi(\eta_q(q), \eta_d(d)) \tag{1}$$

When applying Eq. 1 to sparse retrievers, denoted as f_{sr}, both η_q and η_d will always produce sparse vectors. With the intention of capitalizing on the capabilities of accelerators, we designate the scoring function, ϕ, as the dot product, since it is particularly efficient on accelerators when dealing with sparse vectors, resulting in Eq. 2.

$$f_{sr}(q, D) = \arg \text{top-k}\ \eta_d(D) \cdot \eta_q(q) \tag{2}$$

Within the SPARe framework, any existing sparse retrieval method can be executed on accelerators if we can determine the functions η_d and η_q, such that their dot product retains the original retrieval order of f_{sr}. Encouragingly, many conventional sparse retrieval techniques, like BM25 [17], and emerging learned sparse retrieval (LSR) [3,6,15] approaches are congruent with this formulation. Another convention is also to directly apply η_d to the entire collection and cache the results. As an example, we demonstrate in Eq. 3 how BM25 can be adapted to fit the framework outlined in Eq. 2. For the sake of clarity, we assume that both the query and document are encoded in a bag-of-words format, represented as $\vec{q}$ and $\vec{d}$ respectively.

$$\eta_q(q) = \frac{(k_3 + 1) * \vec{q}}{k_3 + \vec{q}},\ \eta_d(d) = IDF(d) * \frac{(k_1 + 1)) * \vec{d}}{k_1 * ((1 - b) + b * \frac{|\vec{d}|}{avgdl}) + \vec{d}} \tag{3}$$

Inspecting Eq. 2, it becomes clear that our hypothesis is that by utilizing highly efficient sparse computation on accelerators, we can compute all query-document scores more rapidly than traditional methods, which employ algorithms such as WAND [2] to compute only a much limited subset of the total scores.

2.1 Implementation Details

SPARe was designed on top of deep learning frameworks, leveraging their robust ecosystem that supports a wide array of processing units including CPU, GPU, and TPU. The adoption of these frameworks ensures that SPARe can benefit from regular updates and advances in sparse kernel capabilities. To enhance its versatility, SPARe's modular architecture enables easy integration of new backends, requiring only the implementation of the respective a Python interface.

SPARe operates in two primary modes: the indexer mode and the searching mode. The indexer mode is responsible for converting a document collection into a sparse format compatible with our framework. On the other hand, the searching mode utilizes the available accelerators to facilitate high-performance retrieval with the pre-processed sparse collection.

Regarding the sparse data representation, the most popular formats are COO (Coordinate format), CSR (compressed sparse row), and CSC (compressed sparse column). COO stores a vector of non-zero values accompanied by two vectors indicating the row and column indices of these values. Conversely, CSR (and CSC) is a more recent format that condenses the row indices by only storing the position in the value vector for the first element of each row, thus making these compressed formats more memory and computationally efficient. In SPARe, although the COO format is supported, CSR and CSC are the main formats due to their benefits.

Beyond the standard sparse inner product kernel offered by the cuSPARSE library, we introduce an experimental iterative algorithm that efficiently computes the inner product between highly sparse CSC matrices and vectors. The pseudocode shown in Algorithm 1 details our approach to iteratively accumulate, in a dense vector, the product between the value of each query token and the values of documents containing that token. In other words, this algorithm decomposes the sparse computation to a series of dense vector-scalar multiplications and accumulations. Another advantage of this approach is the support for lower-precision data types, an area where the existing cuSPARSE is currently lacking optimizations.

Input: C_{csc}, q_{coo}, top_k, $dtype$
Result: $\vec{r}$
$a \leftarrow \vec{0}.\text{as\_dtype}(dtype)$; `/* Zero vector with size of the collection */`
$\vec{\text{ccol\_indices}}, \vec{\text{row\_indices}}, \vec{\text{values}} \leftarrow \text{decompose}(C_{csc})$;
$\vec{\text{s\_idx}}, \vec{\text{e\_idx}} \leftarrow \vec{\text{ccol\_indices}}[q_{coo}.\vec{\text{indices}}], \vec{\text{ccol\_indices}}[q_{coo}.\vec{\text{indices}} + 1]$;
$i \leftarrow 0$;
while $i < |q_{coo}|$ **do**
  $\vec{\text{v\_indices}} \leftarrow \vec{\text{row\_indices}}[\vec{\text{s\_idx}}[i] : \vec{\text{e\_idx}}[i]]$;
  $\vec{\text{v\_values}} \leftarrow (\vec{\text{values}}[\vec{\text{s\_idx}}[i] : \vec{\text{e\_idx}}[i]] * q_{coo}.\vec{\text{values}}[i]).\text{as\_dtype}(dtype)$;
  $a.\text{add\_at\_index}(\vec{\text{v\_indices}}, \vec{\text{v\_values}})$;
  $i \leftarrow i + 1$;
end
$\vec{r} \leftarrow \text{topk}(a, top_k)$;

Algorithm 1: An iterative inner product Algorithm for Sparse CSC Matrices and Vectors. Inputs: CSC-format matrix C_{csc}, COO-format query vector q_{coo}, number of top-ranking documents top_k, and accumulation data type $dtype$

2.2 How to Index

SPARe offers a versatile indexing API. Users can either index directly from text by supplying a tokenizer or from an iterator of pre-tokenized bag-of-words vectors. Additionally, to promote interoperability with other existing search engines, we also offer a direct index conversion mechanism, as showcased in Code 1.

```
1 import spare
2 sparseCSR_collection = spare.SparseCollection.from_bm25_pyserini_index("anserini_index", k1=1.2, b=0.75,
  ↪  dtype=spare.float32, backend="torch")
3 sparseCSR_collection.save_to_file("spare_converted_anserini_bm25"))
```

Code 1: Example of how to index from another framework with SPARe.

2.3 How to Search

At the heart of the search API is the *SparseRetriever*. As depicted in Code 2, this component bridges the gap between the previously constructed sparse collection and the retrieval operations executed on accelerators. By default, SPARe is designed to prioritize the fastest devices available. This means that if a computer has multiple Accelerators, retrieval is executed in a single program multiple data (SPMD) fashion.

```
1 import spare
2 from transformers import AutoTokenizer
3 tokenizer = AutoTokenizer.from_pretrained("bert-base-uncased")
4 bow = spare.BagOfWords(tokenizer, tokenizer.vocab_size)
5 sparse_collection = spare.SparseCollection.load_from_file("spare_wbertTokenizer_bm25", backend="torch")
6 # BM25WeightingModel will be used as default because collection is of type bm25.
7 sparse_retriver = spare.SparseRetriever(sparse_collection, bow)
8 # runs on multiGPU (SPMD), returns the scores, and records the timings of the operation
9 out = sparse_retriver.retrieve(list_of_questions, top_k=1000, return_scores=True, profiling=True)
```

Code 2: Example of how to search with SPARe.

3 Experimental Results

To assess the performance of our SPARe framework, we benchmarked it against the PyTerrier [13], Pyserini [10] and PISA [16] search engines using BM25 and LSR indexes. For uniformity, as each framework employs a Python interface, PISA was accessed via the PyTerrier Python interface [11], hereafter referred to as PISA (PyTerrier). Furthermore, queries were always processed in batch mode, i.e., all questions are submitted simultaneously, thus minimizing any Python overhead. Our evaluation focused on the question throughput (QPS) of each system across six datasets from BEIR [18]. These datasets were selected based on the size of their collections, ranging from Arguana's 8k documents to MSMarco's nearly 9M. We also considered the number of questions in each dataset, choosing only those with at least 1,000 questions to ensure robust time measurements. In our experimental setup, we first indexed all datasets using Pyserini, PyTerrier, and PISA. Then, we converted these indexes into a sparse format with SPARe, using torch as the backend. Regarding SPARe, it was evaluated on both CPU and GPU, referred to as SPARe (CPU) and SPARe (GPU) respectively. Additionally, we also evaluated a SPARe variant that utilizes Algorithm 1, denoted as SPARe $(\mathrm{GPU})_{iter}$, to compute the question-document scores. All the experiments were conducted in a moderate machine with the following specifications: i9-9900K CPU @3.60 GHz, NVIDIA GeForce RTX 2070 8 GB VRAM and 32 GB of RAM.

In Table 1, we present SPARe's performance using BM25 indexes compared to other frameworks. SPARe consistently outperforms PyTerrier and Pyserini. A key observation is the impact of dataset size on SPARe's various versions. In smaller datasets, SPARe's CPU and GPU versions show comparable performance, attributed to costly data transfer times. Notably, SPARe's iterative algorithm outperforms the cuSPARSE kernel on larger datasets, though its performance drops sharply in the Arguana dataset, largely due to the longer questions (200 words in average), which are less suited for this approach. A notable difference in SPARe's performance, when integrated with PyTerrier compared to Pyserini, stems from PyTerrier's slower tokenizer. In our comparison with PISA (PyTerrier), we introduce a new SPARe variant, SPARe $(\text{GPU})_{iter}^{uint8}$, employing 8-bit accumulation in the iterative algorithm. This variant surprisingly yields considerable throughput improvements while maintaining similar retrieval effectiveness. Overall, SPARe's iterative variant demonstrates competitive performance against PISA (16T) in larger datasets, whereas the cuSPARSE SPARe is more effective in smaller datasets.

Table 1. Comparison of QPS and ndcg@1000 across datasets for BM25.

Systems	MSMARCO (8.8M)		HotPotQA (5.2M)		NQ (2.6M)		Quora (0.5M)		SciDocs (25K)		Arguana (8K)	
	QPS	nDCG	QPS	nDCG	QPS	nDCG	QPS	nDCG	QPS	nDCG	QPS	nDCG
Indexes from PyTerrier												
PyTerrier†	10	31.65	8	64.01	10	36.72	22	79.94	8	26.79	7	41.79
SPARe (CPU)	7	31.66	12	64.92	18	36.71	195	79.70	351	26.79	**353**	41.71
SPARe (GPU)	114	31.65	172	64.93	250	36.71	**481**	79.81	**431**	26.79	344	41.71
SPARe $(\text{GPU})_{iter}$	417	31.64	346	63.93	403	36.70	426	79.84	374	26.79	143	41.71
SPARe $(\text{GPU})_{iter}^{fp16}$	**428**	31.62	**365**	63.94	**415**	36.72	430	79.85	387	26.79	130	41.70
Indexes from Pyserini												
Pyserini	26	31.11	15	64.54	31	37.61	56	83.17	64	26.43	41	41.03
$\text{Pyserini}_{16\ threads}$	93	31.12	71	64.54	97	37.62	119	83.17	113	26.43	96	41.03
SPARe (CPU)	6	31.11	11	64.27	16	37.35	278	83.17	931	26.54	1158	41.22
SPARe (GPU)	88	31.13	145	64.27	213	37.37	**2546**	83.17	**1645**	26.54	**1284**	41.22
SPARe $(\text{GPU})_{iter}$	611	31.13	573	64.27	980	37.38	1324	83.18	1084	26.54	169	41.22
SPARe $(\text{GPU})_{iter}^{fp16}$	**801**	31.11	**662**	64.28	**1087**	37.43	1282	83.19	1038	26.54	153	41.23
Indexes from PyTerrier‡												
PISA (PyTerrier)	228	31.45	76	63.73	233	36.68	921	79.92	885	26.80	217	38.40
PISA $(\text{PyTerrier})_{16\ threads}$	**1161**	31.45	591	63.73	1154	36.68	1947	79.92	1751	26.80	1077	38.40
SPARe (GPU)	115	31.65	172	63.93	254	36.71	**2762**	79.81	**1832**	26.79	**2117**	41.71
SPARe $(\text{GPU})_{iter}^{fp16}$	890	31.62	731	63.94	1228	36.73	1735	79.85	1121	26.79	180	41.70
SPARe $(\text{GPU})_{iter}^{uint8}$	1069	31.12	**856**	63.44	**1364**	37.11	1763	79.47	1121	26.84	187	43.30

† *We excluded a multiprocess version of PyTerrier due to the inability to achieve any significant speed improvements.*
‡ *As we were unable to access the PISA tokenizer over Python, we benchmarked PISA and SPARe without the tokenizer, i.e., both queries and documents are fed as bag-of-words that were created by PyTerrier.*

An interesting observation is that traditional retrieval systems distribute workload across multiple threads to enhance performance, which is especially effective when the number of questions exceeds the number of threads. However, it is important to note that the core search algorithms remain difficult to parallelize [8].

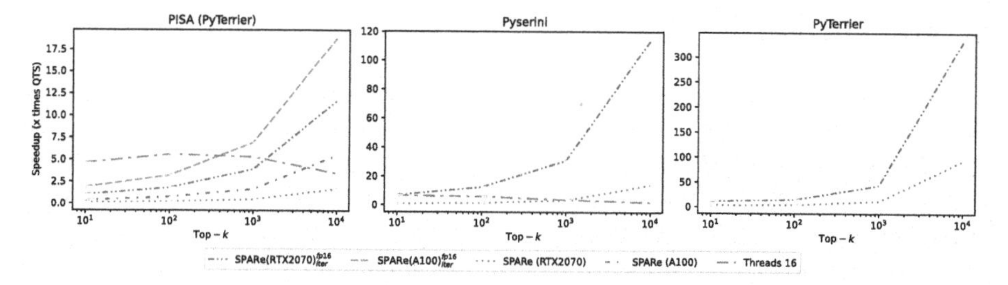

Fig. 1. Speedup performance comparison for MSMARCO dataset: PISA, Pyserini, and PyTerrier single-thread baselines versus SPARe. Note: SPARe results on A100 are exclusive to the PISA benchmark.

Furthermore, to delve deeper into SPARe's capabilities, we suggest an additional benchmark, focusing on question throughputs across different values of top-k. As SPARe's performance is independent of k, gains become evident when retrieving larger ranked lists (over 1000 documents), as depicted in Fig. 1.

Regarding the LSR indexes, we adopted the popular SPLADE++ model [5,12] to encode our documents and questions. For the benchmark, we only consider as baseline the Pyserini$_{16\ threads}$ and PISA(PyTerrier)$_{16\ threads}$, due to the higher computation costs associated with this type of sparse index [14].

Table 2. Comparison of QPS and ndcg@1000 across datasets for SPLADE.

Systems	MSMARCO (8.8M)		HotPotQA (5.2M)		NQ (2.6M)		Quora (0.5M)		SciDocs (25K)		Arguana (8K)	
	QPS	nDCG	QPS	nDCG	QPS	nDCG	QPS	nDCG	QPS	nDCG	QPS	nDCG
Pyserini$_{16\ threads}$	5	51.15	3	72.78	9	58.88	36	85.44	78	27.52	65	43.43
PISA(PyTerrier)$_{16\ threads}$	13	51.17	8	72.79	29	58.90	107	85.43	484	27.54	**396**	43.44
SPARe (GPU)^{fp16}	35	51.11	59	72.80	98	58.96	**982**	85.49	**1833**	27.55	298	43.45
SPARe (GPU)$^{fp16}_{iter}$	**174**	51.13	**155**	72.81	**261**	59.05	271	85.50	244	27.53	80	43.45

Table 2 demonstrates SPARe's superior performance over PISA and Pyserini in this new type of indexes. This advantage is likely due to the SPLADE++ model yielding denser collections, affecting the efficiency of WAND-type algorithms. This is illustrated by the weaker results of these methods in HotPotQA, which saw a 336-fold increase in density, in opposition to the best results in the Arguana dataset, which had the smallest density increase at only two-fold.

4 Limitations

While SPARe exhibits noteworthy performance, it is crucial to acknowledge some inherent limitations. First, the design of the current version of SPARe assumes that the entire sparse collection can be accommodated in memory. Secondly, although we went the extra mile to make sure the experiences were fair, we recognized that some Python/Java overheads may affect the measurements.

5 Conclusion

In this study, we presented SPARe, a Python framework designed for sparse retrieval with accelerator support. Our findings demonstrate SPARe's superior performance over traditional systems like Pyserini and PyTerrier on BM25 indexes, while also closely contending with PISA for larger top-k values. SPARe is particularly effective with the newer, denser LSR indexes, where it significantly outperforms conventional retrieval methods. Furthermore, our iterative algorithm consistently shows substantial speed gains in processing large-scale collections across both sets of experiments.

Acknowledgments. This work was funded by the Foundation for Science and Technology (FCT) in the context of project UIDB/00127/2020. The work was produced with the support of HPC Universidade de Évora with funds from the Advanced Computing Project FCT.CPCA.2022.01. Tiago Almeida is funded by the grant 2020.05784.BD.

References

1. Bell, N., Garland, M.: Efficient sparse matrix-vector multiplication on CUDA. NVIDIA Technical Report NVR-2008-004, NVIDIA Corporation, December 2008
2. Broder, A., Carmel, D., Herscovici, M., Soffer, A., Zien, J.: Efficient query evaluation using a two-level retrieval process, pp. 426–434, November 2003. https://doi.org/10.1145/956863.956944
3. Dai, Z., Callan, J.: Context-aware term weighting for first stage passage retrieval. In: Proceedings of the 43rd International ACM SIGIR Conference on Research and Development in Information Retrieval, SIGIR 2020, New York, NY, USA, pp. 1533–1536. Association for Computing Machinery (2020). https://doi.org/10.1145/3397271.3401204
4. Ding, S., He, J., Yan, H., Suel, T.: Using graphics processors for high performance IR query processing. In: Proceedings of the 18th International Conference on World Wide Web, WWW 2009, New York, NY, USA, pp. 421–430. Association for Computing Machinery (2009). https://doi.org/10.1145/1526709.1526766
5. Formal, T., Lassance, C., Piwowarski, B., Clinchant, S.: From distillation to hard negative sampling: making sparse neural IR models more effective. In: Proceedings of the 45th International ACM SIGIR Conference on Research and Development in Information Retrieval, SIGIR 2022, New York, NY, USA, pp. 2353–2359. Association for Computing Machinery (2022). https://doi.org/10.1145/3477495.3531857
6. Formal, T., Piwowarski, B., Clinchant, S.: SPLADE: sparse lexical and expansion model for first stage ranking. In: Proceedings of the 44th International ACM SIGIR Conference on Research and Development in Information Retrieval, SIGIR 2021, New York, NY, USA, pp. 2288–2292. Association for Computing Machinery (2021). https://doi.org/10.1145/3404835.3463098
7. Gaioso, R., Costa, V., Guardia, H., Senger, H.: Performance evaluation of single vs. batch of queries on GPUs. Concurr. Comput. Practic. Exp. **32**, August 2019. https://doi.org/10.1002/cpe.5474
8. Gaioso, R., Gil-Costa, V., Guardia, H., Senger, H.: A parallel implementation of wand on GPUs. In: 2018 26th Euromicro International Conference on Parallel, Distributed and Network-based Processing (PDP), pp. 10–17 (2018). https://doi.org/10.1109/PDP2018.2018.00011

9. Lin, J.: A proposed conceptual framework for a representational approach to information retrieval. CoRR **abs/2110.01529** (2021). https://arxiv.org/abs/2110.01529
10. Lin, J., Ma, X., Lin, S.C., Yang, J.H., Pradeep, R., Nogueira, R.: Pyserini: a Python toolkit for reproducible information retrieval research with sparse and dense representations. In: Proceedings of the 44th Annual International ACM SIGIR Conference on Research and Development in Information Retrieval (SIGIR 2021), pp. 2356–2362 (2021)
11. MacAvaney, S., Macdonald, C.: A Python interface to PISA! In: Proceedings of the 45th International ACM SIGIR Conference on Research and Development in Information Retrieval, SIGIR 2022, New York, NY, USA, pp. 3339–3344. Association for Computing Machinery (2022). https://doi.org/10.1145/3477495.3531656
12. Macdonald, C., MacAvaney, S.: PyT-SPLADE: PyTerrier plugin for SPLADE models (2023). https://github.com/cmacdonald/pyt_splade
13. Macdonald, C., Tonellotto, N.: Declarative experimentation information retrieval using PyTerrier. In: Proceedings of ICTIR 2020 (2020)
14. Mackenzie, J., Trotman, A., Lin, J.: Wacky weights in learned sparse representations and the revenge of score-at-a-time query evaluation (2021)
15. Mallia, A., Khattab, O., Suel, T., Tonellotto, N.: Learning passage impacts for inverted indexes. In: Proceedings of the 44th International ACM SIGIR Conference on Research and Development in Information Retrieval, SIGIR 2021, New York, NY, USA, pp. 1723–1727. Association for Computing Machinery (2021). https://doi.org/10.1145/3404835.3463030
16. Mallia, A., Siedlaczek, M., Mackenzie, J., Suel, T.: PISA: performant indexes and search for academia. In: Proceedings of the Open-Source IR Replicability Challenge Co-located with 42nd International ACM SIGIR Conference on Research and Development in Information Retrieval, OSIRRC@SIGIR 2019, Paris, France, 25 July 2019, pp. 50–56 (2019). http://ceur-ws.org/Vol-2409/docker08.pdf
17. Robertson, S., Zaragoza, H.: The probabilistic relevance framework: BM25 and beyond. Found. Trends Inf. Retr. **3**(4), 333–389 (2009). https://doi.org/10.1561/1500000019. https://www.nowpublishers.com/article/Details/INR-019
18. Thakur, N., Reimers, N., Rücklé, A., Srivastava, A., Gurevych, I.: BEIR: a heterogeneous benchmark for zero-shot evaluation of information retrieval models. In: Thirty-fifth Conference on Neural Information Processing Systems Datasets and Benchmarks Track (Round 2) (2021). https://openreview.net/forum?id=wCu6T5xFjeJ

DQNC2S: DQN-Based Cross-Stream Crisis Event Summarizer

Daniele Rege Cambrin(✉), Luca Cagliero, and Paolo Garza

Politecnico di Torino, Turin, Italy
{daniele.regecambrin,luca.cagliero,paolo.garza}@polito.it

Abstract. Summarizing multiple disaster-relevant data streams simultaneously is particularly challenging as existing Retrieve&Re-ranking strategies suffer from the inherent redundancy of multi-stream data and limited scalability in a multi-query setting. This work proposes an online approach to crisis timeline generation based on weak annotation with Deep Q-Networks (DQNs). It selects on-the-fly the relevant pieces of text without requiring human annotations or content re-ranking. This makes the inference time independent of the number of input queries. The proposed approach also incorporates a redundancy filter into the reward function to handle cross-stream content overlaps effectively. The ROUGE and BERTScore results achieved on the CrisisFACTS 2022 benchmark are better than those of the best-performing models.

Keywords: Cross-Stream Temporal Summarization · Crisis Management · Timeline Generation · Reinforcement Learning · Text Retrieval

1 Introduction

Automating the extraction of valuable information from large streams of social and news data is particularly relevant to crisis management as could become an active support to disaster-response personnel [12,21]. The problem of generating crisis event summaries of time-evolving data streams has already been addressed by the Temporal Summarization [3] (TS) and Incident Stream [14] (IS) challenges, where the goals are to extract variable-length summaries for a particular event on a set day (TS) and to classify and prioritize information in a single social stream (IS), respectively. More recently, the CrisisFACTS task [13] has focused on retrieving daily timelines from social/news data streams. The main challenges are (1) The contemporary presence of multiple streams of crisis-relevant data (Twitter, Reddit, Facebook, and News), which makes the input collection particularly redundant and complex to summarize; (2) The lack of human annotations on text relevance and topic-level clusters, which limits the scope of supervised techniques; (3) A list of queries about information needed by emergency responders for each emergency event, which increases the complexity compared to single-query tasks. State-of-the-art approaches to the CrisisFACTS task (unicamp [17] and ohmkiz [22]) rely on ColBERT [10] or on the established Retrieve&Re-ranking

N. Goharian et al. (Eds.): ECIR 2024, LNCS 14610, pp. 422–430, 2024.
https://doi.org/10.1007/978-3-031-56063-7_34

approach. This last one entails retrieving the pieces of text relevant to an input query first and then re-rank them. The retrieval step is commonly driven by text relevance scoring functions stored in index structures (e.g., BM25 [20], Dense Passage Retriever [9]) whereas the re-ranking step is commonly based on neural models such as ColBERT [10], DRMM [8], and Conv-KNRM [5]. Both end-to-end ColBERT and Retrieve&Rerank approaches scale linearly with the number of input queries and are not designed for online text processing. Furthermore, it is necessary to incorporate a filtering stage after the re-ranking phase thus retrieving large volumes of redundant content in a multi-stream scenario.

This work proposes a new DQN-based Cross-Stream Crisis event Summarizer. It overcomes the main issues of existing approaches to CrisisFACTS by adopting Deep Reinforcement Learning (DRL). DRL has proved to be effective in both recommender systems [1] and text summarization [4]. The key contributions of the present work are enumerated below.

- **DRL approach for online text retrieval.** Unlike state-of-the-art approaches (e.g., [17]), we rely on DRL. Hence, we do not need to index crisis-relevant data for efficient content retrieval. Conversely, we retrieve relevant texts on the fly from the input streams without any human supervision.
- **Early redundancy filtering.** The DRL reward function already incorporates a redundancy filter to reduce the amount of data processed after retrieval. This improves the efficiency and reduces the complexity of the timeline generation method.
- **Efficient multi-query setting.** Unlike existing approaches to CrisisFACTS, our approach inherently supports multiple queries simultaneously, differently from solutions that are still dependent on the number of input queries [15,24]. This allows the efficient generation of timelines of multi-faceted crisis events, where each query refers to a complementary facet.

Our approach performs better than the state of the art on the CrisisFACTS benchmark. The project code is publicly available for research purposes[1]

.

2 Problem Statement

We address the CrisisFACTS 2022 task first proposed in [13]. The goal is to summarize multiple streams of social and news data describing the same short-lasting crisis event. Let E_a be a set of crisis events about different hazards (see Table 1). We employ cross-validation selecting a subset $E \in E_a$, which will be used to train the model and the remaining events for testing. Each event lasts at least two days within the reference time period T and is described by a set S of streams of textual content, i.e., tweets, Reddit messages, news, and Facebook posts[2]. Stream data consists of timestamped pieces of text te_s^t, where $s \in S$

[1] https://github.com/DarthReca/crisis-dqn Latest access: October 2023.

[2] Due to the restrictions in force to the Facebook data crawler, similar to most peers hereafter we will disregard this stream.

Table 1. CrisisFACTS 2022 events

Event ID	Name	Queries	Texts	Days
001	Lilac Wildfire 2017	52	45578	9
002	Cranston Wildfire 2018	52	25172	6
003	Holy Wildfire 2018	52	25482	6
004	Hurricane Florence 2018	51	180286	15
005	Maryland Flood 2018	48	37598	4
006	Saddleridge Wildfire 2019	52	34480	4
007	Hurricane Laura 2020	51	52561	2
008	Hurricane Sally 2020	51	67632	8

and $t \in T$. The task focuses on a set Q_e of event queries for each $e \in E$ that are of interest to emergency responders. Given e, S, and Q_e, the CrisisFACTS task aims at generating a summary of S that consists of daily crisis timelines reporting the top stories related to Q_e on e. On each timestamp (day) t, the timeline consists of a shortlist of *facts*, sorted by decreasing importance, and described by one or more pieces of text te^t_s.

3 Methodology

We present the DQN-based Cross-Stream Crisis event Summarizer (DQNC2S, in short). It encompasses a three-step process. First, we automatically weakly annotate text candidates based on established extractive QA models to guide the training of the DQN. Secondly, our approach relies on Deep Reinforcement Learning to online retrieve relevant yet non-redundant content suitable for daily summaries. Finally, it applies topic modeling and abstractive summarization to synthesize and rephrase the retrieved texts.

Weak Annotation. The use of extractive Question Answering (QA) models has proved effective in supporting the generation of crisis-related event descriptions [22]. Inspired by the recent use of SQUAD [18] for evidence estimation, we perform a weak annotation step to create an importance score for each query-text pair. Specifically, given a pair of query $q_i \in Q_e$ and a set S of multiple streams related to crisis event e, we leverage Electra[3] and LongFormer[4] to generate tuples $<q_i, te^t_s, CF>$, where CF is the mean confidence score associated by the QA models to the query-text pair. To avoid local model inaccuracies, a tuple is generated only when both QA models provide an answer with a minimal confidence level of 80%. Each text gets a score Sc equal to the number of queries it provides an answer to.

[3] https://huggingface.co/deepset/electra-base-squad2 Latest access: October 2023.
[4] https://huggingface.co/allenai/longformer-large-4096-finetuned-triviaqa Latest access: October 2023.

DQN-Based Text Retrieval. Our retriever is based on a Deep Q-Network [16], which interacts with an environment designed to work online and driven by a single parameter, i.e., the maximum number of retrievable texts. The action space is binary, i.e., $A = 0$ when the text is kept or $A = 1$ when it is discarded. The observation space of size 770 comprises the BERT embedding of the current text (size 768), the remaining percentage of texts P_t that can be chosen (size 1), and the maximum similarity Si_m between the current text and the already kept texts (size 1).

The reward function $\mathcal{R}$ is defined in Eq. (1).

$$\mathcal{R} = \begin{cases} -5 & \text{if } Sc = 0 \wedge A = 0 \\ 1 & \text{if } Sc = 0 \wedge A = 1 \\ N_{Sc} * (1 - Si_m) & \text{if } Sc > 0 \wedge A = 0 \\ -N_{Sc} * (1 - Si_m) & \text{if } Sc > 0 \wedge A = 1 \end{cases} \tag{1}$$

Specifically, if $Sc = 0$, the text te_s^t should be discarded as unlikely to be informative (Sc is equal to zero only if te_s^t does not answer the event queries). If te_s^t is kept, the reward is set to -5 (value chosen after reward shaping); otherwise, the reward is set to 1. If te_s^t has a non-zero score ($Sc > 0$), it should be kept only if it is dissimilar enough to any previously selected texts (according to the cosine similarity between the corresponding embeddings) to limit redundancy. We leverage a normalized score $N_{Sc}=\frac{Sc}{|Q_e|}$ to deal with a variable number of queries per event and then re-scale the product by the maximum similarity score Si_m.

An example of the application of this phase is shown in Fig. 1. In the example reported in Fig. 1, we suppose the input text te_s^t under analysis is irrelevant, and the system already selected two texts (one relevant and another one irrelevant) over a maximum of ten retrievable texts.

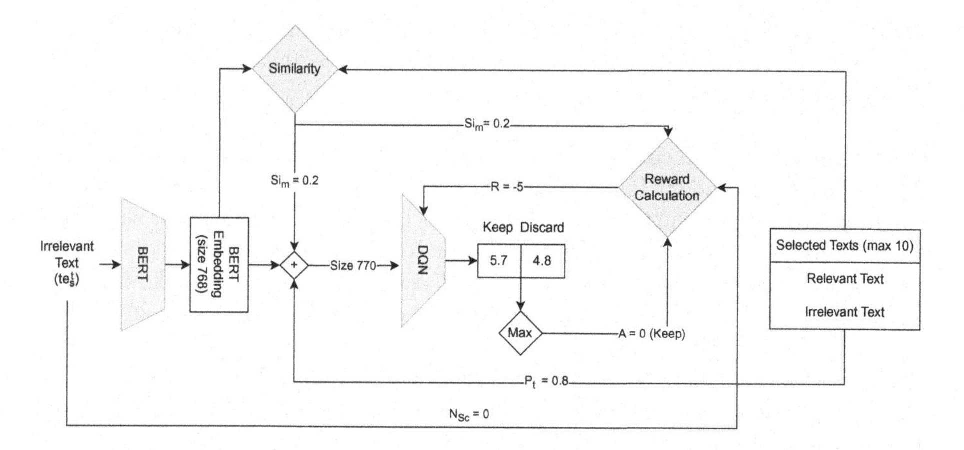

Fig. 1. DQN-based text retrieval. Example of application to an irrelevant text.

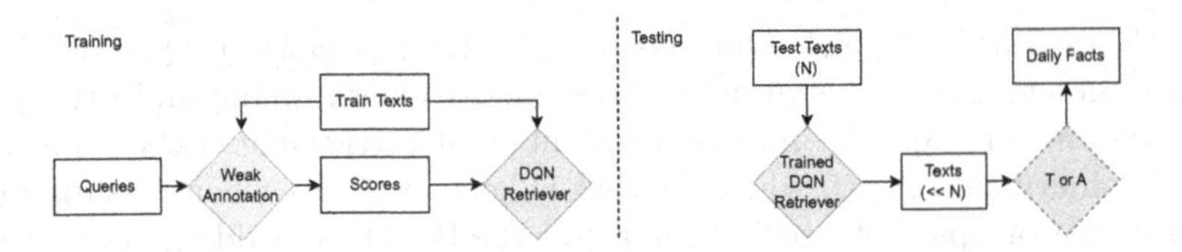

Fig. 2. Pipeline during training on the left and testing on the right. During testing, topic modeling (T) or abstraction (A) are optional.

Topic Modeling and Abstraction. The outcomes of extractive summarization strategies are often suboptimal when the summary consists of several fragments of (potentially incoherent) text, especially while dealing with social data like tweets or Reddit posts. To cope with this issue, we explore the use of state-of-the-art topic modeling and abstractive summarization techniques, i.e., BERTopic [7] and BART-CNN [11]. The former technique groups the retrieved content into subtopics and provides a more faceted description of each fact. The latter reformulates the original text into a concise and more readable form. DQNC2S-T, DQNC2S-A, and DQNC2S-T+A are the system variants incorporating either topic modeling, or abstraction, or a combination of the above.

To generate the crisis event timeline, we produce daily facts. Each fact F corresponds to a different topic and is described by one or more pieces of text. The importance score of text te_s^t of fact F is $I_{te_s^t,F} = Q_T - Q_D$, where Q_T is the Q-value for taking the text, whereas Q_D is the Q-value for discarding it. The higher the gap between the taking/discarding expected rewards, the higher the relevance of the text to the fact description.

The complete pipeline in training and testing can be seen in Fig. 2.

4 Experiments

Dataset. The CrisisFACTS dataset consists of multi-stream data annotated with ground truth summaries extracted from Wikipedia, ICS-209 All-Hazards Dataset [6,23], and NIST annotation.

Evaluation Metrics. In compliance with [13], we evaluated the results in terms of Rouge-2 F1-Score and BERT-Score metrics. They respectively quantify summary coherence from syntactic and semantic perspectives [2]. To make a fair summary comparison, each system shortlists a list of top-k facts in order of decreasing importance, where k is specified by the NIST assessors. Since our work addresses efficiency-related aspects, we also consider the execution time (in seconds).

Baselines. We shortlist the best-performing methods according to the results of the CrisisFACTS 2022 challenge, i.e., unicamp [17] and ohmkiz [22]. They are both Retrieve&Re-ranking strategies. We also consider a state-of-the-art end-to-end model tested by [22], i.e., the ColBERT-based approach [10]. We also include the baselines provided by the CrisisFACTS organizers [13] for completeness.

Table 2. Comparison of mean Rouge-2 (R2) and BERT-Score (BS)

Method	ICS		NIST		Wikipedia		Mean		
	R2	BS	R2	BS	R2	BS	R2	BS	Rank
baseline.run1	0.0418	0.4432	0.1326	0.5565	0.0281	0.5296	0.0675	0.5098	2.4167
baseline.run2	0.0428	0.4427	0.1308	0.5565	0.0281	0.5274	0.0672	0.5089	2.0000
ohm_kiz.ColBERT	0.0497	0.4500	0.1386	0.5460	0.0307	0.5423	0.0730	0.5128	4.6667
ohm_kiz.QACrisis	0.0464	0.4432	0.1471	0.5642	0.0337	0.5448	0.0757	0.5174	6.0833
ohm_kiz.QAasnq	0.0507	0.4477	0.1468	0.5628	0.0362	0.5646	0.0779	0.525	7.1667
unicamp.NM2	**0.0581**	0.4591	0.1338	0.5573	0.0281	0.5321	0.0733	0.5162	5.6667
unicamp.NM1	**0.0581**	0.4591	0.1338	0.5573	0.0281	0.5321	0.0733	0.5162	5.6667
DQNC2S	0.0406	0.4554	**0.1540**	**0.5715**	0.0402	0.5516	0.0783	0.5262	7.6667
DQNC2S-T	0.0513	0.4579	0.1450	0.5667	0.0317	0.5538	0.0760	0.5261	7.5833
DQNC2S-A	0.0412	**0.4596**	0.1538	0.5706	0.0394	0.5538	0.0781	0.528	8.2500
DQNC2S-T+A	0.0452	0.4560	0.1515	0.5709	**0.0420**	**0.5707**	**0.0796**	**0.5325**	**8.8333**

Setup. We run the experiments on an Intel(R) Core(TM) i9-10980XE CPU and P5000 and A6000 GPUs. The Q-Network comprises a mpnet-base-v2 version of SentenceBERT [19] and three linear layers. We employ an Adam optimizer with a constant learning rate of 1e−3 and weight decay of 1e−4. We perform an 8-fold cross-validation on the events. Based on the empirical distribution of the number of facts per day (see Fig. 3), we set the maximum number of texts to 300.

Quantitative Results Overview. As shown in Table 2, DQNC2S performs best in terms of BERTScore for all the summary types, showing maximal semantic similarity with the reference crisis timelines. Its performance is best regarding ROUGE scores for all summaries except for ICS. According to the task organizers [13], the actual incident summaries in ICS are written for the public and from a historical perspective, not for the utility of emergency-response personnel. Thus, the syntactic n-gram overlap is likely less explanatory. DQNC2S-T+A turns out to be, on average, the most effective one. It is particularly effective in generating Wikipedia-like summaries, which provide topic-specific event insights. According to the t-test of statistical significance ($p = 0.08$), DQNC2S-T+A is better than the state-of-the-art method at a significant level of 92%. We have also ranked the solutions from 1 for the worst approach to 11 for the best method for each metric. All versions of DQNC2S are superior to the state-of-the-art.

Inference Time. DQNC2S decides if a piece of text is relevant in 0.0296 ± 0.0037 s for all the N queries. Instead, unicamp [17] requires approximately $N \cdot 0.0752 \pm 0.0380$ s just for the reranking step (disregarding the index creation and the usage of GPT-3). Similarly, ohmkiz [22] requires around $N \cdot 0.0293 \pm 0.0190$ for text re-ranking. The results confirm the higher efficiency of our strategy compared to state-of-the-art methods.

DQN Training and Testing. Figure 3a shows the mean percentage of "take" action per episode. The system exploration phase proceeds until the 500000-th

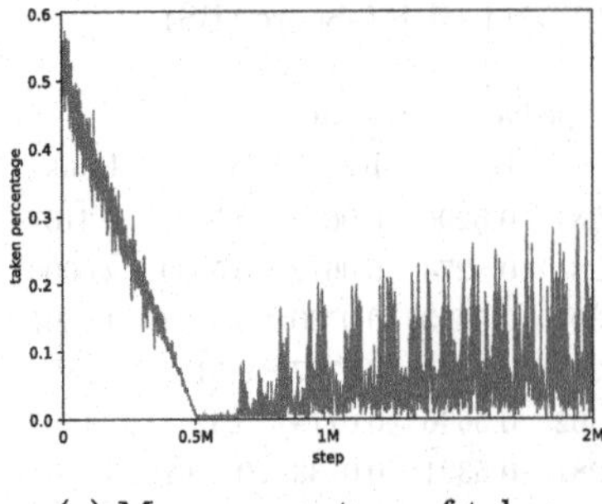

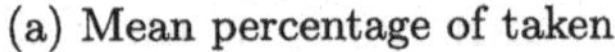

(a) Mean percentage of taken

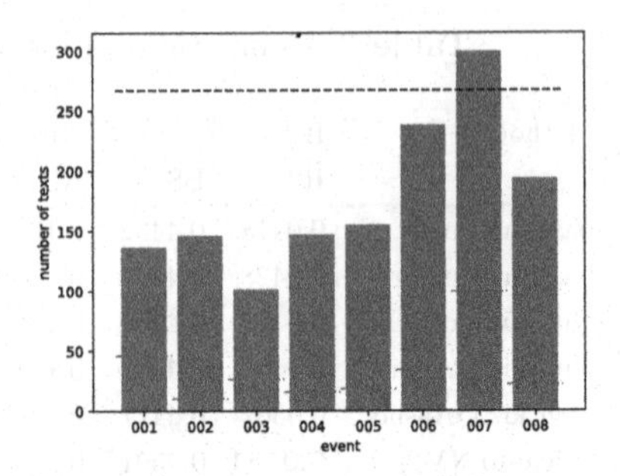

(b) Mean number of selected text

Fig. 3. Mean percentage of "take" action per episode during training (a) and the mean number of retrieved text per event (b). The red and black lines are the mean value and the maximum number of daily NIST facts, respectively. (Color figure online)

step, in which the DQN learns how to shortlist the relevant pieces of text properly. In the exploitation the system learns to take all possible useful texts to maximize the reward, stabilizing its trend. Looking at the number of texts shortlisted at inference time (Fig. 3b) the system rarely fills the whole pool. The mean values of expected (red lines) and retrieved (blue bars) texts are correlated (~ 0.74).

5 Conclusion and Future Work

In this work, we proposed a DRL approach to generate crisis timelines from cross-stream data. We tackled the limitations of existing strategies in (1) Online processing of cross-stream data, avoiding ad-hoc indexing strategies for content retrieval; (2) Extracting salient crisis-relevant information by early filtering redundant content during the text retrieval stage; (3) Efficiently handling multiple event queries, with an inference time independent of the number of input queries. We achieved +0.0063 R2 and +0.0151 BERTScore average improvements compared to the best-performing existing solution. In future work, we plan to leverage Large Language Models to enrich the crisis events representation.

Acknowledgements. This work was partially funded by the SmartData@PoliTO center. This study was carried out within the MICS (Made in Italy - Circular and Sustainable) Extended Partnership and received funding from the European Union Next-GenerationEU (Italian PNRR - M4 C2, Invest 1.3 - D.D. 1551.11-10-2022, PE00000004). Within the FAIR (Future Artificial Intelligence Research) this work also received funding from the European Union Next-GenerationEU (Italian PNRR - M4 C2, Invest 1.3 - D.D. 1551.11-10-2022, PE00000013). This manuscript reflects only the authors' views and opinions, neither the European Union nor the European Commission can be considered responsible for them.

References

1. Afsar, M.M., Crump, T., Far, B.: Reinforcement learning based recommender systems: a survey. ACM Comput. Surv. **55**(7), 1–38 (2022). https://doi.org/10.1145/3543846
2. Antony, D., et al.: A survey of advanced methods for efficient text summarization. In: 13th IEEE Annual Computing and Communication Workshop and Conference, CCWC 2023, Las Vegas, NV, USA, 8–11 March 2023, pp. 962–968. IEEE (2023). https://doi.org/10.1109/CCWC57344.2023.10099322
3. Aslam, J.A., Diaz, F., Ekstrand-Abueg, M., McCreadie, R., Pavlu, V., Sakai, T.: TREC 2014 temporal summarization track overview. In: TREC (2015)
4. Chen, Y.C., Bansal, M.: Fast abstractive summarization with reinforce-selected sentence rewriting. In: Proceedings of the 56th Annual Meeting of the Association for Computational Linguistics (Volume 1: Long Papers), pp. 675–686. Association for Computational Linguistics, Melbourne, Australia, July 2018. https://doi.org/10.18653/v1/P18-1063, https://aclanthology.org/P18-1063
5. Dai, Z., Xiong, C., Callan, J., Liu, Z.: Convolutional neural networks for soft-matching n-grams in ad-hoc search. In: Proceedings of the Eleventh ACM International Conference on Web Search and Data Mining, pp. 126–134, WSDM 2018. Association for Computing Machinery, New York, NY, USA (2018). https://doi.org/10.1145/3159652.3159659
6. Denis, L.S., Mietkiewicz, N., Short, K., Buckland, M., Balch, J.: ICS-209-PLUS - an all-hazards dataset mined from the US National Incident Management System 1999–2014, January 2020. https://doi.org/10.6084/m9.figshare.8048252.v14, https://figshare.com/articles/dataset/ICS209-PLUS_Cleaned_databases/8048252
7. Grootendorst, M.: BERTopic: neural topic modeling with a class-based TF-IDF procedure. arXiv preprint arXiv:2203.05794 (2022)
8. Guo, J., Fan, Y., Ai, Q., Croft, W.B.: A deep relevance matching model for ad-hoc retrieval. In: Proceedings of the 25th ACM International on Conference on Information and Knowledge Management, pp. 55–64, CIKM 2016. Association for Computing Machinery, New York, NY, USA (2016). https://doi.org/10.1145/2983323.2983769
9. Karpukhin, V., et al.: Dense passage retrieval for open-domain question answering. In: Proceedings of the 2020 Conference on Empirical Methods in Natural Language Processing (EMNLP), pp. 6769–6781. Association for Computational Linguistics, Online, November 2020. https://doi.org/10.18653/v1/2020.emnlp-main.550, https://aclanthology.org/2020.emnlp-main.550
10. Khattab, O., Zaharia, M.: ColBERT: efficient and effective passage search via contextualized late interaction over BERT, pp. 39–48. Association for Computing Machinery, New York, NY, USA (2020). https://doi.org/10.1145/3397271.3401075
11. Lewis, M., et al.: BART: denoising sequence-to-sequence pre-training for natural language generation, translation, and comprehension. CoRR abs/1910.13461 (2019). http://arxiv.org/abs/1910.13461
12. Lorini, V., et al.: Social media for emergency management: opportunities and challenges at the intersection of research and practice. In: 18th International Conference on Information Systems for Crisis Response and Management, pp. 772–777 (2021)
13. McCreadie, R., Buntain, C.: CrisisFacts: building and evaluating crisis timelines, pp. 320–339 (2023)
14. McCreadie, R., Buntain, C., Soboroff, I.: TREC incident streams: finding actionable information on social media (2019)

15. McCreadie, R., Santos, R.L.T., Macdonald, C., Ounis, I.: Explicit diversification of event aspects for temporal summarization. ACM Trans. Inf. Syst. **36**(3), 1–31 (2018). https://doi.org/10.1145/3158671
16. Mnih, V., et al.: Human-level control through deep reinforcement learning. Nature **518**(7540), 529–533 (2015)
17. Pereira, J., Fidalgo, R., Lotufo, R., Nogueira, R.: Using neural reranking and GPT-3 for social media disaster content summarization (2023)
18. Rajpurkar, P., Jia, R., Liang, P.: Know what you don't know: unanswerable questions for squad. In: Gurevych, I., Miyao, Y. (eds.) Proceedings of the 56th Annual Meeting of the Association for Computational Linguistics, ACL 2018, Melbourne, Australia, 15–20 July 2018, vol. 2: Short Papers, pp. 784–789. Association for Computational Linguistics (2018). https://doi.org/10.18653/v1/P18-2124, https://aclanthology.org/P18-2124/
19. Reimers, N., Gurevych, I.: Making monolingual sentence embeddings multilingual using knowledge distillation. In: Proceedings of the 2020 Conference on Empirical Methods in Natural Language Processing. Association for Computational Linguistics, November 2020. https://arxiv.org/abs/2004.09813
20. Robertson, S., Zaragoza, H.: The probabilistic relevance framework: BM25 and beyond. Found. Trends Inf. Retr. **3**(4), 333–389 (2009). https://doi.org/10.1561/1500000019
21. Saroj, A., Pal, S.: Use of social media in crisis management: a survey. Int. J. Disaster Risk Reduction **48**, 101584 (2020). https://api.semanticscholar.org/CorpusID:218780990
22. Seeberger, P., Riedhammer, K.: Combining deep neural reranking and unsupervised extraction for multi-query focused summarization. arXiv preprint arXiv:2302.01148 (2023)
23. St. Denis, L.A., et al.: All-hazards dataset mined from the us national incident management system 1999–2020. Sci. Data **10**(1), 112 (2023)
24. Yang, M., Li, C., Sun, F., Zhao, Z., Shen, Y., Wu, C.: Be relevant, non-redundant, and timely: deep reinforcement learning for real-time event summarization. In: Proceedings of the AAAI Conference on Artificial Intelligence, vol. 34, pp. 9410–9417 (2020)

Beneath the [MASK]: An Analysis of Structural Query Tokens in ColBERT

Ben Giacalone(✉), Greg Paiement, Quinn Tucker, and Richard Zanibbi

Rochester Institute of Technology, Rochester, NY 14623, USA
{bsg8294,gfp4857,qt2393,rxzvcs}@rit.edu

Abstract. ColBERT is a highly effective and interpretable retrieval model based on token embeddings. For scoring, the model adds cosine similarities between the most similar pairs of query and document token embeddings. Previous work on interpreting how tokens affect scoring pay little attention to non-text tokens used in ColBERT such as [MASK]. Using MS MARCO and the TREC 2019-2020 deep passage retrieval task, we show that [MASK] embeddings may be replaced by other query and structural token embeddings to obtain similar effectiveness, and that [Q] and [MASK] are sensitive to token order, while [CLS] and [SEP] are not.

Keywords: ColBERT · interpretability · embeddings · relevance scoring

1 Introduction

The ColBERT [4] retrieval model uses BERT [2] to produce token embeddings for document and query passages. Typically, candidate documents are retrieved using dense retrieval on embedded tokens [15,17], and then re-scored using the sum of maximum cosine similarities between each query token embedding and its most similar document token embedding via the *MaxSim* operator. Rescoring improves retrieval effectiveness, and is more interpretable than dense retrieval models that use single vectors (e.g. the BERT [CLS] token), because query tokens contribute individually to document rank scores [3], and token embeddings can be analyzed directly.

Interestingly, not all tokens used in ColBERT's scoring are text tokens. Some are *structural tokens* that mark locations and segments in a token sequence. ColBERT employs a modified BERT model to create contextualized embeddings for *every* document and query token, including structural BERT tokens. Structural tokens include [CLS], which appears at the input start, followed by [Q] or [D] to signify whether a passage is from a query or a document. Text tokens from the query are next, followed by [SEP] after the final text token. Query token sequences shorter than the input size are padded with [MASK] tokens,[1] and

[1] [MASK] was originally devised for BERT to represent a "hidden" input token in its masked token prediction training task.

N. Goharian et al. (Eds.): ECIR 2024, LNCS 14610, pp. 431–439, 2024.
https://doi.org/10.1007/978-3-031-56063-7_35

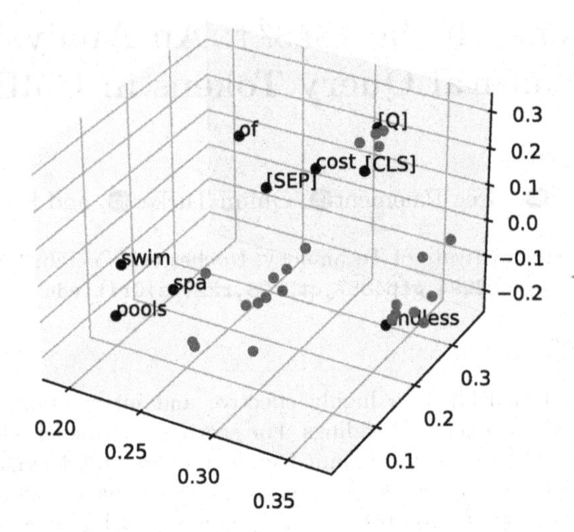

Fig. 1. PCA embeddings for the MS MARCO query "cost of endless pools swim spa". `[MASK]` tokens (red points) cluster around query words and structural tokens (black). (Color figure online)

document token sequences are padded with `[PAD]` tokens. Below are example query and document passage tokenizations with input sizes of 32 and 180 tokens, respectively. Subscripts are used to indicate token position in the input.

Q: `[CLS]`$_1$ `[Q]`$_2$ `cost`$_3$ `of`$_4$ `endless`$_5$ `pools`$_5$ `swim`$_6$ `spa`$_7$ `[SEP]`$_8$ `[MASK]`$_9$ `[MASK]`$_{10}$ `[MASK]`$_{11}$ ... `[MASK]`$_{30}$ `[MASK]`$_{31}$ `[MASK]`$_{32}$
D: `[CLS]`$_1$ `[D]`$_2$ `prices`$_3$ ... `swim`$_{12}$ `spa`$_{13}$ `.`$_{14}$ `[SEP]`$_{15}$ `[PAD]`$_{16}$ `[PAD]`$_{17}$ `[PAD]`$_{18}$ ... `[PAD]`$_{178}$ `[PAD]`$_{179}$ `[PAD]`$_{180}$

Figure 1 shows the token embeddings for the query above.[2] As seen in Fig. 1, `[MASK]` tokens tend to cluster around other query tokens, giving them additional weight [4,13]. The original ColBERT paper suggests `[MASK]` tokens provide a form of query augmentation through term re-weighting and query expansion. Wang et al. [12,13] study a version of ColBERT that performs query expansion using pseudo-relevance feedback, and find that `[MASK]` tokens generally do *not* expand the query by matching terms outside the query, and instead need to add them explicitly. In this way, `[MASK]` tokens primarily weight query tokens by matching query text tokens in documents. Wang et al. [14] also find that for many ColBERT based models, using only query structural tokens for retrieval (`[CLS]`, `[SEP]`, `[Q]`, `[MASK]`) is nearly as effective as using all token embeddings for retrieval, and outperforms using only low IDF query token embeddings.

Previous studies of ColBERT's retrieval behavior have focused on text tokens. In considering why ColBERT's ranking mechanism outperforms standard lexical models such as BM25, Formal et al. [3] focus on query text tokens, and find

[2] Interactive version: https://cs.rit.edu/~bsg8294/colbert/query_viz.html.

that tokens with high Inverse Document Frequency (IDF) produce more *exact* matches in ColBERT query/document token alignments (e.g. (Q:pool,D:pool)) while low IDF terms produce more *inexact* matches (e.g. (Q:is,D:and)). Low IDF token embeddings also tend to shift position more, and removing high IDF tokens perturbs ranking more than removing low IDF tokens. MacAvaney et al. [5] also found a sensitivity for text tokens in ColBERT, with misspellings harming retrieval more than in lexical models. Curiously, they also find ColBERT increasing document scores when *non-relevant* tokens are appended to a document token sequence, while appending *relevant* terms decreases rank scores even after controlling for document length. Perhaps appending relevant terms produces an 'unnatural' token sequence for the embedding language model which interferes with token embedding/contextualization and *MaxSim* scoring.

In this paper, we extend inquiries into how tokens impact retrieval in ColBERT by shifting focus to structural tokens, and [MASK] in particular. In the next Section we present experiments to address the following research questions:

RQ1. Do [MASK] tokens perform more than just term weighting?
RQ2. How sensitive are [CLS], [SEP], [Q], and [MASK] to query token order?

2 Methodology and Experimental Designs

For our experiments, we use the ColBERT v1 model integrated within PyTerrier [6]. The state-of-the-art ColBERT v2 [9] model adds index compression and training with hard negatives and distillation to improve rank quality. Index compression and embedding modifications may alter retrieval candidates and token cosine similarities, and we plan to check this in future work. However, we wish to first study the simpler, original ColBERT model. We also believe insights into the workings of ColBERT v1 and models inspired by it (e.g. the text/image model FILIP [16]) will be beneficial for the research community.

Implementation, Datasets, and Metrics. We use a ColBERT v1 checkpoint from the University of Glasgow trained on passage ranking triples for 44k batches,[3] and run experiments on a server with 4 Intel Xeon E5-2667v4 CPUs, 4 NVIDIA RTX2080-Ti GPUs, and 512 GB RAM. For our experiments, we use two datasets:

1. MS MARCO [7]'s passage retrieval dev set (8.8 million documents, 1 million queries, binary relevance judgements). Each query has at most 1 matching document. We use this dataset for query statistics (e.g. cosine distances between query embeddings).
2. A dataset combining test queries from the TREC 2019 [11] and 2020 [1] deep passage retrieval task (99 queries, graded relevance judgements). Collection documents are the same as MS MARCO. We use this dataset for experiments focused upon retrieval quality.[4]

[3] http://www.dcs.gla.ac.uk/~craigm/ecir2021-tutorial/colbert_model_checkpoint.zip.
[4] Running the TREC test queries takes roughly 15 min to complete using a multi-threaded Rust program: https://github.com/Boxxfish/IR2023-Project.

For the TREC 2019-2020 collection, the relevance scale is between 0 and 3 with a score of 2 considered relevant for metrics using binary relevance (e.g. MAP). We examine relevance scores thresholded at 1, 2, and 3 to see the effect of binarizing at different relevance grades. We use MRR@10 to characterize effectiveness for top results, MAP to characterize effectiveness for complete rankings, and to complement MAP we use nDCG@k measures ($k \in \{10, 1000\}$) to utilize graded relevance labels from the TREC data.

RQ1: Do [MASK] tokens perform more than just term weighting? Figure 1 illustrates how `[MASK]` token embeddings cluster around query terms, which was consistent for MS MARCO queries we examined. As mentioned previously, `[MASK]` tokens have been identified as having two roles in scoring: (1) query term weighting through matching document terms to `[MASK]` tokens with embeddings similar to non-`[MASK]` query tokens, and (2) query expansion through `[MASK]` embeddings shifting toward potentially relevant tokens outside of the query.

In this experiment, we test whether the clustering of `[MASK]` tokens around query tokens indicates that term weighting is the *only* role `[MASK]` tokens actually play in ColBERT scoring. To do this, we replace structural token embeddings in a query with text token embeddings from the same query. This forces ColBERT to perform term weighting: replacing structural token embeddings by their nearest text embeddings cannot introduce new terms or perform "soft weighting" by increasing the weight of multiple query tokens. We use the TREC 2019-2020 collection, and compare four token remapping conditions: (1) no remapping, (2) remapping `[MASK]` tokens to text tokens, (3) remapping all structural token embeddings (`[CLS]`, `[SEP]`, `[Q]`, and `[MASK]`) to text tokens, and finally (4) remapping each `[MASK]` to its most similar embedded text token *or* non-`[MASK]` structural token (i.e. `[CLS]`, `[SEP]`, or `[Q]`). We hypothesize that replacing `[MASK]` embeddings by non-`[MASK]` embeddings in queries will reduce effectiveness by preventing matches with terms that do not appear in the query (i.e. by preventing query expansion).

RQ2: How sensitive are [CLS], [SEP], [Q], and [MASK] to query token order? As shown in the example above, ColBERT begins every query token sequence passed to BERT with the structural tokens `[CLS]` and `[Q]`, followed by the text tokens and the structural token `[SEP]` marking the end of the text tokens, and finally zero or more `[MASK]` tokens to fill out the fixed-size input (e.g. 32 tokens). `[CLS]` is included in BERT's training objective function, and aggregates context across entire query and document passages resulting in a "summary" representation. We thus expect queries with similar wording and intent to produce similar `[CLS]` embeddings, even when the query word order changes. We expect the same pattern to hold for `[SEP]` which terminates every query and document passage. In contrast, we expect `[MASK]` embeddings to vary more than `[CLS]` and `[SEP]` tokens when words are re-ordered, because of their observed clustering around query terms and resulting weighting of individual terms in scoring. We expect `[Q]` embeddings to also vary more than `[CLS]` and `[SEP]`, because `[Q]` is absent in the original BERT training objective.

Table 1. Replacing structural token embeddings with other query token embeddings (TREC 2019-2020, RQ1). Maximum values are in bold; significant differences from "None" are shown with a dagger ($p < 0.05$, Bonferroni-corrected t-tests).

METRIC	STRUCTURAL TOKEN REMAPPING			
	None	**All** `[X]` → **Text**	`[MASK]` → **Text**	`[MASK]` → **Str. & Text**
Binary Relevance				
MAP(rel≥1)	0.447	0.454	†**0.462**	†**0.462**
MRR(rel≥1)@10	**0.930**	0.924	0.929	0.923
MAP(rel≥2)	0.450	0.444	0.454	**0.457**
MRR(rel≥2)@10	**0.851**	0.820	0.835	0.837
MAP(rel≥3)	0.366	0.362	**0.373**	0.372
MRR(rel≥3)@10	0.557	0.560	**0.563**	**0.563**
Graded Relevance				
nDCG@10	0.689	0.685	0.691	**0.694**
nDCG@1000	0.680	0.673	0.683	**0.684**

To study how query word order influences contextualization for query structural tokens, we reorder query text terms prior to contextualization similar to Rau et al. [8]. Randomly shuffling query tokens may alter the meaning of a query, so we limit the permutations considered. Specifically, we transform queries of the form "what is ..." into "... is what", moving the first two text tokens to the end of the query in the opposite order. To further avoid accidental semantic shifting, we only examine queries that are 3–8 tokens long. 12,513 queries in the MS MARCO dev set fit this criteria. As a baseline, we also apply the same reordering pattern to *all* queries 3–8 tokens long, without requiring the first two tokens to be "what is". 68,318 queries in the dev set fit this criteria. For the reasons given above, we hypothesize that `[Q]` and `[MASK]` embeddings will change more than `[SEP]` and `[CLS]` under this reordering. We use the cosine distance to quantify the shift in token embeddings after reordering the query text tokens.

3 Results

RQ1: Do [MASK] tokens perform more than just term weighting? In Table 1 we see replacing embeddings for *all* structural tokens with their closest text token embedding produces non-significant reductions in metrics other than MRR@10 (rel≥3). The two conditions mapping only `[MASK]` have very similar metrics, but surprisingly produce slight increases in MAP and nDCG@10/@1000 over both the "None" and "All" conditions. For MAP(rel≥1), the increase is significant (1.5%). MRR@10(rel≥3) is also slightly higher than standard ColBERT (but not significantly so). These small increases are likely from incorporating additional context through the `[CLS]`, `[Q]`, and `[SEP]` tokens (especially `[CLS]`). This contradicts our hypothesis that remapping `[MASK]` embeddings would harm performance, and is also interesting because `[MASK]` tokens comprise most of the input

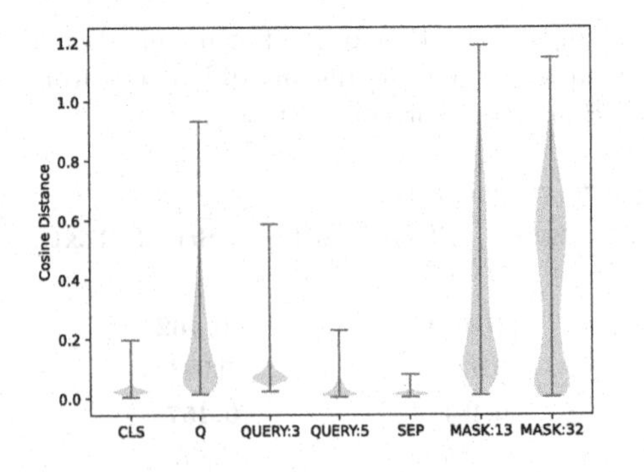

(a) "`what is ...`" → "`... is what`"

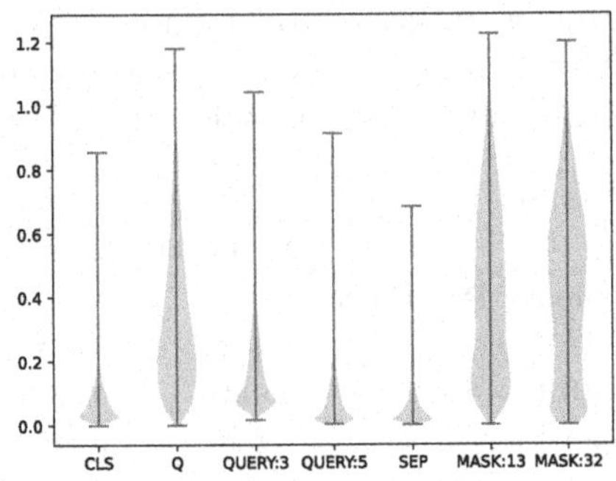

(b) All queries: "`1 2 ...`" → "`... 2 1`"

Fig. 2. Distribution of cosine distance $(1 - \cos(e, e'))$ for token embeddings before and after query token reordering (MS MARCO, RQ2). For brevity not all tokens are shown, but the general trend of higher variance holds for all `[MASK]` tokens. **Left:** Cosine distances for queries starting with "what is". **Right:** Cosine distances without requiring queries to start with "what is". `[QUERY:3]` and `[QUERY:5]` are the first and third text tokens, respectively; `[MASK:13]` represents the `[MASK]` token at position 13, and `[MASK:32]` represents the final `[MASK]` input token at position 32.

for short queries. In other words, its appears that strong retrieval performance with ColBERT is possible even when using only a few text token embeddings, provided that term weighting is taken into account.

RQ2: How sensitive are [CLS], [SEP], [Q], and [MASK] to query token order? In Fig. 2(a), `[QUERY:3]` is the third token and first text token (always "what") while `[QUERY:5]` is the fifth token containing the text token after "what is." We see distinct differences in how cosine distances are distributed for `[CLS]`, `[SEP]`, `[QUERY:3]`, and `[QUERY:5]` versus `[Q]`, `[MASK:13]`, and `[MASK:32]`. The first group shows barely any shift, while the latter group shows large shifts, with higher variation. Figure 2(b) shows results for queries reordered similarly, but without requiring them to start with "what is". For example, *"airplane flights to florida"* produces the somewhat unnatural query *"to florida flights airplane"*. All tokens show larger representational shifts in this condition; however, we again find that `[CLS]`, `[SEP]`, and the `[QUERY:3/5]` text token embeddings vary far less than the `[Q]` and `[MASK]` embeddings.

Despite our efforts to avoid it, some "what is" queries have their meaning altered by our reordering. For example, *"what is some examples homogeneous"* becomes *"some examples homogeneous is what"*, which may change the query from a request for examples to asking for a definition. When we filtered out queries containing the word "example", the variance of `[QUERY:3]` dropped from $8.53 \cdot 10^{-4}$ to $7.73 \cdot 10^{-4}$, while the variance of `[Q]` had less of a proportional drop ($2.07 \cdot 10^{-2}$ to $2.06 \cdot 10^{-2}$), indicating some of the variance of non-`[Q]` or `[MASK]` embeddings may be due to these edge cases.

4 Discussion and Conclusion

To our surprise, replacing [MASK] token embeddings in queries with either their most similar text token embedding in the same query, or the most similar text or non-[MASK] structural token embedding from the query yielded similar effectiveness to standard ColBERT for the TREC 2019-2020 dataset. There is even a small significant increase in MAP when weakly-relevant documents are considered relevant (i.e. MAP(rel$\geq$1)). The differences between mapping [MASK] to only text tokens or to both text and non-MASK structural tokens was statistically insignificant for all metrics observed. So if [MASK] tokens have effects other than term weighting in scoring, their role appears to be minor (RQ1).

This suggests a possible optimization. We can multiply each non-[MASK] query token embedding's score contribution by the number of [MASK] token embeddings most similar to it. Regarding interpretability, using [MASK] only to weights tokens in this manner simplifies the ColBERT scoring model both conceptually and computationally. For short queries, most of the input to ColBERT is [MASK] tokens, and so the number of query tokens to match against document tokens with *MaxSim* may be a fraction of the full token input size. A related approach is described by Tonellotto et al. [10], where query embeddings are dropped after contextualization based on frequency statistics. However, rather than pruning a set number of tokens based on collection frequency, we would use all token embeddings to retrieve candidates, and then weight non-[MASK] query tokens using fewer nearest neighbor lookups during scoring.

However, the question of [MASK] tokens' role in retrieving candidates still remains, as this paper has focused on the final scoring step; all query tokens were used to retrieve candidates in our experiments. How might limiting or removing the use of [MASK] tokens in the first-stage dense retrieval impact performance? We wonder about the small statistically insignificant improvements seen in MAP and MRR for highly relevant documents (rel$\geq$3), as well as nDCG@10, nDCG@1000, and MAP. Are these stable and/or significant in other collections? To better understand [MASK], one possible experiment is appending different numbers of [MASK] tokens to each query. This may reveal whether having fewer [MASK] tokens causes them to move closer to non-[MASK] embeddings, and whether having more [MASK] tokens might improve term weighting.

Regarding the effect of token ordering on contextualized embeddings (RQ2), our findings are consistent with our original hypothesis: [CLS] and [SEP] embeddings do not vary greatly for similar queries with a different token ordering, while [Q] and [MASK] do. The shift in [Q] is the most interesting result here; the model may be treating [Q] similar to another [MASK] token. Some early analysis suggests that a query [CLS] tends to match a document [CLS], a query [SEP] tends to match ending punctuation, and [MASK] tends to match tokens other than [CLS] or [SEP] (see our interactive visualization for ColBERT scoring (See footnote 2)). We have not observed [Q] matching to any specific document tokens.

In the future we would also like to validate our results using ColBERT v2. We believe that our results *should* hold for the newer model – if [MASK]s continue to cluster around query word embeddings, we expect [MASK]s will continue to act as term weights, and the training process in ColBERT v2 should not alter how [Q] is processed. We would also like to extend our evaluation to additional datasets, since we have only focused on MS MARCO and the MS MARCO-derived TREC 2019-2020 datasets in our experiments.

References

1. Craswell, N., Mitra, B., Yilmaz, E., Campos, D.: Overview of the TREC 2020 deep learning track. In: Voorhees, E.M., Ellis, A. (eds.) Proceedings of Text REtrieval Conference (TREC), vol. 1266. NIST Special Publication (2020)
2. Devlin, J., Chang, M.W., Lee, K., Toutanova, K.: BERT: pre-training of deep bidirectional transformers for language understanding. In: Proceedings of North American Chapter of the Association for Computational Linguistics (NAACL), pp. 4171–4186 (2019). https://doi.org/10.18653/v1/N19-1423, https://aclanthology.org/N19-1423
3. Formal, T., Piwowarski, B., Clinchant, S.: A white box analysis of ColBERT. In: Hiemstra, D., Moens, M.F., Mothe, J., Perego, R., Potthast, M., Sebastiani, F. (eds.) Proceedings of European Conference on Information Retrieval (ECIR). LNCS, vol. 12657, pp. 257–263. Springer, Cham (2021). https://doi.org/10.1007/978-3-030-72240-1_23
4. Khattab, O., Zaharia, M.: ColBERT: efficient and effective passage search via contextualized late interaction over BERT. In: Huang, J.X., et al. (eds.) Proceedings of SIGIR, pp. 39–48 (2020). https://doi.org/10.1145/3397271.3401075
5. MacAvaney, S., Feldman, S., Goharian, N., Downey, D., Cohan, A.: ABNIRML: analyzing the behavior of neural IR Models. Trans. Assoc. Comput. Linguist. **10**, 224–239 (2022). https://doi.org/10.1162/tacl_a_00457
6. Macdonald, C., Tonellotto, N., MacAvaney, S., Ounis, I.: PyTerrier: declarative experimentation in Python from BM25 to dense retrieval. In: Proceedings of International Conference on Information & Knowledge Management (CIKM), pp. 4526–4533 (2021). https://doi.org/10.1145/3459637.3482013
7. Nguyen, T., et al.: MS MARCO: a human generated machine reading comprehension dataset. In: Besold, T.R., Bordes, A., d'Avila Garcez, A.S., Wayne, G. (eds.) Proceedings of the Workshop on Cognitive Computation: Integrating Neural and Symbolic Approaches 2016 Co-located with the 30th Annual Conference on Neural Information Processing Systems (NIPS 2016), Barcelona, Spain, 9 December 2016. CEUR Workshop Proceedings, vol. 1773. CEUR-WS.org (2016). https://ceur-ws.org/Vol-1773/CoCoNIPS_2016_paper9.pdf
8. Rau, D., Kamps, J.: The role of complex NLP in transformers for text ranking. In: Proceedings of ICTIR, pp. 153–160 (2022). https://doi.org/10.1145/3539813.3545144
9. Santhanam, K., Khattab, O., Saad-Falcon, J., Potts, C., Zaharia, M.: ColBERTv2: effective and efficient retrieval via lightweight late interaction. In: Carpuat, M., de Marneffe, M.C., Meza Ruiz, I.V. (eds.) Proceedings of North American Chapter of the Association for Computational Linguistics (NAACL), pp. 3715–3734 (2022). https://doi.org/10.18653/v1/2022.naacl-main.272
10. Tonellotto, N., Macdonald, C.: Query embedding pruning for dense retrieval. In: Proceedings of International Conference on Information & Knowledge Management (CIKM), pp. 3453–3457 (2021). https://doi.org/10.1145/3459637.3482162

11. Voorhees, E.M., Ellis, A. (eds.): Proceedings of the Twenty-Eighth Text REtrieval Conference, TREC 2019, Gaithersburg, Maryland, USA, 13–15 November 2019, vol. 1250. NIST Special Publication. National Institute of Standards and Technology (NIST) (2019). https://trec.nist.gov/pubs/trec28/trec2019.html
12. Wang, X., Macdonald, C., Ounis, I.: Improving zero-shot retrieval using dense external expansion. Inf. Process. Manag. **59**(5), 103026 (2022). https://doi.org/10.1016/j.ipm.2022.103026, https://www.sciencedirect.com/science/article/pii/S0306457322001364
13. Wang, X., MacDonald, C., Tonellotto, N., Ounis, I.: ColBERT-PRF: semantic pseudo-relevance feedback for dense passage and document retrieval. ACM Trans. Web **17**(1), 1–39 (2023). https://doi.org/10.1145/3572405
14. Wang, X., Macdonald, C., Tonellotto, N., Ounis, I.: Reproducibility, replicability, and insights into dense multi-representation retrieval models: from ColBERT to Col*. In: Proceedings of SIGIR, pp. 2552–2561. ACM (2023). https://doi.org/10.1145/3539618.3591916
15. Xiong, L., et al.: Approximate nearest neighbor negative contrastive learning for dense text retrieval. In: Proceedings of International Conference on Learning Representations (ICLR). OpenReview.net (2021). https://openreview.net/forum?id=zeFrfgyZln
16. Yao, L., et al.: FILIP: fine-grained interactive language-image pre-training. In: The Tenth International Conference on Learning Representations, ICLR 2022, Virtual Event, 25–29 April 2022. OpenReview.net (2022). https://openreview.net/forum?id=cpDhcsEDC2
17. Zhan, J., Mao, J., Liu, Y., Zhang, M., Ma, S.: RepBERT: contextualized text embeddings for first-stage retrieval. CoRR abs/2006.15498 (2020). https://arxiv.org/abs/2006.15498

A Cost-Sensitive Meta-learning Strategy for Fair Provider Exposure in Recommendation

Ludovico Boratto, Giulia Cerniglia, Mirko Marras(✉), Alessandra Perniciano, and Barbara Pes

University of Cagliari, Cagliari, Italy
{ludovico.boratto,mirko.marras}@acm.org, g.cerniglia1@studenti.unica.it, {alessandra.pernician,pes}@unica.it

Abstract. When devising recommendation services, it is important to account for the interests of all content providers, encompassing not only newcomers but also minority demographic groups. In various instances, certain provider groups find themselves underrepresented in the item catalog, a situation that can influence recommendation results. Hence, platform owners often seek to regulate the exposure of these provider groups in the recommended lists. In this paper, we propose a novel cost-sensitive approach designed to guarantee these target exposure levels in pairwise recommendation models. This approach quantifies, and consequently mitigate, the discrepancies between the volume of recommendations allocated to groups and their contribution in the item catalog, under the principle of equity. Our results show that this approach, while aligning groups exposure with their assigned levels, does not compromise to the original recommendation utility. Source code and pre-processed data can be retrieved at https://github.com/alessandraperniciano/meta-learning-strategy-fair-provider-exposure.

Keywords: Fairness · Equity · Recommendation · Cost-Sensitive · Provider

1 Introduction

Online platforms serve as intermediaries that connect various parties, typically consumers and providers. These platforms frequently rely on recommender systems that employ predictive relevance to match consumers with suitable items offered by providers [17]. Traditionally, optimizing recommendations solely for the benefit of consumers has been the ultimate objective. However, it becomes important to recognize that recommender systems exist within multi-sided environments and, as such, should carefully consider interests and needs of providers, going beyond the scope of consumer-centric concerns only [1].

Notable concerns pertain to the unfair exposure of provider groups in the recommended lists, which goes beyond just legal aspects. It can also be analyzed

N. Goharian et al. (Eds.): ECIR 2024, LNCS 14610, pp. 440–448, 2024.
https://doi.org/10.1007/978-3-031-56063-7_36

from the perspective of a platform's business model [16]. For example, platforms like Coursera and GoFundMe may prioritize ensuring that new instructors or projects receive adequate exposure in the recommendations, even though during the initial period of inclusion in the item catalog, they naturally engaged fewer users than established providers [9]. Mitigating these exposure discrepancies can not only counter a possible discrimination but also bring beyond-utility benefits, such as introducing novelty and diversity [13]. Consequently, supporting provider groups, especially those belonging to minorities, becomes imperative [4,10,11].

Several mitigation methods have been proposed to reorganize the initial lists of a recommender system, aiming to meet specific objectives related to diversification. These methods frequently concentrate solely on the aspect of visibility [19,20]. Visibility, in this context, refers to the fraction of items from a particular provider group in the uppermost segment of the rankings. Conversely, there are other approaches, often introduced within a fairness framework, which strive to guarantee an equitable exposure among provider groups [3,9,18]. However, these methods do not consistently address situations in which items from minority groups end up being ranked at the lower end of the ranking segment.

Our approach in this study stands out in four key ways. First, it addresses position bias in ranked results, which is overlooked in diversification-related fair ranking research [19,20]. Second, in contrast to [3], it does not depend on metrics that require a rank-aware quality assessment, making it more practical for recommender systems. Third, it achieves balancing objectives for user rankings without the need for extensive feedback data from all users, in contrast to [9]. Finally, [15] can control provider group exposure, but its complexity is directly tied to the number of users and items, posing challenges when applied to large datasets commonly encountered in real-world applications. The computational complexity of our approach in this paper does not depend on these factors.

In this paper, we examine a scenario where providers are categorized based on a shared characteristic, and certain provider groups have limited representation of items in the catalog and, subsequently, in the recommendations. As a first novel contribution, we introduce a cost-sensitive meta-learning method to ensure that the level of exposure for provider groups resembles their contribution in the catalog, guided by the principle of equity [3]. As a second novel contribution, we conduct a comprehensive evaluation and discuss our approach with respect to state-of-the-art alternatives, to assess how supporting provider groups influences aspects of utility and beyond-utility on two distinct public data sets.

2 Methodology

2.1 Fairness Objective Definition

Our study centers on the exploration of a recommendation policy having a primary objective: securing an equitable distribution of exposure for provider groups relative to their representation within the catalog. This policy's philosophy aligns with the well-known distributive norm based on equity [3,6,7]. To illustrate the importance of this policy, consider an online course platform

Table 1. In addition to the number of users, we show the number of items in the catalog and the percentage of those provided by females. Then, we place the total number of interactions and of their subset in the train set, and the percentage of train interactions for items from female providers. We also show the percentage of train triplets with a positive/negative item from a female provider, under a random sampling.

Data Set	# Users	Items		Interactions			% Female in Triplets	
		Total	% Female	Total	Train	% Female	Pos. Item	Neg. Item
ML-1M	6,040	3,876	6.30 %	998,539	634,640	3.60 %	3.40 %	6.40 %
COCO	73,022	34,442	12.70 %	520,531	294,271	7.60 %	7.50 %	12.70 %

that hosts some courses provided by male instructors and others sourced from female instructors[1]. With no intervention, the platform may inadvertently favor courses from one group, possibly reinforcing disparities. Following our policy, courses provided by male and female instructors receive proportionate exposure in recommendations, leading to platform that empowers instructors irrespective of gender and reduces any concentration of attention on a given group.

2.2 Data Preparation

To align with benchmarks in prior work [4,15], we consider experimenting with the target policy through our approach in two scenarios, namely movie recommendation and online course recommendation. To this end, we used two publicly available data sets, whose main statistics are described in Table 1. They are notable for the inclusion of providers' gender labels, a rarity in such collections.

MovieLens-1M (ML-1M) [12] encompasses ratings for movies collected from users. Movie directors are traditionally assumed to be the providers in this scenario. We considered the gender labels from [4]. As the main preprocessing steps, we (i) retrieved the directors and their gender from public sources for those movies this information was unknown, (ii) kept only the first director for movies with multiple directors primarily because s/he is typically the most influential in the film-making process (and to prevent duplication of directors in different sets), and (iii) filtered interactions such that items had at least 10 interactions since extreme cold-start scenarios are not our main focus.

COCO [5] consists of ratings assigned to courses by learners, collected from an online course platform. For this benchmark, following prior work, we assumed that the course instructors represent the providers [11]. Similar to the preprocessing steps for ML-1M, to address the issue of missing gender information, we removed instructors and their associated courses, for those instructors we could not retrieve their gender from public sources. Then, to ensure enough input for personalization and avoid extreme cold-start scenarios, we filtered the data such that each course had at least 5 ratings and each learner gave at least 5 ratings.

[1] While gender is by no means a binary construct, to the best of our knowledge, no data set with non-binary gender exists. What we are considering is a binary feature.

On both the considered data sets, we performed a temporal train-test split, allocating the most recent 20% of interactions to the test set, another 20% to the validation set, and the remaining to the train set.

2.3 Pairwise Recommendation Model Creation

Once the two considered datasets were pre-processed, we leveraged them to train a series of recommendation models. In our study, we focused on models optimized through pairwise learning, which forms the basis of several state-of-the-art personalized ranking algorithms [21] and is widely studied in the context of methods pursuing fairness or diversification objectives, e.g., in [4,15]. Specifically, we employed the Bayesian-Personalized Ranking with Matrix Factorization (BPRMF) method, structured around an optimization criterion derived from a Bayesian analysis. It uses the prior probability for the model and a likelihood function that calculates the probability of a user preferring an item i the user previously interacted with over another item j the user did not interacted with. We adopted the BPRMF implementation provided in Elliot for better reproducibility [2].

2.4 Traditional Strategy: Pairwise Training via Random Sampling

To conduct an initial assessment about fairness among provider groups under the equity-based policy, we trained a recommendation model based on BPRMF, by following the traditional training protocol adopted in [21], as an example. To this end, we used a bootstrap sampling with replacement to sample train triplets for a user. Each sample consists of a triplet (u, i, j), where 'u' is the user's index, 'i' is the positive item (i.e., the user interacted with), and 'j' is the negative item (i.e., the user did not interact with). These triplets are created by randomly sampling items according to the interactions in the train set. To initialize the 10-sized latent factor matrices for both users and items within the BPRMF's model architecture, we assigned values uniformly distributed between 0 and 1. The training process ran for 10 epochs, and for optimization, we utilized SGD with a learning rate of 0.001. Detailed (hyper-)parameter values are listed in the source code repository for a better reproducibility.

2.5 Our Proposal: Pairwise Training via Cost-Sensitive Sampling

Motivation. Once the baseline recommendation models were trained on both data sets, we performed an assessment of fairness aimed to uncover potential discrepancies between the representation of different provider groups in the item catalog, the train interactions (and consequently the train triplets), and the provided recommendations. For convenience, let us assume to focus on the model trained on the ML-1M data set. Considering the top-10 recommended items, we observed that only 1.40% of the exposure is given to items coming from a female director. This exposure level goes in contrast with the representation of female providers in the train interactions (3.60%) and in the item catalog (6.30%) - see Table 1. We conjecture that this disparity in representation is rooted in the characteristics of the triplets sampled during training. While the percentage of triplets that involve a positive item from a female provider (3.40%) is close to the representation of female providers in the train interactions (3.60%), the same observation does not hold if we consider the negative items. The percentage of triplets that involve a negative item from a female provider (6.40%) is disproportionately higher than their representation in the interactions and in the catalog. Hence, the model is fed with data that leads to disadvantage female providers more, since the model is optimized to predict a lower relevance for the negative item over the positive one for each triplet.

Intuition. Guided from the abovementioned motivation, we propose to adopt a cost-sensitive learning strategy as a countermeasure. Cost-sensitive learning in data mining encompasses a set of algorithms tailored to account for different costs associated with specific characteristics of the problem to consider [8]. Under this class-cost perspective that, while training a model, more attention is given to training elements with higher associated costs (e.g., rare and valuable classes). While direct methods like CSTree [14] directly integrate costs into learning algorithms, meta-learning automatically transforms existing cost-insensitive models into cost-sensitive ones. It can be seen as a preliminary step that requires to pre-process train data from cost-insensitive learning methods (the BPR-MF model in our study), remaining applicable to any recommendation model.

Proposal. To validate our intuition, we therefore decided to propose and experiment with a novel parametric cost-sensitive meta-learning approach which adjusts the sampling distribution. This approach can regulate the exposure of provider groups in recommendations generated by the BPR-MF model and similar pairwise learning models, based on the contributions of these groups in the

---

Procedure 1. Cost-sensitive meta-learning sampling of train triplets

---

Input: $C \geq 1$: Parametric cost for the majority group; $N \geq 0$: No. of train triplets.
Output: triplets : Generated triplets to train the pairwise recommendation model.

```
p = [(C × 100)/(C + 1), 100/(C + 1)]
for i : trainItems do
    probabilities(i) = isItemFromMajorityClass(i) ? p[0] : p[1]
end for
triplets = [ ]
while N ≥ 0 do
    u = sampleRandomly(listUsers)
    i = sampleRandomly(getUserItems(u)))
    repeat
        j = sampleWithProb(trainItems, probabilities)
    until listItems(u).includes(j)
    triplets.append((u, i, j))
    N = N − 1
end while
```

---

catalog. Procedure 1 outlines the proposed sampling strategy for controlling the distribution of items from female and male providers within the negative item set of triplets. The parameter C alters the distribution, allowing us to reduce or increase instances from the minority group in the negative or positive set and, consequently, influencing the relative exposure of different groups. The vector p includes two elements: (i) the probability of selecting an item from the provider group deserving more emphasis (in the context of the negative set, this represents the majority group), and (ii) the probability of selecting the other group. In essence, by manipulating C, our approach provides control over the balance between female and male provider groups in the triplets, both in the positive and negative item sets. Our hypothesis is that, by aligning the proportions of items in the catalog from female and male providers in both positive and negative sets, the exposure of groups reflects the target fairness level. Note that our approach can be applied also for demographic attributes with more than two classes.

3 Results and Discussion

Given the intuitions we seek to validate, our analysis was focused on examining whether the proposed cost-sensitive meta-learning approach can lead to achieving the objective defined by the equity-based fairness policy and, if so, how it impacts recommendation utility. To this end, we compared the original pairwise model (Baseline, $C = 1$), with our cost-sensitive approach under different cost settings. In Table 2, we monitored recommendation utility (NDCG, with binary relevance scores, base-2 logarithm decay), representation in train data (% Female Interactions, % Female in Triplets), and group exposure in both the top-10 and top-20 recommended lists provided to users (% Female Exposure).

Table 2. Impact of the parameter C on recommended lists (utility, exposure). For each data set, the gray row identifies the setting under which the representation of female items in the positive (Pos. Item) and negative (Neg. Item) item sets is the closest.

Data Set	Item Type	Cost C	NDCG	% Female in Triplets		% Female Exposure	
				Pos. Item	Neg. Item	Top 10	Top 20
ML-1M	–	C = 1.0	0.07	3.4 %	6.4 %	1.4%	2.7%
	NEG	C = 1.2	0.07	3.4 %	5.3 %	2.5%	2.8%
	NEG	C = 2.0	0.07	3.4 %	3.2 %	6.0%	3.9%
	NEG	C = 3.0	0.07	3.4 %	2.2 %	6.8%	4.5%
COCO	–	C = 1.0	0.02	7.5 %	12.7 %	0.0%	0.0%
	NEG	C = 2.0	0.02	7.5 %	6.6 %	0.0%	0.4%
	NEG	C = 3.0	0.02	7.5 %	4.7 %	0.0%	2.2%
	NEG	C = 5.0	0.02	7.5 %	2.9 %	0.0%	2.2%
	POS	C = 1.2	0.02	10.5 %	12.7 %	2.1%	6.9%
	POS	C = 2.0	0.02	10.9 %	12.7 %	5.3%	7.6%
	POS	C = 3.0	0.02	12.3 %	12.7 %	12.2%	11.9%

From the obtained results, we observed that our approach does not affect recommendation utility (NDCG column) in both data sets. In ML-1M, assigning a cost $C = 2.0$ to the majority group (male items in this case) allowed us to have a very close representation of female items in both positive (3.4%) and negative (3.2%) elements of the triplets. As hypothesized, the exposure of items from female providers is also positively affected, reaching 6.0%. This is close to the representation of female providers in the catalog (6.4% - see Table 1). Differently from ML-1M, playing with the negative items during sampling did not allow to reach a similar representation of female items between positive and negative elements of the triplets in COCO. To this end, we applied the cost-sensitive approach to *increase* the female representation in the *positive* set. With this setting, under a cost $C = 3.0$, we achieved such similar representation (12.3% vs 12.7%). Exposure was impacted, reaching 12.2% on the top 10. Again, this is close to the representation of female providers in the catalog (12.7% in Table 1).

In summary, by playing with the parameter C to make the representation of the minority group equal in the positive and negative item sets right before the training phase, pairwise models trained on triplets sampled with our approach can lead to provider groups having an exposure estimate equal or substantially close to their representation in the catalog. No impact is observed in recommendation utility. Therefore, the proposed approach does not suffer from the typical trade-off between the overall recommendation utility of the recommender system and the fairness of exposure across provider groups in the recommended lists. Despite the importance of showing a more consolidated view of the existing methods for provider group fairness, any comparison to other state-of-the-art exposure control methods, such as LFRank [19], FA*IR [20], FOEIR [18], GDE [9], and MMR [15], would not influence the overall assessment of the feasibility and effectiveness of our approach, which is the main focus of this paper. For instance, the method proposed in [15], proven to lead to a better trade-off than all the others the original authors considered, still reports a non-negligible decrease in recommendation utility, between 5% and 20%, under a comparable protocol.

4 Conclusions and Future Work

In this paper, we introduced a cost-sensitive meta-learning approach to tackle the issue of imbalanced class distributions in train data for recommendation. By manipulating the sampling distribution, our method effectively manages provider group exposure in the recommended lists. The meta-learning nature of our approach makes it adaptable to different recommendation methods. Future work will involve applying our method to other pairwise recommendation methods and conducting extensive performance evaluations of effectiveness, fairness, and scalability across diverse domains. Moreover, for a better contextualization, we will perform an extensive comparison to fully assess the relative strengths and weaknesses of the proposed method against existing solutions under a wider range of metrics, including also the required effort to tune their (hyper-)parameters.

Acknowledgement. We acknowledge financial support under the National Recovery and Resilience Plan (NRRP), Miss. 4 Comp. 2 Inv. 1.5 - Call for tender No.3277 published on Dec 30, 2021 by the Italian Ministry of University and Research (MUR) funded by the European Union - NextGenerationEU. Prj. Code ECS0000038 eINS Ecosystem of Innovation for Next Generation Sardinia, CUP F53C22000430001, Grant Assignment Decree N. 1056, Jun 23, 2022 by the MUR.

References

1. Abdollahpouri, H., et al.: Multistakeholder recommendation: survey and research directions. User Model. User Adapt. Interact. **30**(1), 127–158 (2020)
2. Anelli, V.W., et al.: Elliot: a comprehensive and rigorous framework for reproducible recommender systems evaluation. In: Proceedings of the 44th International ACM SIGIR Conference on Research and Development in Information Retrieval, SIGIR, pp. 2405–2414. ACM (2021)
3. Biega, A.J., Gummadi, K.P., Weikum, G.: Equity of attention: amortizing individual fairness in rankings. In: Proceedings of the 41st International ACM SIGIR Conference on Research and Development in Information Retrieval, SIGIR, pp. 405–414. ACM (2018)
4. Boratto, L., Fenu, G., Marras, M.: Interplay between upsampling and regularization for provider fairness in recommender systems. User Model. User Adapt. Interact. **31**(3), 421–455 (2021)
5. Dessì, D., Fenu, G., Marras, M., Reforgiato Recupero, D.: COCO: semantic-enriched collection of online courses at scale with experimental use Cases. In: Rocha, Á., Adeli, H., Reis, L.P., Costanzo, S. (eds.) Trends and Advances in Information Systems and Technologies, vol. 746, pp. 1386–1396. Springer, Cham (2018). https://doi.org/10.1007/978-3-319-77712-2_133
6. Fenu, G., Lafhouli, H., Marras, M.: Exploring algorithmic fairness in deep speaker verification. In: Gervasi, O., et al. (eds.) Computational Science and Its Applications – ICCSA 2020. LNCS, vol. 12252, pp. 77–93. Springer, Cham (2020). https://doi.org/10.1007/978-3-030-58811-3_6
7. Fenu, G., Marras, M., Medda, G., Meloni, G.: Fair voice biometrics: impact of demographic imbalance on group fairness in speaker recognition. In: Proceedings of the 22nd Annual Conference of the International Speech Communication Association, Interspeech, pp. 1892–1896. ISCA (2021)
8. Fernández, A., García, S., Galar, M., Prati, R.C., Krawczyk, B., Herrera, F.: Learning from imbalanced data sets. Springer, Cham (2018). https://doi.org/10.1007/978-3-319-98074-4
9. Gómez, E., Boratto, L., Salamó, M.: Disparate impact in item recommendation: a case of geographic imbalance. In: Hiemstra, D., et al. (eds.) Advances in Information Retrieval: 43rd European Conference on IR Research, ECIR 2021. LNCS, vol. 12656, pp. 190–206. Springer, Cham (2021). https://doi.org/10.1007/978-3-030-72113-8_13
10. Gómez, E., Boratto, L., Salamó, M.: Provider fairness across continents in collaborative recommender systems. Inf. Process. Manag. **59**(1), 102719 (2022)
11. Gómez, E., Zhang, C.S., Boratto, L., Salamó, M., Marras, M.: The winner takes it all: geographic imbalance and provider (un)fairness in educational recommender systems. In: Proceedings of the 44th International ACM SIGIR Conference on Research and Development in Information Retrieval, SIGIR, pp. 1808–1812. ACM (2021)

12. Harper, F.M., Konstan, J.A.: The movielens datasets: history and context. ACM Trans. Interact. Intell. Syst. **5**(4), 19:1–19:19 (2016)
13. Kaminskas, M., Bridge, D.: Diversity, serendipity, novelty, and coverage: a survey and empirical analysis of beyond-accuracy objectives in recommender systems. ACM Trans. Interact. Intell. Syst. **7**(1), 2:1–2:42 (2017)
14. Ling, C.X., Yang, Q., Wang, J., Zhang, S.: Decision trees with minimal costs. In: Proceedings of the Twenty-first International Conference on Machine Learning, ICML. ACM, vol. 69. ACM (2004)
15. Marras, M., Boratto, L., Ramos, G., Fenu, G.: Regulating group exposure for item providers in recommendation. In: Proceedings of the 45th International ACM SIGIR Conference on Research and Development in Information Retrieval, SIGIR, pp. 1839–1843. ACM (2022)
16. Muzellec, L., Ronteau, S., Lambkin, M.: Two-sided internet platforms: a business model lifecycle perspective. Ind. Mark. Manage. **45**, 139–150 (2015)
17. Ricci, F., Rokach, L., Shapira, B. (eds.): Recommender Systems Handbook. Springer, New York (2022)
18. Singh, A., Joachims, T.: Fairness of exposure in rankings. In: Proceedings of the 24th ACM SIGKDD International Conference on Knowledge Discovery and Data Mining, KDD, pp. 2219–2228. ACM (2018)
19. Yang, K., Stoyanovich, J.: Measuring fairness in ranked outputs. In: Proceedings of the 29th International Conference on Scientific and Statistical Database Management, SSDBM, pp. 22:1–22:6. ACM (2017)
20. Zehlike, M., Bonchi, F., Castillo, C., Hajian, S., Megahed, M., Baeza-Yates, R.: Fa*ir: a fair top-k ranking algorithm. In: Proceedings of the ACM on Conference on Information and Knowledge Management, CIKM, pp. 1569–1578. ACM (2017)
21. Zhang, S., Yao, L., Sun, A., Tay, Y.: Deep learning based recommender system: a survey and new perspectives. ACM Comput. Surv. **52**(1), 5:1–5:38 (2019)

Multiple Testing for IR and Recommendation System Experiments

Ngozi Ihemelandu[1](✉) and Michael D. Ekstrand[2]

[1] Boise State University, Boise, ID 83725, USA
ngoziihemelandu@u.boisestate.edu
[2] Department of Information Science, Drexel University, Philadelphia, PA 19104, USA
ekstrand@acm.org

Abstract. While there has been significant research on statistical techniques for comparing two information retrieval (IR) systems, many IR experiments test more than two systems. This can lead to inflated false discoveries due to the multiple-comparison problem (MCP). A few IR studies have investigated multiple comparison procedures; these studies mostly use TREC data and control the familywise error rate. In this study, we extend their investigation to include recommendation system evaluation data as well as multiple comparison procedures that controls for False Discovery Rate (FDR).

Keywords: statistical inference · family-wise error rate · false discovery rate · multiple testing · recommender system · information retrieval

1 Introduction

Effective evaluation of information retrieval and recommender systems requires an assessment of whether the difference in performance metrics observed in an experiment likely represent a real improvement. Statistical tests are designed to fill this gap, assessing the likelihood that an observed improvement could be seen with random chance. Several studies have investigated which significance tests are most appropriate for analyzing evaluation results [10,12–16,18,19,21] mostly examining comparisons between two systems; a few studies [5,6,20] consider comparing more than two systems.

Comparing all pairs of k systems requires $m = k(k-1)/2$ tests, while the significance level α controls the probability of falsely finding significance only for a single test; with m tests, the probability of incorrectly finding a significant difference increases to $1-(1-\alpha)^m$. This is known as the multiple comparison problem (MCP).

Partly supported by the National Science Foundation on Grant 17-51278.

N. Goharian et al. (Eds.): ECIR 2024, LNCS 14610, pp. 449–457, 2024.
https://doi.org/10.1007/978-3-031-56063-7_37

One common approach to addressing the MCP is to adjust the p-value to control the traditional family-wise error rate (FWER). FWER is the probability of making at least one false positive in m experiments. Another approach to adjusting p-values is to control the false discovery rate (FDR) [2]. FDR is the expected proportion of false positive results out of all positive test results.

When we fail to apply appropriate statistical analysis that addresses the MCP, we run the risk of either failing to advance algorithmic methods that should be advanced or falsely identifying interesting results where none exist. Both of these problems can hinder progress in information retrieval research and application both by holding back improvements and by spending time on methods whose observed improvement was a fluke.

Previous IR studies that have investigated the MCP focused on multiple comparison methods that controlled the FWER and used TREC data for their analysis. In this study, we extend our analysis to include procedures that control the FDR in p-value adjustment and to use recommendation system evaluation data. Recommendation system data have a few distinguishing features such as large sample size and high sparsity which induces bias in effectiveness metrics [1]. Our goal is to understand how controlling for each of these different error rate impacts IR and recommendation system evaluation result analysis and enable researchers choose the appropriate test for their analysis.

To that end, we address the following research questions:

- **RQ1** - When systems have equivalent performance (the null hypothesis is true) do procedures that correct for MCP control the family-wise error rate or false discovery rate at a specified α level?
- **RQ2** - How many missed findings does controlling for FWER and FDR lead to respectively?
- **RQ3** - Which method does best in identifying as many actual differences between systems as possible while still maintaining a low false positive rate?

We find that multiple correction tests meets its objective of controlling the FWER and FDR at or below the given α level when the sample size is small (≤ 1000). However, when both the sample size and the number of hypothesis tests are large, they control the error rate at a much higher level than the target α level. We also observe that correction tests that control the false discovery rate instead of the family-wise error rate are more powerful.

2 Related Works and Background

A number of studies [6,12,14,21] have investigated appropriate statistical methods for analyzing IR evaluation results for experiment comparing two systems. A few studies have focused on multiple testing adjustments when comparing more than two systems. Tague-Sutcliffe and Blustein [20] adjusted the p-value using the Scheffe's method [17] and found that only large effect sizes could be detected. Boytsov et al. [5] focused on adjusting p-values for non-parametric statistical procedures and found that the correction procedures found fewer true differences compared to unadjusted tests. Carterette [6] used a single-step method that relied on multivariate Student distribution to adjust the p-value for MCP; like [20] found that small pairwise differences were not detected.

There are two main approaches used in multiple hypothesis testing to correct for multiple comparisons or alpha inflation: controlling for family-wise error rate and controlling for false discovery rate.

2.1 Controlling the Family-Wise Error Rate

The family-wise error rate is the probability of having one or more false positives out of all the hypothesis tests conducted. To guarantee that the probability of having one or more false positives in m tests is α or less, Bonferonni adjusts the significance level of each hypothesis test to α/m. We consider two of the popular procedures in this study: Bonferroni [4] and Holms [9].

2.2 Controlling the False Discovery Rate

The false discovery rate (FDR) is the expected proportion of false positives (or discoveries) among all positives/discoveries. Controlling the FDR instead of the FWER is a more recent approach to addressing the multiple comparison problem. When all hypothesis are true, this error rate is equivalent to the FWER [2] but may not be controlled at the same level otherwise. We investigated the Benjamini & Hochberg (BH) [2] and Benjamini-Yekuteili (BY) [3] multiple correction procedures in this study. BY makes fewer assumptions than BH.

3 Methodology and Data

We study the behavior of multiple-comparison corrections using simulated replications of search experiments on TREC 2013 Web [7] and a recommendation experiment with MovieLens 100K [8]. We adopt the methodology developed by Urbano and Nagler [22], which is based on marginal distributions and inter-system dependencies. This methodology employs evaluation scores from each algorithm run to construct parametric and semi-parametric models. These models represent the marginal distribution of per-topic effectiveness scores for a system, as well as vine copulas that models inter-system dependencies.

The TREC data comes with runs, and for MovieLens, we generated runs using four collaborative filtering recommender algorithms (ALS BiasedMF, ALS ImplicitMF, Item k-NN, and User-based k-NN) in various configurations from the LensKit toolkit. All runs are evaluated using normalized Discounted Cumulative Gain (nDCG), with a cutoff threshold set at 20 for TREC data and 100 for recommender system data.

This combination generates data that is realistic both in terms of score distributions and inter-system correlations. To simulate an experiment with k systems, the process begins by fitting a model to generate replicates:

1. Randomly select k systems and their runs from the original data set.

2. Fit a marginal distribution F_B to each run B. Using code from Urbano and Nagler, candidate distributions take parametric (Truncated Normal, Beta, and Beta-Binomial) and non-parametric (discrete kernel smoothing) forms, and the best-fitting distribution is used.
3. Fit a Vine copula to model the joint score distributions between runs.

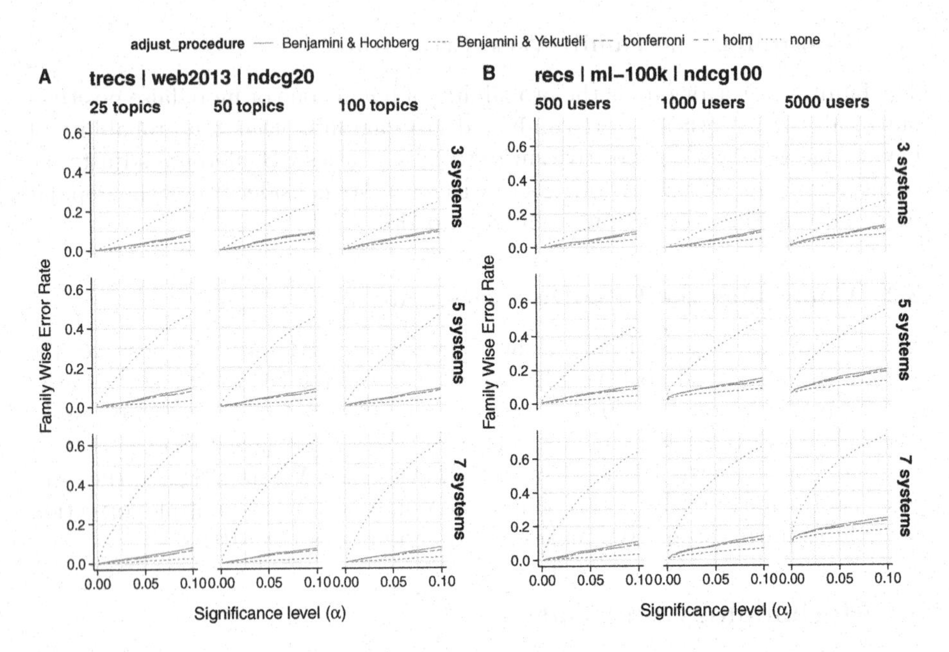

Fig. 1. When systems are equivalent (null hypothesis is true), the error rate is controlled at a significance level (α) by each adjustment test. The control of the error rate is determined by the sample size and number of hypothesis tests. We use all pairwise hypothesis test which determines the number of hypothesis tests. (For example, 5 systems would give 10 hypothesis tests).

We assess significance with pairwise t-tests between systems [11]. To address **RQ1**, which considers scenarios where systems have equivalent performance (thus making the null hypothesis true), we simulate true null hypotheses with all compared systems having the same mean effectiveness. The process involves:

1. randomly selecting a run, denoted as k_0, from the k runs.
2. calculating its mean effectiveness μ_0.
3. transforming the marginal distributions of the remaining $k-1$ runs to have the mean μ_0.

To address **RQ2**, which examines the ability of procedures controlling for FWER and FDR to detect effects when they exist (i.e., when the null hypothesis is false), we simulate scenarios where system E outperforms system B by a specific effect size of δ (effect size δ quantifies the magnitude of the difference in effectiveness between two recommender algorithms, providing insight into the practical or real-world significance of the findings). The simulation steps are as follows:

1. randomly select a run, denoted as k_0 from the k runs.
2. calculate its mean effectiveness, μ_0.
3. transform the marginal distributions of the other runs to have a mean of $\mu_0 + \delta$.

To address **RQ3**, which aims to identify the correction method most effective at detecting actual differences between systems while maintaining a low false positive rate (in scenarios with a mix of true and false null hypotheses), we simulate a mixture of equal and non-equal systems. The simulation process involves:

1. selecting a pair of runs and transforming their marginal distributions to have the same mean, μ_0.
2. choosing a different pair of runs and adjusting their marginal distributions so that the difference in their means is δ.

We then simulate new experimental data with a sample size n:

1. Adjust marginal distributions for selected runs to match experimental condition (all systems equal, one system better and $k - 1$ equal, or a mix of equal and non-equal systems).
2. Draw n topics each with k pseudo-observations from the fitted copula.
3. Apply the adjusted inverse Cumulative Distribution Function (CDF) F_B^{-1} of each system to the corresponding pseudo-observations to get final scores.
4. Use paired t-test to compare all pairs of systems.
5. Correct the t-test's p-values using the selected multiple comparison correction procedure to yield corrected p-values.
6. Calculate the false positives and/or power (as applicable) of comparing the adjusted p-value to the significance threshold α.

We ran 10,000 simulations for each configuration, using $k \in \{3, 5, 7\}$ and n ($n \in \{25, 50, 100\}$ for TREC, $n \in \{500, 1000, 5000\}$ for MovieLens). For the case of one system outperforming several baselines, we used effect sizes of 0.01, 0.05, and 0.1; for a mix, we had pairs of equivalent systems and individual systems with effect size δ over the pairs.

4 Results

Figure 1 shows the case corresponding to **RQ1**, wherein the null hypothesis is true, indicating that all systems exhibit equivalent performance. We observe that when the number of systems and sample size are small, or the number of systems is large and the sample size is small, the correction procedures control the error rate at (or below) the specified α significance level. However, when the sample size is large and the number of systems is large (specific with the

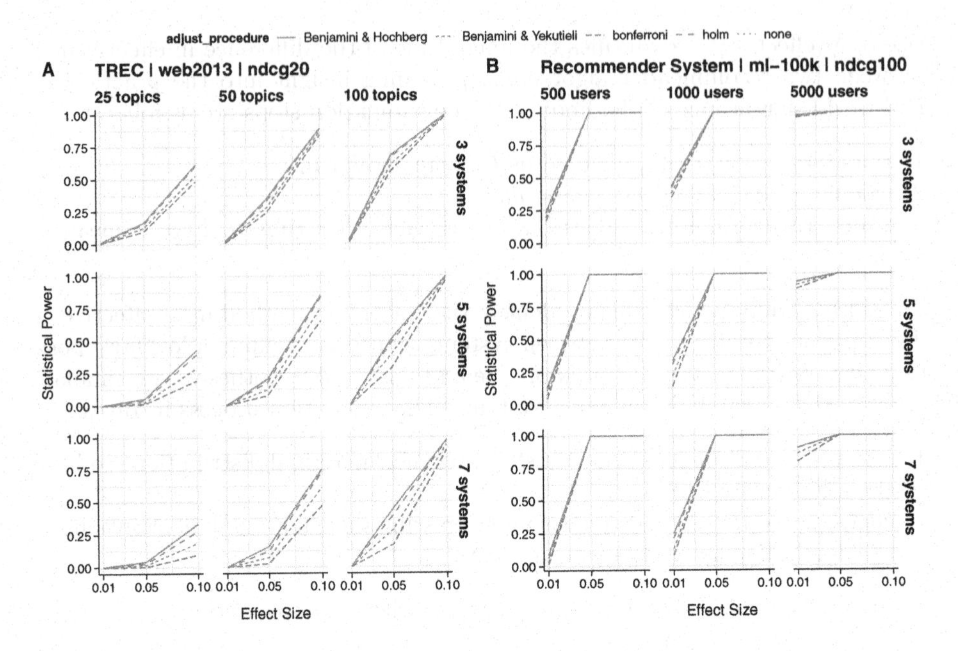

Fig. 2. The ability of adjustment procedures to find 'real' effects (statistical power) while controlling for FWER and FDR at $\alpha = 0.05$.

recommendation dataset), the error rate is inflated, but not nearly as inflated as the uncorrected test.

Figure 2 presents the results for **RQ2**, showing that with the TREC style experiment, when the sample size is small and when many tests are being conducted, the correction procedures impose a fairly severe penalty which reduces the statistical power. This is more pronounced with the Bonferroni test. Holms and Benjamini & Hochberg tests have the most power while the Bonferroni test has the least power. However, with the recommendation experiment, we observe differences in statistical power between the adjustment tests only when both the sample size and effect size are small. For the large sample sizes, there are no differences in the statistical power between the correction procedures.

Figure 3 displays the results for **RQ3**, highlighting the influence of sample size and the number of systems on the balance between error rates and the power of correction procedures. We observe that as the sample size and number of systems increase, the error rate and power also increase, while power decreases and the error rate is maintained at or below the specified threshold when the number of systems increases but the sample size is small.

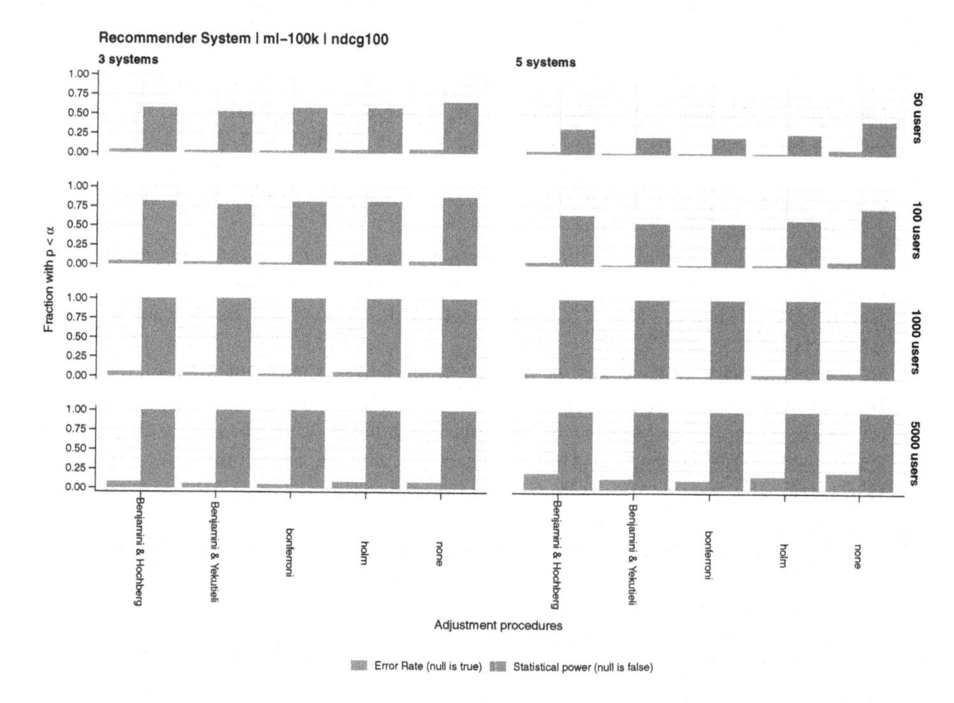

Fig. 3. A mixture of null hypothesis and non-null hypothesis (with effect size = 0.05). The proportion of null hypothesis that are $\leq \alpha$ (error rate) and the proportion of non-null hypothesis that are $< \alpha$ (statistical power).

5 Discussion and Conclusion

We find that corrections for multiple comparisons generally behave as expected in the small sample sizes of TREC-style IR experiments, controlling the error rate below the target significance threshold. However, as the sample size and number of systems increases, as in a large-scale search or recommendation experiment, the corrections no longer keep the overall error rate under the target level, although they still result in far fewer false discoveries than uncorrected pairwise t-tests. Our results also show that while the corrections have differing power and fail to find small effects in small experiments, they recover their power in a larger-scale experiment, and small effects are easily found even with conservative corrections like Bonferonni.

The Benjamini-Yekuteili test showed the lowest error rate in experiments with all systems equivalent, had greater power than the Bonferonni test at small to medium sample sizes, and strikes a balance between power and error in mixed-effect-size experiments. We therefore recommend it as the default correction for comparing multiple systems in an IR experiment. In large-scale experiments, if the computational and conceptual simplicity of the Bonferonni test is preferred, it can be used without meaningful loss in power.

References

1. Bellogín, A., Castells, P., Cantador, I.: Statistical biases in information retrieval metrics for recommender systems. Inf. Retriev. J. **20**, 606–634 (2017)
2. Benjamini, Y., Hochberg, Y.: Controlling the false discovery rate: a practical and powerful approach to multiple testing. J. Roy. Stat. Soc.: Ser. B (Methodol.) **57**(1), 289–300 (1995)
3. Benjamini, Y., Yekutieli, D.: The control of the false discovery rate in multiple testing under dependency. Annals Stat. **29**(4), 1165–1188 (2001)
4. Bland, J.M., Altman, D.G.: Multiple significance tests: the bonferroni method. BMJ **310**(6973), 170 (1995)
5. Boytsov, L., Belova, A., Westfall, P.: Deciding on an adjustment for multiplicity in IR experiments. In: Proceedings of the 36th International ACM SIGIR Conference on Research and Development in Information Retrieval, pp. 403–412 (2013)
6. Carterette, B.A.: Multiple testing in statistical analysis of systems-based information retrieval experiments. ACM Trans. Inf. Syst. **30**(1), 1–34 (2012). https://doi.org/10.1145/2094072.2094076
7. Hagen, M., et al.: Webis at trec 2013-session and web track. In: TREC (2013)
8. Harper, F.M., Konstan, J.A.: The movielens datasets: history and context. ACM Trans. Interact. Intell. Syst. **5**(4), 1–19 (2015)
9. Holm, S.: A simple sequentially rejective multiple test procedure. Scand. J. Statist. 65–70 (1979)
10. Hull, D.: Using statistical testing in the evaluation of retrieval experiments. In: Proceedings of the 16th Annual International ACM SIGIR Conference on Research and Development in Information Retrieval, pp. 329–338 (1993)
11. Ihemelandu, N., Ekstrand, M.D.: Statistical inference: the missing piece of recsys experiment reliability discourse. arXiv preprint arXiv:2109.06424 (2021)
12. Ihemelandu, N., Ekstrand, M.D.: Inference at scale: significance testing for large search and recommendation experiments. In: Proceedings of the 46th International ACM SIGIR Conference on Research and Development in Information Retrieval (SIGIR 2023) (2023)
13. Jones, K.S., Willett, P.: Readings in Information Retrieval. Morgan Kaufmann (1997)
14. Parapar, J., Losada, D.E., Presedo-Quindimil, M.A., Barreiro, A.: Using score distributions to compare statistical significance tests for information retrieval evaluation. J. Am. Soc. Inf. Sci. **71**(1), 98–113 (2020)
15. Rijsbergen, C.V.: Van. Information Retrieval, vol. 2. Butterworths (1979)
16. Savoy, J.: Statistical inference in retrieval effectiveness evaluation. Inf. Process. Manag. **33**(4), 495–512 (1997)
17. Scheffé, H.: A method for judging all contrasts in the analysis of variance. Biometrika **40**(1–2), 87–110 (1953)
18. Smucker, M.D., Allan, J., Carterette, B.: A comparison of statistical significance tests for information retrieval evaluation. In: Proceedings of the Sixteenth ACM Conference on Information and Knowledge Management, pp. 623–632 (2007)
19. Tague-Sutcliffe, J.: The pragmatics of information retrieval experimentation, revisited. Inf. Process. Manag. **28**(4), 467–490 (1992)
20. Tague-Sutcliffe, J., Blustein, J.: A Statistical Analysis of the Trec-3 Data, pp. 385–385. NIST Special Publication SP (1995)

21. Urbano, J., Lima, H., Hanjalic, A.: Statistical significance testing in information retrieval: an empirical analysis of type I, type II and type III errors. In: Proceedings of the 42nd International ACM SIGIR Conference on Research and Development in Information Retrieval, pp. 505–514 (2019)
22. Urbano, J., Nagler, T.: Stochastic simulation of test collections: evaluation scores. In: Proceedings of the 41st International ACM SIGIR Conference on Research and Development in Information Retrieval, pp. 695–704 (2018)

Can We Predict QPP? An Approach Based on Multivariate Outliers

Adrian-Gabriel Chifu[1], Sébastien Déjean[2], Moncef Garouani[4], Josiane Mothe[3,5](✉), Diégo Ortiz[5], and Md Zia Ullah[6]

[1] Aix Marseille Université, Université de Toulon, CNRS, LIS, Marseille, France
adrian.chifu@univ-amu.fr

[2] IMT, UMR5219 CNRS, UPS, Univ. de Toulouse, Toulouse, France
sebastien.dejean@math.univ-toulouse.fr

[3] IRIT, UMR5505 CNRS, Université de Toulouse, INSPE, UT2J, Toulouse, France
josiane.mothe@irit.fr

[4] IRIT, UMR 5505 CNRS, Université Toulouse Capitole, UT1, Toulouse, France
moncef.garouani@irit.fr

[5] IRIT, UMR, 5505 Toulouse, France
diego.ortiz@irit.fr

[6] Edinburgh Napier University, Edinburgh, UK
m.ullah@napier.ac.uk

Abstract. Query performance prediction (QPP) aims to forecast the effectiveness of a search engine across a range of queries and documents. While state-of-the-art predictors offer a certain level of precision, their accuracy is not flawless. Prior research has recognized the challenges inherent in QPP but often lacks a thorough qualitative analysis. In this paper, we delve into QPP by examining the factors that influence the predictability of query performance accuracy. We propose the working hypothesis that while some queries are readily predictable, others present significant challenges. By focusing on outliers, we aim to identify the queries that are particularly challenging to predict. To this end, we employ multivariate outlier detection method. Our results demonstrate the effectiveness of this approach in identifying queries on which QPP do not perform well, yielding less reliable predictions. Moreover, we provide evidence that excluding these hard-to-predict queries from the analysis significantly enhances the overall accuracy of QPP.

Keywords: Information Retrieval · Query performance prediction · QPP · Post-retrieval features · Multivariate outlier detection

1 Introduction

A search engine aims to process and answer any user query by retrieving relevant documents. However, the performance of a particular search engine can vary substantially depending on the specific queries it encounters [15,20].

Query performance prediction (QPP) addresses the crucial task of predicting how effective a system will be on a given query [7,8,35,37]. This problem is of

N. Goharian et al. (Eds.): ECIR 2024, LNCS 14610, pp. 458–467, 2024.
https://doi.org/10.1007/978-3-031-56063-7_38

paramount importance for two primary reasons. Firstly, when a query is expected to be difficult, it may be necessary to adopt a specialized, albeit potentially costly, approach. Conversely, when a query is predicted to be easy, a simpler and more cost-effective method can be employed [2,11,14].

QPP accuracy is typically assessed by measuring the correlation between the predicted and the actual performance values [12,19,31,35]. This approach is sound as long as the underlying assumptions and conditions for applying correlation measurements are respected [3].

Systems vary in their approach to processing a given query. This variability encompasses processes such as automatic query reformulation, the choice of the weighting/matching function (BM25 [32], a Language model [36], or an LLM-based model [9,16,26,38], for example), and the application of document re-ranking models [25]. Consequently, different systems will perform differently on the same queries [14]. QPP is thus considered with regard to the system it predicts the performance for. This implies that a QPP predictor should adapt its behavior to suit a particular system. This could be a reason why post-retrieval QPPs, which use the results of the search, tend to be more accurate than pre-retrieval ones, which rely solely on the query and the set of documents [21,22].

We found for example that *LemurTF_IDF*, a post-retrieval QPP corresponding to a Letor feature [6,8] has a higher correlation with the actual Average precision (AP) obtained with the LGD weighting function [10] than with the JS weighting function [1] (resp. 0.522 and 0.504 Pearson correlation). On the other hand, two predictors will behave differently on a particular system (Fig. 1).

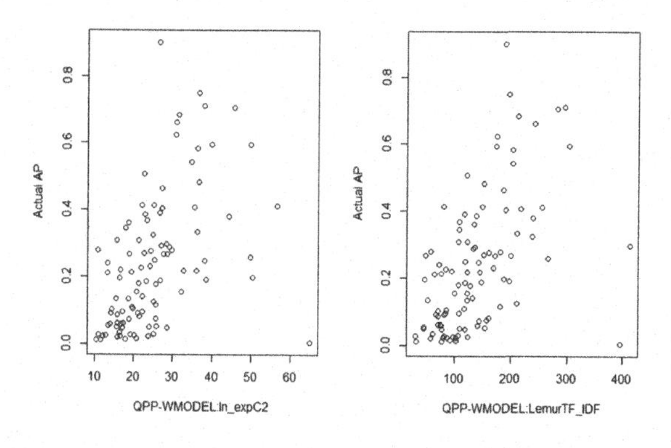

Fig. 1. QPPs behave differently on the same system and set of queries (TREC78). Predicted AP (X-axis) and actual AP (Y-axis) obtained with LGD weighting function and their Pearson correlation. Left side In_expC2 (ρ=0.484) - Right side LemurTF_IDF QPP (ρ=0.552).

In Fig. 1, we can see that the correlation is relatively weak, which is accurately reflected by the 0.484 (resp. 0.552) Pearson correlation values. Current QPP models lack accuracy. Single features, even post-retrieval ones could not demonstrate high correlation with actual performance [7,18,21]; even the combination of predictors, which is out of the scope of this paper, has not been very successful [8,13,27,31,33,37].

In this paper, we aim to analyze performance prediction deeper. Past studies showed that predictors do not work well when considering all queries. We hypothesize that prediction accuracy can significantly vary among queries, and our objective is to identify the queries for which the predictor fails to estimate the effectiveness. In other words, we want to predict the predictability of performance and address the following research question: Is it possible to predict the queries for which a QPP can provide an accurate effectiveness estimation for a given system? To put it differently, can we concentrate on those queries that are more likely to yield accurate predictions and perhaps automatically disregard queries that are more challenging to predict?

Some queries can be considered outliers (*abnormally* easy or difficult); similarly, predictions may have abnormal values. We hypothesize that the queries with abnormal prediction values are difficult to predict or get unreliable predictions. To tackle this problem, we consider multivariate outlier detection as a means to identify these hard-to-predict queries. Since a given QPP may behave differently to estimate the effectiveness of a system, the idea is to consider multiple QPPs in the query identification phase.

We consider several effectiveness measures and several benchmark collections. We show that our hypothesis could pave the way for a new research direction on QPP on accuracy predictability.

2 Identification of Difficult to Predict Queries by TRC

Here, we hypothesize that some predictions are outliers because the queries are difficult to predict. We aim to identify those queries for which we anticipate the QPP will not be accurate.

Johnson defines an outlier as an observation in a data set that appears to be inconsistent with the remainder of that set of data [24]. Univariate methods consider each variable independently, so only observations that appear odd for that variable are detected, while in the case of multivariate outlier detection, the interactions among different variables are compared. Multivariate outliers are a combination of unusual scores on several variables [30], and the idea is to detect the observations that are located relatively far from the center of the data distribution [5]. To identify outliers, we thus consider here multivariate outliers and identify the queries for which predictions are abnormal for different QPPs.

Mahalanobis distance is a common criterion for multivariate outlier detection. Applied to QPP, the Mahalanobis distance for a given query q_i, from a set of n queries $Q = \{q_1, q_2, \dots, q_n\}$, can be defined as follows:

$$D_M\left(q_i, Q\right) = \sqrt{(q_i - \overline{q})^T V_n^{-1} (q_i - \overline{q})}, \quad (1)$$

where the query vector q_i is composed of m ($n >> m$) predictor values $q_i = \left(p_1^i, p_2^i, \dots, p_m^i\right)$, $\overline{q}$ is the mean vector of the queries, and V_n^{-1} is the inverse of the covariance matrix of the queries. The superscript T denotes the transpose of the vector. We used the Transformed Rank Correlations (TRC) for multivariate outlier detection function implemented in R in this paper [4][1]. Instead of working with the raw data, which might be skewed or have a non-normal distribution, the TRC method works with the ranks of the data. This involves sorting the data from smallest to largest and using their position in this sorted order as their value. It then calculates the Spearman rank correlation or a similar non-parametric correlation measure between all pairs of variables. This helps in understanding the underlying structure and relationships in the data, which is crucial for identifying outliers in a multivariate context. The TRC method calculates a distance measure for each observation in the multivariate space. This distance is typically based on rank correlations and reflects how far away each observation is from the central trend of the data. It then determines a threshold for these distances (default value is set to the 0.95 quantile of F-distribution). Observations with a distance greater than this threshold are flagged as outliers.

3 Data

In this study, we use TREC adhoc collections: TREC78 with 528K documents and 100 queries (351 – 450) and WT10G which consists of 1.692M documents and 100 queries (451 – 550). We use Average Precision (AP@1000) [34] and normalized Discounted Cumulative Gain (nDCG@20) [23] as measures to evaluate the search engine performance, which are common in adhoc retrieval. Like previous studies on QPP [7,8,13,31,35,37], we use Pearson correlation as a measure to evaluate the accuracy of the QPP, in addition to plots as recommended in statistics. Pearson correlation measures the linear correlation between the predicted effectiveness of queries and their actual effectiveness. It indicates how well the QPP method predictions align with the actual query performances. Although it may not be a perfect measure [28], we kept it in this study. Alternative measures will be used in future work. We will consider other measures such as Mean Squared Error (MSE), Mean Absolute Error (MAE) or its normalized variant. The former measures the average of the squares of the errors between predicted and actual query effectiveness. The latter measures the average of the absolute differences between predicted and actual values. It is less sensitive to outliers compared to MSE. These two measures could help in understanding the magnitude of the prediction errors.

We consider a series of systems that treat the same queries over the same set of documents. In this study, we construct systems using Terrier [29]. These systems differ on several factors: the scoring function employed and the variant of the query reformulation module utilized, if any. We report the results on the best system according to the considered collection and performance measures, the best in terms of the average effectiveness over the set of queries. We also used a reference system based on LM Dirichlet [36].

In this study, we consider the first four Letor features which have been shown as among the most accurate for single feature QPP [6,8]. Letor features have been

[1] https://cran.r-project.org/web/packages/modi/modi.pdf.

initially used for retrieved document re-ranking [6]. Letor features are associated with each (query, retrieved document) pair[2]. To obtain a single value for each query for a given letor feature, the values are aggregated over the documents. We used maximum as the aggregation function which has been shown as the most accurate for QPP [8]. The four features we kept are LemurTF_IDF, In_expC2, InB2, and InL2, aggregated using the maximum function. We also consider four state-of-the-art QPP: Normalized query commitment (NQC) [35], Unnormalized query commitment (UCQ) [35], QF (query feedback) [37], and WIG [37] .

These features will be used individually to predict system performance. Although methods that combine features have been shown more accurate, here the objective is to conduct a first study on how to detect queries that are difficult to predict. We though opted for simple prediction models (single feature); we will analyze the proposed method for more complex models. Complex models that involve several features need to be trained. Here since we are not training a model, we do not need to split the data into train and test.

4 Results

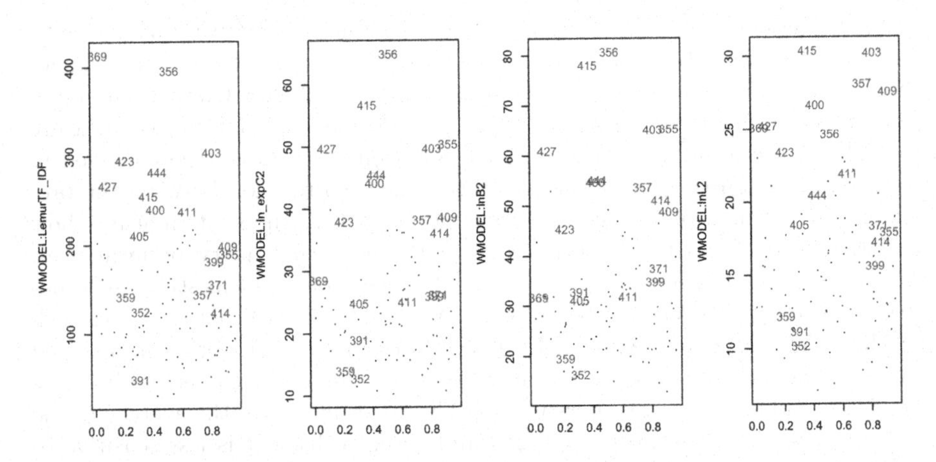

Fig. 2. Outlier queries are different according to the QPP. Queries detected as outliers by our method are in red. The four plots correspond to the four QPPs. The X-axis is not meaningful. Y-axis corresponds to the predicted values by the considered Letor QPP. QQP values are calculated for the TREC78 collection.

Outlier queries are different according to the considered QPP, which shows the importance of using multivariate outlier detection (Fig. 2). For example, query 403 is a clear outlier for InL2 QPP (top right of the right-side sub-figure) but not for the other QPPs. This implies that, if we used a univariate method such as LemurTF_IDF in isolation, we would not have identified this query as

[2] using Terrier (see http://terrier.org/docs/v5.1/learning.html).

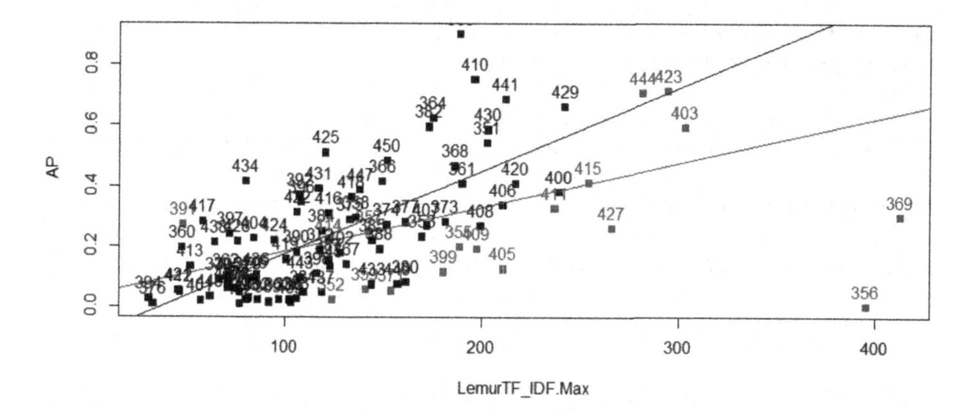

Fig. 3. Outlier queries are correctly identified. Queries that have been detected as outliers by our method are displayed in red. X-axis represents the QPP values and Y-axis is the actual AP. Here the values are calculated for LemurTF_IDF on the TREC78 collection and a model based on the LGD weighting function. 18 out of 100 queries were identified as outliers. The red line is the regression when all the topics are considered while the black one is after the outliers were removed.

one that is difficult to predict, even though it truly is (Fig. 3). Similarly, consider Query 369 (shown at the top left in the left-side sub-figure). It clearly stands out as an outlier when evaluated using the LemurTF_IDF QPP, but does not exhibit the same outlier behavior when assessed by the other three QPPs. These two queries (360 and 403) are indeed not well predicted (Fig. 3). In addition, we can see that query 356 has also been accurately identified as an outlier, as well as some other queries (Fig. 3). If these queries were not considered when calculating the correlation, the correlation would be higher.

We can observe that the correlation when considering the outliers only is very weak (Table 1, "LemurTF_IDF - Outliers only" row, left-part). This result was expected since we want to remove these difficult to predict queries. On the contrary, when the outlier queries are removed, the correlation between the QPPs and the actual effectiveness measures is much higher (Table 1 "No Outliers" rows). Note that if we use univariate outlier detection using LemurTF_IDF predictor only, the correlation decreases (0.658 compared to 0.700 here) (Table 1 "Univariate" rows). Results are consistent across the QPPs for TREC78 collection. This consistency remains valid for the best predictor LemurTF_IDF also for WT10G. We also evaluated the results considering other reference systems (NN-based model will be included in future work) and the results were also consistent (e.g., using In_expB2 weighting function). Results are also consistent when considering the other 4 QPP from the literature. Here we consider one single reference system (LM-based), WT10G and TREC Robust collections (Table 1, right side part).

Table 1. Pearson correlation is consistently better when queries our method detects as difficult to predict are removed. Correlation between actual effectiveness (either AP or ndcg) on TREC78 and WT10G. We report the number of outlier queries detected, the correlation when outliers are removed using univariate and multivariate methods and when all the queries are considered for the 4 Letor features and 4 other features. The reference system rows indicate the effectiveness on average for the considered measure and set of queries; this is the system that obtained the highest effectiveness measure over a set of system configurations we tried. A unilateral test of significance for the difference between the correlations with "No Outliers" and "All" were performed using the cocor R package [17]. A * after a correlation indicates that the p-value of the test is lower than 0.05.

Collection	WT10G	WT10G	TREC78	TREC78
Measure	NDCG	AP	NDCG	AP
Reference system	0.444	0.236	0.524	0.238
Outliers	16	16	19	18
LemurTF_IDF - Univariate	0.337	0.342	0.544	0.658
LemurTF_IDF - Outliers only	0.206	0.292	0.095	0.350
LemurTF_IDF - No Outliers	**0.438**	**0.468**	**0.601***	**0.700***
LemurTF_IDF - All	0.365	0.393	0.381	0.522
In_expC2 - Univariate	0.423	0.368	0.607	0.631
In_expC2 - No Outliers	0.391	0.350	**0.607***	**0.635**
In_expC2 - All	0.425	0.371	0.418	0.484
InB2 - Univariate	0.329	0.286	0.542	0.536
InB2 - No Outliers	0.286	0.214	0.530	**0.543**
InB2 - All	0.336	0.274	0.372	0.416
InL2 - Univariate	0.264	0.341	0.380	0.426
InL2 - No Outliers	0.258	0.347	**0.458**	**0.491**
InL2 - All	0.340	**0.353**	0.398	0.446

Collection	WT10G	WT10G	TREC78	TREC78
Measure	NDCG	AP	NDCG	AP
Ref system	0.4528	0.187	0.5235	0.2538
Outliers	24	24	42	42
NQC - No Outliers	**0.330**	**0.310***	**0.246**	**0.227**
NQC - All	0.097	0.051	0.219	0.174
UQC - No Outliers	**0.350**	**0.351**	**0.382**	**0.340**
UQC - All	0.206	0.151	0.333	0.283
WIG - No Outliers	-0.015	-0.007	0.059	-0.020
WIG - All	0.077	0.041	0.002	-0.103
QF - No Outliers	**0.357**	**0.369**	**0.493**	**0.438**
QF - All	0.283	0.281	0.469	0.403

5 Conclusion

Studies on QPP generally focus on prediction accuracy [7,8,13,18,31,35,37], but seldom on the difficulty of that prediction. QPP is clearly a difficult task, since current predictors are not very accurate. In this study we show that difficult to predict queries can be detected. We used multivariate outlier detection for that; it has the advantage to consider different QPPs to detect the queries for which the prediction may not be accurate. We also show that removing these automatically detected queries, we have a higher accuracy of the predictor. That means that we know that the predictor is not accurate for some queries that we can identify. This result pave the way to a new research direction: the prediction of the accuracy of the prediction. Some predictor may be accurate for certain queries and not for others. In future work, we will re-examine the results on more benchmark collections and reference retrieval systems. In this study we considered a single feature at a time to make the prediction. In future work, we will consider trained models that combine several features. That will also allow us to consider methods that allow models to abstain from making a prediction when they are not sufficiently confident.

References

1. Amati, G.: Frequentist and Bayesian approach to information retrieval. In: Lalmas, M., MacFarlane, A., Rüger, S., Tombros, A., Tsikrika, T., Yavlinsky, A. (eds.) Advances in Information Retrieval. LNCS, vol. 3936, pp. 13–24. Springer, Heidelberg (2006). https://doi.org/10.1007/11735106_3
2. Amati, G., Carpineto, C., Romano, G.: Query difficulty, robustness, and selective application of query expansion. In: McDonald, S. Tait, J. (eds.) Advances in Information Retrieval. LNCS, vol. 2997, pp. 127–137. Springer, Heidelberg (2004). https://doi.org/10.1007/978-3-540-24752-4_10
3. Anscombe, F.J.: Graphs in statistical analysis. Am. Stat. **27**(1), 17–21 (1973)
4. Béguin, C., Hulliger, B.: Multivariate outlier detection in incomplete survey data: the epidemic algorithm and transformed rank correlations. J. R. Stat. Soc. Ser. A Stat. Soc. **167**(2), 275–294 (2004)
5. Ben-Gal, I.: Outlier detection. In: Data Mining and Knowledge Discovery Handbook, pp. 131–146 (2005)
6. Cao, Z., Qin, T., Liu, T.Y., Tsai, M.F., Li, H.: Learning to rank: from pairwise approach to listwise approach. In: ICML, pp. 129–136 (2007)
7. Carmel, D., Yom-Tov, E.: Estimating the Query Difficulty for Information Retrieval. Morgan & Claypool Publishers (2010)
8. Chifu, A.G., Laporte, L., Mothe, J., Ullah, M.Z.: Query performance prediction focused on summarized letor features. In: The 41st International ACM SIGIR Conference on Research and Development in Information Retrieval, pp. 1177–1180 (2018)
9. Clark, K., Luong, M.T., Le, Q.V., Manning, C.D.: Electra: pre-training text encoders as discriminators rather than generators. arXiv preprint arXiv:2003.10555 (2020)
10. Clinchant, S., Gaussier, E.: Information-based models for ad hoc IR. In: Proceedings of the 33rd International ACM SIGIR Conference on Research and Development in Information Retrieval, pp. 234–241 (2010)
11. Cronen-Townsend, S., Zhou, Y., Croft, W.B.: Predicting query performance. In: International ACM SIGIR Conference on Research and Development in Information Retrieval, pp. 299–306 (2002)
12. Datta, S., MacAvaney, S., Ganguly, D., Greene, D.: A'pointwise-query, listwise-document'based query performance prediction approach. In: Proceedings of the 45th International ACM SIGIR Conference on Research and Development in Information Retrieval, pp. 2148–2153 (2022)
13. De, R., Grivolla, J., Jourlin, P., de Mori, R.: Automatic classification of queries by expected retrieval performance. In: ACM SIGIR 2005 Workshop on Predicting Query Difficulty-Methods and Applications (2005)
14. Déjean, S., Mothe, J., Ullah, M.Z.: Studying the variability of system setting effectiveness by data analytics and visualization. In: Crestani, F., et al. (eds.) Experimental IR Meets Multilinguality, Multimodality, and Interaction. CLEF 2019. LNCS, vol. 11696, pp. 62–74. Springer, Cham (2019). https://doi.org/10.1007/978-3-030-28577-7_3
15. Deveaud, R., Mothe, J., Ullah, M.Z., Nie, J.Y.: Learning to adaptively rank document retrieval system configurations. ACM Trans. Inf. Syst. **37**(1), 3 (2018). https://doi.org/10.1145/3231937
16. Devlin, J., Chang, M.W., Lee, K., Toutanova, K.: Bert: pre-training of deep bidirectional transformers for language understanding. arXiv preprint arXiv:1810.04805 (2018)

17. Diedenhofen, B., Musch, J.: cocor: a comprehensive solution for the statistical comparison of correlations. PLoS ONE **10**(4), e0121 (2005)
18. Faggioli, G., et al.: Query performance prediction for neural IR: are we there yet? In: Kamps, J., et al. (eds.) Advances in Information Retrieval: 45th European Conference on Information Retrieval, ECIR 2023. LNCS, vol. 13980, pp. 232–248. Springer, Cham (2023). https://doi.org/10.1007/978-3-031-28244-7_15
19. Faggioli, G., Zendel, O., Culpepper, J.S., Ferro, N., Scholer, F.: smare: a new paradigm to evaluate and understand query performance prediction methods. Inf. Retriev. J. **25**(2), 94–122 (2022)
20. Harman, D., Buckley, C.: Overview of the reliable information access workshop. Inf. Retrieval **12**(6), 615 (2009)
21. Hauff, C., Hiemstra, D., de Jong, F.: A survey of pre-retrieval query performance predictors. In: ACM CIKM, pp. 1419–1420 (2008)
22. Jafarzadeh, P., Ensan, F.: A semantic approach to post-retrieval query performance prediction. Inf. Process. Manag. **59**(1), 102746 (2022)
23. Järvelin, K., Kekäläinen, J.: Ir evaluation methods for retrieving highly relevant documents. ACM SIGIR Forum **51**(2), 243–250 (2017)
24. Johnson, R.A., Wichern, D.W., et al.: Applied Multivariate Statistical Analysis. Prentice Hall, Upper Saddle River (2002)
25. Liu, T.Y., et al.: Learning to rank for information retrieval. Found. Trends® Inf. Retriev. **3**(3), 225–331 (2009)
26. Liu, Y., et al.: Roberta: a robustly optimized bert pretraining approach. arXiv preprint arXiv:1907.11692 (2019)
27. Mizzaro, S., Mothe, J., Roitero, K., Ullah, M.Z.: Query performance prediction and effectiveness evaluation without relevance judgments: two sides of the same coin. In: The 41st International ACM SIGIR Conference on Research and Development in Information Retrieval, pp. 1233–1236 (2018)
28. Mothe, J.: On correlation to evaluate QPP. In: Proceedings of the QPP++ 2023: Query Performance Prediction and Its Evaluation in New Tasks Workshop co-located with 45th European Conference on Information Retrieval (ECIR), vol. 3366, pp. 29–36 (2023). https://www.scopus.com/inward/record.uri?eid=2-s2.0-85152226741&partnerID=40&md5=a5dea754c0149bc72f573d7f89457568
29. Ounis, I., Amati, G., Plachouras, V., He, B., Macdonald, C., Johnson, D.: Terrier information retrieval platform. In: Losada, D.E., Fernaindez-Luna, J.M. (eds.) Advances in Information Retrieval. ECIR 2005. LNCS, vol. 3408, pp. 517–519. Springer, Heidelberg (2005). https://doi.org/10.1007/978-3-540-31865-1_37
30. Peña, D., Prieto, F.J.: Multivariate outlier detection and robust covariance matrix estimation. Technometrics **43**(3), 286–310 (2001)
31. Raiber, F., Kurland, O.: Query-performance prediction: setting the expectations straight. In: Proceedings of the 37th International ACM SIGIR Conference on Research and Development in Information Retrieval (SIGIR 2014), pp. 13–22. ACM, New York (2014). https://doi.org/10.1145/2600428.2609581
32. Robertson, S., Zaragoza, H., et al.: The probabilistic relevance framework: Bm25 and beyond. Found. Trends Inf. Retr. **3**(4), 333–389 (2009)
33. Roy, D., Ganguly, D., Mitra, M., Jones, G.J.F.: Estimating Gaussian mixture models in the local neighbourhood of embedded word vectors for query performance prediction. Inf. Process. Manag. **56**(3), 1026–1045 (2019)
34. Sakai, T.: On the reliability of information retrieval metrics based on graded relevance. Inf. Process. Manag. **43**(2), 531–548 (2007)
35. Shtok, A., Kurland, O., Carmel, D., Raiber, F., Markovits, G.: Predicting query performance by query-drift estimation. ACM Trans. Inf. Syst. **30**(2), 11 (2012)

36. Zhai, C., Lafferty, J.: A study of smoothing methods for language models applied to information retrieval. ACM Trans. Inf. Syst. **22**(2), 179–214 (2004)
37. Zhou, Y., Croft, W.B.: Query performance prediction in web search environments. In: ACM SIGIR, pp. 543–550 (2007)
38. Zhu, Y., et al.: Large language models for information retrieval: a survey. arXiv preprint arXiv:2308.07107 (2023)

An In-Depth Comparison of Neural and Probabilistic Tree Models for Learning-to-rank

Haonan Tan[1], Kaiyu Yang[1], and Haitao Yu[2](✉)

[1] Graduate School of Comprehensive Human Sciences, University of Tsukuba, 1-2 Kasuga, Tsukuba, Ibaraki 305-0821, Japan
{s2221689,s2321730}@u.tsukuba.ac.jp

[2] Institute of Lilbray, Information and Media Science, University of Tsukuba, 1-2 Kasuga, Tsukuba, Ibaraki 305-0821, Japan
yuhaitao@slis.tsukuba.ac.jp

Abstract. Learning-to-rank has been intensively studied and has demonstrated significant value in several fields, such as web search and recommender systems. Over the learning-to-rank datasets given as vectors of feature values, LambdaMART proposed more than a decade ago, and its subsequent descendants based on gradient-boosted decision trees (GBDT), have demonstrated leading performance. Recently, different novel tree models have been developed, such as *neural tree ensembles* that utilize neural networks to emulate decision tree models and *probabilistic gradient boosting machines* (PGBM). However, the effectiveness of these tree models for learning-to-rank has not been comprehensively explored. Hence, this study bridges the gap by systematically comparing several representative neural tree ensembles (e.g., TabNet, NODE, and GANDALF), PGBM, and traditional learning-to-rank models on two benchmark datasets. The experimental results reveal that benefiting from end-to-end gradient-based optimization and the power of feature representation and adaptive feature selection, the neural tree ensemble does have its advantage for learning-to-rank over the conventional tree-based ranking model, such as LambdaMART. This finding is important as LambdaMART has achieved leading performance in a long period.

Keywords: Learning-to-rank · Gradient-boosted Decision Trees · Neural Tree Ensembles · Probabilistic Gradient Boosting Machines

1 Introduction

In the era of the *information flood*, an enormous volume of search requests are submitted by worldwide users daily. However, to accurately and efficiently provide the requested information, many open challenges are still far from being resolved. For example, a key problem is *ranking*, which has attracted increasing attention in recent years across many fields. The modern approach for solving the ranking problem is to employ machine learning methods to learn the ranking model. The resulting research, known as *learning-to-rank*, has gained significant

N. Goharian et al. (Eds.): ECIR 2024, LNCS 14610, pp. 468–476, 2024.
https://doi.org/10.1007/978-3-031-56063-7_39

attention due to its ability to contribute to and foster diverse research topics, such as *web search* and *dialogue systems*. Specifically, the rich machine learning frameworks, such as support vector machines (SVM) and deep neural networks (DNNs), enable the flexible development of powerful ranking models. As a result, the information retrieval (IR) community has experienced a flourishing development of learning-to-rank methods, such as *pointwise* methods [11–13], *pairwise* methods [16,21,36] and *listwise* methods [7,8,10,18,25,34,38,39,42,43,46–49]. For a comprehensive overview of learning-to-rank, we refer the reader to the works [19,26,30]. This work focuses on the cranfield learning-to-rank framework. Specifically, given a query and a set of documents to be ranked, the desired ranking model (or function) assigns a score to each document. Then, a ranked list of documents is obtained by sorting the documents in descending order of scores. Over the learning-to-rank datasets given as vectors of feature values (hereby referred to as *tabular datasets*), the tree-based ranking models, especially LambdaMART [42] and its variants [6,14,17,27], have shown state-of-the-art performance.

Recently, different novel tree-based models have been proposed. For example, unlike the traditional GBDT-based models that output a point prediction, PGBM [37] outputs a probabilistic prediction to quantify the prediction uncertainty. Given that the shallow GBDT-based methods employ greedy and local optimization procedures for tree construction, making it hard to perform gradient descent-based end-to-end learning, the neural tree ensembles [2,22,33] utilize neural architectures to emulate traditional decision tree behaviors. However, the effectiveness of the aforementioned novel tree models for learning-to-rank has not been comprehensively explored. Taking PGBM, TabNet, and NODE as examples, the authors merely compared the performance over tabular datasets for ranking based on simple mean squared error (MSE) or root MSE rather than detailed evaluation by considering the rank position and the relevance level (e.g., nDCG [20]). Furthermore, recent works [4,28] comprehensively studied tree-based models for ranking, especially from the viewpoint of both effectiveness and efficiency. However, the abovementioned novel tree models are not included.

Motivated by the research gap mentioned above, this work conducted several experiments on two benchmark collections to test the effectiveness of neural and probabilistic tree models for learning-to-rank. The main contributions of our study are as follows:

1. We systematically compare many representative neural tree ensembles, probabilistic tree models, and traditional models on two benchmark learning-to-rank datasets. In particular, we adopt three neural tree models (i.e., TabNet, NODE, and GANDALF) for learning-to-rank, where the same ranking objective function as LambdaMART is utilized. Given the selected probabilistic tree model of PGBM, we also implement LambdaMART by following the same GBDT framework as PGBM instead of only using the implementation included in LightGBM. This enables us to conduct a direct comparison betweem PGBM and LambdaMART.

2. By discussing the pros and cons of each method, we shed new light on the advantages of the newly proposed neural tree models. Particularly, benefiting from end-to-end gradient-based optimization and the power of feature representation and adaptive feature selection, the neural tree ensemble can achieve better performance than the conventional tree-based ranking model, such as LambdaMART.

The remainder of the paper is structured as follows. Section 2 details the experimental setting, and Sect. 3 conducts a series of experiments and discusses the corresponding findings. Finally, Sect. 4 concludes this work.

2 Experimental Setup

This section first details the adopted datasets and the evaluation metrics. Then, it describes the training and configuration of each method to be evaluated, including the parameter selection and fine-tuning.

2.1 Datasets

For effectively evaluating learning-to-rank methods, there are a number of widely used benchmark datasets, such as MSLRWEB30K[1], Yahoo-Set1[2] and Istella[3]. The experiments by Ai et al. [1] have shown that Yahoo-Set1 with 29,921 queries is a relatively easy dataset due to the fact that the 700 features are the outputs of a feature selection where the most predictive features for ranking are kept [9]. Therefore, this work employs MSLRWEB30K and Istella22 [15]. Specifically, each dataset comprises feature vectors extracted from query-document pairs and relevance judgment labels. For more detailed information, e.g., the feature description, the reader is referred to [35] and [15], respectively. The basic statistics of each dataset are reported in Table 1.

Given the above datasets, we use the training data to learn the ranking model, the validation data to select the hyperparameters based on nDCG@1, and the testing data for evaluation. Regarding MSLRWEB30K, previous studies [3,5,41] just used a single fold (i.e., Fold1) for the experimental evaluation. Hence, we use all five folds and conduct a 5-fold cross-validation to reduce the possible impact of overfitting on performance comparison. Finally, we report the ranking performance based on the five-fold average evaluation scores. For all the training data, we filter out the *dumb* queries with no relevant document and limit the minimum number of documents per query to 10.

Besides, we use nDCG to measure the performance, which considers rank position and relevance level. We report the results with different cutoff values 1, 3, 5, 10, 20, and 50 to show the performance of each method at different positions. We observe that the results in terms of nERR are consistent with nDCG, which are not included due to space constraints. The statistical significance is computed with a paired t-test with a significance level of $p < 0.01$.

[1] https://www.microsoft.com/en-us/research/project/mslr/.
[2] https://webscope.sandbox.yahoo.com/catalog.php?datatype=c.
[3] http://quickrank.isti.cnr.it/istella-dataset/.

Table 1. The statistics of adopted datasets.

	MSLRWEB30K	Istella22
#Queries	31,531	57,452
#Docs	3,771,125	8,421,456
#Features	136	220
#Avg relevant docs per query	58.0	4.77
#Docs per query (Min; Avg; Max)	(1; 119.6; 1,251)	(1; 207; 1000)
#Ground-truth labels & distribution	0 1,940,952	11,645,177
	1 1,225,770	155,720
	2 504,958	24,888
	3 69,010	60,383
	4 30,435	27,770

2.2 Methods

This work compares the following methods:

(1) **LambdaRank** and **LambdaMART** represent the traditional learning-to-rank methods. LambdaMART represents the typical GBDT-based ranking models, and we compare two different implementations, namely **LambdaMART(P)**, which is our implementation based on PyTorch[4], and **LambdaMART(L)**, which is the implementation in LightGBM [23]. Since this work focuses on comparing tree ensembles for learning-to-rank, other types of ranking approaches, such as context-aware ranking approaches (e.g., [1,32]) and adversarial ranking approaches (e.g., [40,47]) are not included.

(2) **TabNet**, **NODE**, and **GANDALF** represent neural tree ensembles that utilize neural networks to emulate decision tree models. LambdaRank and LambdaMART optimize the same objective function that imposes a weight on the pairwise loss of RankNet [44] by multiplying the absolute change in terms of a specific evaluation metric (e.g., $|\Delta nDCG|$). Thus, we use the same objective function for these neural tree ensembles, allowing us to understand the differences when deploying different machine learning frameworks.

(3) **PGBM** represents the tree model that outputs a probabilistic prediction. The objective function for PGBM is MSE based on the point estimate, where a document's relevance score is assumed to follow the Gaussian distribution. The exploration of optimizing the same objective function as LambdaRank is considered a future work.

2.3 Configuration

All methods and baseline approaches (excluding LambdaMART(L)) were implemented and trained using PyTorch v1.12 on an Nvidia A100 GPU (80 GB

[4] https://pytorch.org/.

memory). The general training settings include an $L2$ regularization (decay rate: 1×10^{-4}) and the Adam optimizer (learning rate: 1×10^{-3})[5] LambdaRank involves a 5-layer feed-forward neural network with 100 units in each hidden layer and a ReLU activation. In LambdaMART(L), the parameters are set as follows: learning rate: 0.05, num_leaves: 400, min_data_in_leaf: 50, min_sum_hessian_in_leaf: 200. The parameters in LambdaMART(P) are the same as in LambdaMART(L). For TabNet, the parameters are n_a: 128, n_d: 128, n_steps: 5, and gamma: 1.0. Regarding PGBM, the following parameters are adopted, bins: 256, leaves: 128, trees: 100. In NODE, the parameters are: layer_dim: 1024, num_layers: 2, tree_dim: 2, depth: 6. Finally, in GANDALF the parameters are: GFLU Stages: 3, GFLU Dropout: 0.25, Init Sparsity: 0.0.

3 Results and Analysis

Table 2 reports the overall performance of the tested approaches on each dataset. The best result is indicated in bold, while the second and third best results are denoted in underline and italics, respectively. In order to highlight the performance differences concerning the traditional GBDT-based ranking model, namely LambdaMART, the results that significantly differ from LambdaMART(L) are marked with an asterisk *.

Table 2. The performance of the compared methods on MSLR-WEB30K and Istella22.

MSLR-WEB30K						
Model	**nDCG@1**	**nDCG@3**	**nDCG@5**	**nDCG@10**	**nDCG@20**	**nDCG@50**
LambdaRank	0.4633*	0.4424*	0.4438*	0.4576*	0.4780*	0.5132*
LambdaMART(L)	0.4905	0.4694	0.4725	**0.4892**	**0.5105**	**0.5393**
LambdaMART(P)	0.4704*	0.4484*	0.4514*	0.4684*	0.4898*	0.5238*
PGBM	0.4724*	0.4511*	0.4544*	0.4708*	0.4930*	0.5207*
NODE	_0.4779*_	_0.4574*_	_0.4597*_	_0.4763*_	_0.4985*_	_0.5309*_
TabNet	0.4649*	0.4465*	0.4500*	0.4678*	0.4907*	0.5243*
GANDALF	__0.4954*__	__0.4720*__	__0.4728*__	0.4870*	0.5068*	0.5373*
Istella22						
LambdaRank	0.6154	0.7427*	0.8107*	0.8444*	0.8155*	0.770
LambdaMART(L)	**0.6808**	**0.8308**	**0.8695**	**0.8748**	**0.8358**	**0.7821**
LambdaMART(P)	0.6322*	0.7497*	0.8204*	0.8509*	0.8211*	0.7762*
PGBM	_0.6356*_	_0.7523*_	_0.8212*_	_0.8513*_	_0.8225*_	_0.7778*_
NODE	0.6267	0.7476*	0.8185*	0.8487*	0.8197*	0.7755*
TabNet	0.6299*	0.7495*	0.8180*	0.8499*	0.8205*	0.7756*
GANDALF	0.6500*	0.7621*	0.8289*	0.8582*	0.8290*	0.7842*

[5] The source code for reproducing the results: https://github.com/wildltr/ptranking/.

The performance of the conventional learning-to-rank approaches, namely LambdaRank and LambdaMART, reveals that: (1) Both LambdaMART(L) and LambdaMART(P) achieve significantly better performance than LambdaRank on the adopted two datasets. This again demonstrates the superior performance of LambdaMART over LambdaRank due to the effective ensemble of weak prediction models, which is consistent with the previous studies [41,45]. (2) Benefiting from the algorithmic and engineering optimizations of LightGBM, LambdaMART(L) demonstrates a further enhanced performance than LambdaMART(P), which uses the native histogram-based algorithm by employing global equal density histogram binning to bin continuous features into discrete bins.

Regarding the performance of each neural tree model in Table 2, GANDALF shows significantly better performance than NODE and TabNet. Although GANDALF, NODE, and TabNet follow similar ensemble learning, they rely on different strategies for feature representation learning and feature selection. Specifically, TabNet uses an attention mechanism at the instance level. NODE uses learnable feature masks with the α-entmax function [31] within the context of the oblivious trees. GANDALF uses the α-entmax function as feature masks but introduces learnable sparsity with the gating mechanism.

Next, we compare the three types of tree-based ranking models: the conventional GBDT-based model LambdaMART, the probabilistic GBDT-based model PGBM, and the neural tree model GANDALF. Table 2 highlights that: (1) PGBM achieves competitive performance as LambdaMART(P). Since PGBM and LambdaMART(P) are implemented based on the same GBDT framework, their key difference is that PGBM considers the prediction's uncertainty. This suggests that it is possible to achieve probabilistic estimates without compromising point performance. Moreover, algorithmic and engineering optimizations are required to achieve competitive performance as LambdaMART(L). (2) GANDALF shows significantly better performance than LambdaMART(P) and PGBM on both datasets. Even compared with LambdaMART(L), GANDALF performs significantly better over MSLR-WEB30K. Since LambdaMART(L) benefits from the algorithmic and engineering optimizations of LightGBM, it is reasonable to say that the neural tree ensemble does have its advantage over the conventional tree-based ranking model, especially benefiting from both end-to-end gradient-based optimization and the power of feature representation and adaptive feature selection.

4 Conclusion

This paper conducted several experiments to investigate the effectiveness of current representative neural tree ensembles and probabilistic tree models on the learning-to-rank task. By imposing the same ranking objective function as LambdaMART over the adopted neural tree models (i.e., TabNet, NODE, and GANDALF) and using the same ensemble structure of PGBM for LambdaMART, we are able to investigate the advantages and disadvantages of the newly proposed neural tree ensembles and probabilistic tree model in the context of learning-to-rank. Also, our results indicate that the neural tree ensemble can achieve better performance than the conventional tree-based ranking model, especially benefiting from both end-to-end gradient-based optimization and the power of feature representation and adaptive feature selection. Since learning-to-rank has demonstrated significant value in a variety of fields, we believe that our work shed new light on the advantages of the newly proposed neural tree models.

For future work, the following practical issues are worthy to be investigated. First, in terms of scalability, we did not conduct an in-depth comparison on the newly proposed neural tree ensembles and probabilistic tree model. However, we do note that a number of prior studies [24,29] are quite inspiring. Second, we plan to extend our comparison study by exploring different datasets, such as MS MARCO, which consists of raw text queries and documents.

Acknowledgments. This research has been supported by JSPS KAKENHI Grant Number 19H04215.

References

1. Ai, Q., Bi, K., Guo, J., Croft, W.B.: Learning a deep listwise context model for ranking refinement. In: The 41st International ACM SIGIR Conference on Research & Development in Information Retrieval, pp. 135–144. SIGIR '18, Association for Computing Machinery, New York, NY, USA (2018)
2. Arik, S.O., Pfister, T.: TabNet: attentive interpretable tabular learning. Proc. AAAI Conf. Artif. Intell. **35**(8), 6679–6687 (2021)
3. Bruch, S., Han, S., Bendersky, M., Najork, M.: A stochastic treatment of learning to rank scoring functions. In: Proceedings of the 13th WSDM, pp. 61–69 (2020)
4. Bruch, S., Lucchese, C., Nardini, F.M.: Efficient and effective tree-based and neural learning to rank. Found. Trends® Inf. Retrieval **17**(1), 1–123 (2023)
5. Bruch, S., Zoghi, M., Bendersky, M., Najork, M.: Revisiting approximate metric optimization in the age of deep neural networks. In: Proceedings of the 42nd SIGIR, pp. 1241–1244 (2019)
6. Bruch, S.: An alternative cross entropy loss for learning-to-rank. In: Proceedings of the Web Conference 2021, pp. 118–126. WWW '21, Association for Computing Machinery, New York, NY, USA (2021)
7. Burges, C.J.C., Ragno, R., Le, Q.V.: Learning to rank with nonsmooth cost functions. In: Proceedings of NeurIPS, pp. 193–200 (2006)
8. Cao, Z., Qin, T., Liu, T.Y., Tsai, M.F., Li, H.: Learning to Rank: from pairwise approach to listwise approach. In: Proceedings of the 24th ICML, pp. 129–136 (2007)
9. Chapelle, O., Chang, Y.: Yahoo! learning to rank challenge overview. In: Chapelle, O., Chang, Y., Liu, T.Y. (eds.) Proceedings of the Learning to Rank Challenge. Proceedings of Machine Learning Research, vol. 14, pp. 1–24. PMLR, Haifa, Israel (2011)
10. Chapelle, O., Le, Q., Smola, A.: Large margin optimization of ranking measures. In: NIPS workshop on Machine Learning for Web Search (2007)
11. Chu, W., Ghahramani, Z.: Gaussian processes for ordinal regression. J. Mach. Learn. Res. **6**, 1019–1041 (2005)

12. Chu, W., Keerthi, S.S.: New approaches to support vector ordinal regression. In: Proceedings of the 22nd ICML, pp. 145–152 (2005)
13. Cossock, D., Zhang, T.: Subset ranking using regression. In: Proceedings of the 19th Annual Conference on Learning Theory, pp. 605–619 (2006)
14. Dato, D., et al.: Fast ranking with additive ensembles of oblivious and non-oblivious regression trees. ACM Trans. Inf. Syst. **35**(2) (2016)
15. Dato, D., MacAvaney, S., Nardini, F.M., Perego, R., Tonellotto, N.: The istella22 dataset: bridging traditional and neural learning to rank evaluation. In: Proceedings of the 45th International ACM SIGIR Conference on Research and Development in Information Retrieval, pp. 3099–3107 (2022)
16. Freund, Y., Iyer, R., Schapire, R.E., Singer, Y.: An efficient boosting algorithm for combining preferences. J. Mach. Learn. Res. **4**, 933–969 (2003)
17. Ganjisaffar, Y., Caruana, R., Lopes, C.V.: Bagging gradient-boosted trees for high precision, low variance ranking models. In: Proceedings of the 34th International ACM SIGIR Conference on Research and Development in Information Retrieval, pp. 85–94. SIGIR '11, Association for Computing Machinery, New York, NY, USA (2011)
18. Guiver, J., Snelson, E.: Learning to rank with SoftRank and Gaussian processes. In: Proceedings of the 31st SIGIR, pp. 259–266 (2008)
19. Guo, J., et al.: A deep look into neural ranking models for information retrieval. Inf. Process. Manag. (2019)
20. Järvelin, K., Kekäläinen, J.: Cumulated gain-based evaluation of IR techniques. ACM Trans. Inf. Syst. **20**(4), 422–446 (2002)
21. Joachims, T.: Training linear SVMs in linear time. In: Proceedings of the 12th KDD, pp. 217–226 (2006)
22. Joseph, M., Raj, H.: GANDALF: gated adaptive network for deep automated learning of features. arXiv:2207.08548 [cs.LG] (2022)
23. Ke, G., et al.: LightGBM: a highly efficient gradient boosting decision tree. In: Proceedings of NeurIPS, pp. 3149–3157 (2017)
24. Ke, G., Xu, Z., Zhang, J., Bian, J., Liu, T.Y.: DeepGBM: a deep learning framework distilled by GBDT for online prediction tasks. In: Proceedings of the 25th ACM SIGKDD International Conference on Knowledge Discovery & Data Mining, pp. 384–394. KDD '19, Association for Computing Machinery, New York, NY, USA (2019)
25. Lan, Y., Zhu, Y., Guo, J., Niu, S., Cheng, X.: Position-aware ListMLE: a sequential learning process for ranking. In: Proceedings of the 30th Conference on UAI, pp. 449–458 (2014)
26. Lin, J., Nogueira, R., Yates, A.: Pretrained transformers for text ranking: BERT and beyond. arxiv.org/abs/2010.06467 (2020)
27. Lucchese, C., Nardini, F.M., Perego, R., Orlando, S., Trani, S.: Selective gradient boosting for effective learning to rank. In: The 41st International ACM SIGIR Conference on Research & Development in Information Retrieval, pp. 155–164. SIGIR '18, New York, NY, USA (2018)
28. Lyzhin, I., Ustimenko, A., Gulin, A., Prokhorenkova, L.: Which tricks are important for learning to rank? In: Proceedings of the 40th International Conference on Machine Learning, ICML'23, JMLR.org (2023)
29. Nardini, F., Rulli, C., Trani, S., Venturini, R.: Distilled neural networks for efficient learning to rank. IEEE Trans. Knowl. Data Eng. **35**(05), 4695–4712 (2023)
30. Onal, K.D., Zhang, Y., Altingovde, I.S.: Others: neural information retrieval: at the end of the early years. J. Inf. Retrieval **21**(2–3), 111–182 (2018)

31. Peters, B., Niculae, V., Martins, A.F.T.: Sparse sequence-to-sequence models. In: Proceedings of the 57th Annual Meeting of the Association for Computational Linguistics, pp. 1504–1519. Association for Computational Linguistics, Florence, Italy (2019)
32. Pobrotyn, P., Bartczak, T., Synowiec, M., Bialobrzeski, R., Bojar, J.: Context-aware learning to rank with self-attention. CoRR (2005)
33. Popov, S., Morozov, S., Babenko, A.: Neural oblivious decision ensembles for deep learning on tabular data. CoRR abs/1909.06312 (2019)
34. Qin, T., Liu, T.Y., Li, H.: A general approximation framework for direct optimization of information retrieval measures. J. Inf. Retrieval **13**(4), 375–397 (2010)
35. Qin, T., Liu, T.Y., Xu, J., Li, H.: LETOR: a benchmark collection for research on learning to rank for information retrieval. Inf. Retrieval J. **13**(4), 346–374 (2010)
36. Shen, L., Joshi, A.K.: Ranking and reranking with perceptron. Mach. Learn. **60**(1–3), 73–96 (2005)
37. Sprangers, O., Schelter, S., de Rijke, M.: Probabilistic gradient boosting machines for large-scale probabilistic regression. In: Proceedings of the 27th ACM SIGKDD Conference on Knowledge Discovery & amp; Data Mining, pp. 1510–1520 (2021)
38. Taylor, M., Guiver, J., Robertson, S., Minka, T.: SoftRank: optimizing non-smooth rank metrics. In: Proceedings of the 1st WSDM, pp. 77–86 (2008)
39. Volkovs, M.N., Zemel, R.S.: BoltzRank: learning to maximize expected ranking gain. In: Proceedings of ICML, pp. 1089–1096 (2009)
40. Wang, J., et al.: IRGAN: a minimax game for unifying generative and discriminative information retrieval models. In: Proceedings of the 40th International ACM SIGIR Conference on Research and Development in Information Retrieval, pp. 515–524. SIGIR '17, Association for Computing Machinery, New York, NY, USA (2017)
41. Wang, X., Li, C., Golbandi, N., Bendersky, M., Najork, M.: The lambdaloss framework for ranking metric optimization. In: Proceedings of the 27th CIKM, pp. 1313–1322 (2018)
42. Wu, Q., Burges, C.J., Svore, K.M., Gao, J.: Adapting boosting for information retrieval measures. J. Inf. Retrieval **13**(3), 254–270 (2010)
43. Xu, J., Li, H.: AdaRank: a boosting algorithm for information retrieval. In: Proceedings of the 30th SIGIR, pp. 391–398 (2007)
44. Yu, H.: Optimize what you evaluate with: a simple yet effective framework for direct optimization of IR metrics. CoRR abs/2008.13373 (2020)
45. Yu, H.T., Huang, D., Ren, F., Li, L.: Diagnostic evaluation of policy-gradient-based ranking. Electronics **11**(1) (2022)
46. Yu, H.T., Jatowt, A., Joho, H., Jose, J.M., Yang, X., Chen, L.: WassRank: Listwise document ranking using optimal transport theory. In: Proceedings of the Twelfth ACM International Conference on Web Search and Data Mining, pp. 24–32. WSDM '19, Association for Computing Machinery, New York, NY, USA (2019)
47. Yu, H.T., Piryani, R., Jatowt, A., Inagaki, R., Joho, H., Kim, K.S.: An in-depth study on adversarial learning-to-rank. Inf. Retr. **26**(1) (2023)
48. Yu, H.T.: Optimize what you evaluate with: search result diversification based on metric optimization. Proc. AAAI Conf. Artif. Intell. **36**(9), 10399–10407 (2022)
49. Yue, Y., Finley, T., Radlinski, F., Joachims, T.: A support vector method for optimizing average precision. In: Proceedings of the 30th SIGIR, pp. 271–278 (2007)

Answer Retrieval in Legal Community Question Answering

Arian Askari(✉), Zihui Yang, Zhaochun Ren, and Suzan Verberne

Leiden University, Leiden, The Netherlands
{a.askari,z.ren,s.verberne}@liacs.leidenuniv.nl,
zoeyangzihui@gmail.com

Abstract. The task of answer retrieval in the legal domain aims to help users to seek relevant legal advice from massive amounts of professional responses. Two main challenges hinder applying existing answer retrieval approaches in other domains to the legal domain: (1) a huge knowledge gap between lawyers and non-professionals; and (2) a mix of informal and formal content on legal QA websites. To tackle these challenges, we propose CE_{FS}, a novel cross-encoder (CE) re-ranker based on the fine-grained structured inputs. CE_{FS} uses additional structured information in the CQA data to improve the effectiveness of cross-encoder re-rankers. Furthermore, we propose *LegalQA*: a real-world benchmark dataset for evaluating answer retrieval in the legal domain. Experiments conducted on LegalQA show that our proposed method significantly outperforms strong cross-encoder re-rankers fine-tuned on MS MARCO. Our novel finding is that adding the question tags of each question besides the question description and title into the input of cross-encoder re-rankers structurally boosts the rankers' effectiveness. While we study our proposed method in the legal domain, we believe that our method can be applied in similar applications in other domains.

Keywords: Legal Answer Retrieval · Legal IR · Data collection · Fine-grained structured cross-encoder

1 Introduction

As an established problem in information retrieval, answer[1] retrieval [6] has been studied in a variety of domains, including healthcare [8], social media [6,29], and programming [31]. Community question answering (CQA) platforms provide common sources for answer retrieval [5,22,23], e.g., Stackoverflow[2] for finding answers of programming questions [5]. In the legal domain, users[3]

[1] We interchangeably use the word answer and response to refer to the content written by the professional lawyer.
[2] https://stackoverflow.com/.
[3] We refer to the person who posts a question as a user or questioner throughout this paper.

N. Goharian et al. (Eds.): ECIR 2024, LNCS 14610, pp. 477–485, 2024.
https://doi.org/10.1007/978-3-031-56063-7_40

seek legal advice from professionals, and timely access to accurate answers is important. Legal language, often considered a sub-language due to its distinct characteristics, emphasizes the importance of answer retrieval in the legal domain [10,25,27]. However, answer retrieval has not been addressed in the legal domain yet.

In this paper, we propose the task of legal answer retrieval. We start by analyzing the effectiveness of the widely used two-stage ranking pipeline [3], which consists of an efficient retriever, e.g., BM25 [21], to retrieve a shortlist of documents from the collection, and a re-ranker [18] to increase the effectiveness of the initial ranking. On this premise, we propose cross-encoder$_{CAT}$ (CE$_{CAT}$), which uses cross-encoders for optimizing the re-ranking stage by concatenating the query and the candidate document in the input [3]. Given a new question and the corresponding initial ranked list produced by BM25, the CE$_{CAT}$ aims to effectively reorder the candidate documents from the initial ranked list to locate the most relevant responses on top of the ranked list. We fine-tune a cross-encoder re-ranker, called CE$_{FS}$, based on our novel **f**ine-grained **s**tructured inputs to learn the relevance between a pair of a question and its best answer based on structured information from the data; such a model would be able to effectively re-rank prior existing relevant responses where the best answer is not provided [30].

Our proposed method is inspired by recent studies that have shown task-level input modification could improve the effectiveness of cross-encoder re-rankers. For instance, BERT-FP [11] has shown that adding splitter tokens between each utterance of a dialogue can improve the effectiveness of BERT-based re-rankers. We are also motivated by the fact that, in legal CQA, question tags consist of important legal terms related to the legal question and using them can potentially bridge the knowledge gap between lawyer content and questioner content. It is noteworthy to mention that question tags are dynamically generated by the online platform that we have used in this paper.

For the first attempt, we release a new benchmark dataset, namely LegalQA, in this paper. Each question in LegalQA has a corresponding best answer selected by the questioner[4]. It consists of 9,846 questions and 33,670 answers responses by identified lawyers, organized in train, validation, and test splits. Our experiments on LegalQA show that CE$_{FS}$ significantly outperforms regular and strong fine-tuned cross-encoder re-rankers such as MiniLM-MSMARCO [24] which is a highly effective cross-encoder re-ranker trained on MS MARCO [17].

Our contributions are as follows: (1) We formulate the task of answer retrieval in the legal domain and release LegalQA. As far as we know, this is the first benchmark test collection for legal answer retrieval. (2) We evaluate the applicability of both probabilistic and existing effective fine-tuned cross-encoder re-rankers on legal answer retrieval. (3) We propose CE$_{FS}$ taking into account the different elements of a legal question with a fine-grained structured input that significantly outperforms regular fine-tuning cross-encoder re-rankers and the strong cross-encoder re-ranker trained on MS MARCO.

[4] https://github.com/arian-askari/AnswerRetrieval-Legal.

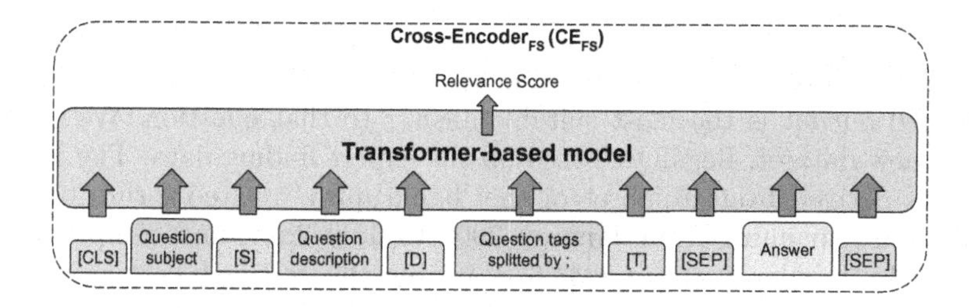

Fig. 1. Fine-grained structured input of CE_{FS}

2 Related Work

The objective of answer retrieval is to find the most relevant response given a question, which aligns with the core retrieval objective [5,6,22,23,29–31]. Therefore, the effective CE_{CAT} re-rankers are suitable to be invested in this task as they have shown high effectiveness in addressing various core retrieval tasks [1,12,18,20]. Several studies have shown the impact of modifying the regular input of CE_{CAT} could improve their effectiveness such as [2,3,7,11]. However, there are no studies that investigated the usage of question tags to improve cross-encoder re-rankers in community question-answering systems. The most relevant work to this study is the recent work by [15] that investigates the effectiveness of existing methods on legal CQA data. However, the responses in that work are written by legal forum users rather than identified lawyers, and they do not improve the effectiveness of existing methods for this particular domain. They also included the fine-tuned MiniLM on MS Marco [24] in their study and showed its poor performance in the legal domain. Furthermore, Martinez-Gil et al. [16] explore potential future directions for legal answering systems in a survey on legal systems designed for lawyers or law students including statute law retrieval and task legal textual entailment tasks. In contrast, our focus is on answer retrieval in a legal community question-answering system with a combination of legal language by lawyers and everyday language by questioners who ask questions.

3 Methods

We present our method, CE_{FS}, as an effective cross-encoder based re-ranker for the question answering retrieval in the legal domain. Our retrieval pipeline consists of first-stage retrieval and re-ranking. For a user's questions, the first-stage retrieval returns a set of initial responses from the dataset. These results are then re-ranked in the second step to improve the effectiveness of retrieval.

Collected Dataset for the Task. We create the LegalQA dataset by using the question URLs shared by [4] to investigate expert finding in the legal domain. In the context of the expert finding task, the input consists solely of specific question tags, and the output is a list of experts about that specific question tag. In contrast, in answer retrieval, the standard input is the question's content, and the desired output is the most suitable answer to that question. We therefore create a new dataset, LegalQA based on the expert finding data. The LegalQA dataset is derived from a subset of the 'bankruptcy' related forum on Avvo[5] in California, spanning from January 2008 to July 2021, and comprises 9,846 questions by regular users and 33,670 answers by certified lawyers. Notably, lawyers' profiles on Avvo are associated with their real names, distinguishing them from regular users. The questions are categorized, and each category, such as 'bankruptcy,' includes questions with different category tags, for instance, 'bankruptcy homestead exemption'. Following [4], we determine a question's best answer based on whether it is selected as the most helpful by the questioner or if it receives "lawyer agree" votes from at least three certified lawyers, which is distinct from the "helpful" upvotes provided by other users. To facilitate model training and evaluation, we split the dataset into three subsets: 70% for training, 10% for validation, and 20% for testing, based on the chronological order of questions. To ensure the dataset's integrity and eliminate duplicates, we perform a lexical similarity analysis using Levenshtein distance [14] on all the questions[6]. This analysis identified 50 pairs of questions with more than a 90% overlap. In cases of high overlap between question pairs, we retain the longer question and discard the shorter one. Additionally, we reassign the responses from the removed question to the retained one.

Baselines. For the first-stage retrieval, we employ BM25 as the baseline method. Additionally, we compare BM25 to a lexical-based probabilistic first-stage retriever named LMD [19] to assess the effectiveness of BM25 compared to another lexical-based retrieval. We leave out investigating the effectiveness of Transformer-based first-stage retrievers, e.g., dense retrievers, since we focus on improving the effectiveness of the re-ranking stage of answer retrieval in the legal domain. We leave further analysis on Transformer-based first-stage retrievers to future work. For the re-ranking, we use cross-encoder re-rankers (CE_{CAT}) in different settings: (1) a fine-tuned cross-encoder on MS MARCO; (2) a fine-tuned cross-encoder on the LegalQA training set; (3) a pre-trained cross-encoder as the zero-shot baseline.

BM25 and LMD. We use BM25 as first stage retriever, a commonly used ranking function that efficiently retrieves a set of documents from the full document collection based on word overlap [9,21].

CE_{CAT}. We use CE_{CAT} as a strong re-ranker. The query q1:m and answer content a1:n sequences are concatenated with the [SEP] token, and the [CLS] token representation computed by CE is scored with a single linear layer W in the CECAT ranking model: $CE_{CAT}(q_{1:m}, a_{1:n}) = CE([CLS]\, q\, [SEP]\, a\, [SEP]) * W$. We evaluate fine-tuned both MS MARCO-trained CE_{CAT}, MiniLM-MSMARCO

[5] https://www.avvo.com/topics/bankruptcy.

[6] We use the implementation available at https://github.com/seatgeek/thefuzz.

Table 1. Effectiveness results. † denotes a statistically significant improvement of $\mathbf{CE_{FS}}$ over the second most effective re-ranker, CE_{CAT} fine-tuned on the LegalQA training set. Statistical significance was measured with a paired t-test ($p < 0.001$) with Bonferroni correction for multiple testing. All re-rankers used top-1000 retrieved answers by BM25 as the initial ranking.

Training source	Model	MAP@1k	R@1k	R@100	R@10	R@2	R@1
First-stage retrievers							
—	BM25	.120	**.542**	.354	.192	.113	.069
—	LMD	.080	.540	.321	.153	.840	.050
Cross-encoder re-rankers							
MS MARCO	CE_{CAT}	.109	**.542**	.341	.173	.101	.087
LegalQA (**ours**)	CE_{CAT}	.236	**.542**	.495	.381	.204	.181
LegalQA (**ours**)	$\mathbf{CE_{FS}}$ **(ours)**	**.270**†	**.542**	**.524**†	**.428**†	**.261**†	**.209**†

[24], and LegalQA-trained CE_{CAT} following by the above design. We use MiniLM [26] as the cross-encoder model thorough all of the experiments.

Proposed method: Cross-encoderFS (CEFS).

We propose CE_{FS} in order to capture the relevance within the question and best answer based on a fine-grained structural-based input representation tailored for cross-encoder re-rankers. In Fig. 1, we present the input representation of CE_{FS}, which is formally explained as follows:

$$CE_{FS}(q_{1:m}, a_{1:n}) = CE([CLS]\, q_{Subject}\, [S]\, q_{Description}\, [D]\, q_{Tags}\, [T]\, [SEP]\, a\, [SEP]) * W \quad (1)$$

Here, the [S], [D], and [T] tokens serve as separators, i.e., splitter tokens, for different parts of the question, namely Subject, Description, and Category tags, respectively. The input representation of CE_{FS} is designed to take into account different aspects of the retrieval context. The novelties brought by CE_{FS} can be summarized in two key aspects:

- *Structured Input.* CE_{FS} employs structured input by dividing the question into distinct sections - the subject, description, and tags. These sections are separated by splitter tokens. Such structuring not only facilitates a more comprehensive representation of the query but also emphasizes the importance of individual sections to the re-ranker since the cross-encoder in CE_{FS} can comprehend different aspects of the information and assign varying levels of importance to each section of the question.
- *Question Tags.* CE_{FS} takes the question tags into account in a straightforward yet effective manner by incorporating them into the cross-encoder re-ranker's input. Each question tag is separated by a semicolon. The motivation behind these additions is to equip the re-ranker with more comprehensive knowledge about the query, enabling it to grasp both an overview of the legal question's topic and its detailed category tags.

4 Experiments and Results

Experimental Setup. We use the Huggingface library [28] for the cross-encoder re-ranking training and inference. We add the splitter tokens into the tokenizer of the cross-encoder. Following prior work [12] we use Cross-Entropy loss [32], training batch size of 32, and Adam [13] optimizer with a learning rate of $7*10^{-6}$ for all cross-encoder layers, regardless of the number of layers trained.

Table 2. Results of the ablation study on CE_{FS}.

Model	MAP@1k	R@100	R@10	R@1
CE_{FS} w/o [T] splitter and query tags	.252	.510	.405	.163
CE_{FS} w/o [S] splitter and question subject	.239	.498	.379	.154
CE_{FS} w/o [D] splitter and question description	.191	.461	.315	.119
CE_{FS}	**.270**	**.524**	**.428**	**.209**

Ranking Quality. Table 1 illustrates the effectiveness of lexical-based first-stage retrievers and cross-encoder re-rankers. For re-ranking, our proposed method, CE_{FS}, demonstrates significant improvements over both CE_{CAT} fine-tuned on the LegalQA dataset and CE_{CAT} trained on MS MARCO, referred to as MiniLM-MSMARCO [24]. E.g., in terms of MAP@1k, CE_{FS} achieves 0.270 vs. 0.236 for CE_{CAT} and 0.109 for MiniLM-MSMARCO. The higher effectiveness of CE_{FS} over the CE_{CAT} confirms the effectiveness of the proposed method, and the low performance of MiniLM-MSMARCO on LegalQA confirms the challenges of the legal domain since MiniLM-MSMARCO achieves a three times higher effectiveness on MS MARCO dataset in terms of MAP@1k. Among the first-stage retrievers, BM25 outperforms LMD in terms of initial ranking effectiveness. The relatively low effectiveness of BM25 reveals a noticeable difference in lexical word overlap between the question and the best answer. This difference serves as an indicator of a knowledge gap between the questioner and the lawyer, resulting in a higher occurrence of lexical mismatches between the question and the relevant answer. Achieving a higher overall effectiveness in this task is a potential area for improvement in future works.

Ablation Study. We do an ablation study on the CE_{FS} to analyze to what extent each section of the fine-grained structured input of CE_{FS} has an impact on the effectiveness of CE_{FS}. As shown in Table 2, the effectiveness of CE_{FS} is highest when we use all of the splitter and corresponding contents. We see that query tags and [T] splittor have less impact on the effectiveness and question description and [D] has the highest impact. Question subject and [S] have the second-highest impact. This suggests that although the query tags have a role in improved effectiveness, question subject and description have still a large impact on the effectiveness.

5 Conclusion

For investigating answer retrieval in legal community question answering, we created a dedicated dataset, called LegelQA, which we divide into training, validation, and test sets. We use this dataset to train neural retrieval models, introducing a novel and highly effective re-ranker called CE_{FS}, and take a fine-grained structured approach to leverage the information available in the legal domain, demonstrating significantly higher effectiveness over common strong cross-encoder re-rankers. We investigate the impact of each part of the fine-grained structured input within our method and highlight the significant role played by question tags in improving retrieval effectiveness besides the most important part of the question which is question description. While our method is initially proposed for answer retrieval within the legal domain, we foresee its potential application in answer retrieval for community question-answering systems across various domains where question tags are provided alongside each question post. For future work, our data, LegalQA, facilitates other tasks such as legal response generation.

Acknowledgments. This work was supported by the EU Horizon 2020 ITN/ETN on Domain Specific Systems for Information Extraction and Retrieval (H2020-EU.1.3.1., ID: 860721).

References

1. Abolghasemi, A., Verberne, S., Azzopardi, L.: Improving BERT-based query-by-document retrieval with multi-task optimization. In: Hagen, M., et al. (eds.) ECIR 2022. LNCS, vol. 13186, pp. 3–12. Springer, Cham (2022). https://doi.org/10.1007/978-3-030-99739-7_1
2. Askari, A., Abolghasemi, A., Aliannejadi, M., Kanoulas, E., Verberne, S.: Closer: conversational legal longformer with expertise-aware passage response ranker for long contexts. In: The 32nd ACM International Conference on Information and Knowledge Management (CIKM 2023). ACM (2023)
3. Askari, A., Abolghasemi, A., Pasi, G., Kraaij, W., Verberne, S.: Injecting the BM25 score as text improves BERT-based re-rankers. In: Advances in Information Retrieval, pp. 66–83. Springer Nature Switzerland, Cham (2023). https://doi.org/10.1007/978-3-031-28244-7_5
4. Askari, A., Verberne, S., Pasi, G.: Expert finding in legal community question answering. In: Hagen, M., et al. (eds.) Advances in Information Retrieval, pp. 22–30. Springer International Publishing, Cham (2022). https://doi.org/10.1007/978-3-030-99739-7_3
5. Atkinson, J., Figueroa, A., Andrade, C.: Evolutionary optimization for ranking how-to questions based on user-generated contents. Expert Syst. Appl. **40**(17), 7060–7068 (2013)
6. Bian, J., Liu, Y., Agichtein, E., Zha, H.: Finding the right facts in the crowd: factoid question answering over social media. In: Proceedings of the 17th International Conference on World Wide Web, pp. 467–476 (2008)

7. Boualili, L., Moreno, J.G., Boughanem, M.: MarkedBERT: integrating traditional IR cues in pre-trained language models for passage retrieval. In: Proceedings of the 43rd International ACM SIGIR Conference on Research and Development in Information Retrieval, pp. 1977–1980 (2020)
8. Budler, L.C., Gosak, L., Stiglic, G.: Review of artificial intelligence-based question-answering systems in healthcare. Wiley Interdisc. Rev.: Data Min. Knowl. Disc. **13**(2), e1487 (2023)
9. Chen, T., Zhang, M., Lu, J., Bendersky, M., Najork, M.: Out-of-domain semantics to the rescue! zero-shot hybrid retrieval models. In: Hagen, M., et al. (eds.) ECIR 2022. LNCS, vol. 13185, pp. 95–110. Springer, Cham (2022). https://doi.org/10.1007/978-3-030-99736-6_7
10. Haigh, R.: Legal English. Routledge (2018)
11. Han, J., Hong, T., Kim, B., Ko, Y., Seo, J.: Fine-grained post-training for improving retrieval-based dialogue systems. In: Proceedings of the 2021 Conference of the North American Chapter of the Association for Computational Linguistics: Human Language Technologies, pp. 1549–1558 (2021)
12. Hofstätter, S., Althammer, S., Schröder, M., Sertkan, M., Hanbury, A.: Improving efficient neural ranking models with cross-architecture knowledge distillation. arXiv preprint arXiv:2010.02666 (2020)
13. Kingma, D.P., Ba, J.: Adam: a method for stochastic optimization. arXiv preprint arXiv:1412.6980 (2014)
14. Levenshtein, V.: Binary codes capable of correcting deletions, insertions and reversals. In: Soviet Physics-Doklady, vol. 10, pp. 707–710 (1966)
15. Mansouri, B., Campos, R.: FALQU: finding answers to legal questions. arXiv preprint arXiv:2304.05611 (2023)
16. Martinez-Gil, J.: A survey on legal question-answering systems. Comput. Sci. Rev. **48**, 100552 (2023)
17. Nguyen, T., et al.: Ms MARCO: a human generated machine reading comprehension dataset. In: CoCo@ NIPs (2016)
18. Nogueira, R., Cho, K.: Passage re-ranking with BERT. arXiv preprint arXiv:1901.04085 (2019)
19. Ponte, J.M., Croft, W.B.: A language modeling approach to information retrieval. In: ACM SIGIR Forum, vol. 51, pp. 202–208. ACM New York, NY, USA (2017)
20. Rau, D., Kamps, J.: The role of complex NLP in transformers for text ranking. In: Proceedings of the 2022 ACM SIGIR International Conference on Theory of Information Retrieval, pp. 153–160 (2022)
21. Robertson, S.E., Walker, S.: Some simple effective approximations to the 2-poisson model for probabilistic weighted retrieval. In: SIGIR'94, pp. 232–241. Springer, London (1994). https://doi.org/10.1007/978-1-4471-2099-5_24
22. Roy, P.K., Ahmad, Z., Singh, J.P., Alryalat, M.A.A., Rana, N.P., Dwivedi, Y.K.: Finding and ranking high-quality answers in community question answering sites. Glob. J. Flex. Syst. Manag. **19**, 53–68 (2018)
23. Roy, P.K., Saumya, S., Singh, J.P., Banerjee, S., Gutub, A.: Analysis of community question-answering issues via machine learning and deep learning: state-of-the-art review. CAAI Trans. Intell. Technol. **8**(1), 95–117 (2023)
24. Sentence-BERT: cross-encoder for ms MARCO: ms-marco-minilm-l-12-v2 (2023). https://huggingface.co/cross-encoder/ms-marco-MiniLM-L-12-v2
25. Tiersma, P.M.: Legal language. University of Chicago Press (1999)
26. Wang, W., Wei, F., Dong, L., Bao, H., Yang, N., Zhou, M.: MiniLM: deep self-attention distillation for task-agnostic compression of pre-trained transformers. Adv. Neural. Inf. Process. Syst. **33**, 5776–5788 (2020)

27. Williams, C.: Tradition and change in legal English: verbal constructions in prescriptive texts, vol. 20. Peter Lang (2007)
28. Wolf, T., et al.: Huggingface's transformers: state-of-the-art natural language processing. arXiv preprint arXiv:1910.03771 (2019)
29. Xiong, W., et al.: TWEETQA: a social media focused question answering dataset. arXiv preprint arXiv:1907.06292 (2019)
30. Yang, W., et al.: End-to-end open-domain question answering with BERTserini. arXiv preprint arXiv:1902.01718 (2019)
31. Yen, S.J., Wu, Y.C., Yang, J.C., Lee, Y.S., Lee, C.J., Liu, J.J.: A support vector machine-based context-ranking model for question answering. Inf. Sci. **224**, 77–87 (2013)
32. Zhang, Z., Sabuncu, M.: Generalized cross entropy loss for training deep neural networks with noisy labels. In: Advances in Neural Information Processing Systems, vol. 31 (2018)

News Gathering: Leveraging Transformers to Rank News

Carlos Muñoz[1], María José Apolo[2], Maximiliano Ojeda[1,2], Hans Lobel[1], and Marcelo Mendoza[1(✉)]

[1] Pontificia Universidad Católica de Chile, Vicuña Mackenna 6840, Santiago, Chile
{carlos.munoz,halobel,marcelo.mendoza}@uc.cl

[2] Universidad Técnica Federico Santa María, Vicuña Mackenna 3939, Santiago, Chile
{maria.apolo,maximiliano.ojeda}@sansano.usm.cl

Abstract. News media outlets disseminate information across various platforms. Often, these posts present complementary content and perspectives on the same news story. However, to compile a set of related news articles, users must thoroughly scour multiple sources and platforms, manually identifying which publications pertain to the same story. This tedious process hinders the speed at which journalists can perform essential tasks, notably fact-checking. To tackle this problem, we created a dataset containing both related and unrelated news pairs. This dataset allows us to develop information retrieval models grounded in the principle of binary relevance. Recognizing that many Transformer-based models might be suited for this task but could overemphasize relationships based on lexical connections, we tailored a dataset to fine-tune these models to focus on semantically relevant connections in the news domain. To craft this dataset, we introduced a methodology to identify pairs of news stories that are lexically similar yet refer to different events and pairs that discuss the same event but have distinct lexical structures. This design compels Transformers to recognize semantic connections between stories, even when their lexical similarities might be absent. Following a human-annotation assessment, we reveal that BERT outperformed other techniques, excelling even in challenging test cases. To ensure the reproducibility of our approach, we have made the dataset and top-performing models publicly available.

Keywords: News gathering · Document ranking · News IR

1 Introduction

Determining if a group of news articles relates to the same event poses a significant challenge. We tackle this problem by utilizing a comprehensive dataset

Supported by the Millennium Institute for Foundational Research on Data (ANID ICM grant ICN17-002) and ANID PIA National Center of Artificial Intelligence, grant FB210017. Mr. Mendoza acknowledges funding by ANID Fondecyt grant 1200211.

N. Goharian et al. (Eds.): ECIR 2024, LNCS 14610, pp. 486–493, 2024.
https://doi.org/10.1007/978-3-031-56063-7_41

of paired news articles. These articles were retrieved from Crowdtangle, a versatile media tracking tool. Our dataset consists of news items that share the same content but employ different writing styles. This method gave us a robust dataset, ideal for refining our models. Drawing from the principle of binary relevance, we examined several Transformer-based models [11] to discern between related and unrelated news pieces. With this dataset, we optimized various text encoding/decoding techniques, producing models adept at news matching, which can be instrumental for news ranking. Our experimental results demonstrate the capability of our models to recognize related news articles.

The primary contribution of our study is the creation of a dataset specifically designed to optimize Transformers, enhancing their ability to detect semantically related news. **To imbue Transformers with this capability, we identified pairs of unrelated news with lexical connections and pairs of related news devoid of lexical links. This design compels the base model to recognize semantic connections between stories, even when their lexical similarities might be absent.** Our findings indicate that this approach proves especially beneficial in the news landscape.

2 Related Work

In the 2013 version of SemEval, a task closely aligned with semantic matching of news articles was introduced [2]. Two specific tasks were highlighted: CORE, which involves sentence pairs, and TYPED, where semi-structured sentence pairs are deemed related if they align on metadata elements such as title, author, or description. From the 2013 SemEval dataset, a total of 750 pairs were drawn from Newswire headlines. The subsequent 2014 version expanded the domain of semantic matching into a multilingual dimension [1]. Articles from the 'International' section of Google News - which curates content from various news platforms - were parsed. Following human annotation, 480 sentence pairs showcasing high similarity were manually derived. The 2017 SemEval iteration is renowned as the Semantic Textual Similarity Benchmark (STS-B) since it is the most comprehensive, encompassing paired items in diverse formats, including those like video and image captions [3].

Pre-trained language models have become fundamental in text mining, with models such as BERT [4] - rooted in the Transformer encoder - delivering exceptional performance across a myriad of tasks. A prevalent technique for tasks centered on sentence pairs is cross-encoding, wherein a symbol delineates the paired sentences. While effective, another strategy involves using bi-encoders [10], which, despite offering superior results, can be computationally demanding. To address this, Humeau et al. [7] introduced poly-encoding, an enhancement of the transformer designed for multiple-sentence encoding. Empirical studies highlight that poly-encoders surpass bi-encoders in performance and are more efficient than cross-encoders. In a subsequent development, Reimers et al. [8] presented Sentence-BERT (SBERT), a Siamese network-based sentence embedding model tailored for sentence pair tasks, which when assessed on multiple SentEval tasks, outclassed competing state-of-the-art methodologies. More recently, Gao et al. [5] unveiled SimCSE, a contrastive learning-based sentence embedding technique. Evaluated under both unsupervised and supervised frameworks, SimCSE utilizes pre-trained BERT models, excelling in predicting textual entailments from premises and distinguishing between related and unrelated sentence pairs.

3 A Parallel Corpus for News Gathering

We introduce a methodology designed to construct a massive parallel dataset specifically designed for news articles. This methodology is intended to optimize the process of developing models for ranking news articles.

In Fig. 1, we illustrate the three key stages of our methodology.

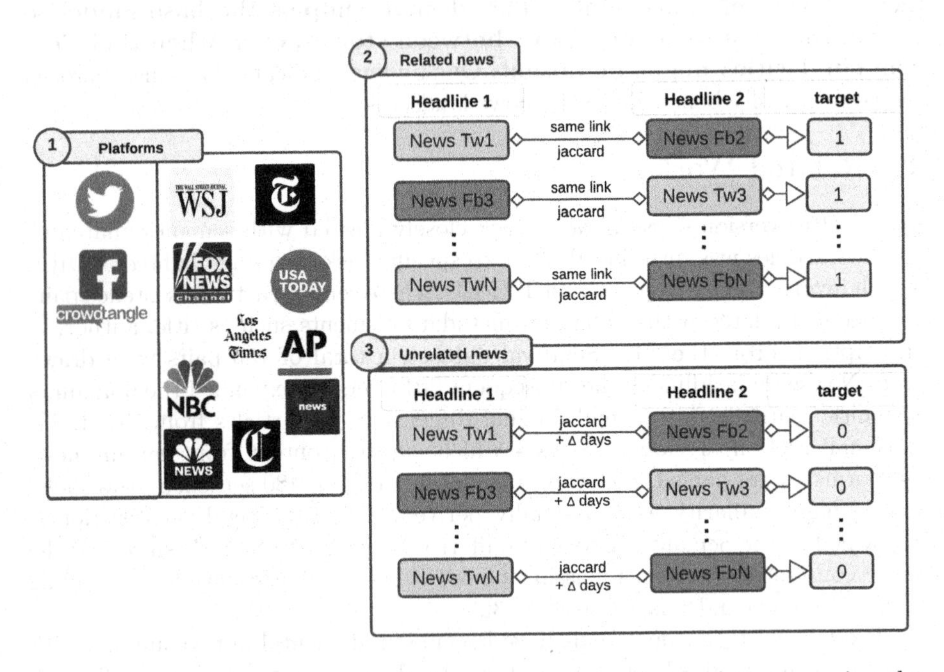

Fig. 1. Schematic representation of the primary phases involved in generating the dataset. Fb and Tw denote Facebook and Twitter (now X), respectively.

1. **Extraction of News Media Posts**: To create our news collection, we employed two platforms: Twitter (Academic API) and Facebook (Crowdtangle). We gathered a diverse range of mainstream news media posts spanning from 2019 to 2022. The selected news outlets included prominent names such as The New York Times, San Francisco Chronicle, National Broadcasting, and Yahoo! News, among other news outlets. We extracted headlines, source website links, and publication dates.

2. **Related News**: We determine pairs of news articles that pertain to the same event but exhibit differences in their wording. Initially, we identified pairs of posts that directed readers to the same news outlet's URL from Twitter and Facebook, respectively. From this collection of paired news articles, we identified lexically dissimilar related news using Locality Sensitive Hashing (LSH) with a Jaccard coefficient threshold of at most 0.5.
3. **Unrelated News**: We determine pairs of news articles that exhibit lexical similarity but pertain to distinct events. A robust method for establishing this partition involves identifying pairs of news articles with significant lexical overlap, yet they cover dissimilar events. To identify such lexically related pairs, we employed Locality Sensitive Hashing (LSH) with a Jaccard coefficient of at least 0.5. Among all lexically similar pairs, we specifically sought those with a time gap of at least 60 d between their publication dates.

We identify a collection of **hard examples** from the dataset. These challenging instances consist of **related news with either an extremely low lexical connection (Jaccard coefficient of at most 0.1) or unrelated news with a strong lexical connection (Jaccard coefficient of at least 0.9)**. Examples that do not meet this definition are categorized as **soft examples**.

The dataset comprises 130,709 pairs, consisting of 52,192 pairs of unrelated news articles (40%) and 78,517 pairs of related news articles (60%). During the dataset preparation, 86% of the data was allocated for training purposes, while 12% was set aside for validation. For human-based annotation, two distinct partitions were created, corresponding to 0.6% and 1% of the total dataset, respectively. The first human-based testing partition, referred to as the **soft test**, is a subset of the dataset that excludes hard examples. The second partition, known as **hard test** includes hard pairs that were initially identified during the dataset construction phase.

Both testing partitions –hard and soft test– underwent human annotation to guarantee the highest level of label reliability. We leveraged the crowdsourcing platform Clickworker[1] to perform the data annotation, employing a majority voting system. A total of 149 participants were tasked with categorizing pairs as related, unrelated, or uncertain. To enhance precision, we introduced gold standard cases, which constituted a subset of results validated by humans. Only pairs with a minimum of three agreements out of four responses were included in the final testing datasets. A summary of the corpus is depicted in Table 1.

4 Experiments

We trained a total of six models using our dataset. Our approach was divided into two primary categories. The first category used transformer-encoder methods, specifically BERT [4], SBERT [8], and SimSCE [5]. On the other hand, the second category was centered on Transformer-decoder techniques.

[1] https://www.clickworker.com/.

Table 1. The parallel news corpus for news ranking.

Partition	Unrelated	Related	Total	%
Train	43,352	69,480	112,832	86,3%
Val	7,688	8,061	15,749	12,0%
Soft-test	362	472	834	0.6%
Hard-test	790	504	1,294	1,0%
	52,192	78,517	130,709	

For BERT encoder models, we utilized BERT base [4] to encode the headlines. In BERT's implementation, each pair of sentences was ingested using cross-encodings, where a special token separated the two sentences in each sample. For SBERT and SIMCSE, we adhered to their conventional encoding methods. These three models underwent training on the training set computing the errors during training over validation data. The SIMCSE model was fine-tuned based on the original pre-trained version called sup-simcse-roberta-base [5]. For all these methods, the base models were fine-tuned with a batch size of 32 pairs, over 3 epochs, a learning rate of 5e-5, and a weight decay of 0.01.

Turning our attention to the second category, we aligned three large language models using prompts. For each language model alignment, we employed a standardized prompt format. This format comprised three segments: a task definition (*'Below is an instruction that describes a task and two sentences that provides further context. Write a response that appropriately completes the request. You have two sentences related to news headlines. The task is to determine if both sentences refer to the same news or not. Provide the output as 1 if both sentences refer to the same news, 0 otherwise.'*), an input section presenting the two headlines for comparison, and an output section indicating the expected outcome (0 for unrelated pairs, and 1 otherwise). The prompts were curated using data from the validation set. To facilitate efficient fine-tuning using these prompts, we used two specific strategies: LORA [6] to align the GPT-J-6B model[2] [12], and the Adapter [13] to align Falcon 7B[3] and Alpaca 7B[4] [9].

We assessed the performance of the six models using the testing partition (human-labeled data). The results are presented using Precision (**P**), Recall (**R**), and F_1 measures. The results are broken down by class, with "R" indicating related pairs and "U" denoting unrelated pairs. Additionally, these metrics are provided at both micro and macro levels for comprehensive analysis. Results in Table 2 show on bold fonts the best performances.

The findings indicate that BERT outperforms its competitors on both segments of the test set. Additionally, it is observed that methods based on encoders surpass the performance of aligned LLMs on the dataset.

[2] https://huggingface.co/EleutherAI/gpt-j-6b.
[3] https://huggingface.co/tiiuae/falcon-7b.
[4] https://huggingface.co/tloen/alpaca-lora-7b.

Table 2. Results for binary relevance on the testing partition.

		Soft-test			Hard-test					Soft-test			Hard-test		
		P	**R**	F_1	**P**	**R**	F_1			**P**	**R**	F_1	**P**	**R**	F_1
BERT	U	0.79	0.94	0.86	0.88	0.79	0.84	GPT-J-6B	U	0.69	0.57	0.62	0.75	0.58	0.65
	R	0.95	0.81	0.87	0.72	0.84	0.77		R	0.71	0.81	0.76	0.49	0.68	0.57
	Micro	0.86	0.86	**0.86**	0.81	0.81	**0.81**		Micro	0.71	0.71	0.71	0.65	0.62	0.62
	Macro	0.87	0.88	**0.87**	0.80	0.82	**0.81**		Macro	0.70	0.69	0.69	0.62	0.63	0.61
SBERT	U	0.92	0.66	0.77	0.84	0.85	0.84	Falcon-7B	U	0.69	0.51	0.59	0.68	0.19	0.30
	R	0.78	0.96	0.83	0.76	0.74	0.75		R	0.69	0.83	0.76	0.40	0.85	0.54
	Micro	0.83	0.83	0.83	0.81	0.81	0.81		Micro	0.69	0.69	0.68	0.57	0.44	0.38
	Macro	0.85	0.81	0.81	0.80	0.79	0.80		Macro	0.69	0.67	0.67	0.54	0.52	0.42
SimSCE	U	0.66	0.55	0.60	0.79	0.87	0.83	Alpaca-7B	U	0.51	0.95	0.66	0.70	0.92	0.79
	R	0.69	0.78	0.74	0.76	0.64	0.70		R	0.89	0.29	0.44	0.74	0.38	0.50
	Micro	0.68	0.68	0.69	0.78	0.78	0.78		Micro	0.58	0.58	0.58	0.71	0.71	0.71
	Macro	0.67	0.67	0.66	0.78	0.76	0.76		Macro	0.70	0.62	0.55	0.72	0.65	0.65

A large scale evaluation on the wild. To evaluate the ability of models to rank news "on the wild", we conducted an experiment using an extensive collection of news articles. We utilized Google News to retrieve **top stories** from a diverse range of sources published in the US across categories such as Business, Entertainment, Health, Science, Sports, Technology, US, and World. The collection encompasses news from July 23, 2022, to July 30, 2022,

The news items were grouped into pairs. These pairs were formed by pairing news articles published on the same day. All pairs were manually annotated as either related or unrelated by human annotators. Each pair was reviewed by four annotators. Only those pairs that matched in at least three out of the four annotations were selected for the dataset. The remaining pairs were discarded. In total, the dataset comprises 32,100 pairs.

To rank news articles, we implemented the following method: For each news article we retrieve all news articles published on the same day and pair them with the original article, as depicted in Fig. 2.

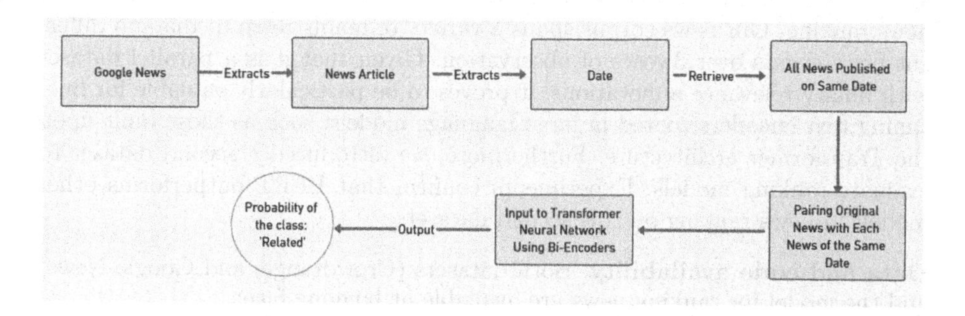

Fig. 2. Ranking news with transformers using a pairwise approach.

We then use bi-encoders to rank the news articles that are in the second position of each pair, assessing their relevance to the first article in the pair. The organization of each list produced by the encoders is based on the probabilities derived for the related class. The results of this experiment are shown in Fig. 3.

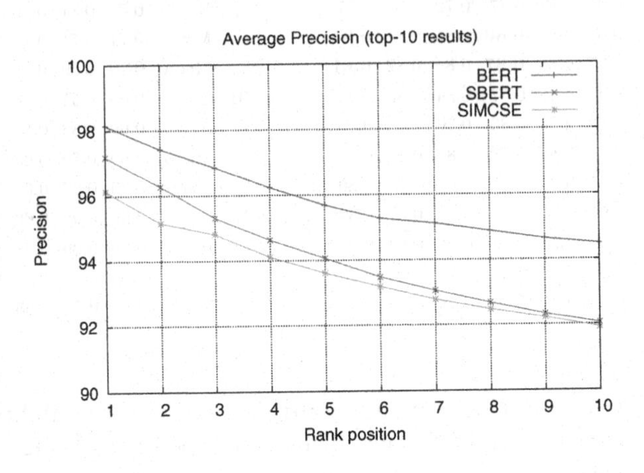

Fig. 3. Average precision for top-10 results retrieved using BERT, SBERT and SIMCSE.

Figure 3 illustrates the average precision curve for the top-10 results across the three models. The methods BERT, SBERT, and SIMCSE achieved overall precision rates of 89.85%, 84.07%, and 86.29% respectively. The figure illustrates that BERT outperforms the other models, and the gap between these models becomes more pronounced with the increase in rank position. Conversely, the performance of SBERT and SIMCSE is almost identical, with the gap between these two models narrowing as the rank position increases.

5 Conclusion

We have introduced datasets to fine-tune transformer models for the task of news ranking. Our news corpus spans a variety of mainstream media and different news events over 3 years of observation. Given that it is a parallel dataset with binary relevance annotations, it proves to be particularly valuable for fine-tuning text encoders rooted in large language models, such as those built upon the Transformer architecture. Furthermore, we introduced a second dataset to evaluate ranking models. Experiments confirm that BERT outperforms other models in news ranking on this second dataset.

Data and code availability. Both datasets (Crowdtangle and Google News) and the model for ranking news are available at hugging face:

1. Crowdtangle: https://huggingface.co/datasets/cmunhozc/usa_news_en
2. Google News: https://huggingface.co/datasets/cmunhozc/google_news_en
3. News ranking model: https://huggingface.co/cmunhozc/news-ranking-ft-bert

References

1. Agirre, E., et al.: Semeval-2014 task 10: multilingual semantic textual similarity. In: Proceedings of the 8th International Workshop on Semantic Evaluation, SemEval@COLING 2014, Dublin, Ireland, August 23–24, 2014. pp. 81–91. The Association for Computer Linguistics (2014)
2. Agirre, E., Cer, D.M., Diab, M.T., Gonzalez-Agirre, A., Guo, W.: *sem 2013 shared task: semantic textual similarity. In: *SEMEVAL (2013)
3. Cer, D.M., Diab, M.T., Agirre, E., Lopez-Gazpio, I., Specia, L.: Semeval-2017 task 1: semantic textual similarity multilingual and crosslingual focused evaluation. In: Proceedings of the 11th International Workshop on Semantic Evaluation, SemEval@ACL 2017, Vancouver, Canada, August 3–4, 2017, pp. 1–14. The Association for Computational Linguistics (2017)
4. Devlin, J., Chang, M., Lee, K., Toutanova, K.: BERT: pre-training of deep bidirectional transformers for language understanding. In: Proceedings of the 2019 Conference of the North American Chapter of the Association for Computational Linguistics: Human Language Technologies, NAACL-HLT 2019, Minneapolis, MN, USA, June 2–7, 2019, Volume 1 (Long and Short Papers), pp. 4171–4186. Association for Computational Linguistics (2019)
5. Gao, T., Yao, X., Chen, D.: SimCSE: Simple contrastive learning of sentence embeddings. In: Proceedings of the 2021 Conference on Empirical Methods in Natural Language Processing, EMNLP 2021, Virtual Event/Punta Cana, Dominican Republic, 7–11 November, 2021, pp. 6894–6910. Association for Computational Linguistics (2021)
6. Hu, E.J., et al.: LoRA: low-rank adaptation of large language models. In: The Tenth International Conference on Learning Representations, ICLR, Virtual Event, April 25–29, 2022. OpenReview.net (2022)
7. Humeau, S., Shuster, K., Lachaux, M., Weston, J.: Poly-encoders: architectures and pre-training strategies for fast and accurate multi-sentence scoring. In: 8th International Conference on Learning Representations, ICLR 2020, Addis Ababa, Ethiopia, April 26–30, 2020 (2020)
8. Reimers, N., Gurevych, I.: Sentence-BERT: sentence embeddings using siamese BERT-networks. In: Proceedings of the 2019 Conference on Empirical Methods in Natural Language Processing and the 9th International Joint Conference on Natural Language Processing, EMNLP-IJCNLP 2019, Hong Kong, China, November 3–7, 2019, pp. 3980–3990. Association for Computational Linguistics (2019)
9. Touvron, H., et al.: LLaMA: open and efficient foundation language models. CoRR abs/2302.13971 (2023)
10. Urbanek, J., et al.: Learning to speak and act in a fantasy text adventure game. In: Proceedings of the 2019 Conference on Empirical Methods in Natural Language Processing and the 9th International Joint Conference on Natural Language Processing, EMNLP-IJCNLP 2019, Hong Kong, China, November 3–7, 2019, pp. 673–683. Association for Computational Linguistics (2019)
11. Vaswani, A., et al.: Attention is all you need. In: Advances in Neural Information Processing Systems 30: Annual Conference on Neural Information Processing Systems 2017, December 4–9, 2017, Long Beach, CA, USA, pp. 5998–6008 (2017)
12. Wang, B., Komatsuzaki, A.: GPT-J-6B: a 6 billion parameter autoregressive language model. https://github.com/kingoflolz/mesh-transformer-jax (2021)
13. Zhang, R., et al.: LlaMA-adapter: efficient fine-tuning of language models with zero-init attention. CoRR abs/2303.16199 (2023)

GenRec: Large Language Model for Generative Recommendation

Jianchao Ji, Zelong Li, Shuyuan Xu, Wenyue Hua, Yingqiang Ge, Juntao Tan, and Yongfeng Zhang(✉)

Department of Computer Science, Rutgers University, New Brunswick, NJ 08854, Canada
{jianchao.ji,zelong.li,shuyuan.xu,wenyue.hua, yingqiang.ge,juntao.tan,yongfeng.zhang}@rutgers.edu

Abstract. In recent years, Large Language Models (LLMs) have emerged as powerful tools for diverse natural language processing tasks. However, their potential for recommender systems under the generative recommendation paradigm remains relatively unexplored. This paper presents an innovative approach to recommender systems using Large Language Models (LLMs) purely based on raw text data, i.e., using item name or title as item IDs rather than creating meticulously designed user or item IDs. More specifically, we present a novel LLM for Generative Recommendation (GenRec) method that utilizes the expressive power of LLM to directly generate the target item to recommend, rather than calculating the ranking score for each candidate item one by one as in traditional discriminative recommendation. GenRec uses LLM's understanding ability to interpret context, learn user preferences, and generate relevant recommendations. Our proposed approach leverages the vast knowledge encoded in Large Language Models to accomplish recommendation tasks. We formulate specialized prompts to enhance the ability of LLM to comprehend recommendation tasks. Subsequently, we use these prompts to LoRA-fine-tune the LLaMA backbone LLM on the user-item interaction data represented by raw text (using raw item name or title as the item's ID) to capture user preferences and item characteristics. Our research underscores the potential of LLM-based generative recommendation in revolutionizing the domain of recommendation systems and offers a foundational framework for future explorations in this field. We conduct extensive experiments on benchmark datasets, and the experiments show that our GenRec method achieves better results on large datasets. Code and data are are open-source at GitHub (https://github.com/rutgerswiselab/GenRec).

Keywords: Large Language Model · Recommender Systems · Natural Language Processing · Generative Recommendation

1 Introduction

Large Language Models (LLMs) have made a particularly significant milestone in this technological evolution. These LLMs, designed to understand and gen-

N. Goharian et al. (Eds.): ECIR 2024, LNCS 14610, pp. 494–502, 2024.
https://doi.org/10.1007/978-3-031-56063-7_42

erate human-like text, have revolutionized numerous applications, from search engines to chatbots, and have facilitated more natural and intuitive interactions between humans and machines. This paper seeks to explore a relatively new and promising application of these models in recommender systems. Recommender systems have become an integral part of our digital experience. They are the unseen force guiding us through the vast amounts of data, suggesting relevant products on e-commerce websites, recommending movies on streaming platforms, or proposing what news to read or videos to watch. The primary aim of these systems is to predict the individual user preferences and enhance user experience and engagement.

Traditionally, recommender systems have been built around methods such as collaborative filtering [11,19,21], content-based filtering [22,24], and hybrid approaches [1,2,18]. Collaborative filtering leverages user-item interactions, making suggestions based on patterns found in the behavior of similar users or items. On the other hand, content-based filtering uses item features to recommend similar items to those a user has previously interacted with. Hybrid methods attempt to combine the strengths of these two approaches to overcome their limitations.

Despite the progress made with these traditional techniques, there still have some significant challenges. For instance, both collaborative and content-based filtering can hardly handle the issue of data sparsity, given that most users interact with only a small fraction of the total items available. Additionally, because of the computational complexity of processing large interaction matrices, these models often struggle to scale effectively with the growth of users and items.

The integration of language-based LLMs into recommender systems presents an exciting opportunity to address these challenges [3,26]. These models can learn and understand complex patterns in human language, which allows for a more nuanced interpretation of user preferences and a more sophisticated generation of recommendations. A significant number of the prevailing LLM-based recommendation models are trained using meticulously designed user and item IDs [9,15,16,26,27]. These approaches demonstrate an important direction towards LLM-RecSys alignment, since they have the advantage of encoding the crucial collaborative information into the user or item IDs, which helps LLMs to align the content and collaborative information and better learn the relationship between users and items or between items and items, thus enhancing the LLM-based recommendation performance. However, creating effective user or item IDs is not a trivial task, which requires meticulously designed techniques such as sequential indexing, collaborative indexing, content-based indexing, or hybrid indexing [9].

In this paper, we propose a novel pure-text-based large language model for generative recommendation (GenRec). GenRec directly uses the textual item name or title as the ID for the item, eliminating the need to create specifically designed IDs for each item. One of the primary benefits of the GenRec model is that it capitalizes on the rich, descriptive information inherently contained within the item names, which often contain features that can be semantically analyzed, enabling a better understanding of the item's potential relevance to the user. This could potentially provide more accurate and personalized recommendations, thereby enhancing the overall user experience. The key contributions of this paper can be summarized as follows:

- We highlight the promising paradigm of LLM-based generative recommendation, which directly generates the name of the target item to recommend, rather than traditional discriminative recommendation, which has to calculate a ranking score for each and every candidate item one by one and then sorts them for deciding which to recommend.
- We introduce a novel approach, GenRec, to enhance the generative recommendation performance by properly incorporating the textual information into the generative recommendation model.
- We also illustrate the efficacy of GenRec on practical recommendation tasks, underscoring its prospective abilities for a wider scope of applications.

2 Related Work

The use of Large Language Models (LLMs) for recommender systems has gained significant attention recently [8,14]. These models exhibit great potential in the understanding and modeling of user-item interactions, exploiting rich semantics and long-range dependencies present in user activity data.

The pioneering work of P5 [3,26] illustrates the feasibility of using large language models for generative recommendation. It fine-tunes a large language model backbone to create a unified system capable of handling various recommendation tasks, such as sequential recommendation, direct recommendaitaon, rating prediction, explanation generation, etc. This innovative approach highlighted the effectiveness of LLMs in handling multi-task learning in the recommendation context. To enhance the Large Language Model's comprehension of a user's behavior history, some researchers have attempted to incorporate collaborative information into the training process [9,15,16]. For example, collaborative indexing creates item IDs from the user-item or item-item collaborative information based on spectral collaborative learning [9] or graph collaborative learning [16]. Each item is assigned an ID, ensuring that items of that share similar user behavior also share similar IDs. These IDs are then utilized to represent the items during the training process and to generate recommendations.

These recent studies by P5 and item ID creation methods [3,9,13,15,16,27] provide compelling evidence that, with suitable fine-tuning, LLMs can exhibit remarkable performance in recommendation tasks. This advancement underscores the adaptability and potential of LLMs in recommendation domains. Beyond this, some researchers extend LLMs to address cold start or zero-shot problems [6,20,25], where traditional models often falter due to the lack of historical data or prior information. By leveraging the inherent knowledge encoded within their extensive training data, LLMs can generate meaningful insights and predictions even in these challenging situations. This capability is significantly enhanced with carefully crafted prompts, which guide the models to focus on relevant aspects of the problem at hand, thereby improving the quality and relevance of the output. Such findings highlight the impact of LLMs in areas where data scarcity or the need for immediate insights presents substantial obstacles.

Besides employing Large Language Models (LLMs) as recommender systems, several researchers also explored the use of LLM as auxiliary tools within a traditional recommender system. This approach involves utilizing LLMs as feature encoders and scoring functions [12]. By doing so, LLMs can process and interpret user data, extracting nuanced features that traditional methods might overlook. These features are then integrated into recommendation algorithms, enhancing its accuracy and relevance. Additionally, LLMs can be used to score and rank recommendations, leveraging their advanced natural language processing capabilities to better align suggestions with user preferences and behaviors. This application of LLMs represents a significant shift from their conventional use, opening new avenues for more personalized and efficient recommender systems.

However, the potential of LLMs to understand and generate purely text-described item IDs as recommendations has not been fully explored. In this paper, we propose a novel approach to purely text-ID-based generative recommendation, leveraging the latest advances in LLMs. We aim to address some of the limitations of previous works and push the boundaries of what is possible in the realm of LLM-based recommender systems.

3 The GenRec Method

Our proposed GenRec framework for LLM-based generative recommendation is simple and effective. The architecture of the proposed framework is illustrated in Fig. 1. Given a user's item interaction sequence, the large language model for Generative Recommendation (GenRec) will format the text-based item names or titles with a prompt. This reformatted sequence is subsequently employed to fine-tune a Large Language Model (LLM). The fine-tuned LLM can then predict subsequent items the user is likely to interact with. In our paper, we select the LLaMA [23] language model as the backbone and use Low-Rank Adaptation (LoRA) [7] for fine-tuning. However, our framework retains flexibility, allowing for seamless integration with any other LLM, thus broadening its potential usability and adaptability.

3.1 Sequence Generation

The initial component of GenRec is a generative function, tasked with producing various sequences that encapsulate user interests. To enhance the model's comprehension of the recommendation task, we have devised multiple prompts that facilitate sequence generation. Take Fig. 2 as an example, we use the user's movie watching history as the training data and use this information to format the training sequence. The sequence consists of three parts: instruction, input and output. The instruction element outlines the specific task of movie recommendation, for which we have created several directives to enhance the LLM's

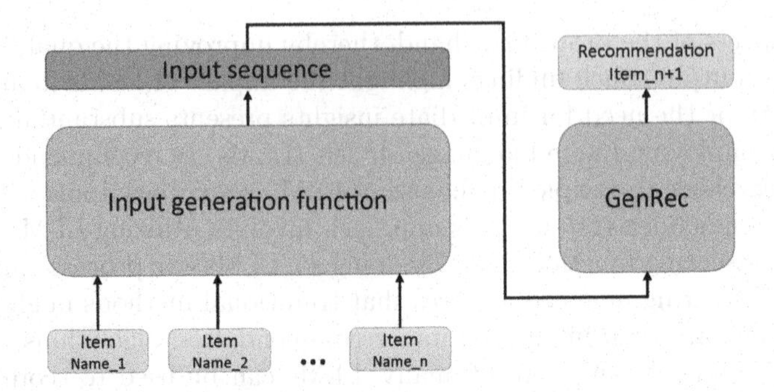

Fig. 1. An illustration of GenRec's simple architecture. Our model generates an input sequence based on the interaction history. Then the model predicts the next item the user may interact with.

comprehension of the ongoing recommendation task. The input represents the history of the user's interactions, excluding the most recent instance. The output is the latest interaction in this record. The primary task for the LLM here is to accurately predict this final interaction.

Interaction history: Pinocchio (1940), Legends of the Fall (1994), Once Were Warriors (1994), In the Name of the Father (1993), Shadowlands (1993), Heavenly Creatures (1994), Quiz Show (1994), In the Line of Fire (1993)
Recommendation Prompt Example:
Instruction: Given the movie viewing habits, what is the most probable movie they will choose to watch next?
input: Pinocchio (1940), Legends of the Fall (1994), Once Were Warriors (1994), In the Name of the Father (1993), Shadowlands (1993), Heavenly Creatures (1994), Quiz Show (1994)
output: In the Line of Fire (1993)

Fig. 2. GenRec prompt and data. GenRec converts the interactive history to a training sequence consisting of instruction, input and output.

Refer to Fig. 2 for an illustration. This figure represents how we utilize a user's history of watched movies as interaction data. Given the prompt "Based on the movie viewing habits, what is the most likely movie they will select to watch next?" and the provided input, we then allow GenRec to generate the subsequent output.

3.2 Training Strategy

In this paper, we use the LLaMA large language model [23] as the backbone for the training of GenRec. The LLaMA model is pre-trained on an expansive language corpus, offering a valuable resource for our intended purpose of efficiently capturing both user interests and item content information. However, it is important to note that the memory requirements for GPU to fine-tune LLaMA, even the smallest 7-billion parameter version, are pretty substantial.

To circumvent this challenge and conserve GPU memory, we adopt the Low-Rank Adaptation (LoRA) [7] method for fine-tuning and inference tasks over the LLaMA-7b model within the scope of this study. By this measure, we have achieved a significant reduction in the GPU memory requirements. With this optimized approach, we can fine-tune the LLaMA-LoRA model on a single GPU with a memory capacity of 24GB.

4 Experiments

We conduct extensive experiments on two real-world datasets from Amazon [17] and MovieLens [4], respectively, to evaluate the performance of our proposed GenRec approach on recommendation tasks.

4.1 Baseline Methods

In the following, we introduce the baseline models used in this research. These baselines serve as foundational benchmarks against which the performance and efficacy of our proposed methods are evaluated. By introducing these established methods, we aim to provide a clear context and point of reference for understanding the relative strengths and advancements of our contributions.

GRU4Rec [5]: The GRU4Rec model utilizes a session-based recommendation strategy, harnessing the Gated Recurrent Unit (GRU) to discern user preferences. Within a session, this model draws on past preferences as context to predict the subsequent item that a user may interact with.

SASRec [10]: The Self-attentive Sequential Recommendation (SASRec) method incorporates the self-attention technique into its sequential recommendation framework, appropriately applying the usage of both Markov Chains and RNN-driven methods.

P5 [3,26]: The Pre-train, Personalized Prompt, and Predict Paradigm (P5) for LLM-based generative recommendation, which incorporates an array of templates for input and target sequences throughout the training process. This approach dissolves the boundaries between different tasks, promoting a more fluid and integrated training procedure. It has showcased noteworthy performance in the domain of sequential recommendation tasks, underlining its effectiveness and applicability. In this work, we compare with the sequential recommendation performance of P5.

4.2 Performance Comparison

As we can see in Table 1, P5 gains better performance on Amazon Toys dataset, while our GenRec approach has significantly better performance on MovieLens-25M dataset. The possible reasons behind the different performances could be attributed to the distinct nature of the datasets: the MovieLens-25M dataset, unlike Amazon Toys dataset, contains a much richer amount of interaction information due to its larger scale, which provides a more robust understanding of the user's preferences and behavior, thus likely leading to more accurate recommendations even without the need to create meticulously designed item IDs. This observation implies that when the training data is abundant enough, then LLM may be able to directly learn the user-item collaborative information into the purely name- or title-based item IDs by re-optimizing the word embeddings. However, when the amount of training data is not large enough, it would be necessary to create meticulously designed item IDs such as sequential IDs or collaborative IDs [9] so as to "pre-encode" the collaborative information into the IDs for better facilitating LLMs to learn from the training data.

Table 1. Experimental results on Normalize Discounted Cumulative Gain (NDCG@k) and Hit Ratio (HR@k). Bold numbers represent best performance. We use * to indicate that the performance is significantly better than baselines. The significance is at 0.05 level on paired t-test.

Methods	MovieLens 25M				Amazon Toys			
	HR@5	NDCG@5	HR@10	NDCG@10	HR@5	NDCG@5	HR@10	NDCG@10
GRU4Rec	0.0439	0.0241	0.0753	0.0312	0.0125	0.0076	0.0171	0.0105
SASRec	0.0517	0.0344	0.0878	0.0408	0.0184	0.0124	0.0218	0.0142
P5	0.0688	0.0464	0.1040	0.0577	**0.0239***	**0.0145***	**0.0411***	**0.0201***
GenRec	**0.1034***	**0.0716***	**0.1311***	**0.0837***	0.0190	0.0136	0.0251	0.0157

5 Conclusion

This paper proposes GenRec, a Large Language Model approach for Generative Recommendation based on textual item name or title as IDs. By focusing on the semantic richness of item names as input, GenRec promises personalized and contextually relevant recommendations. Our practical demonstrations highlight GenRec's efficacy and point towards its adaptability across different recommendation scnarios. Furthermore, the flexibility of the GenRec framework facilitates integration with any Large Language Model, hence widening its sphere of potential utility. In terms of future work, there are several directions to explore. We intend to refine the generation of sequences by developing more sophisticated prompts, which could further enhance the model's understanding of recommendation tasks. Additionally, we plan to extend our research to incorporate more complex user interaction data, such as ratings or reviews, which could provide deeper insights into user behavior and preferences. Another direction is to see how GenRec works with different Large Language Models, since we are curious about the benefits and downsides of using different models.

Acknowledgement. The work was supported in part by NSF IIS-2046457 and IIS-2007907. Any opinions, findings, conclusions or recommendations in this material are those of the authors and do not necessarily reflect those of the sponsors.

References

1. Basilico, J., Hofmann, T.: Unifying collaborative and content-based filtering. In: Proceedings of the Twenty-First International Conference on Machine Learning, p. 9 (2004)
2. Burke, R.: Hybrid recommender systems: survey and experiments. User Model. User-Adap. Inter. **12**, 331–370 (2002)
3. Geng, S., Liu, S., Fu, Z., Ge, Y., Zhang, Y.: Recommendation as language processing (RLP): a unified pretrain, personalized prompt & predict paradigm (p5). In: Proceedings of the 16th ACM Conference on Recommender Systems, pp. 299–315 (2022)
4. Harper, F.M., Konstan, J.A.: The MovieLens Datasets: history and context. ACM Trans. Interact. Intell. Syst. (TIIS) **5**(4), 1–19 (2015)
5. Hidasi, B., Quadrana, M., Karatzoglou, A., Tikk, D.: Parallel recurrent neural network architectures for feature-rich session-based recommendations. In: Proceedings of the 10th ACM Conference on Recommender Systems, pp. 241–248 (2016)
6. Hou, Y., et al.: Large language models are zero-shot rankers for recommender systems. arXiv preprint arXiv:2305.08845 (2023)
7. Hu, E.J., et al.: LoRA: low-rank adaptation of large language models. arXiv preprint arXiv:2106.09685 (2021)
8. Hua, W., Li, L., Xu, S., Chen, L., Zhang, Y.: Tutorial on large language models for recommendation. In: Proceedings of the 17th ACM Conference on Recommender Systems, pp. 1281–1283 (2023)
9. Hua, W., Xu, S., Ge, Y., Zhang, Y.: How to index item IDs for recommendation foundation models. SIGIR-AP (2023)
10. Kang, W.C., McAuley, J.: Self-attentive sequential recommendation. In: 2018 IEEE International Conference on Data Mining (ICDM), pp. 197–206. IEEE (2018)
11. Konstan, J.A., Miller, B.N., Maltz, D., Herlocker, J.L., Gordon, L.R., Riedl, J.: GroupLens: applying collaborative filtering to Usenet news. Commun. ACM **40**(3), 77–87 (1997)
12. Li, J., Zhang, W., Wang, T., Xiong, G., Lu, A., Medioni, G.: GPT4Rec: a generative framework for personalized recommendation and user interests interpretation. arXiv preprint arXiv:2304.03879 (2023)
13. Li, L., Zhang, Y., Chen, L.: Prompt distillation for efficient LLM-based recommendation. In: Proceedings of the 32nd ACM International Conference on Information and Knowledge Management, pp. 1348–1357 (2023)
14. Li, L., Zhang, Y., Liu, D., Chen, L.: Large language models for generative recommendation: a survey and visionary discussions. arXiv:2309.01157 (2023)
15. Lin, X., Wang, W., Li, Y., Feng, F., Ng, S.K., Chua, T.S.: A multi-facet paradigm to bridge large language model and recommendation. arXiv:2310.06491 (2023)
16. Mei, K., Zhang, Y.: LightLM: a lightweight deep and narrow language model for generative recommendation. arXiv:2310.17488 (2023)

17. Ni, J., Li, J., McAuley, J.: Justifying recommendations using distantly-labeled reviews and fine-grained aspects. In: Proceedings of the 2019 Conference on Empirical Methods in Natural Language Processing and the 9th International Joint Conference on Natural Language Processing (EMNLP-IJCNLP), pp. 188–197 (2019)
18. Pazzani, M.J.: A framework for collaborative, content-based and demographic filtering. Artif. Intell. Rev. **13**, 393–408 (1999)
19. Resnick, P., Iacovou, N., Suchak, M., Bergstrom, P., Riedl, J.: GroupLens: an open architecture for collaborative filtering of netnews. In: Proceedings of the 1994 ACM Conference on Computer Supported Cooperative Work, pp. 175–186 (1994)
20. Sanner, S., Balog, K., Radlinski, F., Wedin, B., Dixon, L.: Large language models are competitive near cold-start recommenders for language-and item-based preferences. In: Proceedings of the 17th ACM Conference on Recommender Systems, pp. 890–896 (2023)
21. Sarwar, B., Karypis, G., Konstan, J., Riedl, J.: Item-based collaborative filtering recommendation algorithms. In: Proceedings of the 10th International Conference on World Wide Web, pp. 285–295 (2001)
22. Son, J., Kim, S.B.: Content-based filtering for recommendation systems using multiattribute networks. Expert Syst. Appl. **89**, 404–412 (2017)
23. Touvron, H., et al.: LlaMA: open and efficient foundation language models. arXiv preprint arXiv:2302.13971 (2023)
24. Van Meteren, R., Van Someren, M.: Using content-based filtering for recommendation. In: Proceedings of the Machine Learning in the New Information Age: MLnet/ECML2000 workshop, vol. 30, pp. 47–56. Barcelona (2000)
25. Wang, L., Lim, E.P.: Zero-shot next-item recommendation using large pretrained language models. arXiv preprint arXiv:2304.03153 (2023)
26. Xu, S., Hua, W., Zhang, Y.: OpenP5: benchmarking foundation models for recommendation. arXiv:2306.11134 (2023)
27. Zheng, B., Hou, Y., Lu, H., Chen, Y., Zhao, W.X., Wen, J.R.: Adapting large language models by integrating collaborative semantics for recommendation. arXiv:2311.09049 (2023)

Author Index

N. Goharian et al. (Eds.): ECIR 2024, LNCS 14610, pp. 503–505, 2024.
https://doi.org/10.1007/978-3-031-56063-7

Printed in the United States
by Baker & Taylor Publisher Services